ALLE ZEIT WACH
1842

Zeev Berger

Satellite Hydrocarbon Exploration

Interpretation and Integration Techniques

With 267 Figures, Including 43 Color Plates

Springer-Verlag
Berlin Heidelberg New York
London Paris Tokyo
Hong Kong Barcelona
Budapest

Dr. Zeev Berger
Berger Enterprises Ltd.
P. O. Box 72054
1600-90 Av. SW
T2V 5HO Calgary
Alberta, Canada

ISBN 3-540-57348-8
Springer-Verlag Berlin Heidelberg New York
ISBN 0-387-57348-8
Springer-Verlag New York Berlin Heidelberg

Library of Congress Cataloging-in-Publication Data. Berger, Zeev. Satellite hydrocarbon exploration: interpretation and integration techniques/Zeev Berger. p. cm. Includes bibliographical references and index. ISBN 3-540-57348-8. – ISBN 0-387-57348-8 (U.S.) 1. Petroleum – Prospecting – Remote sensing. I. Title. TN24.P4B47 1994 622'.1828 – dc20 94-8948 CIP

Typesetting: Appl, Wemding
SPIN: 10129945 32/3130-5 4 3 2 1 0
Printed on acid-free paper

Dedicated to the memory
of my mentor, Dr. Jacob Aghassy,
and his wife, Ahuva,
both dear friends who lost their lives
in a tragic airplane accident
over the skies of Pittsburgh.

Preface

Opening Remarks

The debut of satellite imaging systems on board Landsat I in 1972 was a technological advance of considerable interest to earth scientists in general and exploration geologists in particular. Two major uses were anticipated for the satellite data. First, it was expected to replace the traditional aerial photograph that had proven to be useful for mapping geological structures, whether well exposed at the surface or obscured by thick vegetative and soil coverage. In addition, it was predicted that the spectral information provided by the imaging systems could be used to directly detect hydrocarbons from space.

Nearly three decades of satellite image application have proven that the technology aptly fulfills the first role. Satellite image interpretation is now considered to be a conventional exploration tool, used routinely for regional geological mapping and identification of prospect scale structures. We have also learned that the key to successful application of this technology is in its integration with other exploration tools such as seismic, well, gravity and magnetic data. However, there is no evidence to date that imaging systems from space can provide any reliable surface information that directly indicates the accumulation of hydrocarbons at depth.

Nearly all the available remote sensing textbooks are designed to provide comprehensive information on remote sensing technology. To an exploration geologist, such books contain too much technical information on the systems themselves rather than on the specific techniques for analyzing the images. Furthermore, these books generally lack three ingredients which would make them truly beneficial to the explorationist: (1) examples of satellite image interpretation that are supported by surface and, more importantly, subsurface controls; (2) demonstrations of a systematic approach to the interpretation and integration of satellite image data as part of an exploration program; and (3) synthesis of the processes that lead to the structural, geomorphic and spectral signatures which are manifested on satellite imagery data.

This book aims to fill that gap. It is based on experience gained in the past 14 years by me and other members of the remote sensing and the structural analysis research groups at Exxon Production Research Company. Explorationists from various Exxon affiliates which have used image data to support hydrocarbon exploration have also contributed. The examples used here, therefore, are taken directly from Exxon's case studies and training material. The reader must bear in mind that some of the examples which are illustrated here have been modified to some extent for the sake of simplicity as well as for proprietary reasons. Unless specified, the interpretation presented here does not necessarily reflect the present understanding by Exxon of the area described, but rather the writer's and his colleagues' interpretation at the time the studies were done.

Throughout the text, attempts are made to recognize the contributions of individuals. However, assigning credit to one specific source is not always possible. The major contributing authors also are credited in the acknowledgment.

Organization of the Book

The book is divided into two major parts. The first is designed to introduce the reader to the fundamental principles of image technology, interpretation and integration. The first chapter describes the various satellite imaging systems, their operation and surface mapping capabilities. The second chapter discusses common methods of computer enhancement of satellite imagery data. The next three chapters describe the typical diagnostic surface features that are used to detect the expression of, first, well-exposed and, then, buried or obscured structures. Detection of basement warp structures, which are a unique subgroup of buried structures, receives its own chapter because it requires specialized integration techniques. Chapter 6 deals with stereo image production, viewing and interpretation techniques. The whole range of interpretation tools are used in Chapter 7 to demonstrate a step-by-step approach to the integration of satellite image interpretation into a conventional hydrocarbon exploration program. Chapter 8 covers several additional aspects of image interpretation including direct detection of hydrocarbons, photostratigraphic mapping of outcrops and logistic applications (marine and on-shore).

The second part of the book (Chaps. 9–14) provides additional examples from six areas representing a wide range of structural complexities and surface conditions. These examples, which are constrained with a considerable amount of surface and subsurface control, can be used for further training and testing of the concepts illustrated in the first part.

A glossary of terms is provided at the end of the book followed by an Appendix containing a summary of symbols used for image interpretation in this book. Reviewing these terms and symbols in advance should prove beneficial to the reader. Also included is a brief list of major suppliers of imagery data and ways to contact them for further inquiries.

Acknowledgments

First and foremost, this book would never have been written without the influence of two of my former professors, who introduced me to the beauty of geology, geomorphology and field observations – Dr. Akiva Flexer and Dr. Jacob Aghassy. This book was reviewed by H.R. Hopkins and D.C. Peters. Selected portions were reviewed by F.F. Sabins, Jr., D. Simpson, H.R. Lang, R. Welch, M.A. Chapman, S.H. Lingrey and M. Covey. Their suggestions and contributions are greatly appreciated. Thanks are due to the Exxon Production Research Company (EPRco) and several Exxon affiliates for support of this effort and for allowing me to publish their data. Special thanks to Imperial Oil Resources Limited, especially the drafting department, for their continuous support, to Intera Inc. for providing radar data and to EOSAT and SPOT for releasing imagery data for publication. Special contributions to this book were made by several individuals from the remote sensing and structural analysis group at EPRco including F.V. Corona, R.L. Brovey, T.E. Townsend, H.R. Hopkins, T.P. Harding, R.L. Dodge, D.W. Phelps, C.C. Wielchowsky, D.M. Davidson, B.F. Merembeck, R.L. Kite and J.R. Kyle. Imagery products were made by R.L. Vernon and J.E. Firey. Many of the drafted maps were made by S.D. Adharsingh. Thanks, also to H.R. Hopkins for his enthusiastic support in assisting in the publication of this book. Finally, this book would have never been finished without the contribution of my son, Shai, who edited, drafted and rephrased several portions of this book, particularly those which cover his areas of expertise in physics and engineering.

Financial Support

Financial support for the production of the color plates was given by

- Intera Information Technologies, Ltd., Calgary, Canada.
- The Earth Observation Satellite Company (EOSAT), USA.
- SPOT Image, France.

Calgary, 1994 Zeev Berger

Contents

Part 1

Fundamentals of Remote Sensing Technology, Interpretation and Integration

Chapter 1 Imaging Systems

1.1 Introduction

The basic principle behind all satellite imaging systems is similar: an orbiting platform containing imaging equipment gathers data from a specified section of the earth and transmits the data to an earth-bound receiving station where it is analyzed. Imaging systems can be divided into those that use framing and those that use scanning techniques. The former method captures an area as one image using a photographic plate or other recording medium. This is the technique used in traditional aerial and space photography. The vast majority of space-based imaging systems now in operation are scanning systems, where one or more detectors collect data by sweeping over the target in parallel scan lines. Figure 1.1 shows a schematic of such a system and illustrates some important scanning-system terminology.

Scanning systems are further categorized as being passive or active. Passive systems employ various kinds of detectors that sense varying intensities and frequencies of light reflected from the Earth's surface. Active systems actually emit their own radiation and portray surface features from the reflected waves. The two major passive scanning satellites extensively used for exploration are the French SPOT satellite and the US Landsat satellite (which includes the multispectral scanner, or MSS, and the thematic mapper, or TM, imaging systems). Important active scanning satellites which are, or have been, in use include SEASAT, SIR-A, SIR-B (shuttle imaging radar projects) and ERS-1 (European remote sensing satellite). These systems represent the most common space-based imaging equipment in use today, and will be discussed in detail in the following chapters.

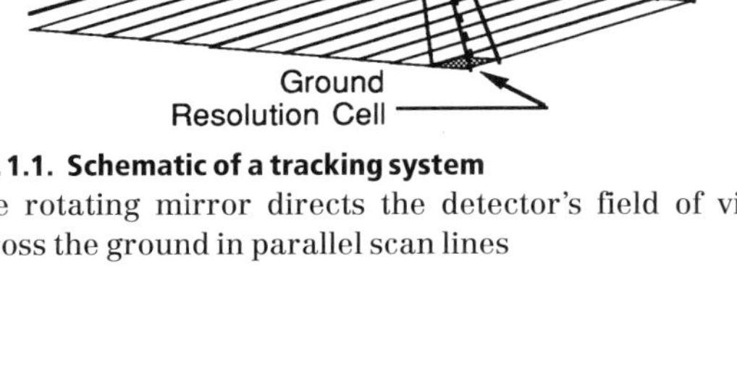

Fig. 1.1. Schematic of a tracking system
The rotating mirror directs the detector's field of view across the ground in parallel scan lines

Scanning imaging systems record data as discrete picture elements, called pixels, each of which represents the ground resolution cell or smallest area of land that is distinguishable on the image (Fig. 1.2). The pixel size determines the spatial resolution of this form of imagery; objects which are smaller than the pixel size cannot be resolved. The spatial resolution therefore indicates the maximum degree of clarity in an image. Spatial resolution is determined by the physical characteristics of the satellite, such as its altitude and velocity, as well as the characteristics of the imaging system, such as the size of the detector elements.

Another important element that determines the quality of the image is the radiometric resolution. This term describes the ability of the scanning instrument to discriminate between objects of similar brightness (the brightness represents the intensity of the reflected energy recorded for each pixel). An increase in radiometric resolution increases the quality of the image, the contrast between objects and therefore the usefulness of the image to the geologist. Figure 1.3 shows an example of how variations in intensity are treated in a computer system. Each pixel has been assigned a numerical value (grey level) relative to its brightness. This provides the basis for further enhancement and balancing of the image by computer, as will be discussed in Chapter 2.

Satellite imagery has numerous advantages over its predecessor, aerial photography. Foremost, the satellite imaging systems provide synoptic views of large areas, near-global repeated coverage, low

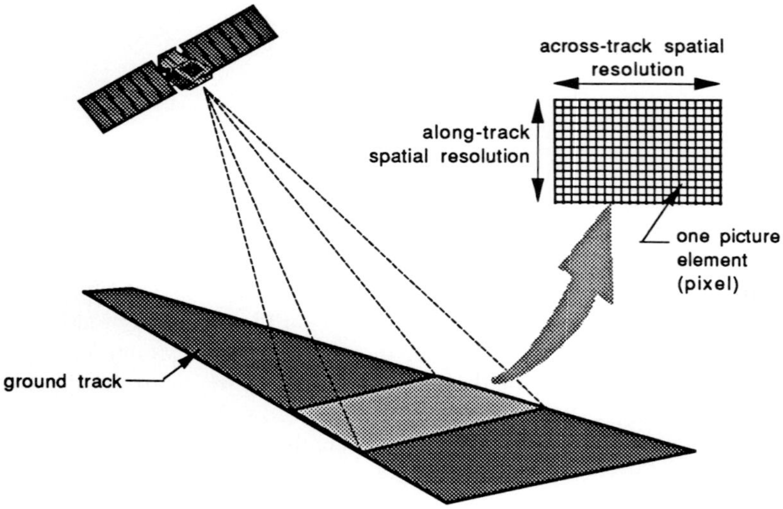

Fig. 1.2. Spatial resolution
The pixels are the fundamental units that form each image. Their size determines the spatial resolution or clarity

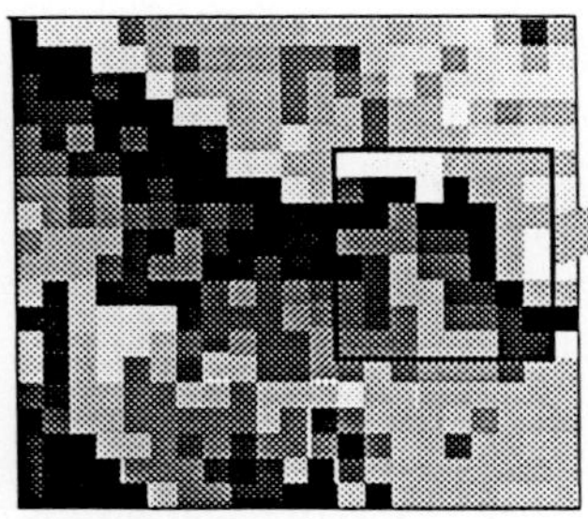

02	06	06	15	17	19	14
35	64	62	05	67	12	17
65	64	45	68	64	64	63
43	42	43	35	35	66	35
59	55	35	42	45	63	35
54	35	35	35	40	44	67
44	42	39	35	35	55	56

Fig. 1.3. The image as a digital matrix
The digital matrix of satellite data illustrated with a subscene from Death Valley, California

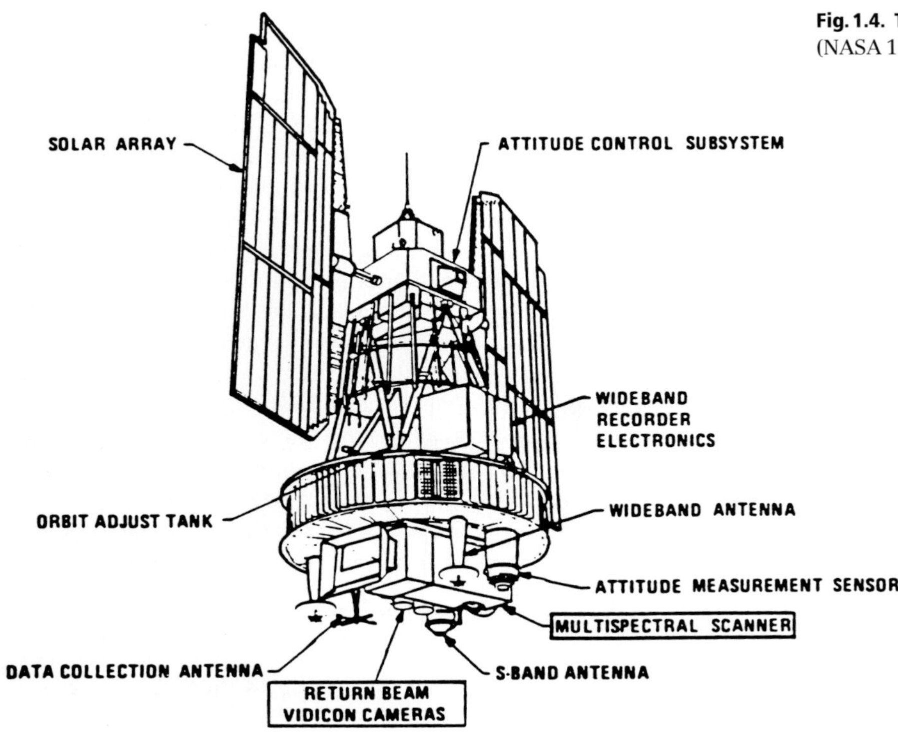

Fig. 1.4. The Landsat satellite
(NASA 1976)

cost and the ability to obtain data rapidly in a digital format. This last characteristic means that satellite data can easily be processed, enhanced and integrated with other data sets by computer. The main disadvantage of satellite imagery compared to aerial photography is its limited clarity. In addition, satellite data traditionally have been limited to monoscopic viewing only. These limitations, however, have both been remedied in recent years; the former with the advent of new high-resolution imaging systems and the latter with the development of the SPOT satellite with stereo capabilities.

1.2 The Multispectral Scanning System

The MSS was launched by NASA in 1972 on board the Landsat 1 satellite, and the program was continued with Landsats 2, 3, 4 and 5 (Fig. 1.4). Although it was not the only imaging system on board the first Landsat satellites, it was the one most commonly used for hydrocarbon exploration. The MSS also can be credited with introducing the principles that were used by all subsequent satellite-based imaging systems.

All of the Landsat satellites follow near-polar, sun-synchronous orbits. In other words, the satellite orbits pass longitudinally over the poles (or close to them). The orbits are synchronized with the sun so that the southbound half of the orbit is always made on the illuminated side of the planet, and the northbound pass on the dark side (Fig. 1.5).

This synchronization with the sun allows the satellite to make repeated passes over any given area at the same local time, so that the sun's angle of elevation remains constant for all images of the area (except for seasonal changes in the sun's elevation). For example, Landsat 1 crosses 40 ° N lat. at 9:30 a. m. every orbit. Because of this consistent illumination, successive passes over adjoining areas can be combined into high-quality, uniform mosaics. Landsat completes 14 orbits a day, each one slightly more westward (as the Earth has rotated to the east) than the last; the entire globe is covered in about 18 days. The only areas not covered by Landsat are latitudes greater than 81 °. The satellite orbits at an altitude of 920 km and has a field of view of 185 × 185 km.

The MSS has six sensors that sweep in parallel lines in the scan direction (perpendicular to the direction of flight). Each sensor records reflectance values in one of the four bandwidths. This informa-

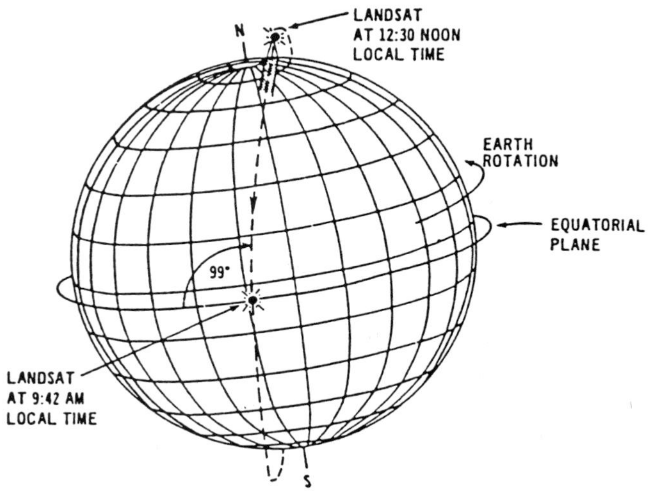

Fig. 1.5. Landsat's orbit
Orbital inclination is 99 ° (near-polar). Sun synchronization allows the satellite to cross each latitude at the same local time on every orbit. Southbound passes are always made on the illuminated side of the planet; northbound on the dark side. (NASA 1976)

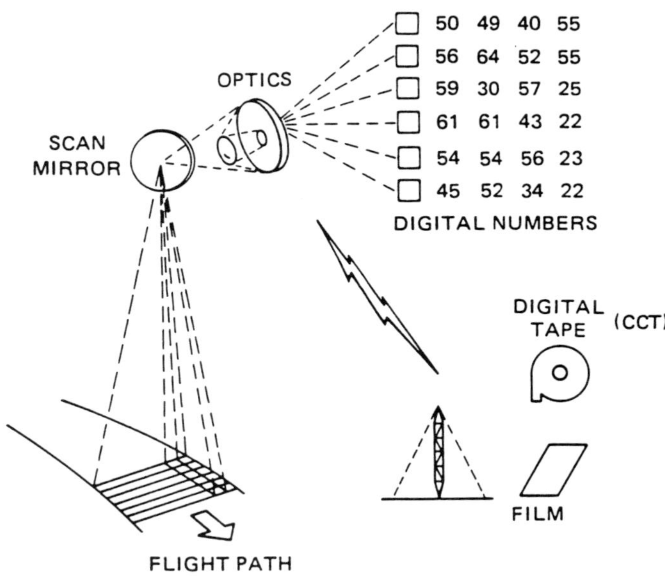

Fig. 1.6. The Landsat satellite system
The MSS uses a scanning instrument to record reflected light from the surface of the Earth in four different wavelength bands. Six lines are scanned across the ground perpendicular to its southbound trajectory and recorded digitally. Each digital element represents an 80 m^2 ground element. The data is then transmitted to an earthbound receiving station

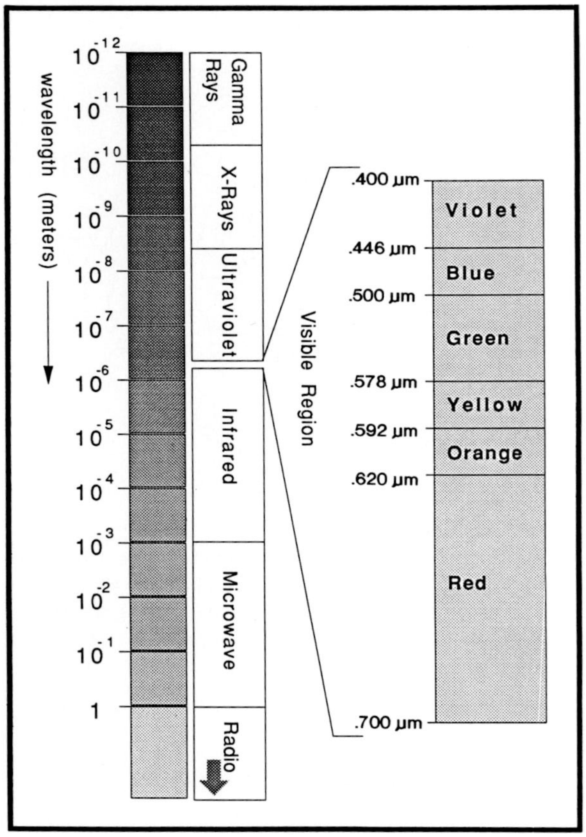

Fig. 1.7. The electromagnetic spectrum

tion is compiled and prepared for transmission to receiving stations (Fig. 1.6). The main advantage of the MSS over traditional photography is that its sensors register reflectance values[1] in several different frequency bands (bandwidths) along the electromagnetic spectrum; thus, the name "multispectral". Figure 1.7 shows the electromagnetic (EM) spectrum with the visible portion enlarged. The range of bandwidths for an imaging system is referred to as its spectral resolution. The MSS uses four bands ranging from 0.5 µm (visible light) to 1.1 µm. Four data sets are therefore created (each one capturing the variations in reflectance within one bandwidth for the scene) then composited and finally enhanced by computer, as discussed in Chapter 2, to create the final image.

1.2.1 The Advantages of Multispectral Imaging

The perceived color of an object is determined by the wavelengths of light which it absorbs. Starting with sunlight, almost the full range of the EM spectrum is available. After specific ranges of frequency are absorbed by an object, the remaining frequencies are reflected and combine to form the color[2]. By

[1] Strictly, reflectance refers to the ratio of incident light to reflected light in a specific band. However, it is often used loosely as a qualitative term for "radiance". Radiance measures the degree of incident light in watts per square meter.

[2] Frequency and wavelength are often used interchangeably when referring to the EM spectrum because they are related by the simple relationship: (freq) × (wavelength) = (the speed of light). Thus, one measure of an EM wave can easily be converted to another.

Fig. 1.8. Frequency response curves for several crucial materials
Near-infrared frequencies allow easy differentiation between the three basic components of the Earth's surface. *UV* Ultraviolet; *B* blue; *G* green; *R* red. (After Estes 1985) ▼

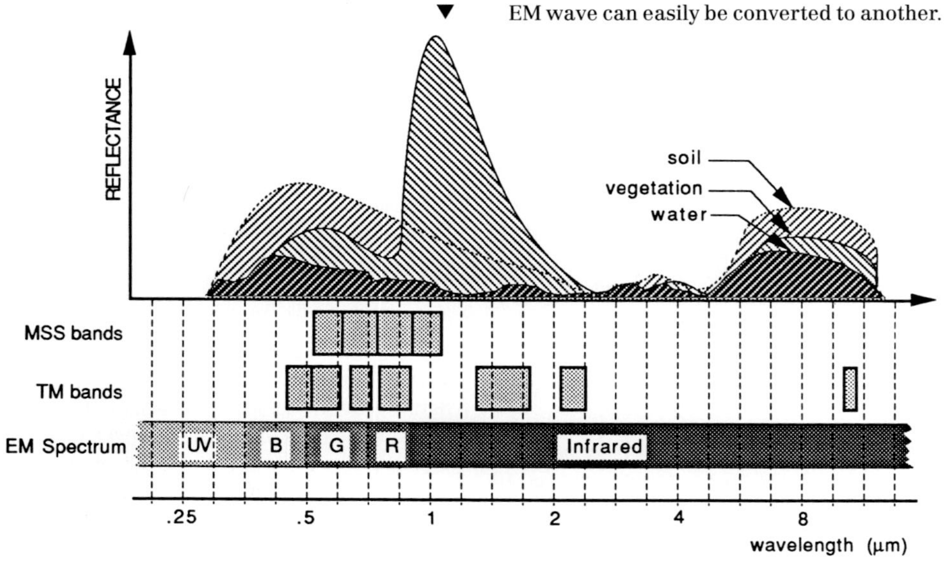

limiting discussion to just three bandwidths, red, green and blue (which can be combined to cover all of the visible spectrum), it is possible to achieve an intuitive feel for this concept. For example, grass and trees absorb red and blue light (a function of the chemical chlorophyll) and therefore appear green. Water absorbs most wavelengths above 0.5 µm and therefore appears blue since blue is the only range left in the visible spectrum.

This concept can then be expanded from three samplings to a continuous curve comparing the intensity of reflected light to frequency. It can be seen that every material has a characteristic curve that can be used to uniquely identify it. This is called the spectral response curve. Figure 1.8 shows the spectral response curve for several materials. Intuitively, the vegetation curve peaks between 0.5 and 0.6 µm, representing the green portion of the spectrum.

Further interpretation of this figure reveals much more information. For example, vegetation also peaks between 0.7 and 0.8 µm, although this range is not in the visible region. If one needed to differentiate between vegetation and water, scanning equipment might not be able to discern the small differences in reflectance within the visible spectrum, but could discriminate between the two in the near-infrared (IR) range where the difference is much greater.

The bandwidths used on the MSS are given in Table 1.1 along with their respective wavelengths. Taking the spectral response curves for several materials, one can simplify the graph by looking at the

Table 1.1. The Landsat MSS bands

These numbers are used to refer to the MSS bands on Landsat 1, 2 and 3. On Landsat 4 and 5, they are referred to as 1, 2 and 3 instead of 4, 5 and 6. The original numbering system allotted bands 1, 2 and 3 to an old system no longer in use

MSS band	Wavelength (µm)	Spectral range
4	0.5–0.6	Green
5	0.6–0.7	Red
6	0.7–0.8	Infrared
7	0.8–1.1	Infrared

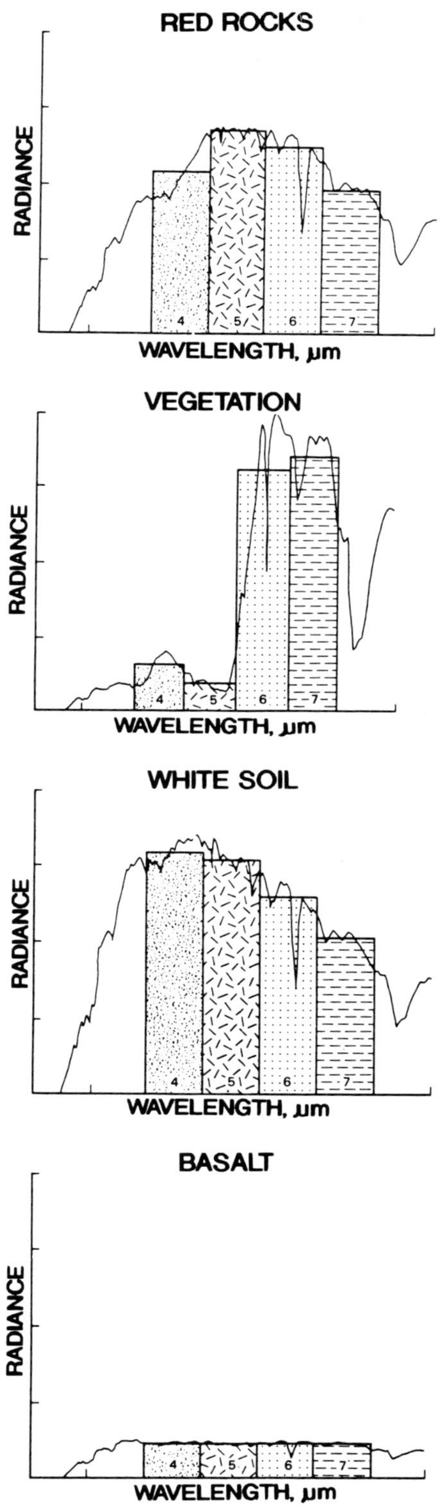

Fig. 1.9. Bar graphs showing frequency response in MSS bandwidths

Note how the reflectance levels follow naturally from the nature of the material. Basalt, a dark rock type, has reflectance values that are in the low range, whereas white soil reflects much of the incoming light. Also note how vegetation has a high infrared response. (Townsend and Dodge 1983, EPRco internal report)

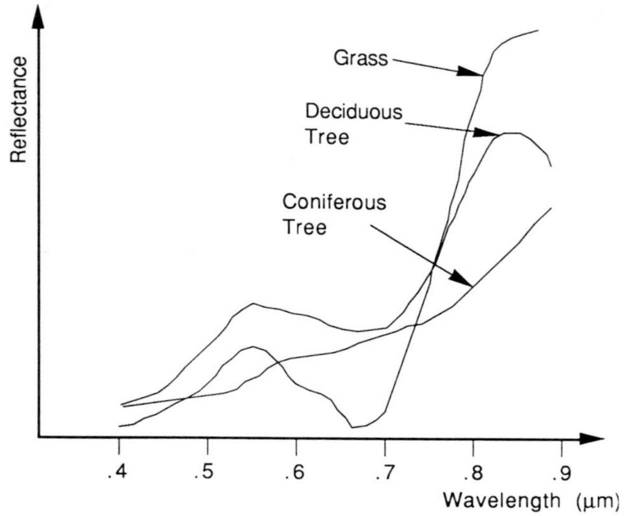

Fig. 1.10. Frequency response curves for three different kinds of vegetation
Note their similar response in the 0.5–0.6 μm range (visible green) but varying responses in the 0.7–0.85 μm range. (After Holz 1985)

radiance level of particular frequencies. This time however, the four bandwidths of the MSS are used instead of the colors of the visible spectrum (Fig. 1.9). Using graphs such as the ones shown in this figure, geologists can compile a table of particular materials and their radiance in each of the bandwidths.

The range of frequencies which is useful for distinguishing one particular material is its diagnostic reflectance range. For example, it was seen above (in Fig. 1.8) that a good diagnostic reflectance range for vegetation was in the near-infrared. Spectral response signatures can be applied even more precisely, even differentiating between different types of vegetation. Figure 1.10 clearly shows that, to accomplish this, the diagnostic range required is between 0.7 μm and 0.8 μm. A critical assessment of an imaging system's capabilities is made by comparing the diagnostic reflectance range of materials of interest with the bandwidths of the system.

1.2.2 A Word of Caution

It is important to note that, after developing the concept of multispectral imagery, the analogy to human vision is not precisely accurate. One must be careful to remember that, although it is common to use words that have meaning in terms of human sight, these terms have different (or at least expanded) meanings in satellite imaging technology. For example, radiometric resolution is often referred to as the ability to discern different "brightness" levels. In fact, it truly indicates the sensitivity of the sensors to different reflectance intensities in each band. For the MSS, the bar graphs in Fig. 1.9 could only be adjusted by increments of 1/64 of 100 % reflectance because the radiometric resolution is 64. Similarly, the term "color" must not be used carelessly since it implies color as perceived by human vision. Scanning systems can only return digital information indicating the intensity levels in different frequencies. If color appears on an image, it must be remembered that it is artificially produced by assigning a particular hue to a bandwidth (these techniques will be discussed in Chap. 2). In fact, the MSS is incapable of producing true color images, since it has no spectral band corresponding to the blue end of the spectrum. (This lack of a bandwidth in the blue range is deliberate, as the frequency coincides with a region of extreme atmospheric absorption which would lead to poor image quality.) This problem was solved with the next implementation of the imaging system on board Landsat, the thematic mapper or TM.

1.3 The Thematic Mapper

The Landsat thematic mapper (TM) is the second generation imaging system on board Landsat. TM introduced several important improvements over the MSS. The system was first launched in 1982 on board Landsat 4 and later on Landsat 5 in 1984. The nominal altitude of Landsat 4 and 5 is 705 km – considerably lower than the previous satellites. This difference permits an improvement in spatial resolution from 80 × 80 m to 30 × 30 m. Radiometric resolution was also increased from 64 to 256 grey levels.

1.3.1 Changes in Spectral Range

The spectral resolution was improved in several ways with TM: (1) seven spectral bands as opposed to the four of MSS are used; (2) TM's spectral range is expanded to include more of the visible and reflected IR regions of the EM spectrum; (3) the addition of blue spectral data (TM band 1) enables the production of normal color composite images; and (4) TM has a thermal infrared band.

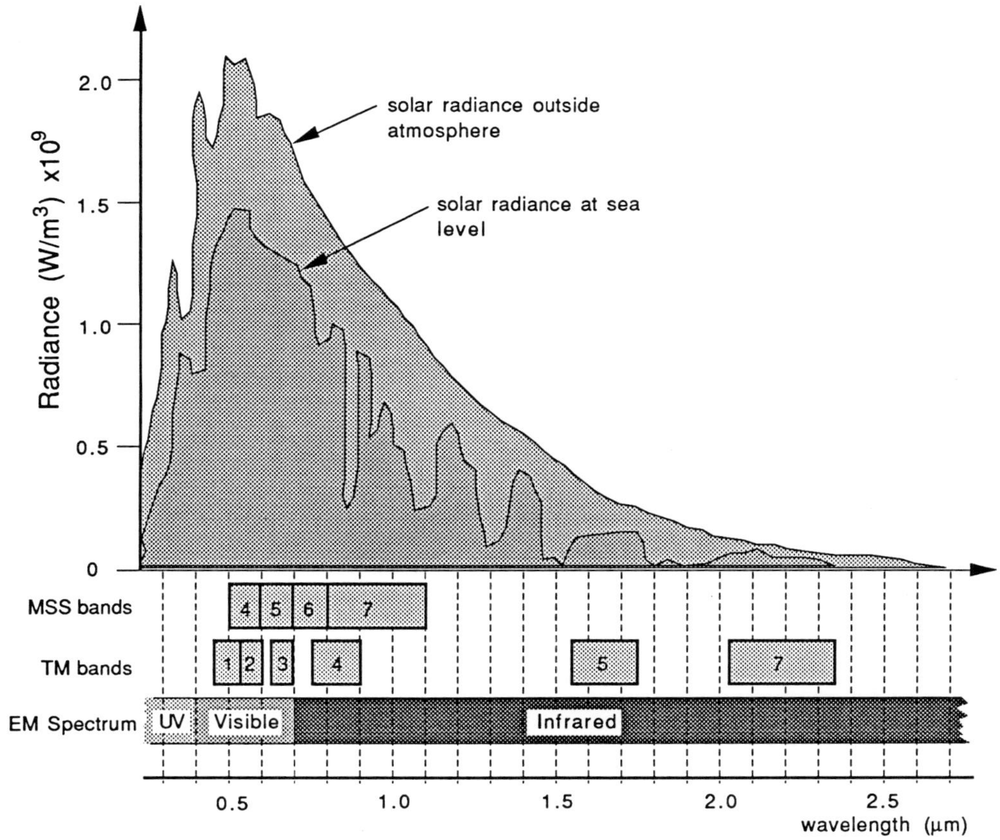

Fig. 1.11. A comparison of MSS and TM bands with radiance levels
The bandwidths of spaceborne scanning systems are designed to take advantage of atmospheric windows where radiation reflecting back from the Earth is not scattered by the atmosphere. Note how TM bands 1, 5 and 7 cover ranges of the spectrum not covered by MSS bands. (After Townsend and Dodge 1983)

Figure 1.11 compares MSS and TM bandwidths. Note that TM bands 1, 5 and 7 are ranges not covered by MSS. Band 1 achieves the strongest penetration through water and, therefore, is useful for bathymetric applications as well as for creating true-color images. Recall that the MSS has no bandwidth in the blue range. Note also how the bandwidths of both satellites are placed to take advantage of "atmospheric windows" or regions of the spectrum not disrupted by atmospheric attenuation.

Figure 1.12 compares MSS and TM bands to the diagnostic ranges of several materials. Lithological mapping applications benefit greatly from bands 5 and 7, as the diagnostic reflectance ranges of both clay and carbonate minerals fall in these bands[3]. Another important expansion of the spectral range is provided by the thermal infrared band (band 6) of the TM, which collects data from 10.4 to 12.5 µm. The main purpose of this band, which is limited in its spatial resolution to 120 m, is estimation of soil moisture content and detection of areas with unique thermal signature.

1.3.2 Advantages of Increased Spatial Resolution and Spectral Range

The significant improvement in surface geological mapping capabilities using TM (as compared to MSS) data is illustrated in Fig. 1.13. The first pair of images demonstrates the increase in the level of structural detail due to TM's improved spatial resolution. In this case, the images show the surface ex-

[3] Clay minerals have another diagnostic range around 1.5 µm but, as the figure shows, this coincides with an area of atmospheric attenuation.

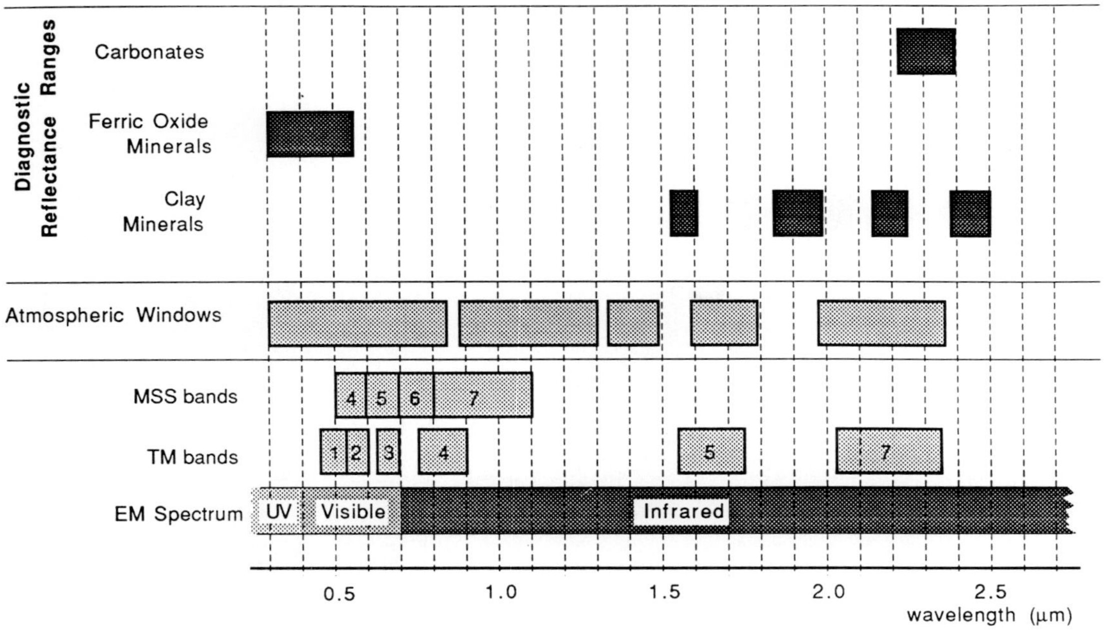

Fig. 1.12. Comparison of diagnostic reflectance ranges with Landsat bandwidths
The TM band 7 is ideally suited for the identification of clay minerals because of its proximity to the diagnostic reflectance range of the material. Note that, although clay minerals have other diagnostic ranges, these coincide with ranges of atmospheric absorption. (After Townsend 1984)

pression of a well-exposed wrench fault system and related oblique anticlines in Death Valley, California. TM provides better definition of the fault line trace and its related diagnostic triangular facets as well as better surface expression of the oblique anticlines.

The middle images show the surface expression of a subtle breached fold in the Central Basin Platform in West Texas. Here, the image benefits not only from increased spatial resolution but also from the addition of a spectral range that highlights the different lithostratigraphic units which crop out around this breached fold. The recognition of these units (marker beds) allows the interpreter to recognize the attitude (dip and strike) of the bedrock units which are exposed at the crest of this fold. Note that the corresponding MSS image does not reveal these marker beds and therefore does not allow such interpretation to be made.

Finally, the third pair of images shows a profound improvement in the surface mapping capabilities of TM as compared to MSS in humid areas where surface expression of structures is hampered by lack of bedrock exposure and extensive man-made features. The image shows the surface expression of structurally controlled geomorphic features reflecting the presence of a buried fault zone which crosscuts the Trinity River in East Texas, near Palestine. Cross-cutting faults are expressed as short linear features that cause deflection of the Trinity River channel and abrupt changes in the active floodplain area. However, other linear features shown also reflect the presence of major highways, roads and edges of cultivated areas in this region. Only with the improved spatial resolution and spectral range of TM can one distinguish between these two similar linear features and make a reliable structural interpretation.

The potential usage of TM's thermal infrared band (band 6) for detection of structurally related moisture conditions is demonstrated in the imagery data from the Mexia-Talco fault system in East Texas (Fig. 1.14). Here, the downthrown side of individual faults within this extensional fault system creates topographic depressions which are charac-

Fig. 1.13. A comparison between MSS and TM imagery data ▶
Three different geological settings are shown: **a** and **b** a wrench fault system in Death Valley, California; **c** and **d** a breached fold in the Central Basin platform of West Texas; **e** and **f** a buried fault system across the Trinity River in East Texas. *FA* Fold axis; *TF* triangular facets; *MB* marker beds; *BF* buried fault; *DV* deflected valley, *MM* man-made linear feature; *FLT* fault line trace

MSS
TM
FLT
FLT
FA
0
3Km.
a
FLT
FLT
TF
FA
0
3Km.
b
FA
0
3Km.
c
MB
FA
MB
0
3Km.
d
BF
DV
MM
DV
BF
0
3Km.
e
BF
DV
MM
BF
MM
DV
DV
BF
0
3Km.
f

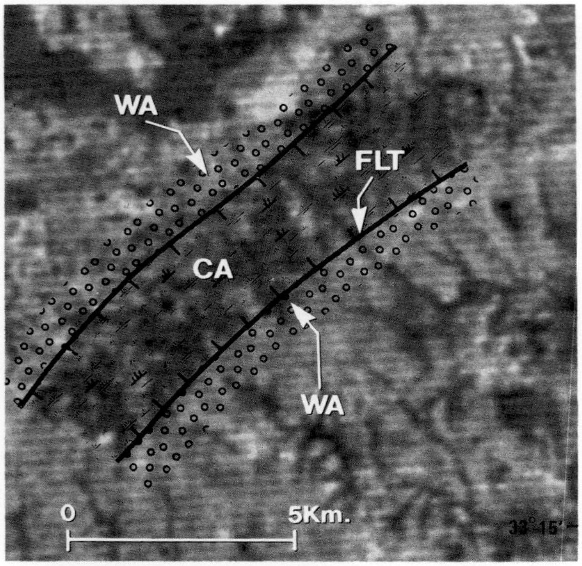

Fig. 1.14. TM thermal imagery (band 6)
Scene is of the Commerce area, northeast Texas, showing changes in thermal properties across faults which are attributed to ground moisture conditions and related vegetative cover. *WA* Warm areas; *CA* cold areas; *FLT* fault line trace (with normal separation)

terized by humid conditions and increases in vegetative cover. These areas are relatively colder than the surroundings and thus are reflected by the thermal image data as darker features. The spatial resolution of this band, however, is quite limited for detailed analysis of individual surface features.

As shown by the above examples, TM imagery provides superior mapping capabilities and provides opportunities for interpretations that cannot be made from MSS images. These advantages have proven to outweigh the increased cost of acquiring TM images and, thus, TM has currently completely replaced the MSS for most types of detailed geological applications. However, MSS mosaics and images are still widely used for cursory mapping of geological structures.

1.4 The SPOT Satellite

The SPOT imaging system, launched in 1986 by the French National Space Agency (CNES), compliments data provided by the Landsat satellites. SPOT uses high-resolution visible (HRV) systems in a passive imaging method. The HRV records data by focusing reflected energy onto a linear array of charge-coupled detectors (CCD's, the imaging system used in modern camcorders.) The SPOT satellite can operate in two modes. In the first, 6000-CCD's are used to analyze a 10 × 10 m ground cell and to record a single panchromatic image in the 0.51 to 0.73 μm band. The panchromatic bandwidth is rather large, covering over half of the visible spectrum. The second mode is multispectral and uses three arrays of 3000 CCD's each for a ground cell of 20 × 20 m. The three bands of the multispectral mode cover roughly the same range as the MSS. Table 1.2 summarizes the parameters of SPOT.

Table 1.2. SPOT system characteristics

Characteristic	Multispectral mode	Panchromatic mode
Spectral bands	0.50–0.59 μm 0.61–0.68 μm 0.79–0.89 μm	0.51–0.73 μm
Spatial resolution	20 m	10 m
Radiometric resolution	256	256

1.4.1 Off-Nadir Viewing Capabilities

Another feature of the SPOT satellite which makes it particularly useful is its capability for off-nadir imaging. The satellite is able to acquire images up to 27° east or west of its path by tilting its imaging system as illustrated in Figs. 1.15 and 1.16.

This capability allows SPOT to make images of the same area on successive orbits, which provides considerable flexibility in scheduling times of acquisition. Without it, the opportunity to scan the exact same area would arise only 26 days later – the time SPOT takes to complete the full cycle of its near-polar orbit.

Off-nadir capability allows SPOT to re-image areas of activity frequently, which is useful for monitoring dynamic events such as floods or growth cycles of crops. Although such frequency is not necessary for geological mapping purposes, the repeated coverage does provide the geologist with stereo mapping capabilities, a truly valuable asset. If two images of the same area are offset by a relatively small angle, a parallax is introduced and the combined data can be processed in such a manner as to reveal three-dimensional features of the area.

Figure 1.17 shows a comparison between TM and SPOT images over the Todd oil field of the Cen-

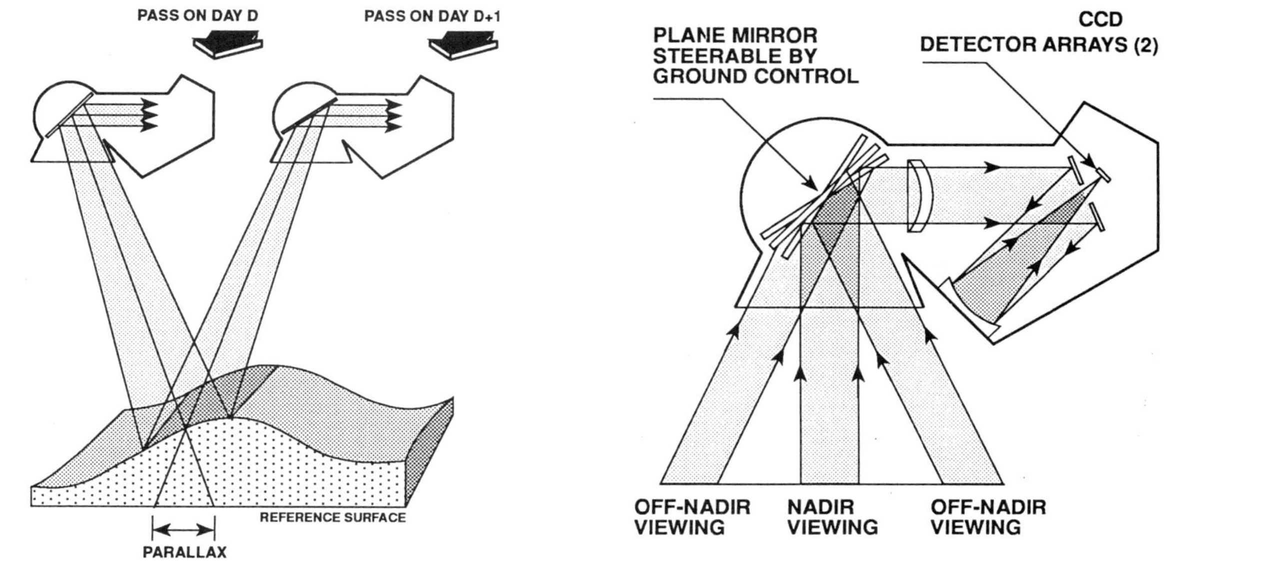

Fig. 1.15. Off-nadir capabilities of SPOT
Because of the angle of SPOT's imaging mirror, which is adjustable up to 27°, the satellite can acquire an image of the same area on successive orbit passes with varying incidence angles. The number of observations possible is determined by the latitude. For example, localities at the equator can be imaged 7 times; at latitude 45°, 11 times. (SPOT 1984)

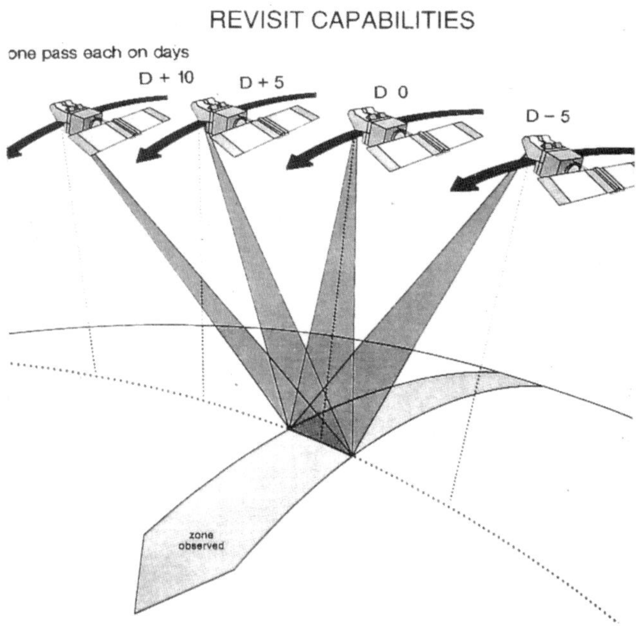

Fig. 1.16. SPOT's revisit capability
By tilting its imaging system to either side of its ground path, SPOT is able to revisit localities several times during its 26-day cycle. (SPOT 1984)

tral Basin Platform in West Texas. The field is manifested at the surface as a subtle fold which is outlined by a radial drainage pattern. The recognition of this field on imagery data requires detailed mapping of drainage patterns and careful analysis of the outcropping units, including dip and strike measurements. Clearly, more detailed drainage features are detectable on the SPOT image. More importantly, lithostratigraphic units, not evident on the TM image, can be mapped from the SPOT image and later measured for dip and strike from SPOT stereo data. The stereo capabilities of SPOT are particularly useful for mapping geological structures that represent the two extreme end members of structural deformation: highly deformed areas, such as fold and thrust belts, and regions of low deformation such as stable platforms and foreland basins (see Chap. 6).

Table 1.3 summarizes and compares the capabilities of the three passive imaging devices discussed so far.

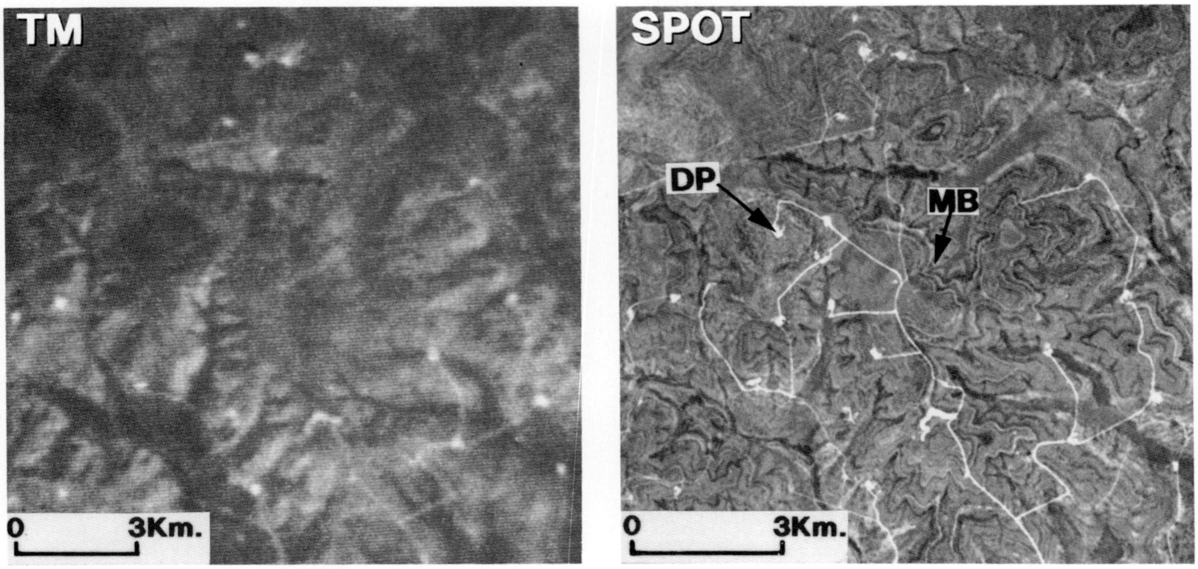

Fig. 1.17. Comparison of TM and SPOT images of the Todd oil field, West Texas
A significantly greater level of detail of drainage patterns and outcrop units is depicted on the 10-m resolution SPOT image. *MB* Marker beds; *DP* drilling pads

1.5 Radar Imagery

In addition to the arsenal of passive scanning devices discussed, geologists now have access as well to spaceborne active sensor systems that use radar techniques for imaging. The use of radar imaging systems is by no means a new concept; airborne radar has been a frequently used tool for geological interpretation. Table 1.4 shows the wavelengths used for radar imaging, with the bands used in geology highlighted. Note that these wavelengths are considerably longer – beyond thermal infrared – than the ones used in the imaging systems previously mentioned.

1.5.1 Background

Radar (an acronym for radio detection and ranging) operates by emitting electromagnetic radiation in the microwave and radio range in short pulses. After transmitting each pulse, the radar antenna switches from a transmitting mode to a receiving mode, and records the reflected waves from the

Table 1.3. Comparison of Landsat and SPOT instrument characteristics

	MSS	TM	Spot[a]
Spatial resolution, m	79	30 (bands 1–5,7) 120 (band 6)	20/10
Radiometric resolution	64	256	256/256
Spectral bands	4	7	3/1
Width of path imaged, km	185	185	60/60

[a] Multispectral/panchromatic mode.

Table 1.4. Radar wavelengths used in remote sensing
The letter codes are a remnant from radar's early use as a military tool in WW II when they were invented for security purposes

Band designation	Wavelength (cm)
Ka	0.8–1.1
K	1.1–1.7
Ku	1.7–2.4
X[a]	2.4–3.8
C[a]	3.8–7.5
S[a]	7.5–15.0
L[a]	15.0–30.0
P	30.0–100.0

[a] Bands commonly used in remote sensing.

pulse. When the pulse reflects off of an object, the radar waves are altered according to features of the object. The variations in those waves are recorded by the radar system and then processed into an image.

Radar instruments are grouped into three categories. The simplest implementation is an altimeter which emits a signal and then determines altitude from careful timing of its return. Typical applications include oceanwide current analysis and wave climatology. Scatterometers form the second group. They measure the degree to which the surface of the area of interest scatters the emitted signal. Scatterometers find use in mapping global wind patterns. The final category is the imaging radar system which combines information of return times and scattering to produce an image of the surface. This type of system is the one of principal interest to the geologist and will be discussed for the remainder of the chapter.

Originally, radar systems were used on board specially fitted aircraft. The radar was aimed downwards and off to the side. For this reason, these systems have been referred to as side-looking airborne radar (SLAR). The direction in which the radar beams are emitted is called the "look direction" or "range direction". The direction in which the plane is moving is called the "azimuth direction". The same techniques, and the same terminology, are applied to radar systems on space platforms.

The information that is recorded by a radar imaging system can be presented in graphic form, as shown in Fig. 1.18.

The horizontal axis measures distance in the look direction, and is determined by the time the echo of a particular feature takes to return. Because the radar waves are traveling at the speed of light, multiplying the time the pulse takes to return by this constant speed gives the distance to the feature. The image tone is determined directly from the intensity of the echo off of a particular feature. Each swath, as shown in Fig. 1.18, captures a strip of land in the look direction.

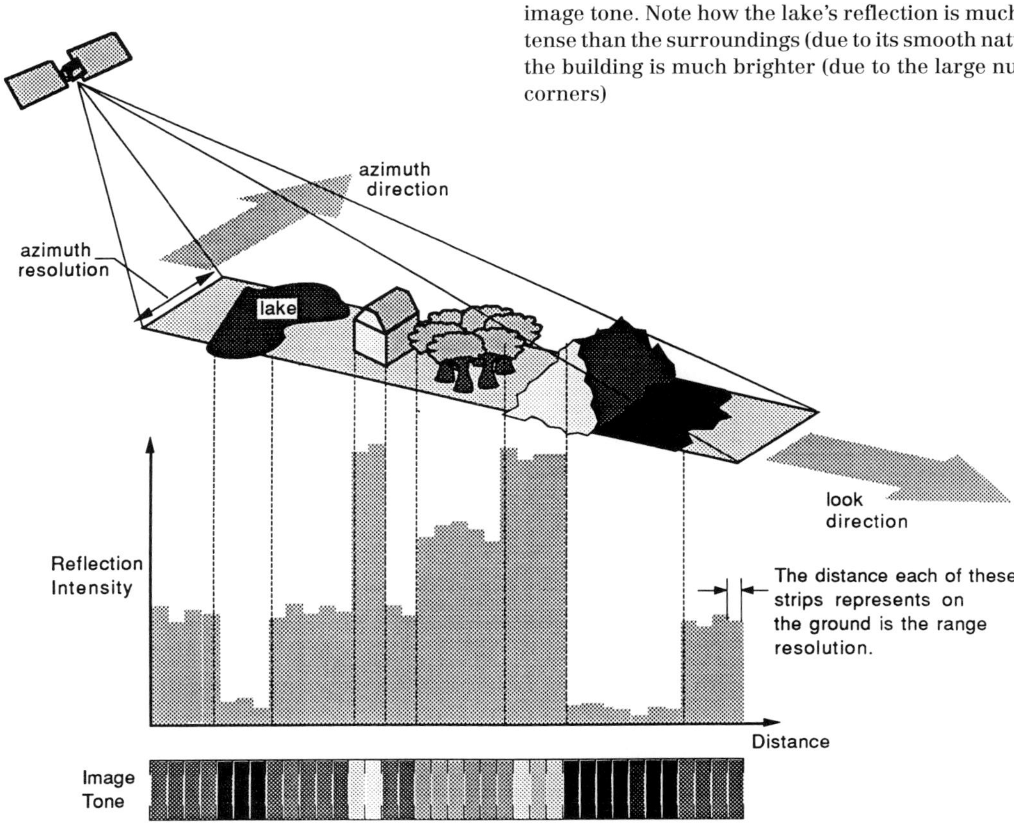

Fig. 1.18. Schematic of a radar imaging system
The transmitter sends out a pulse and then, functioning as a receiver, records the time taken for the echoes to return. These times are translated to distance and then correlated with the relative intensity of the echo which becomes the image tone. Note how the lake's reflection is much less intense than the surroundings (due to its smooth nature) and the building is much brighter (due to the large number of corners)

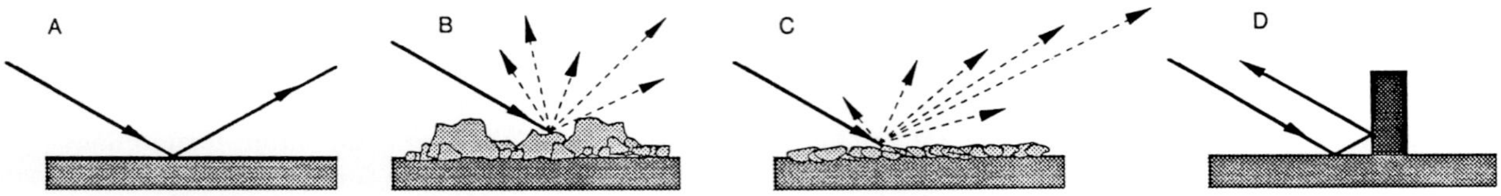

Fig. 1.19. Reflection mechanisms of different surfaces
Reflection intensity varies from **A** none, due to a very flat surface, to **D** total, due to a corner reflection. Intermediate situations are shown in **B** and **C**

1.5.1.1 Factors Affecting Reflection Intensity

The reflection intensity is a complex function depending on both terrain elements such as surface roughness, dielectric properties, orientation of surface features and on system characteristics such as the wavelength and depression angle. Those considerations which are directly related to the surface mapping capabilities of radar imaging as applied to hydrocarbon exploration and as compared to the capabilities of passive systems are discussed below. Other aspects of radar imaging technology and interpretation techniques can be found in Drury (1987), Ford et al. (1980, 1983), Sabins (1987), and other publications that are listed at the end of this chapter.

Surface roughness is by far the most significant factor contributing to the reflectance (and hence to the image tone). As Fig. 1.19 shows, smooth specular surfaces reflect rays at the same angle but in the opposite direction (A). None of the signal is returned to the receiver and, therefore, the feature is characterized by a black appearance on the image. A surface with many orthogonal corners has a high probability of reflecting radiation back to the reflector; thus such surfaces (including buildings, cars, and other man-made objects) show a bright signature (D). The question can be thought of in terms of probability: rougher surfaces have a higher chance of returning radiation to the emitter and thus appear brighter. Table 1.5 gives the typical radar response of several surface elements.

Although easy to understand in a qualitative way, the roughness value of a surface is hard to analyze quantitatively. One must remember that topographic relief (the overall changes in slope, measured usually in meters) does not determine the signature. Rather the small-scale variations in the surface, measured perhaps in centimeters, are the determining factor. Thus, measurement must take into account loose soils, rocks, vegetation and even the relative size and orientation of the leaves on trees. These types of measurements elude mathematical description, but it has been proven that they may be approximated by the average height of the surface feature above ground level (called the vertical relief). Tables have been compiled listing various materials with their respective vertical relief values. For example, floodplains have a value of 0.2 cm, sand and fine gravel have a value of 1 cm and coarse gravel has a value of 12 cm (e. g. Sabins 1987).

Table 1.5. Various terrain features and their radar signatures

Terrain feature	Image signature	Image tone	Cause of signature
Steep slopes and scarps facing *toward* antenna	Highlights	Bright	Much energy is reflected back to antenna
Steep slopes and scarps facing *away* from antenna	Shadows	Very dark	No energy reaches terrain; hence there is no return to antenna
Vegetation	Diffuse surfaces	Intermediate	Vegetation scatters energy in many directions, including returns to antenna
Bridges and cities	Corner reflectors	Very bright	Intersecting planar surfaces strongly reflect energy toward antenna
Calm water, pavement, dry lake beds	Specular surfaces	Very dark	Smooth, horizontal surfaces totally reflect energy, with angle of reflectance opposite to angle of incidence

Radar reflectance characteristics are further complicated by the presence of "radar shadow" or regions where the direct line from the receiver is blocked by a feature (for example, the far side of a mountain). No radiation reaches such a region and it therefore appears black on the image. There is no systematic way to tell if a black region on an image is caused by a specular surface or by a radar shadow; only careful interpretation allows one to distinguish the two. As illustrated later in this section, radar shadows can be both harmful and helpful, depending on their extent and the situation.

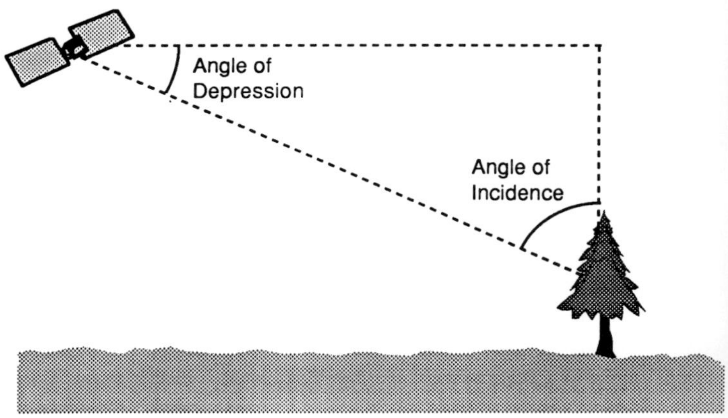

Fig. 1.20. Depressions vs incidence angle
The incidence angle can be calculated by subtracting the depression angle from 90° and vice versa if the ground is assumed flat. If this assumption is not made, then there is no correlation between the two angles

1.5.1.2 Factors Affecting Resolution – Range Resolution

The resolution of a radar system must be given with two measurements – the range resolution and the azimuth resolution. Referring back to Fig. 1.18, the azimuth resolution can be thought of as the width of the "strip" and the range resolution as the smallest division along the strip. Thus, the area of an individual ground resolution cell is given by the product of the azimuth and range resolution.

The range resolution is directly related to the length of the radar pulse. This is because reflections from two objects must be received at separate times for the two objects to be resolved. More importantly, range resolution is affected by the angle of depression (Fig. 1.20). Specifically, the resolution decreases proportionately as depression angle increases. Thus, optimum resolution is achieved from the lowest flight path possible. Recall, however, that the discussion of radar shadows revealed that low flight paths lead to large loss of data due to shadowing. These two effects must therefore be carefully balanced to suit the needs of the survey.

Note that range resolution is independent of the emitter's distance to the object. This fact allows spaceborne systems to use similar techniques with regards to range resolution as land or airborne systems. (This is not the case with azimuth resolution.) One way of increasing the range resolution is to shorten the pulse length. However, doing so decreases the amount of energy sent by the transmitter and therefore reduces the clarity of the returned signals[4]. The balance between these two effects is yet another factor to consider when designing a radar survey.

1.5.1.3 Factors Affecting Resolution – Azimuth Resolution

Azimuth resolution is determined solely by the width of the "illuminated" strip or swath on the ground. Because the radar beam spreads out as it moves further away from the transmitter, azimuth resolution deteriorates for distant objects. The problem is particularly acute for space-based platforms due to their large separation from the ground swath.

Azimuth resolution may be improved by decreasing the wavelength. (Recall, however, that this leads to lower penetration ability.) The radar beam can also be narrowed by using a longer antenna. This method is the best approach to improving resolution, but clearly there are physical limitations to the length of an antenna (particularly on an airborne system). An electronic system to simulate a longer antenna has been devised to remedy this problem. Almost all space-based radar systems use this technique, which is referred to as synthetic aperture radar (SAR)[5].

[4] The energy level of each pulse can be maintained by shortening the wavelength as pulse length is shortened. Doing so, however, limits the penetrating capability of the pulse.

[5] SAR is also called Doppler radar because it relies on the Doppler effect, the shifting of frequency due to relative motion, for its operation. SAR systems monitor the return signals from an object for the entire time it is in range and use the shifting of the return signal's frequency to interpolate more accurately the object's position.

1.5.2 Advantages of Mapping with Radar Imagery

The basic interpretation techniques of geological structures using radar imagery are more or less similar to those used in the interpretation of other remotely sensed data. Radar does, however, provide some unique opportunities to the interpreter due to its active nature. By the same token, radar images contain some systematic distortions and inherent disadvantages that must be considered. Examples in this section will focus on the advantages gained by using radar, and the following section will discuss the pitfalls and limitations of geological interpretation with radar.

1.5.2.1 Cloud Penetration

The most important capability of radar systems lies in their ability to operate independently of visible conditions allowing useful images to be made at night, during rain and through heavy clouds. Except for thermal IR wavelengths, none of these feats can be performed with a passive imaging system. Figure 1.21 illustrates the critical role that radar plays in mapping features in the tropics. It shows a comparison between an SIR-A radar image and an MSS image. For a study conducted of the Irian Jaya in New Guinea, in 1983, the MSS image was the best available at the time and still had a large percentage of cloud cover, whereas a single pass over the region with the space radar provided an excellent picture of the region. (SIR-A made only one global pass; see discussion at the end of this chapter.) It is interesting to note that the radar imagery provides detailed information on the topographic expression of geological features in this region even though the area is completely covered by dense forest. This additional advantage of using radar is illustrated well in the next example.

MSS IRIAN JAYA

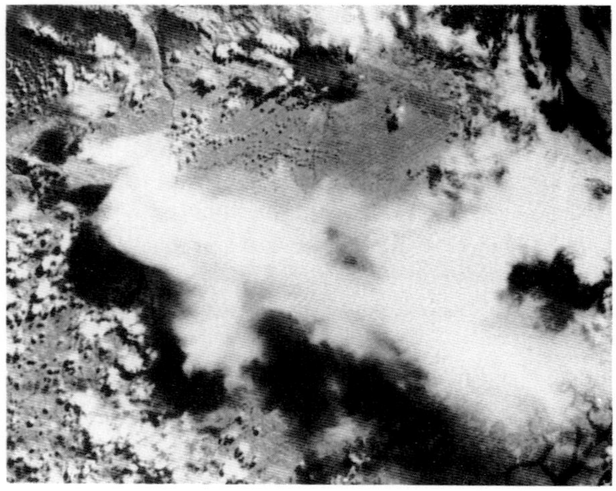

SIR-A IRIAN JAYA

ILLUMINATION

N 0 20 Km

Fig. 1.21. Example of cloud penetration with radar
The scene is from the rain forest of the Irian Jaya fold and thrust belt. This location is almost continuously obscured by heavy cloud cover. (Prepared by R. L. Dodge)

1.5.2.2 "Treetop Geology"

Figure 1.22 shows a comparison between space radar imagery and MSS in the heavy forests of Mexico and Guatemala. The MSS is completely "flat", offering no information on surface structures, whereas the radar image clearly shows the folded belt in this region. In both of the preceding examples, the radar images have captured the surface topography of the region by imaging the surface created by the tree-

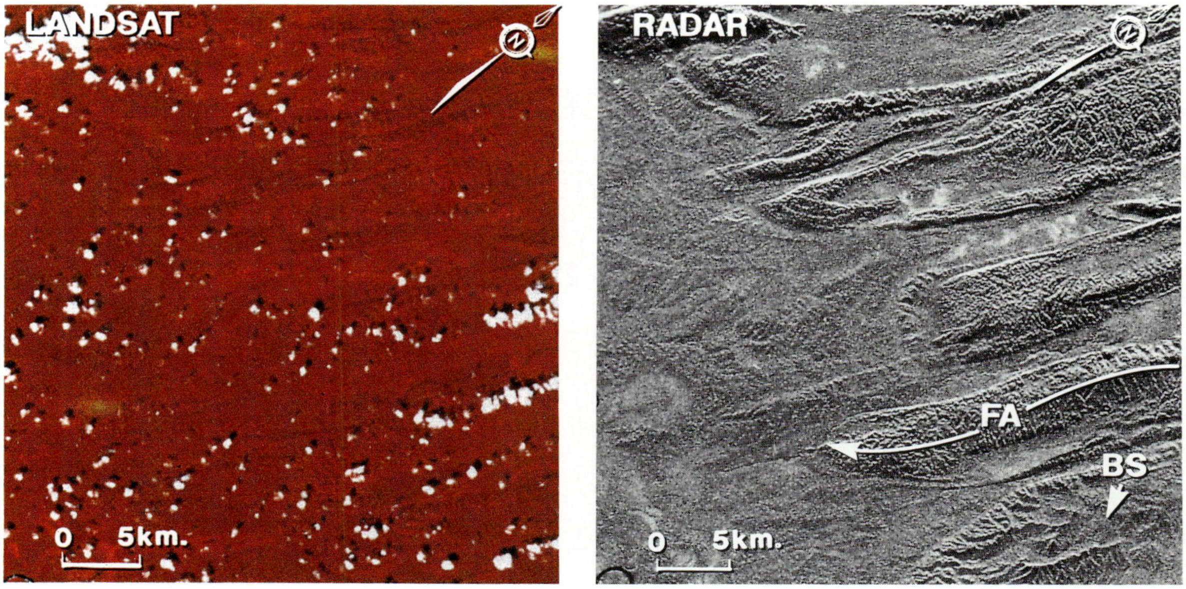

Fig. 1.22. Comparison between satellite and radar data in the heavy forests of Mexico and Guatemala
The satellite image is quite "flat", whereas the radar provides excellent details on "treetop" topography which can be used for geological mapping in this area. *FA* Fold axis; *BS* breached structure. (Prepared by R. L. Dodge; radar data courtesy of Intera)

tops. In other words, the image is not of the actual ground surface but indirectly indicates topographic variations. This capability of the radar is directly related to its active system which provides highly accurate distancing information (from platform to object). This concept in radar mapping can be referred to as "treetop geology". The same principles can also be used in any case where the surface expression of geological structure is masked by uniform vegetation. In other words, trees are not required for the treetop geology idea to work. Furthermore, this concept can be implemented for the detection of small-scale geological structures and other subtle topographic features. An outstanding example of this phenomenon is shown Fig. 1.23 which shows a comparison between a satellite image (MSS) and airborne radar from the Canadian Foothills west of Calgary, Alberta. As illustrated previously, the satellite image is quite flat showing very little information on surface topography and related geological features. The "flatness" on the MSS image is a direct result of the homogeneity of the region (which is partly due to extensive vegetative and soil cover; this aspect is further illustrated in Fig. 1.24 and the related discussion on surface roughness). In other words, there is little variation in the spectral signature of the materials captured by the image. In contrast, the radar image captures the topographic expressions of fold and thrust faults of the foothills as well as subtle topographic expression of surficial deposits related to glacial features. Needless to say, the ability of the interpreter to separate surface features related to geological structures from those created by surficial deposits is a crucial element in reliable interpretation of remote sensing data.

1.5.2.3 Surface Roughness Detection

The sensitivity of the radar to variations in surface roughness, as compared to passive satellite imaging systems, provides some unique advantages for mapping small-scale geological and geomorphological features. The most common usage of this phenomenon is for lithological mapping in heavily vegetated areas; this concept was nicely demonstrated by Sabins (1983, 1987) using examples from Irian Jaya, New Guinea. Another outstanding example of this phenomenon has been illustrated by McCauley et al. (1982) who showed the presence of ancient buried drainage systems in the Sahara Desert using radar imagery. Two possible mechanisms were proposed for the ability of the radar imaging system to capture the presence of this buried drainage

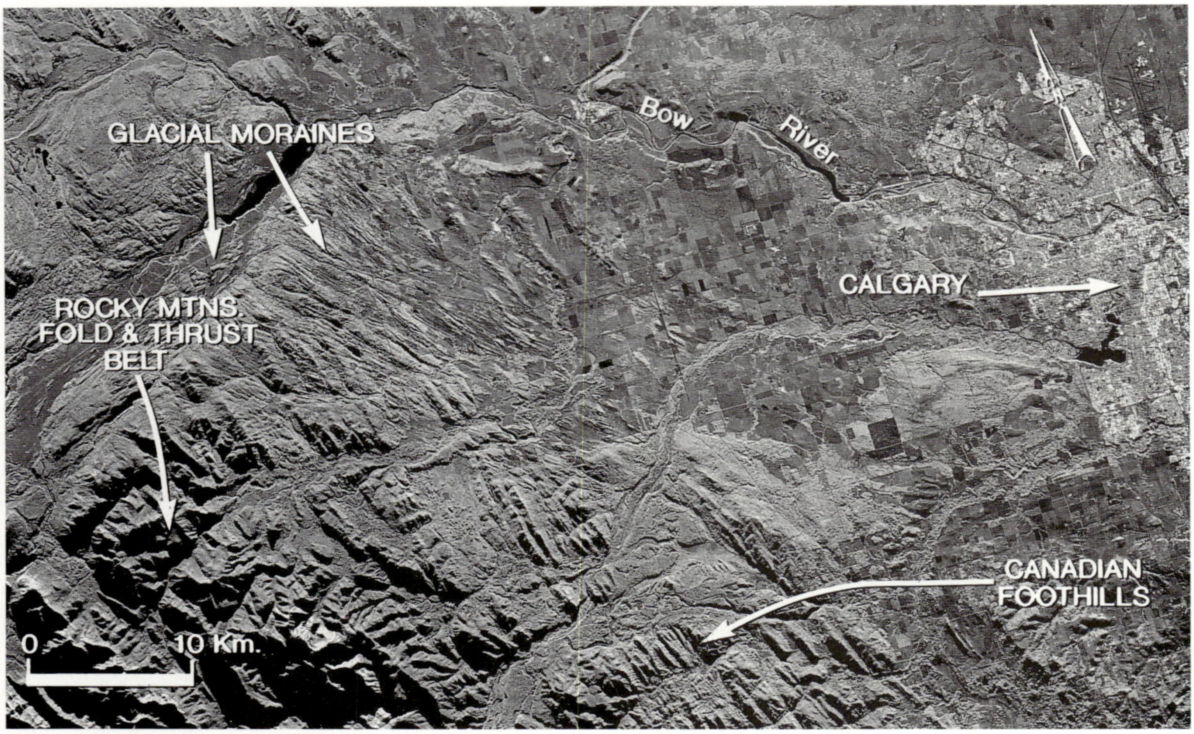

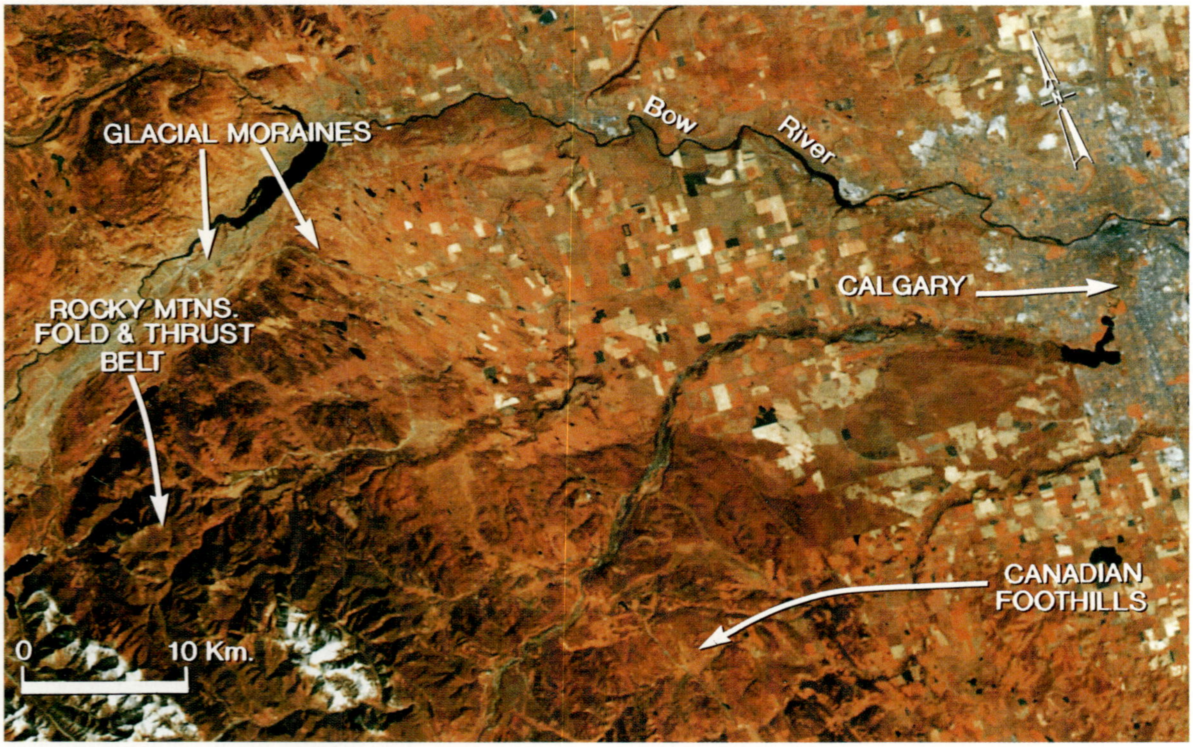

Fig. 1.23. Satellite imagery (MSS) and airborne radar Scene is of the Canadian Foothills, west of Calgary, Alberta. The radar image *(top)* shows detailed expression of the Canadian fold and thrust belt as well as glaciated features. These features are not apparent on the MSS image because of the homogeneity of the surface cover and lack of bedrock exposure. (Radar data courtesy of Intera, Calgary)

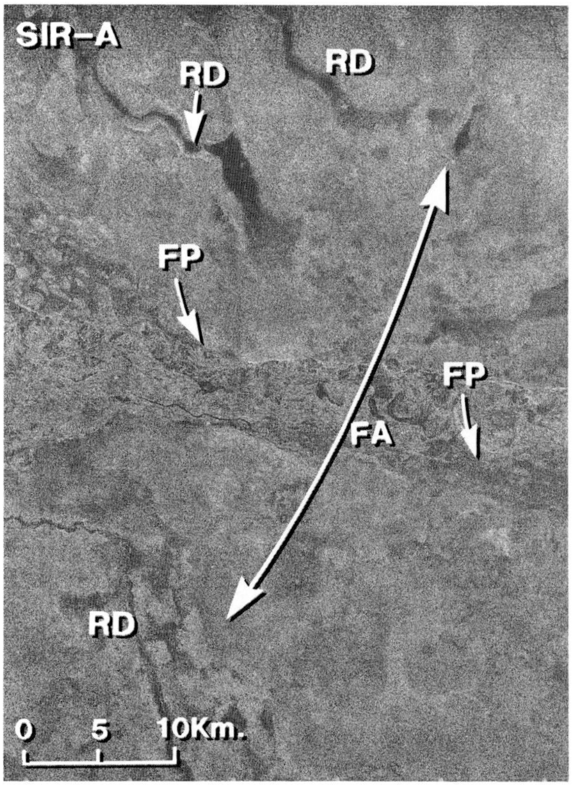

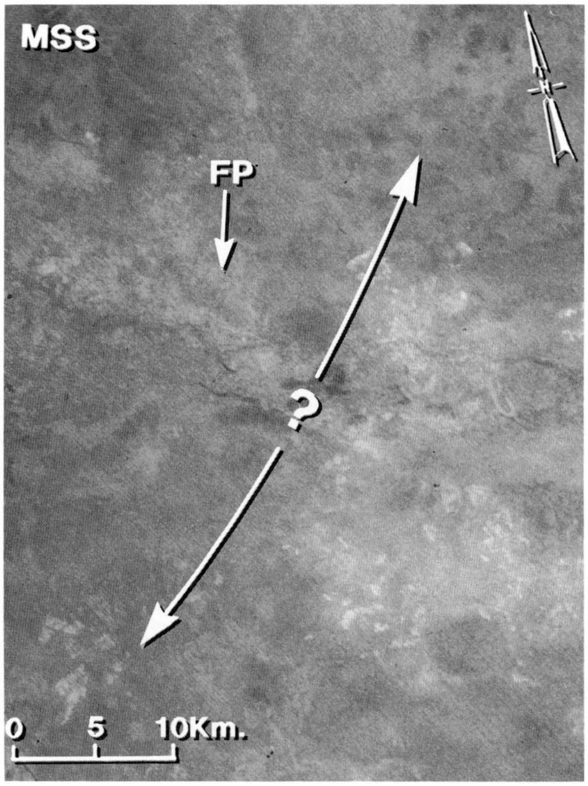

Fig. 1.24. Satellite and radar images from central Africa
Here, the ability of radar to image structurally related features in areas which typically show "flat" responses on satellite images is illustrated. A surface expression of an obscured subtle fold is depicted on SIR-A radar by its influence on drainage patterns and floodplain configuration. These features are much more evident on radar than they are on the satellite imagery data. *RD* Radial drainage; *FP* floodplain; *FA* fold axis

system. One theory holds that the patterns result from radar penetration and backscatter from subsurface variations in grain size. The alternative theory is that the radar is picking up subtle variations in surface roughness related to the old stream courses. Figure 1.24 shows how a similar phenomenon can be used to detect the subtle expressions of geological structures in remote areas in this case, a subtle fold in central Africa. The fold is recognized by the presence of an inactive (or buried) drainage system that outlines its topographic expression and by abrupt changes in the width of the floodplain of the valley that crosses the crest of the anticline. None of these features are evident on the passive system image of the same area.

1.5.2.4 Preferred Look Directions

Because of the directional nature of radar's imaging systems, the expression of geological features can either be enhanced or suppressed according to their orientation relative to the look direction. Features that trend nearly perpendicular to the look direction are enhanced the most, and parallel trending features are suppressed the most. Therefore, radar surveys can be designed to maximize the expression of desired features by choosing appropriate look directions. In addition, varying the look direction can provide a unique perspective of an area with multiple structural directions. An example of this capability is given by Harris (1991) and shows the surface expression of radar from the Cobequid/Chedabucto fault system in Nova Scotia (Fig. 1.25). As illustrated with a series of radar images of the study area with changing look direction (Fig. 1.26), different structural and surficial features of the area are depicted on each image and can be used to map these features at a high level of detail. With prior knowledge of the dominant structural and topographic grain of the area of interest, one could carefully design radar surveys with look directions that

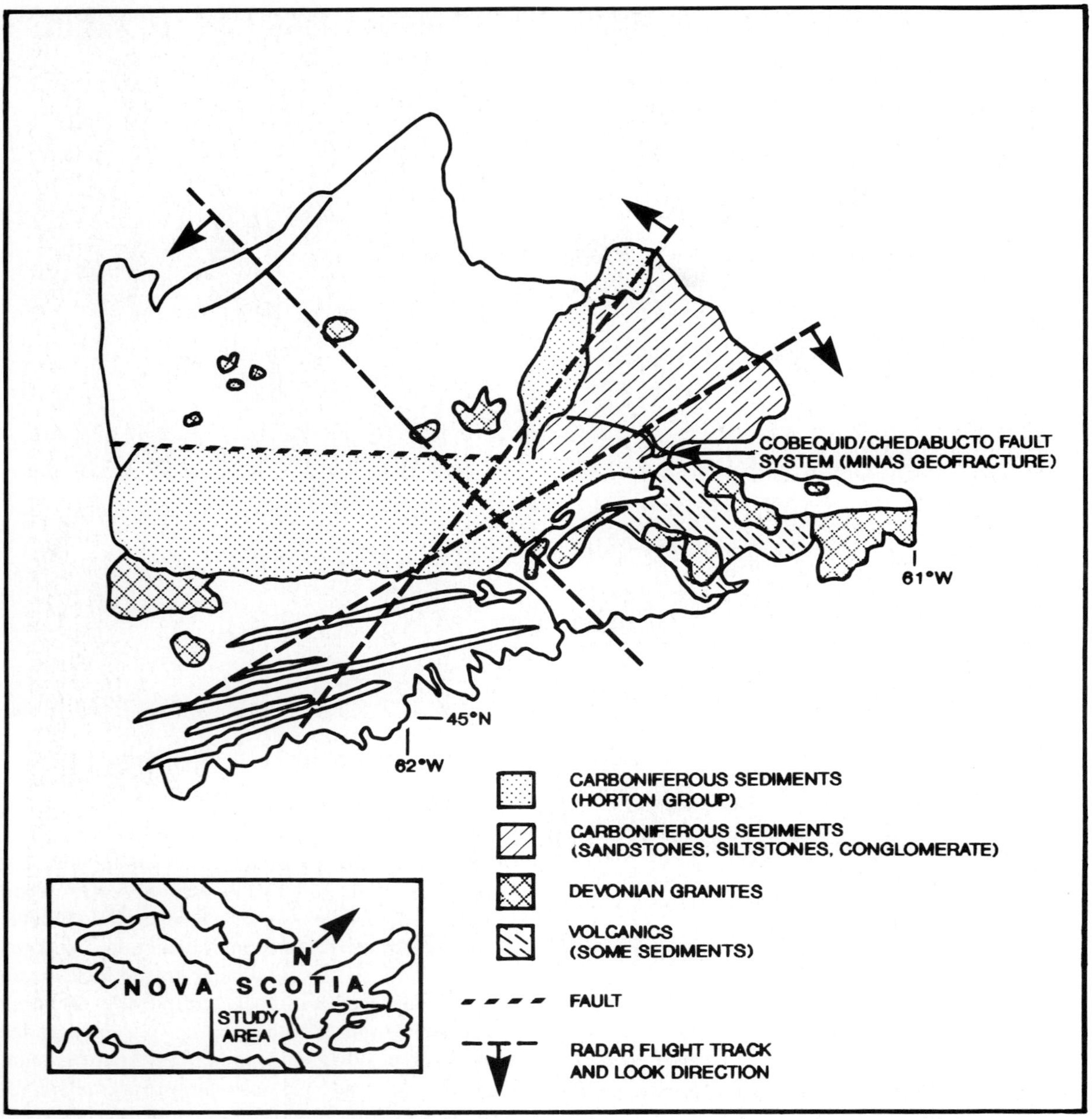

Fig. 1.25. Generalized geologic map of the Cobequid/Chedabucto fault system, Nova Scotia
Flight pass and look directions of airborne radar surveys are shown. (After Harris 1991)

Fig. 1.26. A set of radar images with different look and pass directions ▶
Different features are highlighted by each pass in accordance with their orientation relative to the look direction. *FT* Flight track; *LD* look direction. (After Harris 1991)

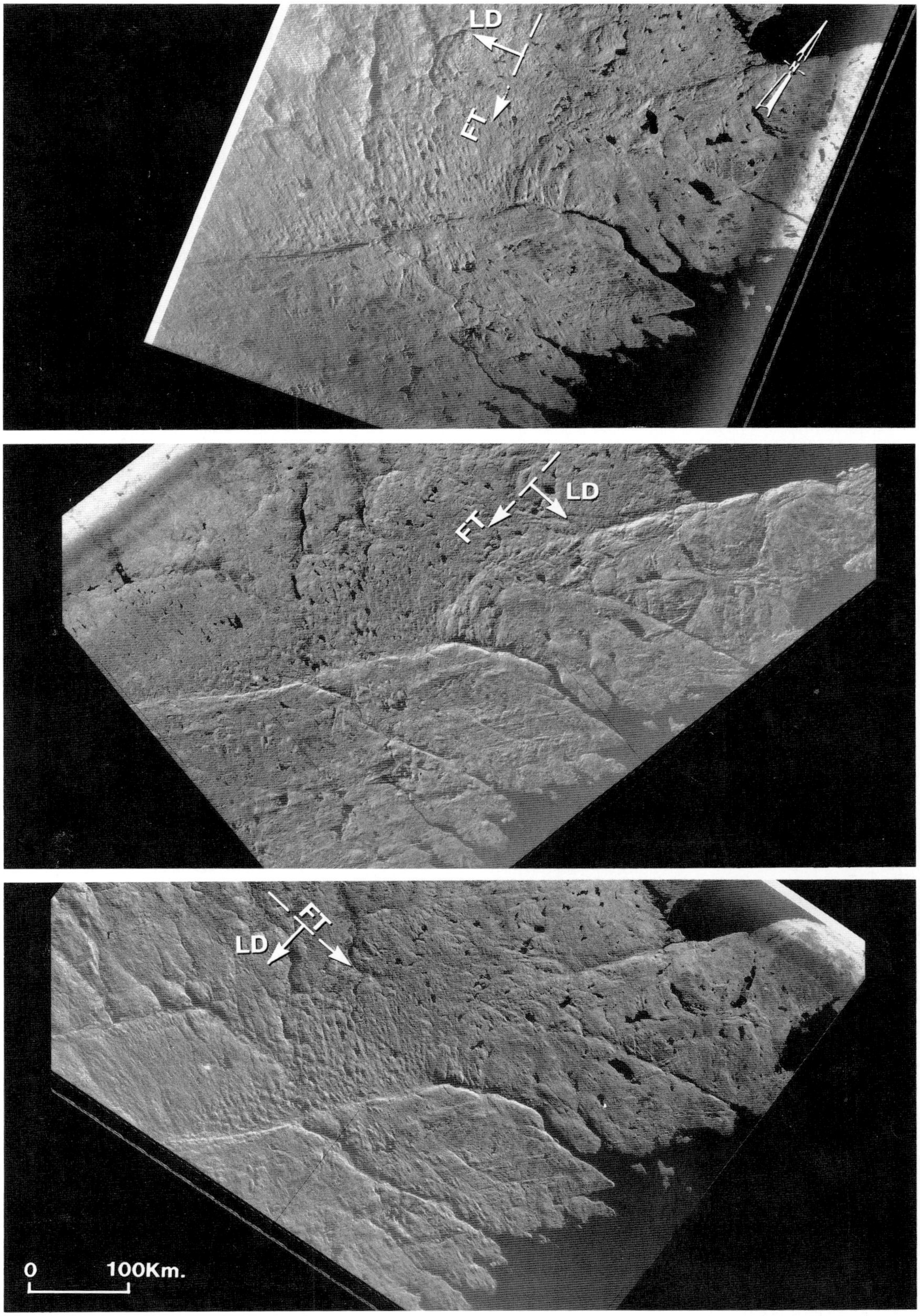
LD
FT
LD
FT
LD
FT
0
100Km.

would optimize the imaging of critical structural features.

1.5.2.5 Stereo Radar Imaging of Geological Structures

As in the case of SPOT stereo capabilities, the availability of stereo radar is particularly useful to geologists to map structures that are either highly deformed or consist of extremely low angle dipping strata. This subject will be extensively covered in Chapter 6 which discusses the principles of stereo mapping as a whole. A specific example of stereo radar imagery as compared to a monoscopic TM image is shown with an example from the Canadian Foothills in British Columbia in Fig. 1.27. A subtle linear topographic feature is nicely captured by the radar data and appears to reflect the surface expression of a wrench fault and related oblique anti-

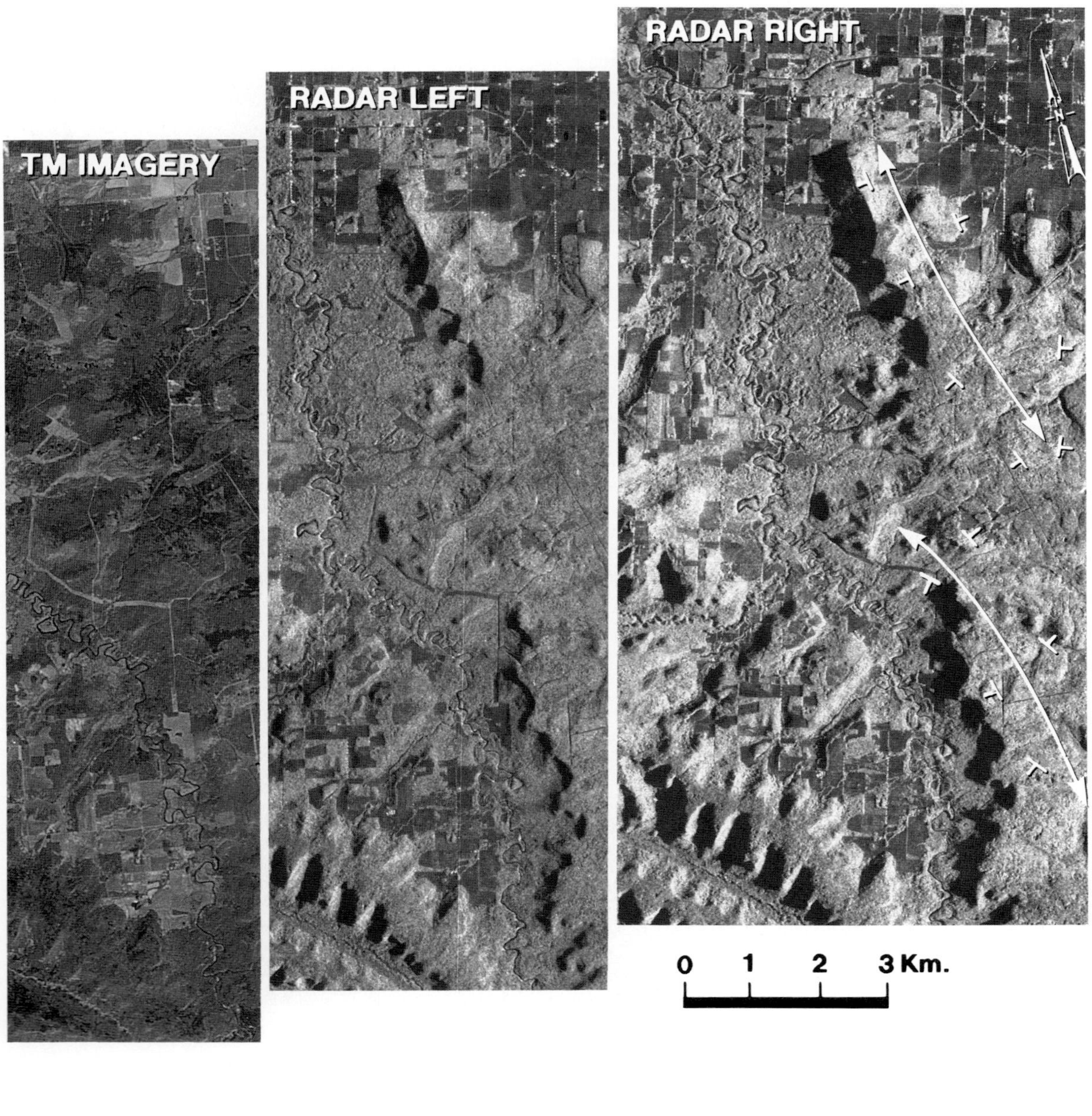

Fig. 1.27. Comparison between a stereo pair of airborne radar and satellite imagery
Scene shows a subtle linear, topographic ridge in the Canadian Foothills, British Columbia. Stereo measurements, which were obtained with a Kelsh stereo plotter, indicate that the ridge is formed by the alignment of two oblique surface anticlines that probably represent the expression of a reactivated wrench fault system. Stereo can be viewed with a pocket stereoscope. (Radar data courtesy of Intera)

clines. This feature is totally obscured on the TM image. This example actually encompasses other advantages of radar images discussed above. In fact, the area is covered by heavy vegetation, lacks bedrock exposure, and the preferred look direction is being used.

1.5.3 Problems with Radar Imagery

1.5.3.1 Slant Range Images

There are several problems unique to radar imaging systems. The first of these results from the peculiar geometry of the radar image. Because radar systems capture images by looking sideways, and distances are determined by return travel times, the scale of the image becomes distorted with increasing distance from the platform (Fig. 1.28). The image perceived by the imaging system is called the slant-range image. It can be corrected by mapping points with a hyperbolic transformation (Fig. 1.29).

1.5.3.2 Layover

High-relief features, such as mountains, produce a unique distortion. Because the radar wave will reach the top of the mountain before the bottom, the top will appear closer than it actually is. If a vertical wall were imaged, it would appear to be leaning over towards the platform. This effect, called layover, is shown in Fig. 1.30. A common effect of layover is to make symmetric mountains resemble flatirons, which may introduce a rather large interpretation error. The layover effect is most severe close to the platform. Layover depends on the relief of the terrain but also increases with depression angle – another effect to consider when deciding on a depression angle. Layover and its distorting effect should be of great concern to geological application of space radar since its sensors are positioned considerably higher than airborne systems and consequently have a steeper angle of incidence. An ERS-1 image from the Northwest Territories (Fig. 1.31) shows the Franklin Mountains to be extremely tilted towards the look direction, exhibiting exaggerated flatirons.

1.5.4 Spaceborne Radar Systems in Use

Launched in June of 1978, Seasat was the first radar satellite to be deployed by NASA. Seasat was originally designed to study oceanic phenomena such as roughness, current patterns, and sea ice conditions, but has also proved useful for terrain observations. Unfortunately, a power failure ended its radar imaging capabilities in mid-October of 1978 after only 99 days of operation so that the database of Seasat images is relatively small. Despite Seasat's

Fig. 1.28. Slant-range vs. ground-range image
The distortion of a sideways looking radar causes scale further from the system to be compressed

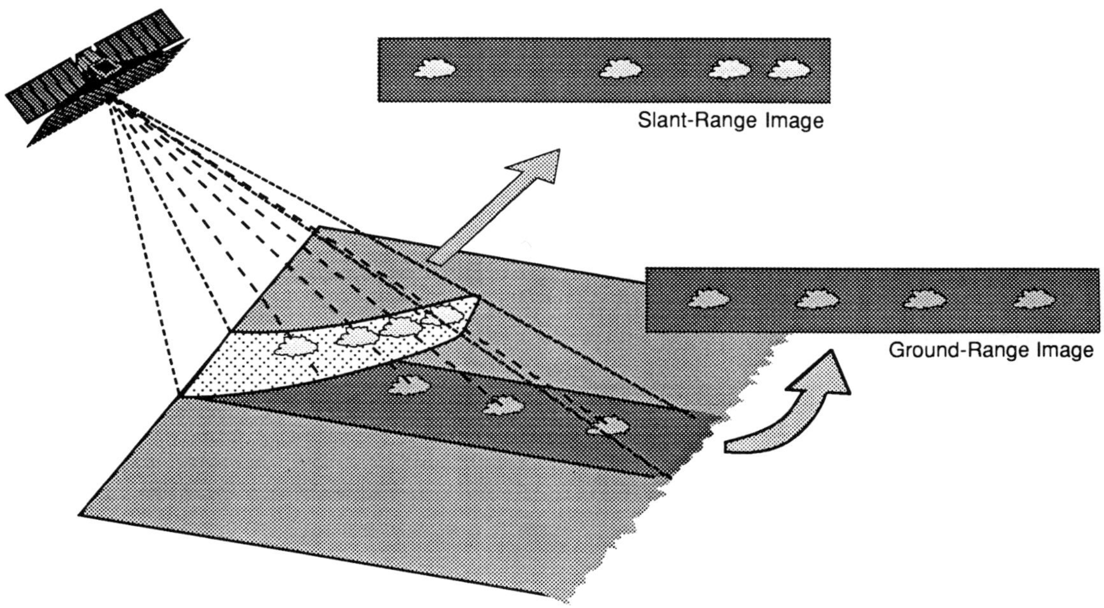

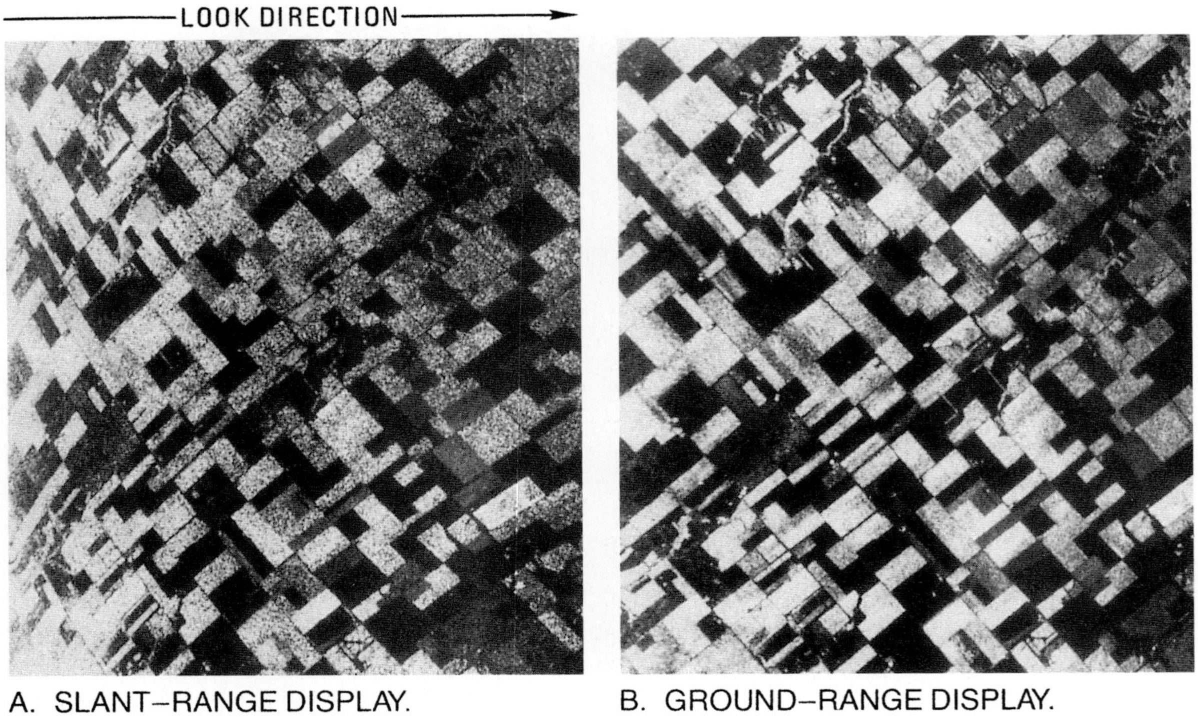

Fig. 1.29. Example of slant-range correction
An airborne radar of central Illinois which was corrected from a slant-range display into a ground-range display. Each image covers a ground area of 10 × 10 km. (After Sabins 1987; images courtesy J. P. Ford, Jet Propulsion Laboratory)

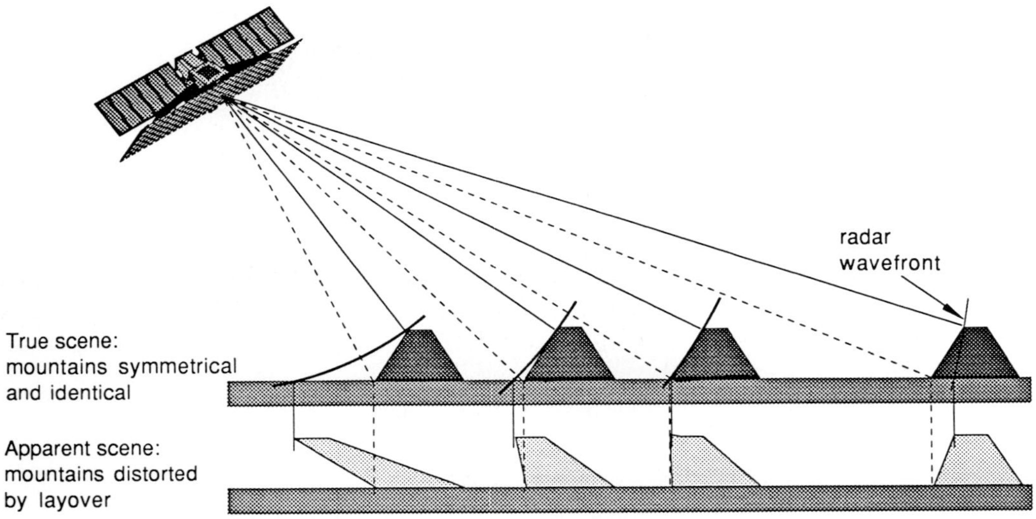

Fig. 1.30. The layover effect
Although all mountains are identical and symmetrical the ones in the near range appear to have steeper fore slopes than they do in reality. The effect decreases in the far range

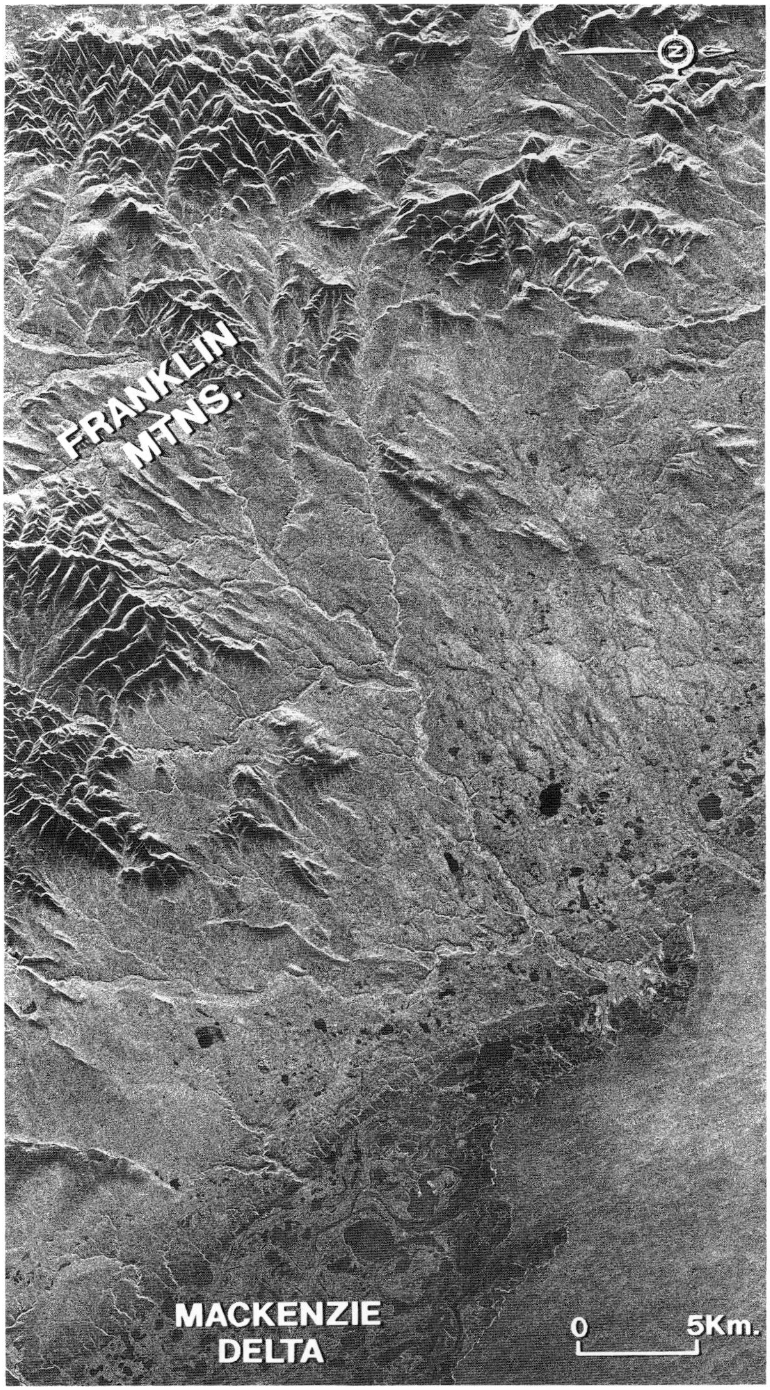

Fig. 1.31. ERS-1 image of the Franklin Mountains, Northwest Territories

The mountains are artificially tilted towards the look direction showing an exaggerated manifestation of flatirons

Table 1.6. Characteristics of common radar satellites used for geological exploration

Characteristics	Seasat (78)	SIR-A (81)	SIR-B (84)	ALMAZ-1 (87)	ERS-1 (91)	JERS-1 (92)
Instruments carried	SAR	SAR	SAR	SAR Radiometric scanner	SAR Scatterometer Altimeter	SAR Optical Imaging
Wavelength	L-band	L-band	L-band	S-band	C-band	L-band
Spatial resolution	25 m	38 m	25 m	15–30 m	30 m	18 m
Nominal altitude	790 km	250 km	225 km	300 km	785 km	568 km
Nominal swath width	100 km	50 km	40 km	20 km	100 km	75 km

short lifetime, it was a significant project that paved the way for future spaceborne radar systems. NASA carried out two other spaceborne radar experiments in 1981 and 1984. The Shuttle Imaging Radar units known as SIR-A and SIR-B were carried on board the space shuttle and remained inside the cargo bay during the imaging process. SIR-A and SIR-B followed different orbits but used similar wavelengths. SIR-B matches Seasat's 25-m resolution, while SIR-A has a less clear image with 38-m resolution. SIR-B had the added ability to change its depression angle while in flight. This allowed both an evaluation of the effect of depression angle on radar returns and the production of stereo radar images. The ALMAZ 1 radar satellite, under direction of the Soviet space agency, Glavkosmos, completed a 2-year mission in 1989. Its data was classified until recently when a joint venture between Glavkosmos and the Space Commerce Corporation (USA) made the archived images commercially available. Almaz 2 was launched in 1991. The new version, as well as the old, has an angle of incidence selectable from 30° to 60°. The ERS-1 (European remote sensing satellite) was launched in 1991 and represents the first such effort from the 13-member nations making up the European Space Agency. The main objectives of the 3-year mission include meteorology, sea-state forecasting and monitoring of seaice. The satellite has on-board recorders that can retain information from the altimeter and scatterometer. The imaging radar system, however, generates too much data for on-board storage and is transmitted to a receiving station. This means that ERS-1 can only provide radar images when it is within the range of one of its receiving stations. These overlapping ranges cover most of North and South America, Asia, Europe and Australia (Attema 1991).

JERS-1 (also named "FUYO-1") is the first radar satellite entry from Japan. Launched in February of 1992, it is the first to include both SAR and optical sensors. Its main purpose is geological mapping but it will also gather data related to agriculture, forestry, fishery, coastal monitoring and disaster prevention. The optical sensor has eight bands ranging from 0.52 to 2.4 μm. JERS-1 encountered technical problems after launch including difficulty in deploying its SAR antenna, but preliminary evaluations of data are now available (even though SAR images are still marred by distortions caused by the difficulties). Despite its problems, JERS-1 has several advantages including a more shallow look angle than other satellites. This feature decreases geometric distortion due to layover (Nishidai 1993; Nemoto et al. 1991). The characteristics of the six radar systems discussed above are summarized in Table 1.6.

1.5.5 Future Spaceborne Radar Systems

Canada plans to launch its own radar satellite, Radarsat, with sponsorship from NASA, in 1995 for a 5-year mission. The project was designed to provide better monitoring capabilities for the northern Canadian regions with sea-ice conditions a particular concern. Because the higher latitude regions are often cloud covered and receive little or no sunlight during the winter, a radar satellite was deemed the best solution. Radarsat will carry only one single-band SAR but will be capable of several imaging modes with varying resolutions, swath widths and incidence angles. Because of extensive development in beam steering capabilities, Radarsat will be able to use off-nadir imaging to capture any area in Canada within 3 days. Regions at the equator can be imaged every 4 days and regions above 79° lat. are covered on a daily basis. Radarsat will also include an on-board SAR data recorder which will allow it complete global coverage. The SAR antenna will be oriented to the north of the satellite's ground track so that images can be obtained of all latitudes including the north pole. Moreover, two periods during the mission are designated for Antarctic mapping requiring the satellite to be rotated 180° so that the SAR antenna points south of the satellite (Raney 1991).

NASA, in cooperation with the German and Italian space agencies, planned another shuttle radar project for early 1994 called SIR-C/X-SAR. SIR-C will acquire images simultaneously with L-band (24 cm) and C-band (5.6 cm), while X-SAR will use X-band radar (3 cm). Two subsequent flights are planned for different seasons to enable data acquisitions under different environmental conditions. The altitude for imaging will be 215 km and 50 h of data should be acquired during each mission. Data will be processed on board as well as relayed to earthbound stations. NASA is also developing a real-time SAR ground processor that will provide high spatial resolution but limited corrections for spatial and radiometric distortions. The output will be a continuously produced strip representing a ground speed of approximately 7 km/s. (Evans et al. 1993). Also, in the 1994, the European Space Agency plans to launch an ERS-2 satellite.

1.5.6 Evaluation of Space Radar

There is great anticipation in the geological community for the arrival of the new generation of space radar imagery data which will offer, for the first time, complete global coverage. Airborne radar companies are also curious to evaluate these data sets in order to ascertain whether it will replace the role of their instruments. A complete evaluation of the new generation of space radar is not available yet, but the first set of ERS-1 and ALMAZ images as well as a small sample of JERS-1 images provides some initial information for its geological mapping capabilities. A comparison between ERS-1 space radar and airborne radar is provided in Fig. 1.32. The two upper images show examples from the rugged region of Vancouver Island, whereas the bottom two compare data from the low-relief terrain of the exposed Canadian Shield. There are two fundamental differences between the airborne and spaceborne radar. One is related to differences in spatial resolution and the other to geometric distortions. Regarding spatial resolution, it is evident from the examples shown that the airborne radar provides significantly sharper images. This difference may not be significant for reconnaissance (regional) geological mapping but may present some limitation for site-specific investigations that require detail mapping of lithological units, faults and other key geological features. For example, the airborne radar from the Canadian Shield shows significantly more detail on the orientations and densities of fractures than the spaceborne radar. Regarding geometric distortion, it appears that the layover effect creates a serious problem for geological mapping in the rugged mountains of the Brooks Peninsula on the west coast of Vancouver Island. The symmetric mountain ridges of these heavily dissected igneous and metamorphic rocks appear on ERS-1 as a series of asymmetric ridges characterized by the development of large-scale hogbacks and related flatirons which falsely create landform characteristics of highly folded and heavily eroded sedimentary rock units. Moreover, the illuminated sides of the mountains have been shortened by the layover effect, making geological mapping virtually impossible.

The serious shortcoming in mapping mountainous areas with spaceborne radar was illustrated in Fig. 1.31, which showed the exaggerated flatiron characteristics of mountain belts. This limitation should be of great concern to explorationists because a significant portion of the unexplored potential in the heavily forested tropics lies within the rugged terrain of highly folded sedimentary strata. Initial imagery from JERS-1, which has a smaller depression angle, illustrates that the layover effect can be reduced to manageable levels (Fig. 1.33).

Overall, the introduction of systematic space radar coverage should significantly enhance the role of both air- and space-based radar imaging systems. At this point, it appears that the new generation of space-radar imaging systems will provide reconnaissance mapping tools for large areas, while detailed interpretation of key areas is most likely to be continued with the aid of airborne radar (in the same manner that stereo photographs are used to compliment the interpretation of passive satellite imaging systems). This is particularly true because a large portion of the previously unmapped oil-rich areas lies within the tropical regions where passive imaging systems have not been effective so far.

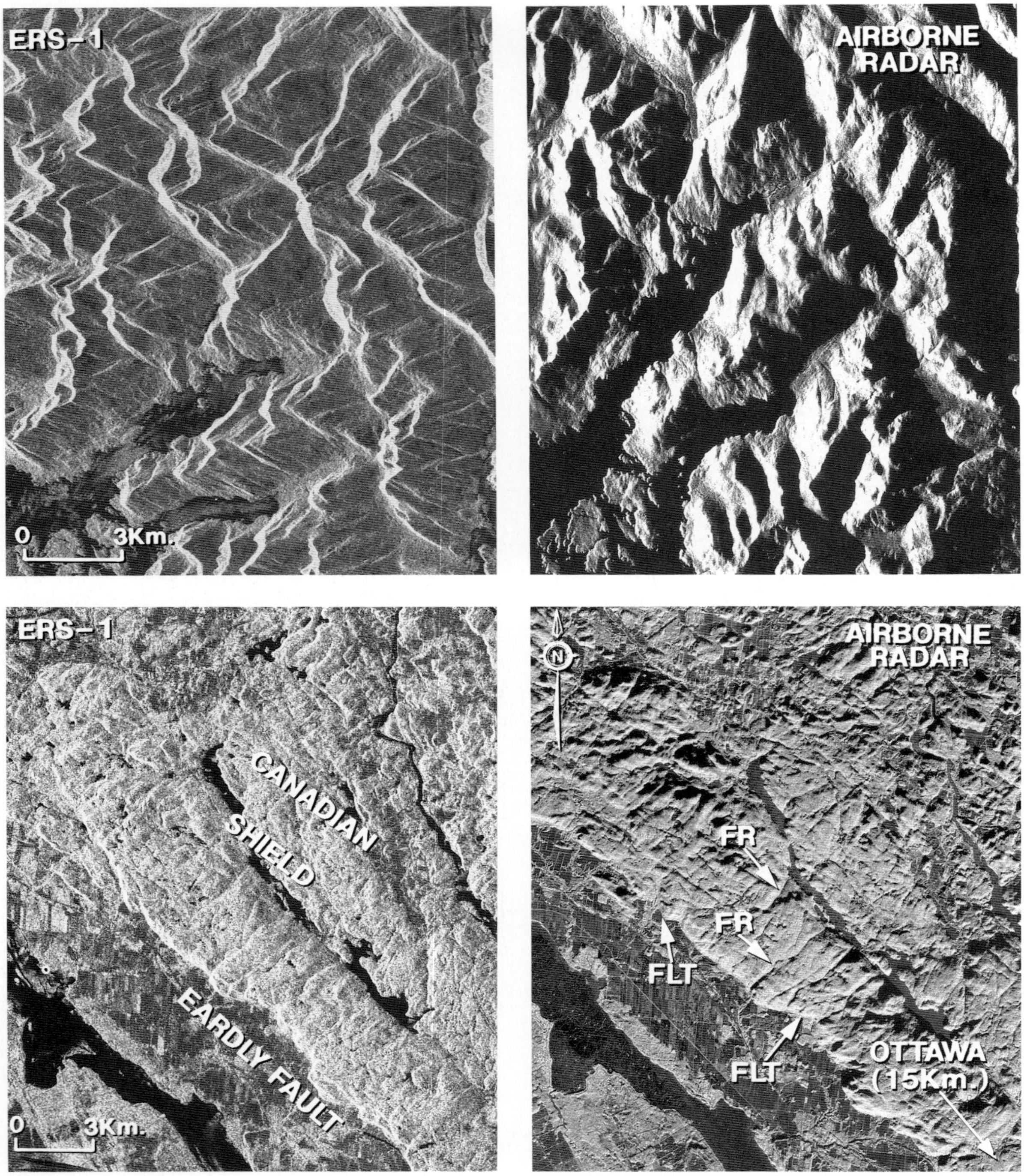

Fig. 1.32. Comparison of ERS-1 and airborne radar
Shown are two data sets from the Vancouver Island area *(upper pair)* and the exposed Canadian Shield near the city of Ottawa *(lower pair)*. *FLT* Fault line trace; *FR* fractures. (Radar data courtesy of Intera)

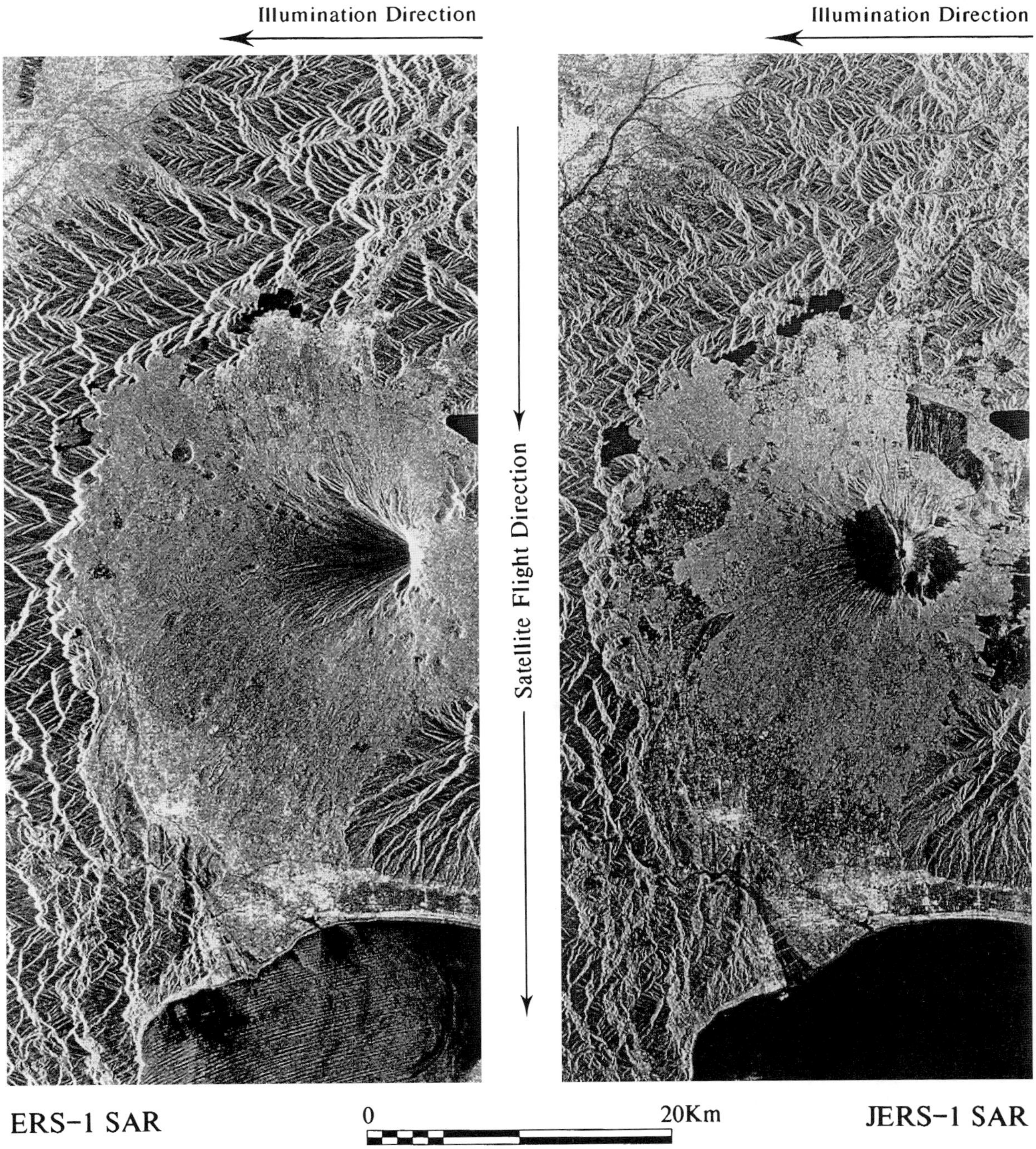

Fig. 1.33. Comparison of JERS-1 and ERS-1 radar
Shown are two radar images of Mt. Fuji in Japan. The ERS-1 image shows a severe amount of layover to the extent that the rim of the caldera is obscured. Moreover, the terrain around the mountain shows severe distortion with exaggerated flatirons. JERS-1 appears to rectify some of these problems. (After Nishidai 1993)

1.6 Summary

Passive imaging systems work by detecting electromagnetic radiation reflected from the Earth and converting it into a digital format which can be processed by computer. Multispectral systems detect radiation in several frequency ranges, or bandwidths, along the EM spectrum; this enhances the interpretability of the image.

Reflectance values from each of the bandwidths can be compared to known spectral response curves and then correlated with specific materials. The range most useful for distinguishing a material is called its diagnostic reflectance range. For effective multispectral studies, the diagnostic reflectance range should be matched to the bandwidths of the imaging system. The effectiveness of an imaging system is measured by its spatial resolution or clarity, its spectral resolution or range of frequencies, and its radiometric resolution, which is the degree of sensitivity to brightness.

There are several space-based imaging systems that provide such data. The Landsat satellites carry the multispectral scanning system (MSS) and the thematic mapper (TM) system. The TM, a more recent device, exhibits greater spatial and spectral resolution than the MSS. The SPOT satellite can operate with extremely high spatial resolution at the expense of its multispectral capabilities and is also capable of off-nadir imaging which allows it to make images of the same area during successive orbits. The off-nadir capability also makes possible stereo imaging where two slightly offset images are superimposed to allow three-dimensional evaluation of data.

Active imaging systems work by emitting their own radiation and detecting its reflections. The advantages of radar use include the ability to penetrate cloud cover, produce stereo imagery and infer topological information from variations in tree or vegetative cover. Disadvantages include systematic distortions such as layover and radar shadowing.

References and Further Reading

Attema Evert PW (1991) The active microwave instrument on board the ERS-1 satellite. Proc IEEE, vol 79 (6): 791–799

Bernstein R (1976) Digital image processing of earth observation sensor data. IBM J Res Dev 20: 40–57

Bidwell TC, Mitchell CA (1975) Author index to published ERTS-1 reports. US Dept Commerce, National Technical Information Service, Document PB 248294, Springfield, VA

Blom RG, Daily MI (1982) Radar image processing for rock-type discrimination. IEEE Trans Geosci Remote Sens GE-20: 343–51

Blom RG, Crippen RE, Elachi C (1984) Detection of subsurface features in Seasat radar images of Deans Valley, Mojave Desert, California. Geology 12: 346–349

Bueche, FJ (1980) Introduction to physics for scientists and engineers. McGraw-Hill, New York, pp 593–612

Christner DG (1940) Todd Ranch (Oil) Discovery, Crockett County, Texas. AAPG Bull 24, (6): 1126–1127

Craib KB (1972) Synthetic aperture SLAR systems and their application for regional resources analysis. In: Shahrokhi F (ed) Remote sensing of earth resources, vol 1. University of Tennessee Space Institute, Tullahoma, Tenn, pp 152–178

Drury SA (1987) Image interpretation in geology. Department of Earth Sciences, The Open University, Boston

Eliason EM, Chavez PS, Soderblom LA (1974) Simulated "true color" images from ERTS data. Geology 2: 231–234

Evans DL, Lang HR (1985) Techniques for multi-sensor image analysis. Proc 18th Int Symp on Remote sensing of the environment, Paris, 1987

Evans DL, Elachi C, Stofan ER, Holt B, Way JB, Kobrick M, Ottl H, Pampaloni P, Vogt M, Wall S, van Zyl J, Schier M (1993) The Shuttle imaging radar-C and X-SAR mission. EOS Transactions, American Geophysical Union (March 30)

Estes R (1985) Remote sensing fundamentals. In: Holz RK (ed) The surveillant science. John Wiley, New York, pp 12–27

Flawn PF (1968) Palestine sheet. Geologic atlas of Texas. Bureau of Economic Geology, Univ Texas, Austin, Texas, scale 1 : 250 000

Ford JP, Blom RG, Bryan MG, Daily MI, Dixon TH, Elachi C, Xenos EC (1980) Seasat views North America, the Caribbean, and Western Europe with imaging radar. JPL Publ, Pasadena, CA, pp 80–67

Ford JP, Cimino JB, Elachi C (1983) Space shuttle Columbia views the world with imaging radar – the SIR-A experiment. JPL Publ, Pasadena, CA, pp 82–95

Freden SC, Gordon F (1983) Landsat satellites. In: Colwell RN (ed) Manual of remote sensing, 2nd edn. Amer Soc Photogramm, Falls Church, VA, pp 517–578

Gess GJ, Chrowicz B, Becue B, Curnelle R, Deroin JP, Huger J, Perrin G, Ronfola D (1986) Methodology for the use of SPOT imagery in petroleum exploration. In: SPOT 1 image utilization assessment, results. Centre National d'Etudes Spatiales, Toulouse, pp 811–819

Goetz AFH (1976) Remote sensing geology – Landsat and beyond. Caltech/JPL Conf on Image processing technology, data sources and soft ware for commercial and scientific purposes, JPL Publ, Pasadena, CA, SP 43–30, pp 8–1 to 8–8

Gugan DJ, Dowman IJ (1988) Topographic mapping from SPOT imagery. Photogramm Eng Remote Sens 54 (10): 1409–1414

Harris JR (1991) Mapping of regional structure of eastern Nova Scotia using remotely sensed imagery. In: Implications for regional tectonics and gold exploration. Can J Remote Sens 17 (2): 122–136

Holz RK (ed) (1985) The surveillant science: remote sensing of environment. John Wiley and Sons, New York

Hunt GR, Salisbury JW (1970) Visible and near-infrared spectra of minerals and rocks – I silicate minerals. Mod Geol 1: 283–300

Hunt GR (1980) Electromagnetic radiation: the communication link in remote sensing. In: Siegal BS, Gillespie AR (eds) Remote sensing in geology. New York, John Wiley and Sons, Chap 2

Jensen H, Graham LC, Porcello LJ, Leith EN (1977) Side-looking airborne radar. Sci Am 237: 84–95

McCauley JF, Schaber GG, Breed CS, Grolier MJ, Haynes CV, Issawi B, Elachi C, Blom RG (1982) Subsurface valleys and geo-archeology of the eastern Sahara revealed by Shuttle radar. Science 218: 1004–1020

Moore RK (ed) (1983) Imaging radar systems. In: Colwell RN (ed) Manual of remote sensing, 2nd edn. Am Soc Photogramm, Falls Church, VA, pp 429–74

NASA (1976) Landsat data users handbook. Goddard Space Flight Center, Doc No 76SDS-4258, Greenbelt, MD

Nassau K (1980) The causes of color. Sci Am 39 (4): 124–154

Nemoto Y, Nishino H, Ono M, Mitzutamari H, Nishikawa K, Tanaka K (1991) Japanese earth resources Satellite-1 synthetic aperture radar. Proc IEEE 79 (6): 800–809

Nishidai T (1993) Early results from Japan's Earth Resources Satellite (JERS-1) Int J Remote Sens 14 (9): 1825–1833

Padgham CA, Saunders JE (1975) The perception of light and color. Bell, London

Pravdo SH, Huneycutt B, Holt BM, Held DN (1982) Seasat synthetic-aperture radar data users' manual JPL Publ, Pasadena, CA, pp 82–90

Raney K (1991) RADARSAT. Proc IEEE 79 (6): 839–848

Sabins FF (1969) Thermal infrared imagery and its application to structural mapping in southern California. Geol Soc Am Bull 80: 397–404

Sabins FF (1973) Geologic interpretation of radar and space imagery of California. AAPG Bull 57: 802 (Abstr)

Sabins FF (1980) Interpretation of thermal infrared images. In: Siegal BS, Gillespie AR (eds) Remote sensing in geology, John Wiley and Sons, New York, pp 275–295

Sabins FF (1983) Geologic interpretation of Space Shuttle radar images of Indonesia. AAPG Bull 64: 612–628

Sabins FF (1987) Remote sensing: principles and interpretation. WH Freeman, San Francisco

Salisbury JW, Hunt GR (1974) Remote sensing of rock-type in the visible and near infrared. Environmental Research Institute of Michigan, Proc 9th Int Symp on Remote sensing of the environment, Ann Arbor, pp 1953–1958

Schanda E (ed) (1976) Remote sensing for environmental sciences. Springer, Berlin Heidelberg New York

Siegel BS, Gillespie AR (1980) Remote sensing in geology. John Wiley and Sons, New York

SPOT Image Corporation (1984) SPOT data user's handbook. Toulouse, France

Taranik JV (1987) First results of international investigation of the applications of SPOT 1 data to geologic problems, mineral and energy exploration. In: SPOT-1 image utilization assessment, results. Centre National D'Etudes Spatiales, Toulouse, pp 701–708

Townsend TE, Dodge RL (1983) Techniques for geologic mapping based on the specral component of Landsat imagery. Exxon Production Research Company, Internal Rep

Townsend TE (1984) The significance of iron oxides and clays in petroleum and minerals exploration. Exxon Production Research Company, Internal Rep

Wadge G, Dixon TH (1984) A geological interpretation of Seasat-SAR imagery of Jamaica. J Geol 92: 561–81

Wolfe EW (1971) Thermal IR for geology. Photogramm Eng 37: 43–52

Chapter 2 Digital Image Manipulation

2.1 Introduction

As discussed in Chapter 1, most satellite images are in the form of digital data. Data is therefore stored as a series of digital numbers (DNs), each representing the intensity value for a particular pixel. Each pixel covers one resolution cell of terrain (whose size is determined by the spatial resolution).

This format is necessary for the imaging system to be able to transmit its data to earthbound receiving stations. Also, it allows the imaging system to record data quickly and store it conveniently. The digital format also opens the door to the whole realm of computer manipulation techniques. Because the image exists as a simple array of numbers within the computer, various mathematical procedures can be performed to make the data more usable.

The techniques of computer manipulation are not limited to digital data. Any image can be changed into a computer-compatible form by means of a digitizing system. Such a system is illustrated in Fig. 2.1.

The output of this process is a matrix of DNs representing the intensity values of the original image similar in format to data recorded by a scanning system. Thus, computer techniques can be applied not only to images from scanning systems but also to aerial photographs and other data such as gravity, magnetics, structure, isopach and topographic maps.

A procedure also exists for turning digital data back into an image. Using a similar principle, a photographic plate is exposed to a moving light source that crosses the plate in a pattern that matches the one used by the scanning system on the terrain. The intensity of the light source varies with the DN for each particular pixel, thus varying the intensity with which the photographic plate is exposed.

It should be noted that scanning systems do not actually produce images. Raw data from a scanning system such as Landsat is simply a string of numbers. Before the image can be "seen", it must be computer processed and turned into a viewable format. It is rare, as well, that data are directly turned into the image. Rather, extensive computer manipulation is usually involved in the conversion. Figure 2.2 summarizes the digital image acquisition process.

Computer manipulation techniques can be divided into four categories:

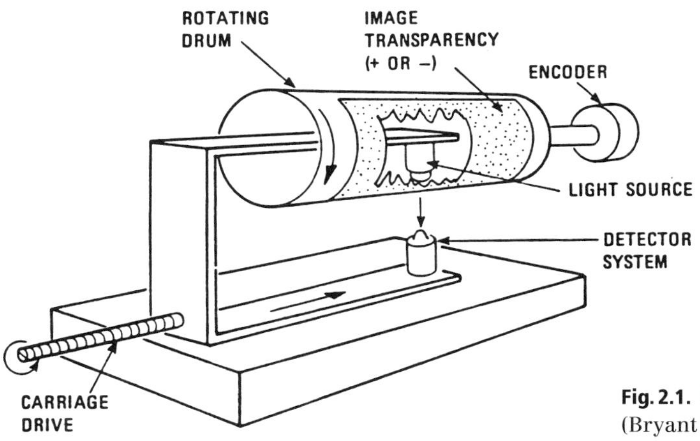

Fig. 2.1. System for digitizing image data (Bryant 1974)

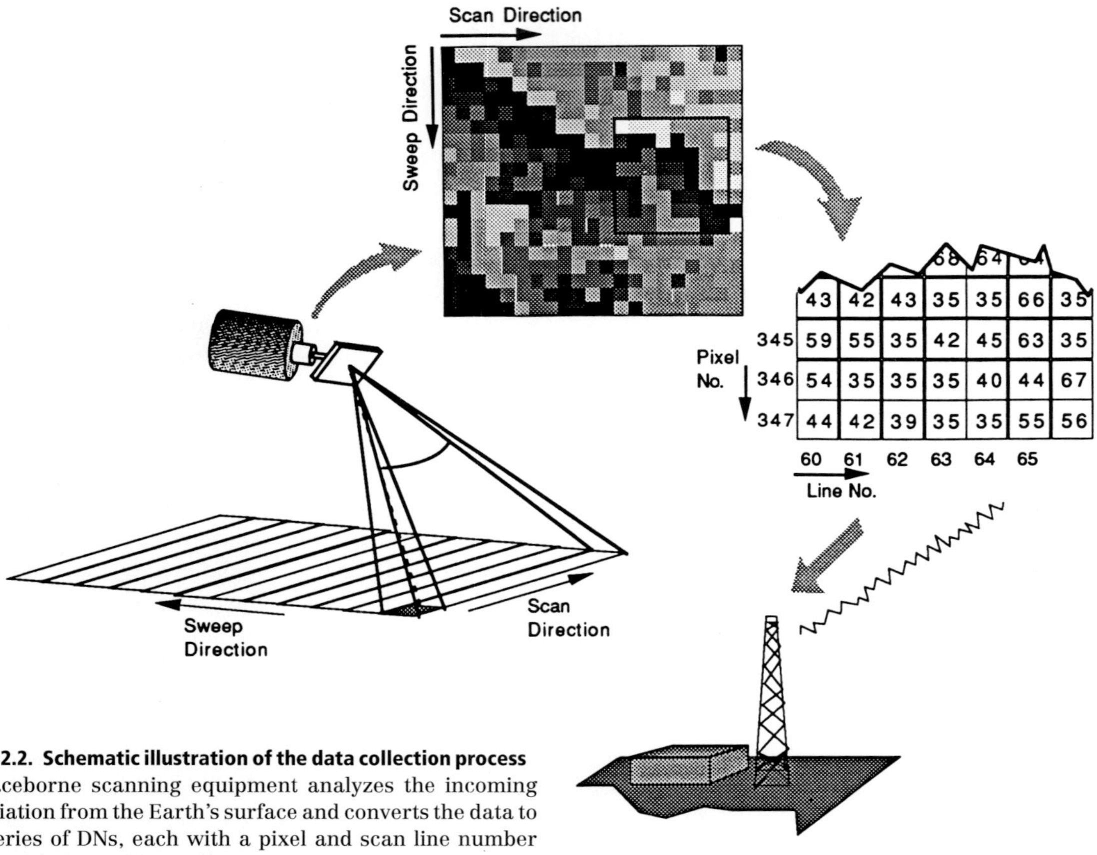

Fig. 2.2. Schematic illustration of the data collection process
Spaceborne scanning equipment analyzes the incoming radiation from the Earth's surface and converts the data to a series of DNs, each with a pixel and scan line number that correlates the pixel to a position on the ground. The pixel information is then transmitted to an earthbound receiving station

1. **Image Restoration.** This involves processes to remove systematic errors in the data such as noise or geometric distortion. Common procedures include restoring periodic line dropouts, filtering random noise, and correcting for atmospheric scattering. Because the majority of processes in the category are carried out before interpretation begins, these processes are of little concern to the interpreter and will not be discussed here. Bernstein (1983), Billingsly (1983) and Drury (1987) provide a more detailed description of these concepts.

2. **Image Enhancement.** This includes any alteration of the data that improves the interpretability of the image information. These procedures include contrast stretching, edge enhancement, false-color compositing and intensity-hue-saturation transformations. The techniques relevant to geological mapping will be discussed below. A more comprehensive treatment of this subject matter can be found in Anuta (1977), Chavez and Bauer (1982), Condit and Chavez (1979), Gillespie (1980) and Sabins (1987).

3. **Information Extraction.** Techniques such as supervised and unsupervised classification use the power of the computer to aid geological interpretation. These procedures will be discussed briefly as they are fairly complex and very application-specific. Other methods include density slicing, principal-component separation and pattern recognition. For further discussion, see Drury (1987), Gillespie (1980), Sabins (1987) and Schowengerdt (1983).

4. **Image Merging.** Interpretation can be aided by using the computer to manipulate and merge several data sets together interactively. This allows the interpreter to view simultaneously as many combinations of merged data sets as desired before arriving at the final image best suited for interpretation. Image merging techniques are very widely used in exploration today.

2.2 Imagery Enhancement Techniques

2.2.1 Contrast Stretching

A common way to examine the data from an image is called a DN distribution histogram, shown in Fig. 2.3.

Here, the DNs are plotted against the frequency with which they occur. (One can see that this is a graph from an image with a radiometric resolution of 256.) In this typical example, the most common intensity values are in the middle range, while the upper and lower values have few or no pixels at all.

These histograms can be analyzed on their own to reveal information about the image. For example, the "sharpness" of the peak (if there is one) indicates the relative contrast in the image. A peak with steep sides indicates an image with very low contrast, whereas a broad peak suggests that a large portion of the image is spread out along a wider range of intensity values. Multiple peaks suggest several different terrain features with very distinct spectral signatures.

With a histogram showing the distribution for just one bandwidth, no conclusions can be drawn about terrain types. However, correlating histograms of several different bandwidths can lead to direct identification of surface features. This is done most conveniently with multidimensional histograms, and this concept leads directly to unsupervised classification which will be discussed later in this chapter.

The sensors on board the Landsat satellite are calibrated to register different brightness values ranging from black basalt plateaus to white sea ice under a wide range of lighting conditions. It is rare that one individual scene will utilize the full range of the brightness scale. A more common result is shown in Fig. 2.4 where 96 % of the data has a DN range of 10 to 45. This utilizes only 14 % of the grey scale which has 256 values. The term "contrast stretching" is used to describe a process in which the lower values (representing 1 %) of this image are assigned to black (0), the upper values (3 %) are assigned to white (255) and the remaining pixel values are distributed evenly between the extremes. Although 4 % of the data in the top and bottom ranges is lost, the overall clarity of the picture is improved and interpretation of surface features is now much easier.

The original scene of the Wind River Basin (Fig. 2.4 a) was quite blurry and "flat", providing little structural and stratigraphic information. This scene will be used several times throughout this chapter to illustrate different enhancement techniques. In viewing this image, the reader should focus on three key features: (1) the triangular-shaped floodplain of the major tributary that forms along the outermost exposed hogback of the basin's margins; (2) the small, breached Dallas Anticline; and (3) a plunging nose of a fold southwest of the Dallas Anticline.

The value below which all DNs are set to zero is appropriately called the cut-off point. (The top value, above which all values become maximum, is sometimes called the saturation cutoff point.) A common way of setting the cutoff is to identify the intensity value of shadow areas and set all DNs of that value or below to zero. In any case, raising the cutoff point and lowering the saturation cutoff will increase contrast but cause more information to be lost. This compromise must be made individually for each image, taking into consideration the interpretation needs.

A more advanced method of this compromise is to select only the areas of the histogram of interest for

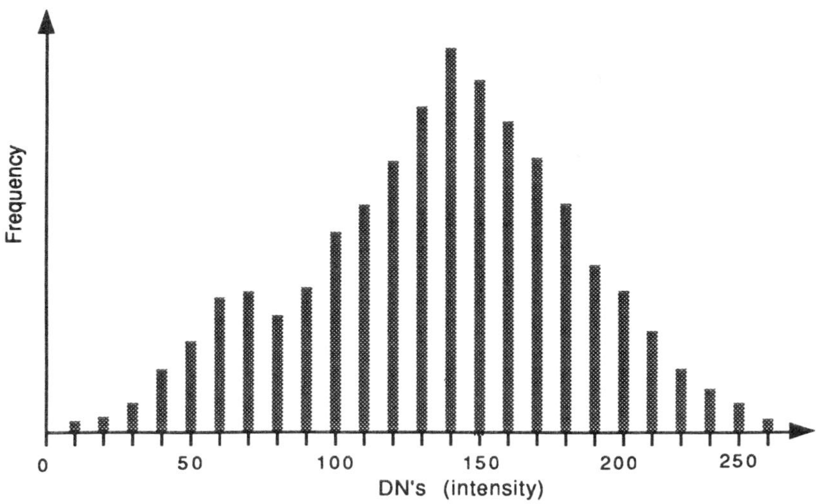

Fig. 2.3. A DN distribution histogram
A simple graph that is commonly used to describe the distribution of intensity values of imagery data

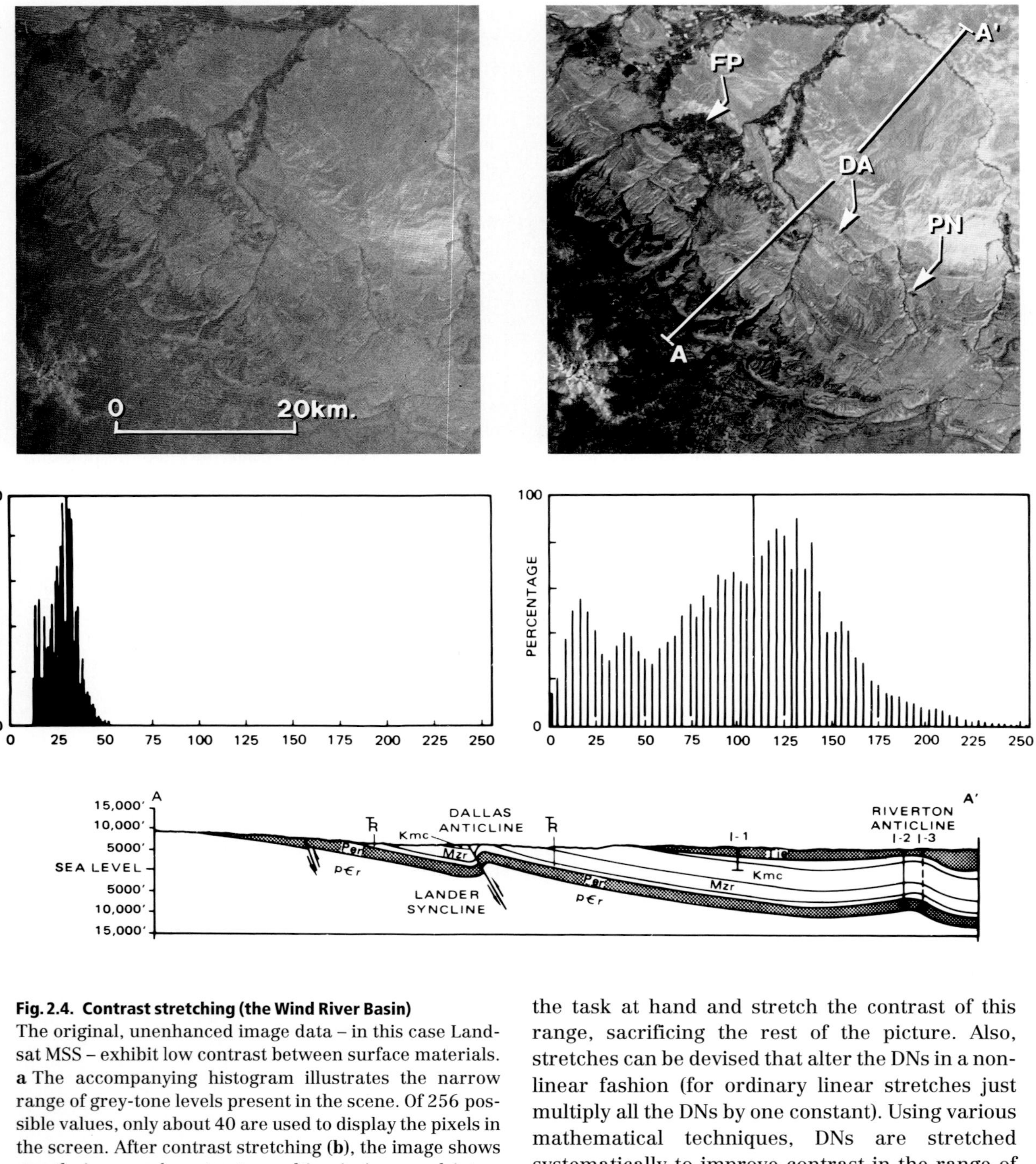

Fig. 2.4. Contrast stretching (the Wind River Basin)
The original, unenhanced image data – in this case Landsat MSS – exhibit low contrast between surface materials. **a** The accompanying histogram illustrates the narrow range of grey-tone levels present in the scene. Of 256 possible values, only about 40 are used to display the pixels in the screen. After contrast stretching (**b**), the image shows greatly increased contrast, resulting in improved interpretability. Here, nearly the entire range of grey levels is used to display the pixels in the scene. *FP* Floodplain; *DA* Dallas Anticline; *PN* plunging nose. (After Townsend and Dodge 1983)

the task at hand and stretch the contrast of this range, sacrificing the rest of the picture. Also, stretches can be devised that alter the DNs in a nonlinear fashion (for ordinary linear stretches just multiply all the DNs by one constant). Using various mathematical techniques, DNs are stretched systematically to improve contrast in the range of interest.

2.2.2 Standard False-Color Imagery

A widely used method in the interpretation of Landsat and other multispectral satellite data involves assigning a specific color to each spectral band and

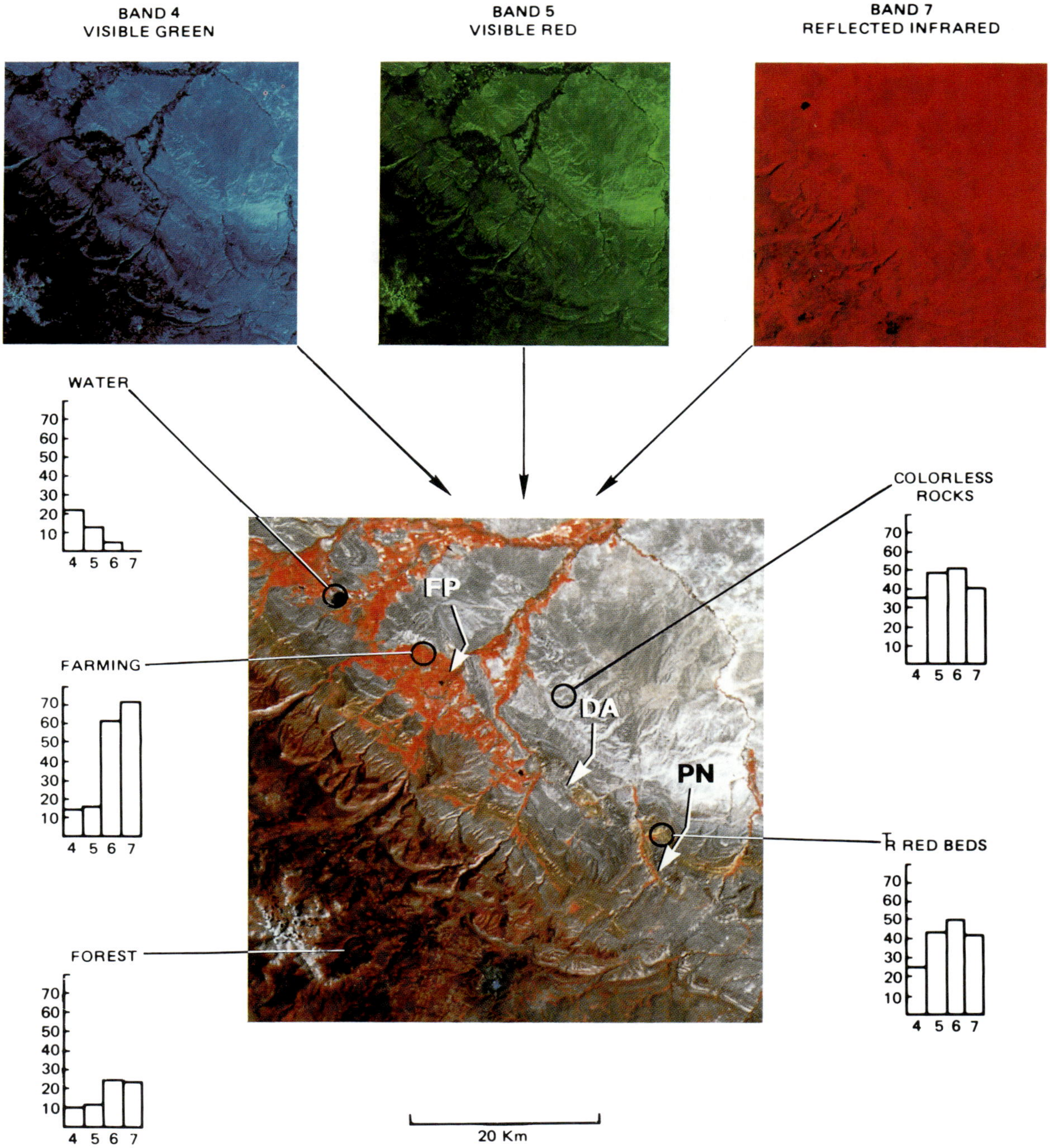

Fig. 2.5. Standard false-color (SFC) image (the Wind River Basin)

The bar graphs illustrate the response of the four separate Landsat bands as the total number of digital data counts (out of a total of 256 possible digital numbers) recorded for specific surface materials. *FP* Floodplain; *DA* Dallas Anticline; *PN* plunging nose. (After Townsend and Dodge 1983) then combining the bands to produce a full-color image. The standard false-color (SFC) assignment is shown in Fig. 2.5. Blue, green and red are assigned to bands 4, 5 and 7 of the MSS, respectively, and then the three monochromatic images are overlaid to form an SFC composite.

The color designations are arbitrary, but an analysis of the relationship between the colors and the bandwidths they represent can bring about an understanding of the color scheme in the composite. By examining the accompanying bar graphs in Fig. 2.5, which show relative responses in each band by various materials, this relationship is easy to see. For example, water appears blue on the SFC composite because the reflectance of water is relatively high in band 4 but almost zero in band 7: the composite has a large blue component, little green and no red. Vegetation typically has high infrared reflectance relative to its visible reflectance, so forests appear red on the SFC image. The grey and white areas of the image represent material that has uniform reflectance across all three bands, and this behavior corresonds to rocks or soils.[1]

In this case, the standard false-color designations provide imagery data with clear definition of the eroded flanks of the Wind River Basin by highlighting the alternating resistant and nonresistant bedrock units and related flatirons. The nonresistant bedrock units, in most cases, appear as unique features containing yellow and green colors (from Triassic redbeds) and red colors (from valleys that are covered by soil and related vegetation). Because these two features often occur in the same area, it is not easy to separate them on the SFC image. To do this, more sophisticated techniques will be introduced.

An interpreter is not limited, however, to the color assignments of the standard false-color composite. The assignments can be picked, rather, according to the interpretation needs of the image. Table 2.1 shows the MSS and TM bands along with the features each is best for detecting. Color assignments can be devised from this table. For example, the standard false-color scheme is good for general purposes, but has the limitation of hiding surface details with the various shades of red in areas of high vegetation. In these cases, MSS band 7 is often assigned the color green to contrast the natural

[1] Actually, the color assignments for SFC were chosen in an attempt to mimic airborne infrared photography that was used extensively before satellite data became widely available

Table 2.1. Detectability of different surface features with TM and MSS bands

Band	Wavelength range	Remarks	Used to detect
TM 1	0.45–0.55 µm (visible blue)	Needed for true color composites	Bathymetry in clear water, soil type discrimination
MSS 4 (TM 2)	0.5–0.6 µm (visible blue-green)	Good water penetration	Bathymetry in turbid water, vegetation vigor, ferric content in rocks and soils
MSS 5 (TM 2&3)	0.6–0.7 µm (visible orange-red)	Minor water penetration	Bare ground, culture, ice and snow mapping
MSS 6 (TM 3&4)	0.7–0.8 µm (red-near IR)	Some cloud penetration	Vegetation, culture
MSS 7	0.8–1.1 µm (near IR)	Maximum MSS cloud penetration	Vegetation, moisture variations
TM 5	1.55–1.75 (mid IR)	Only way to identify clay and carbonates	Soil, vegetation, moisture content
TM 7	2.08–2.35 µm (mid IR)	Only way to identify clay and carbonates	Clay minerals, hypothermal alteration
TM 6	10.14–12.5 µm (Thermal IR)	Cloud/ vegetation	Surface temperature, soil moisture

green reflectivity of vegetation with its strong infrared response. With MSS band 4 assigned to red, good contrast is developed to distinguish vegetation type and density. Experience shows, however, that the best way to select a color scheme is for the interpreting geologist to interactively view different color assignments and select the imagery best suited for the project at hand.

2.2.3 Intensity, Hue and Saturation Transformations

A system has been developed where the color of any pixel can be represented by three criteria: (1) intensity, representing the brightness value (with a sensitivity determined by the radiometric resolution); (2) hue, indicating the dominant wavelength of color (given in a numeric system where 0 represents red with the numbers increasing to 255 along the visible spectrum); and (3) saturation, representing the purity of the color (0 saturation is completely "impure" where all wavelengths are evenly distributed and appears greyish, whereas increasing numbers represent clearer and more intense colors). These values together are called the IHS system and can be represented as a cylinder with the main axis measuring intensity, distance from the center measuring saturation, and position along the rim indicating hue (Fig. 2.6).

It is possible to separate each of these values from an image, enhance them separately and then recombine them for a clearer composite. In its simplest form, this process results in a normalized false-color image. In Fig. 2.7, the hue and saturation values have been separated from the intensity values (a and b, respectively). The two were then enhanced separately and recombined to create Fig. 2.7 c. By comparing the results from this process with the standard false-color image (shown in Fig. 2.5), one can realize how this enhancement technique often can provide critical information needed for geological mapping. The enhanced imagery clearly separates exposed redbed units (which appear in yellow) from heavily vegetated floodplain areas which now appear in red. Such separation between the redbeds and the floodplain areas was not as clear on the standard false-color imagery.

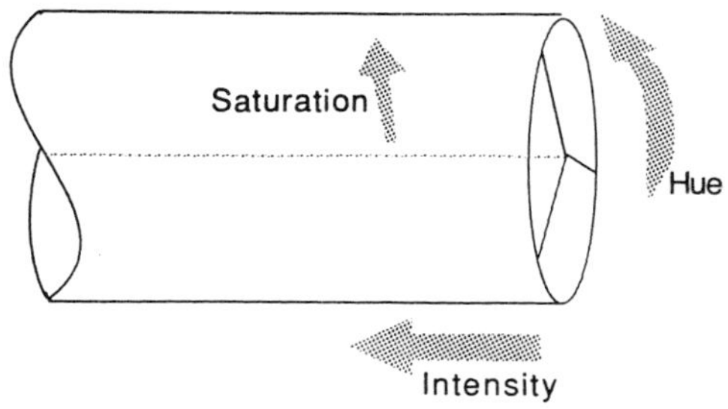

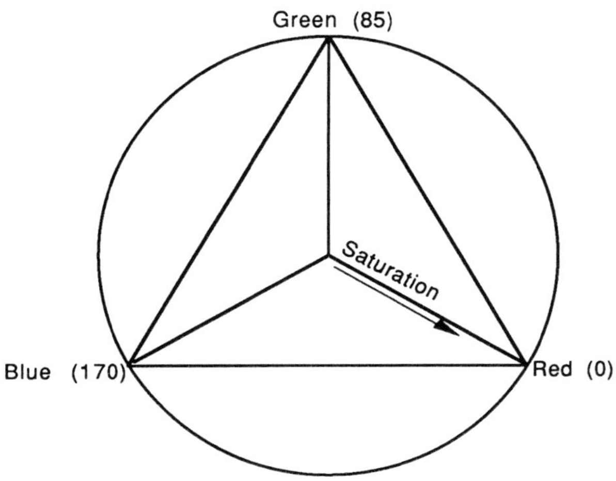

Fig. 2.6. The IHS system
The main axis represents intensity level. Distance from the center of the cylinder represents saturation values. Position along the rim indicates hue values

Fig. 2.7. Normalized false-color enhancement (the Wind River Basin)

a Color components of Landsat MSS image. **b** Brightness components of Landsat image. **c** Normalized false-color image. The color and brightness components are enhanced separately, then recombined to form image **c**. Color assignments for each band are the same as in a standard false-color image: band 4 = blue; band 5 = green; band 7 = red. *FP* Floodplain; *DA* Dallas Anticline; *PN* plunging nose. (After Townsend and Dodge 1983)

2.2.4 Ratio Enhancement

Figure 2.8 shows the profound effect that the angle of illumination has on the brightness data of an image. In trying to identify specific materials by their spectral signature, the discrepancies shown here would make an accurate analysis almost impossible. The shadow effect causes a drastic decrease in the reflectance levels of both bands. However, one can compensate for this by using the fact that the percentage of brightness lost by each band is similar. By dividing the brightness values from band 4 by the values from band 5, a new set of values is created where the effect of shadowing is minimized (note the similarity of the two values in the final column in Fig. 2.8).

Figure 2.9 shows an application of this technique to the imagery data from the Wind River Basin. Here, MSS bands 4 and 5 are ratioed to form the final image. It is important to note that the ratioing process not only minimizes illumination effects but that it also presents the interpreter with an entirely new data set. For example, the bright values on the ratioed image now represent material where band 5 reflectance is much higher than band 4. This correlates to the redbed units which have a considerable ferric oxide content (which has high band 5 values but low band 4 values). Vegetation appears dark on the 5/4 ratio image because of its high reflectance in the band 4 frequencies. Furthermore, the band ratio transformation allows division of the floodplain into different land-use areas, which was not possible with any of the previous transformations.

Fig. 2.8. Suppression of brightness variation with ratio enhancement
Prior to ratioing, the sandstone has a higher reflectance value in sunlight than in shadow; brightness information overpowers color information in this situation. Ratioing suppresses brightness variations caused by topographic effects, thus enhancing color information. Postratio reflectance values for both sandstone outcrops are similar, allowing correlation based on color. This illustration depicts a sunlight illumination angle from the southeast, which is standard for Landsat imagery. (Sabins 1980)

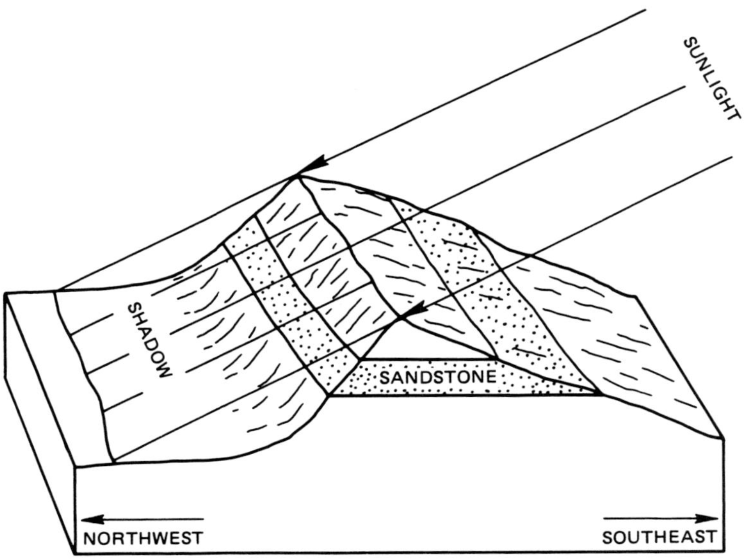

SANDSTONE REFLECTANCE

ILLUMINATION	BAND 4	BAND 5	BAND 4/5
SUNLIGHT	28	42	0.66
SHADOW	22	34	0.65

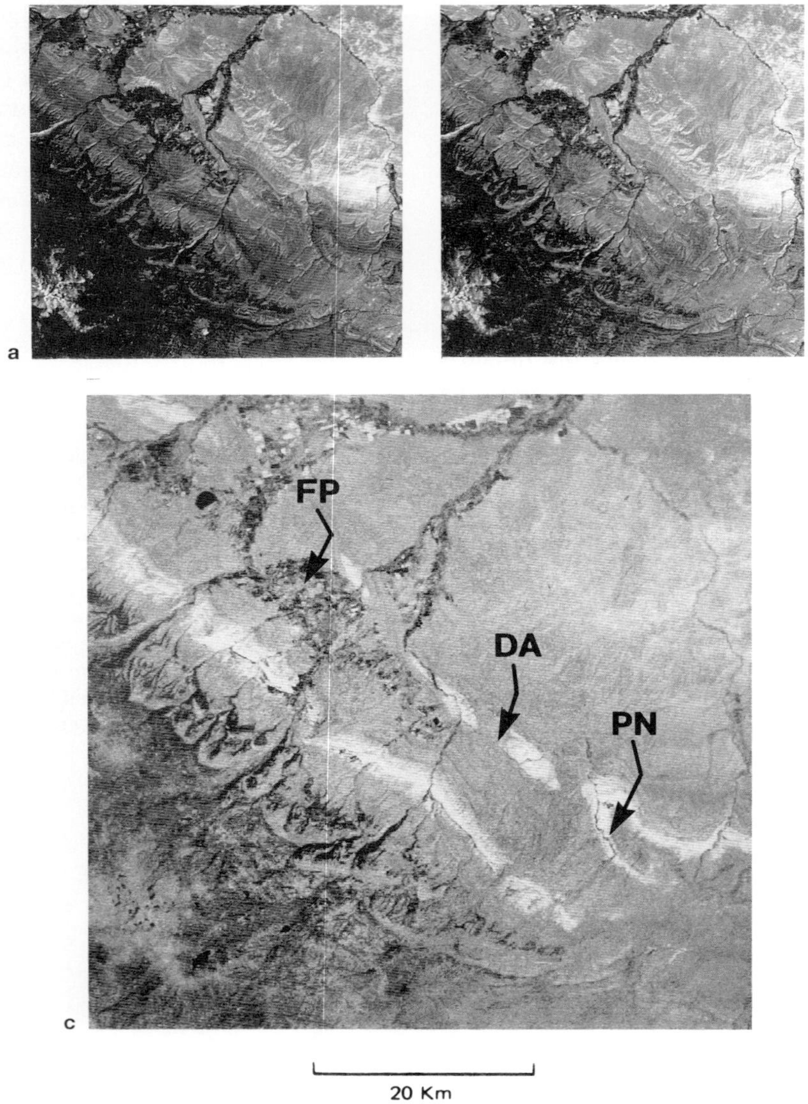

Fig. 2.9. Ratio enhancement (the Wind River Basin)
a Band 4 image; **b** band 5 image; **c** 5/4 ratio image. The *lightest tones* represent high ratio values, which usually delineate rocks with high ferric iron content. *Dark tones* indicate low ratio values and correspond to vegetation. *Intermediate grey tones* represent materials with similar reflectance properties in both bands. *FP* Floodplain; *DA* Dallas Anticline; *PN* plunging nose. (After Townsend and Dodge 1983

2.2.5 Edge Enhancement

Interpretation of satellite imagery often involves the identification of precise boundaries of geological features whether they consist of faults, joints, litho-stratigraphic contacts, or bedding planes. In mapping such features, geologists look for abrupt changes in tones or brightness which define these elements. The interpretation, therefore, is dependent on the precise identification of edges on image data. The limited spatial resolution of many satellite imaging systems is frequently insufficient for the required precision. Mathematical techniques can be used to manipulate the digital image data so that more accurate information on boundaries is revealed.

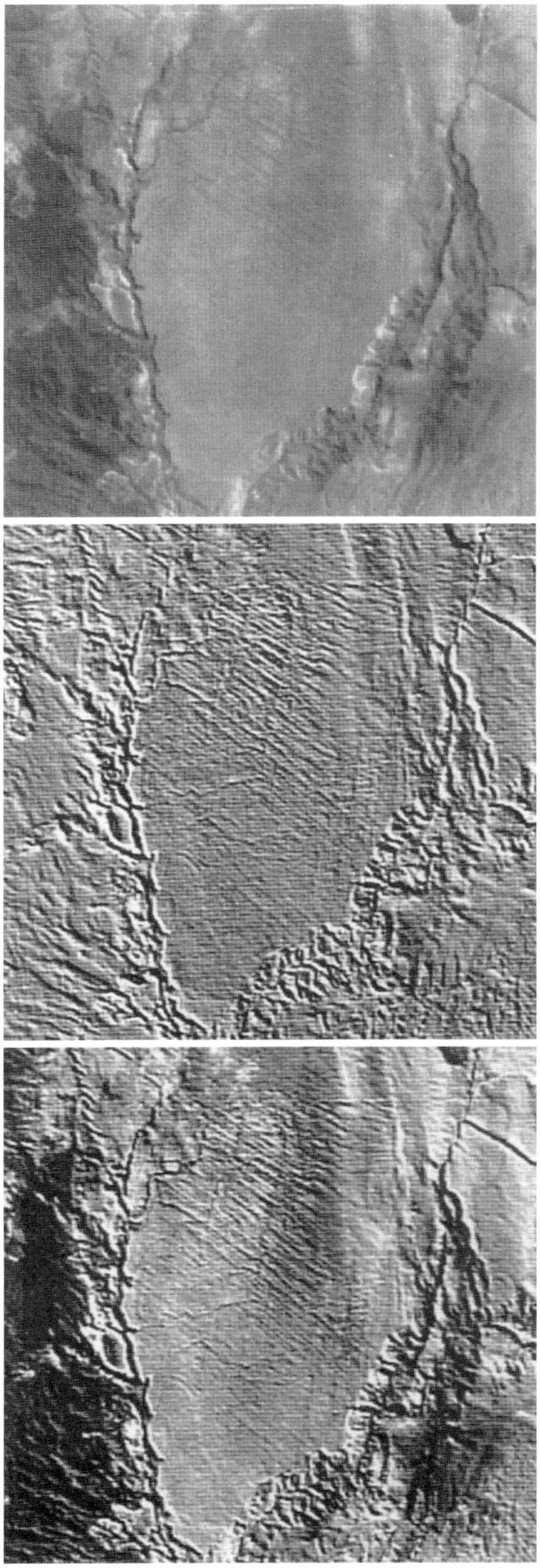

a
b
c

Fig. 2.10. Comparison of nondirectional and directional edge enhancement

Landsat MSS band 5 image in the Altiplano region, Chile. **a** The original image; **b** nondirectional enhancement; **c** directional enhancement of northwest-trending linear features. (Sabins 1987)

Such edge-enhancing techniques can be divided into two categories: directional and nondirectional. In nondirectional edge enhancing, a numerical matrix is applied to each pixel which considers the values of neighboring pixels and reassigns a value so as to maximize edge contrast differences (Fig. 2.10 b). The exact mathematical methods are described more thoroughly in Drury (1987). In directional edge enhancing, a similar method is employed but a matrix that favors the pixels in a particular direction is used. Through directional edge enhancing, the linear features of the landscape that trend in one particular direction can be highlighted for better inspection.

Care must be taken when applying directional edge enhancements because linear features that run near-parallel to the direction of filtering will be suppressed, thus creating a loss of information. With most enhancement techniques, features within 45 ° of perpendicularity with the enhancement direction will be enhanced, while others will be suppressed. A complete directional analysis therefore requires edge enhancement from all eight principal compass directions. This type of procedure is expensive and time-consuming so, as usual, enhancement directions should be chosen by the needs of the interpretation.

2.3 Information Extraction Techniques

2.3.1 Spectral Classification

In previous examples, classification was usually done by correlating the color seen on an image to specific materials. The main limitation of this method is the ability to correlate only three, maybe four, different spectral bands (because there are only three primary colors). As well, this method requires the time and effort of a trained geologist. An alternative approach is to allow the computer to classify materials without human help by comparing the spectral signature of each pixel with a table of known spectral signatures and then assigning a material type.

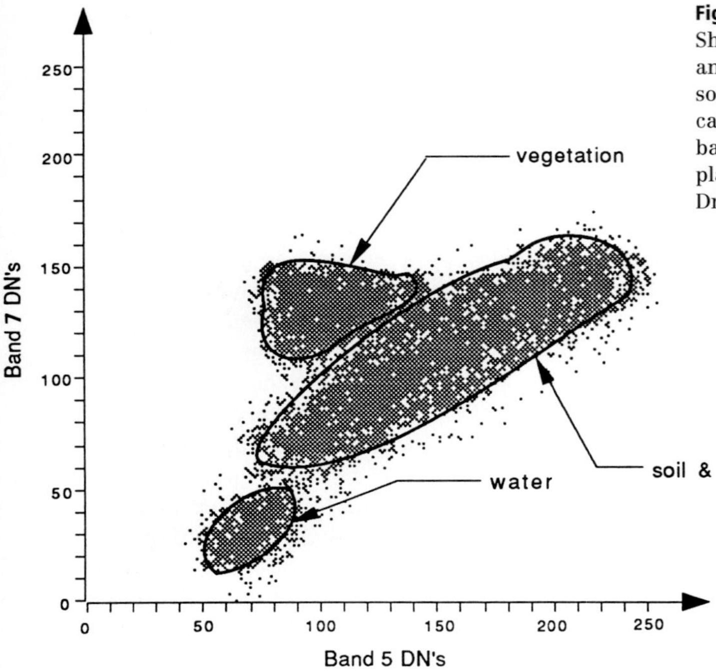

Fig. 2.11. Bivariate plots
Shown are DN values of Landsat MSS band 7 and band 5 of an area that is comprised of a mixture of open water, bare soil and rock, and variable vegetation cover. Note the typical shift of vegetation response towards high values of band 7. The chlorophyll absorbs red (band 5) energy and plant cells reflect infrared (band 7) very strongly. (After Drury 1987)

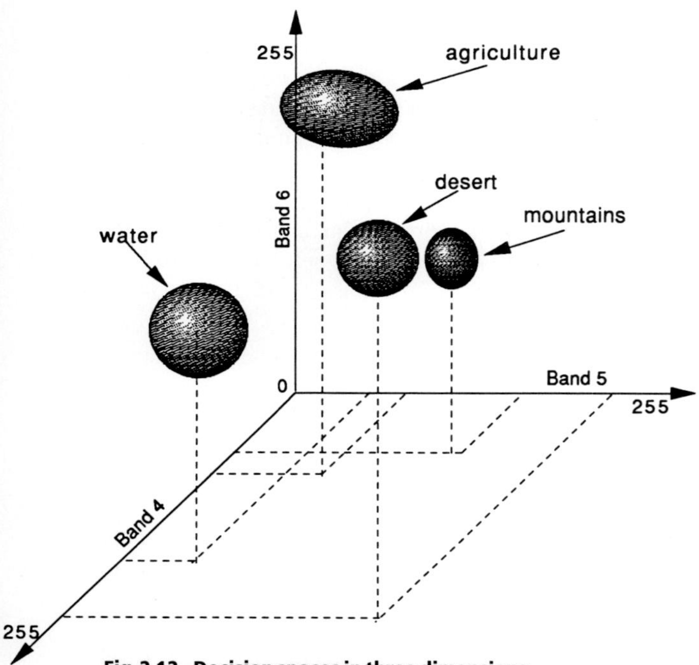

Fig. 2.12. Decision spaces in three dimensions
A three-dimensional spectral response curve and cluster diagram for Landsat MSS data of the Salton Sea and the Imperial Valley. (After Sabins 1987)

This process can most easily be understood by examining a two-dimensional histogram – a bivariate plot – comparing two spectral bands (Fig. 2.11). This plot shows two very distinct patterns with a high degree of correlation between the two bands. A field can be outlined on the graph for each material that encompasses the majority of points. Any point inside the field is then defined as belonging to that material classification. The area of the field is called the decision space. Pixels with signatures falling outside the decision spaces are left as unidentified. Colors then can be assigned to each material and a new map, with colors representing different features, is produced.

This method can then be extended to a comparison of three different bands (Fig. 2.12) which requires a three-dimensional plot. Once again, a high degree of correlation is seen between the bandwidths, this time manifested as three-dimensional clusters. Appropriately, the decision space must be changed to spheres.

In reality, multispectral images can be represented by plots with the number of dimensions matching the number of bands. For example, MSS or TM images can be analyzed by comparing four or seven bands, respectively. This comparison requires graphs in 4- and 7-dimensional space and appropriate higher-dimensional geometric descriptions for the decision spaces. Also, decision spaces need not

be circular or spherical (or the higher dimension equivalents), but rather can take on whatever shape best fits the data of the image.

Fig. 2.13. Supervised classification
Standard false-color MSS image *(upper half)* and supervised classification *(lower half)* of the northeastern flanks of the Wind River Basin. The spectral signatures of the exposed redbed units successfully separated from the spectral signature of the vegetation and are displayed in green color on the imagery of the lower half. *FP* Floodplain; *DA* Dallas Anticline; *PN* plunging nose

2.3.2 Supervised Classification

Spectral classification is divided into two categories, supervised and unsupervised classification. In supervised classification, the interpreter provides the computer with the decision spaces. This process entails analyzing the picture and identifying the important geological materials. Small representative areas of each material are then selected as "training areas", and the computer records the spectral signature of these areas for use as the basis of the decision space. Great care must be taken in selecting the training areas so that the spectral signature is indeed indicative of the general material. Often, more than one training area is taken for each material to cover the range of appearances that the material may have in the scene.

The decision spaces are then adjusted, by the interpreter, to best suit the image. If decision spaces are too large, the computer is less discriminating in identifying materials and the possibility exists for incorrect identification. If decision spaces are too small, large areas of the image will remain unidentified.

In Fig. 2.13, the scene from the Wind River Basin was divided into two halves. The upper half shows an enhanced SFC image, whereas the lower half shows how a supervised classification technique was used to map the distribution of the exposed redbeds in this basin, which in this case were mapped in green color. The supervised classification techniques enable the interpreter to separate the spectral signature of the exposed redbeds from those of the floodplain area along the river which, in the SFC imagery, had a similar spectral response. The objective of obtaining such imagery is twofold: first, the accurate lithological mapping of the redbeds and, second, the use of the redbed outcrop pattern as a guide for mapping exposed folds in the area.

2.3.3 Unsupervised Classification

This form of spectral classification relies even more on the computer. Given a scene, mathematical algorithms are applied to identify materials with distinct features. No training area is provided so the computer must decide independently on the shape and size of the decision spaces.

This method has an advantage of speed but requires very sophisticated software. Frequently, such systems divide the scene into more classifications than are needed because the computer does not consider context in its analysis. An unsupervised classification may identify similar materials that occur in different parts of the scene as different materials because of shadowing or other effects. In this case, a geologist must adjust the decision spaces to arrive at a reasonable interpretive map. Unsupervised classification is most useful where surface conditions are unknown or poorly known.

It should be noted that computerized classification techniques are far from being able to replace a geologist in identifying materials in images. Except under very arid conditions, most geological structures are usually covered to various degrees with vegetation and soil, thus eluding the unsupervised classification scheme. Also, classification systems must deal with the large range of spectral signatures that a single material can have. Images must still be interpreted on an individual basis and, often, a visual inspection must still be used

2.4 Merged Imagery

The availability of images in a digital format and the advent of sophisticated image processing systems allow the coregistration of many different layers of data which can include different satellite images as well as other digital data sets routinely used for exploration (gravity, magnetics, topographic maps, etc.).

The wide range of merged data products that are available can be divided into two categories: those that are presented in a map format (i. e., in two-dimensional mode) and those that are presented with some type of a three-dimensional displaying technique. Representative examples of these two types are shown in Fig. 2.14.

In Fig. 2.14 a, a high-resolution, panchromatic SPOT image (upper right) from the San Rafael Swell in Utah, was merged with an enhanced TM image of the same area to create an image that contains the higher spatial resolution of SPOT and the improved spectral resolution of TM (lower left). The image in Fig. 2.14 b shows a three-dimensional view of the structure and topography of the Eastern Desert, Egypt. The three-dimensional view was achieved by combining digital terrain data with TM imagery, whereas the geological cross section was constructed from surface geologic maps which were coregistered onto the imagery data. The display nicely illustrates the inverse relationships between topography and structures in this region. Without a thorough understanding of these three-dimensional relationships, one could arrive at an incorrect structural interpretation of imagery data. In Fig. 2.14 c and d, geological maps from Death Valley were registered, merged and displayed on the TM image of the area. The merged imagery can be better used for both structural and stratigraphic interpretation.

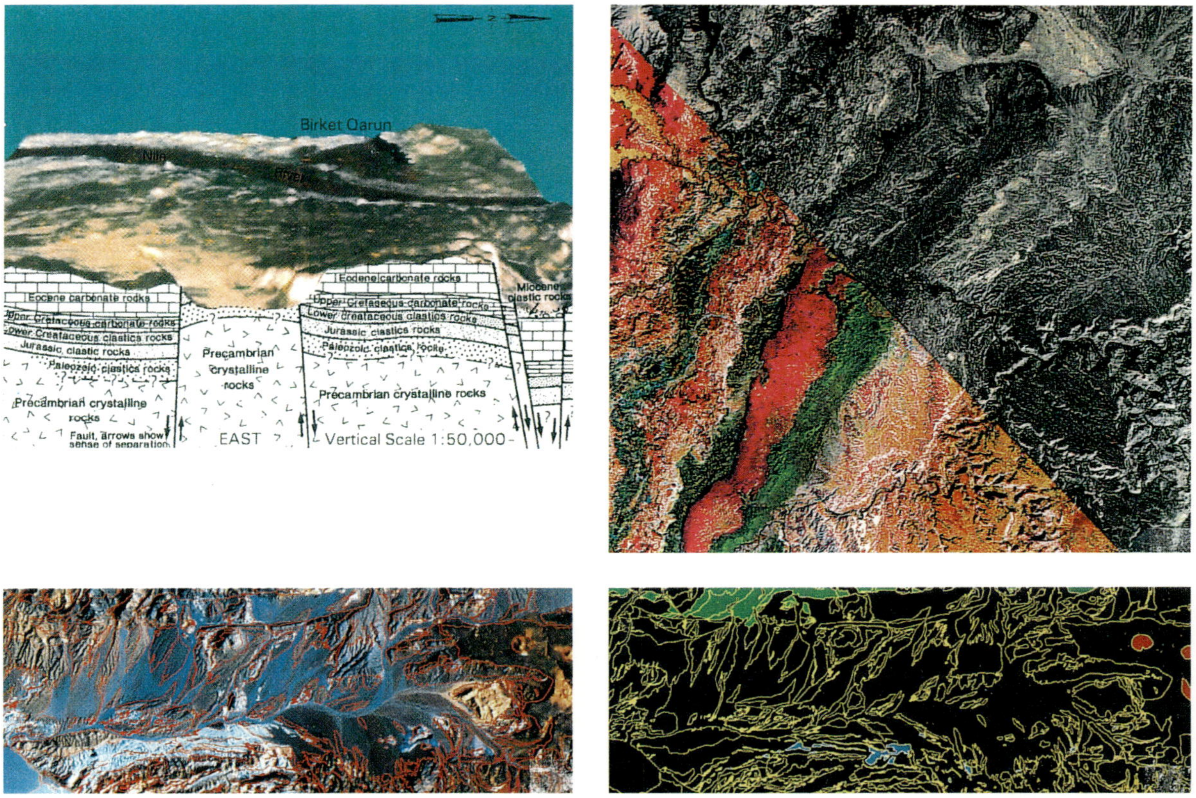

Fig. 2.14. Merged imagery
a A three-dimensional display of satellite imagery, topography and geological cross section from the Eastern Desert, Egypt. **b** A two-dimensional display of a SPOT panchromatic image *(upper right)* merged with TM data *(lower left)* of exposed outcrop units in the San Rafael swell in Utah. **c** and **d** Geological contacts in Death Valley, California, were merged and superimposed on enhanced TM imagery data. (**a** prepared by F. V. Corona and R. L. Brovey; **b** after Hopkins et al. 1987; **c** and **d** prepared by T. E. Townsend)

The process of merging these data sets, including satellite imagery, often requires a considerable amount of data preparation, manipulation, expense and time. One should therefore carefully weigh the benefits of merging such data sets against the cost and time involved. Such considerations are particularly warranted in the dynamic environment of hydrocarbon exploration where data, interpretations, and focus of study are continually changing.

2.5 Summary

Satellite imagery data are recorded in a digital format which can be entered into a computer system for further enhancements and data manipulation techniques. These techniques are divided into four major categories: image restoration, image enhancement, information extraction and image merging. Image restoration involves processes to remove systematic errors in the data such as noise or geometric distortions. Such processes are usually done before the interpretation of the imagery begins and thus their description has not been included in this chapter.

Image enhancement techniques that are most frequently used for geological applications include the following: (1) contrast stretching is designed to maximize the capabilities of the imaging systems to produce clear and high-contrast images; (2) standard false-color imaging involves the production of a colored image where each color represents the spectral signature of a specific band; (3) intensity, hue and saturation transformations provide the capabilities to separately enhance the various components of the imagery and then to combine them back into a single, clearer image. This procedure is often referred to as normalization. (4) Ratio enhancement is designed to eliminate the effect of shadowing on the spectral signature of the image; (5) directional and nondirectional edge enhancement techniques are used to bring out abrupt changes in tone or brightness which usually reflect the presence of geological features such as lithological contacts, faults and fractures or bedding planes.

Information extraction techniques are designed to isolate the spectral signatures of different materials on image data and to map their distributions on imagery by unique colors. Such effort can be done in two ways: by supervising the classification of the entire scene with spectral information that was extracted and analyzed by the interpreter from a small training area, or by an unsupervised method that relies primarily on the computer to identify features or regions with unique spectral characteristics.

The merging of imagery data involves the production of a single image that displays data from several sources. These images are divided into two types: those which display data in a two-dimensional mode and those which portray data in stereo images, block diagrams and perspective views which are presented in a three-dimensional format.

References and Further Reading

Andrews HC, Hunt BR (1977) Digital image restoration. Prentice-Hall, Englewood Cliffs

Anuta PE (1977) Computer assisted analysis techniques for remote sensing data interpretation. Geophysics 42: 468–481

Bernstein R (1976) Digital image processing of earth observation sensor data. IBM J Res Dev 20: 40–57

Bernstein R, Ferneyhough DG (1975) Digital image processing Photogramm Eng 41: 1465–1476

Bernstein R (ed) (1983) Image geometry and rectification. In: Collwell RN (ed) Manual of remote sensing, 2nd edn. Am Soc Photogramm, Falls Church, Virginia, Chap 21

Billingsley FC (1983) Data processing and reprocessing. In: Collwell RN (ed) Manual of remote sensing, 2nd edn. Am Soc Photogram, Falls Church, Virginia, Chap 17

Bryant M (1974) Digital image processing: optronics. Inter Publ 146, Chelmsford, MA

Canas AAD, Barnett AE (1985) The generation and interpretation of false-color composite principal component images. Int J Remote Sens 6: 867–81

Chavez PS, Bauer B (1982) An automatic optimum kernel-size selection technique for edge enhancement. Remote Sens Environ 12: 23–38

Chavez PS, Berlin GL, Sowers LB (1982) Statistical method for selecting Landsat MSS ratios. J App Photograph Eng 8: 23–30

Condit CD, Chavez PS (1979) Basic concepts of computerized digital image processing for geologists. USGS Bull 1462

Cornsweet TN (1970) Visual perception. Academic Press, New York

Daily M (1983) Hue-saturation-intensity split-spectrum processing of Seasat radar imagery. Photogramm Eng 49: 349–55

Drury SA (1987) Image interpretation in geology. Department of Earth Sciences, The Open University, Boston

Eliason EM, Chavez PS, Soderblom LA (1974) Simulated "true color" images from ERTS data. Geology 2: 231–234

Evans DL, Lang HR (1985) Techniques for multi-sensor image analysis. Proc 18th Int Symp on Remote sensing of the environment. The Environemntal Research Institute of Michigan, Ann Arbor, pp 196–215

Freden SC, Gordon F (1983) Landsat satellites. In: Colwell RN (ed) Manual of remote sensing, 2nd edn. Am Soc Photogramm, Falls Church, Virginia, Chap 12

Gillespie AR (1980) Digital techniques of image enhancement. In: Siegal BS, Gillespie AR (eds) Remote sensing in geology. John Wiley and Sons, New York, Chap 6

Goetz AFH, Billingsley FC, Gillespie AR (1975) Applications of ERTS images and image processing to regional geologic problems and geologic mapping in northern Arizona. JPL Tech Rep 32, 1975

Gonzalez RC, Wintz P (1977) Digital image processing. Addison-Wesley, Reading

Holz RK (ed) (1985) The surveillant science: remote sensing of environment. John Wiley and Sons, New York

Hopkins HR, Navail H, Berger Z, Merembeck, BF, Brovey RL, Schriver JS (1987) Structural analysis of the Jura Mountains-Rhine Graben intersection for petroleum exploration using SPOT stereoscopic data. In: SPOT 1 image utilization assessment, results. Centre National d'Etude Spatiales, Toulouse, France, pp 803–810

Hunt GR (1980) Electromagnetic radiation: the communication link in remote sensing. In: Siegal, BS, Gillespie AR (eds) Remote sensing in geology. John Wiley and Sons, New York, Chap 2

Hunt GR, Ashley RP (1979) Spectra of altered rocks in the visible and near infrared. Econ Geol 74: 1613–1629

Hunt GR, Salisbury JW (1970) Visible and near-infrared spectra of minerals and rocks – I silicate minerals. Mod Geol 1: 283–300

Hunt GR, Salisbury JW, Lenhoff CJ (1971) Visible and near-infrared spectra of minerals and rocks – III oxides and hydroxides. Mod Geol 2: 195–205

Keefer WR (1970) Structural geology of the Wind River Basin, Wyoming. USGS Prof Pap 495 D

Pratt WK (1978) Digital image processing. John Wiley and Sons, New York

Rothery DA (1985) Interactive processing of satellite images for geological interpretation – a case study. Geol Mag 12: 57–63

Rowan LC, Wetlaufer PH (1975) Iron-absorption band analysis for the discrimination of iron-rich zones. USGS, Type III Final Rep

Sabins FF (1974) Oil exploration needs for digital processing of imagery. Photogramm Eng 40: 1197–1200

Sabins FF (1980) Interpretation of thermal infrared images. In: Siegal, BS, Gillespie AR (eds) Remote sensing in geology. John Wiley and Sons, New York, Chap 9

Sabins FF (1987) Remote sensing: principles and interpretation. WH Freeman, San Francisco

Salisbury JW, Hunt GR (1974) Remote sensing of rock-type in the visible and near infrared. Proc 9th Int Symp on Remote sensing of environment. The Environmental Research Institute of Michigan, Ann Arbor, pp 1953–1958

Schowengerdt RA (1983) Techniques for image processing and classification in remote sensing. Academic Press, Orlando

Siegal BS, Gillespie AR (1980) Remote sensing in geology: John Wiley and Sons, New York

Townsend TE (1983) Discrimination of iron alteration minerals in remote sensing data. PhD Thesis, Stanford University

Townsend TE (1984) The significance of iron oxides and clays in petroleum and minerals exploration. Exxon Production Research Company, Internal Rep

Townsend TE, Dodge RL (1983) Techniques for geologic mapping based on the spectral component of Landsat imagery. Exxon Production Research Company, Internal Rep

Chapter 3 Image Interpretation Techniques: Exposed Structures

3.1 Definitions and Classification

The ability to recognize and map geological structures from remote sensing data is dependent primarily on two main factors: the level of bedrock exposure of the mapped structures and their magnitude of deformation. These factors determine: (1) the type of imagery data (i. e., monoscopic versus stereoscopic) that is required for structural mapping; (2) the kind of interpretation techniques (i. e., structural versus geomorphic) that must be employed; and (3) the level of integration with other data sets that is needed to constrain the interpretation of the image data (Fig. 3.1).

Exposed structures are recognized and analyzed from image data by the unique expressions of their inclined bedrock strata and fault-line traces. Such analysis can be done either independently from

Fig. 3.1. Classification of geological structures and related interpretation tools
The classification is done on the basis of the magnitude of deformation, level of bedrock exposure and related interpretation tools

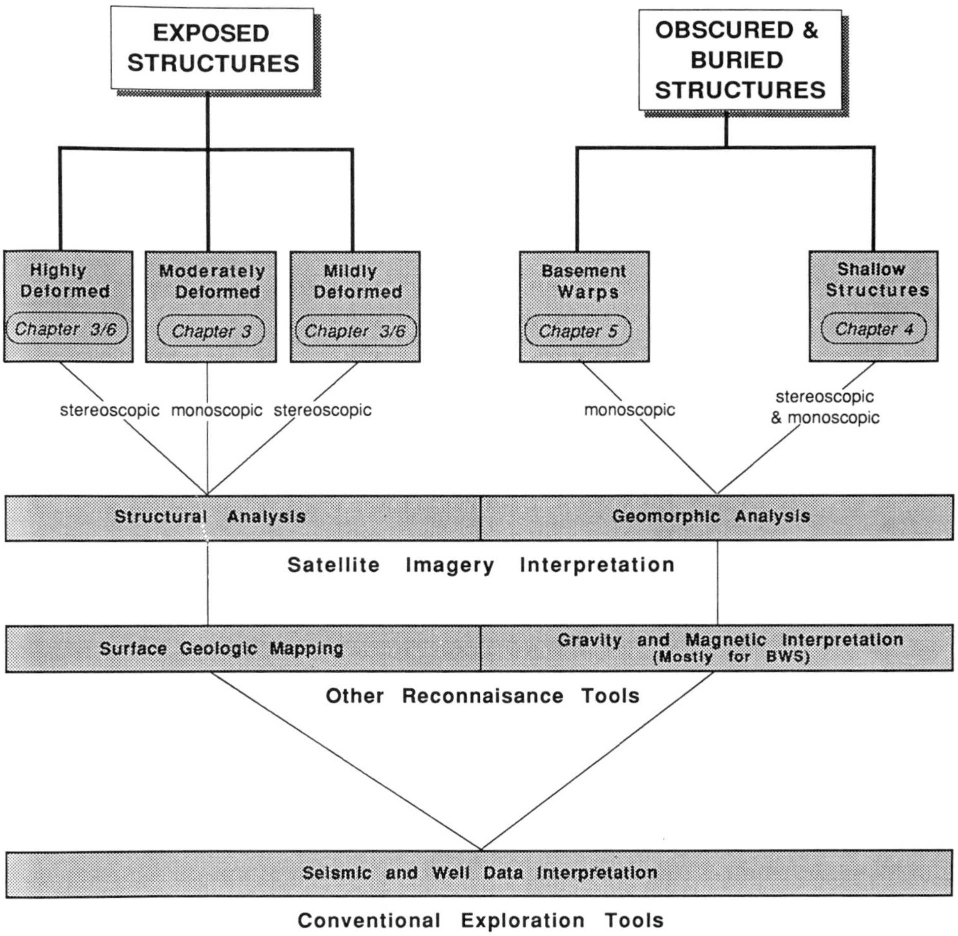

surface or subsurface controls or preferably with some integration of available surface geologic maps and key seismic lines. Exposed structures may be further divided on the basis of their magnitude of deformation into three groups: mildly deformed, moderately deformed, and highly deformed. Moderately deformed structures can be mapped and analyzed with monoscopic imagery because they manifest clear expressions of inclined bedrock strata. The two other groups lack such surface manifestations and usually require the support of stereo data for interpretation.

Geological structures whose outcropping units are either obscured by vegetation and soil or completely buried under consolidated sediments are recognized on imagery data by their subtle influence on the regional topographic setting and related drainage, vegetation and soil moisture patterns. The analysis of such structures requires the integration of all other available subsurface data. Obscured and buried structures may be further divided into two groups which generally reflect their magnitude of deformation and depth of burial. The first group consists of buried faults and fold-related structures which manifest clear expressions on seismic and well data. The second group consists of subtle basement structures which are difficult to constrain by conventional interpretational techniques and are often referred to as basement warp structures (BWSs). Recognition and analysis of the latter group usually involve the integration of gravity and magnetic as well as specially processed and enhanced seismic data. Also, the analysis of BWSs requires reconstruction of paleogeomorphic surfaces that commonly develop over such structures. This process relies heavily on modern-day analogs which are observed on imagery data for both exposed and buried structures.

The present chapter covers the most fundamental of the situations: where structures are exposed and can be mapped without the use of stereo. Subsequent chapters will deal with increasingly more complex conditions. The main objective of this chapter is to introduce the reader to the most important aspect of the analysis of exposed structures, the recognition of structural styles and related hydrocarbon traps. For such identification, it is essential to review some fundamental interpretation techniques of individual folds and faults, their map patterns, typical surface expressions, and related style of deformation.

Appendix A provides an overview of common geologic symbols that will be used in the interpretation of structures in this book. Appendix B gives a list of abbreviations used in the figure captions to describe features on the images.

3.2 Analysis of Exposed Folds

3.2.1 Dip and Strike of Inclined Bedrock Strata

Identification of dip and strike of inclined bedrock strata is fundamental to the analysis of exposed structures with satellite imagery data. Well-exposed folds, domes and basins are usually partially breached by erosion and manifest diagnostic surface features which can be used to recognize the orientation (i.e., dip and strike) of their exposed limbs. The three general criteria which are commonly used for such analysis are illustrated in Fig. 3.2 and can be described as follows:

1. The limbs of exposed folds are expressed as a series of asymmetric ridges (hogbacks) with short, steep slopes on one side and long, gentle slopes on the other. The former reflect the antidip direction and are often referred to as scarp slopes. The latter reflect the dip direction and often are referred to as dip or isoclinal slopes (Fig. 3.2 a).
2. The scarp slopes are characterized by the presence of parallel topographic benches that reflect the presence of alternating resistant and nonresistant lithostratigraphic units. These benches appear on imagery as interrupted slopes. Isoclinal slopes are expressed by a series of triangular-shaped ridges (i.e., flatirons) that always point away from the dip direction (Fig. 3.2 b).
3. The asymmetry of the hogback is also reflected by the drainage pattern. The crest of the cuesta divides two types of drainage systems: those draining the scarp slopes, often referred to as obsequent streams, and those draining the isoclinical slopes, often referred to as subsequent streams. Obsequent streams typically show high frequency and short length, whereas the subsequent streams are longer and less dense (Fig. 3.2 c).

An excellent example of the surface expression of inclined bedrock strata can be found around the exposed limbs of the San Rafael Swell in Utah (Fig. 3.3). The entire swell is surrounded by concentric hogbacks that exhibit various levels of asymmetry. Profound expression of flatirons, obsequent and subsequent stream valleys can be seen in Fig. 3.3 c. Interrupted slopes are most noticeable in Fig. 3.3 e.

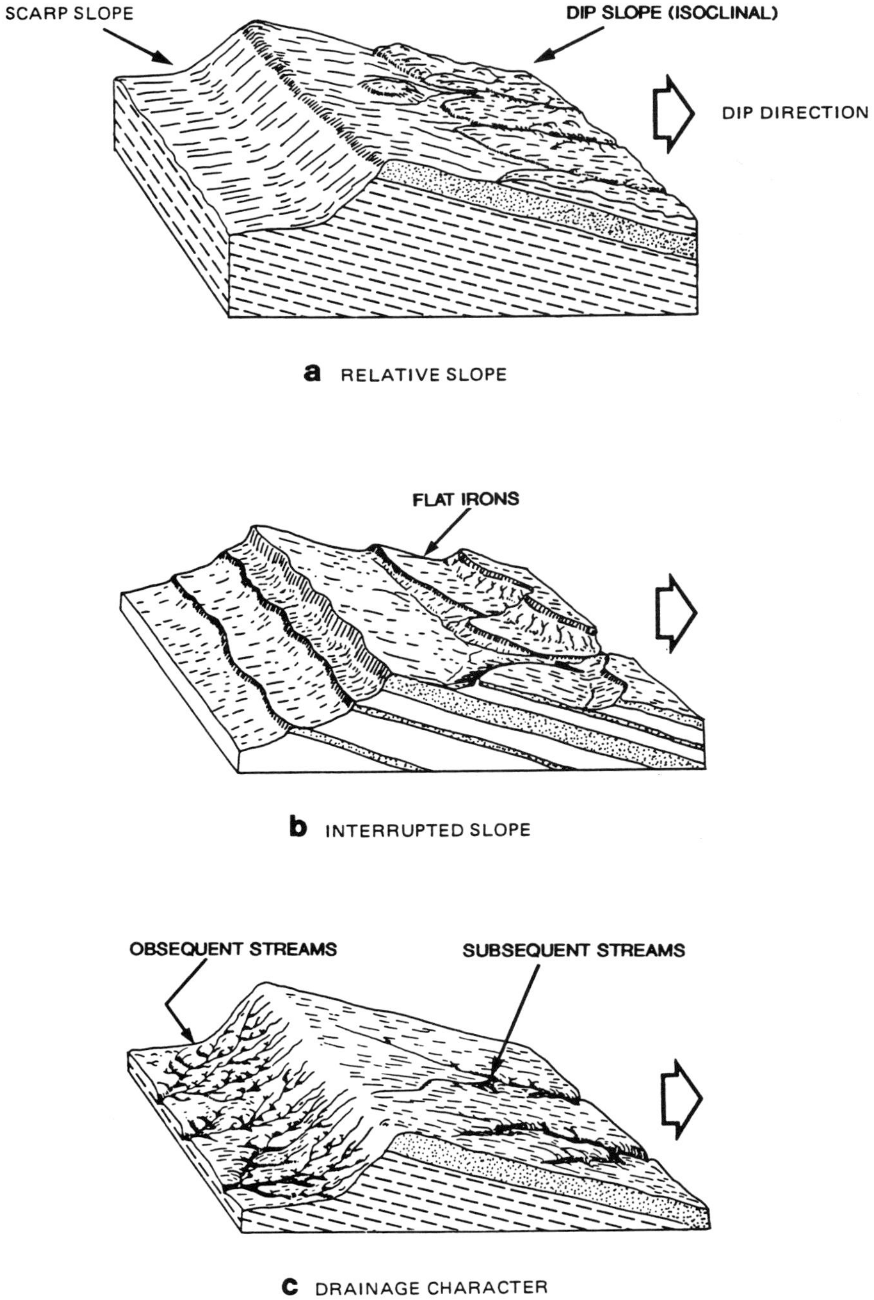

Fig. 3.2. General criteria for determining the dip direction of inclined rock units
(After Miller 1961 by Corona and Wielchowsky 1984)

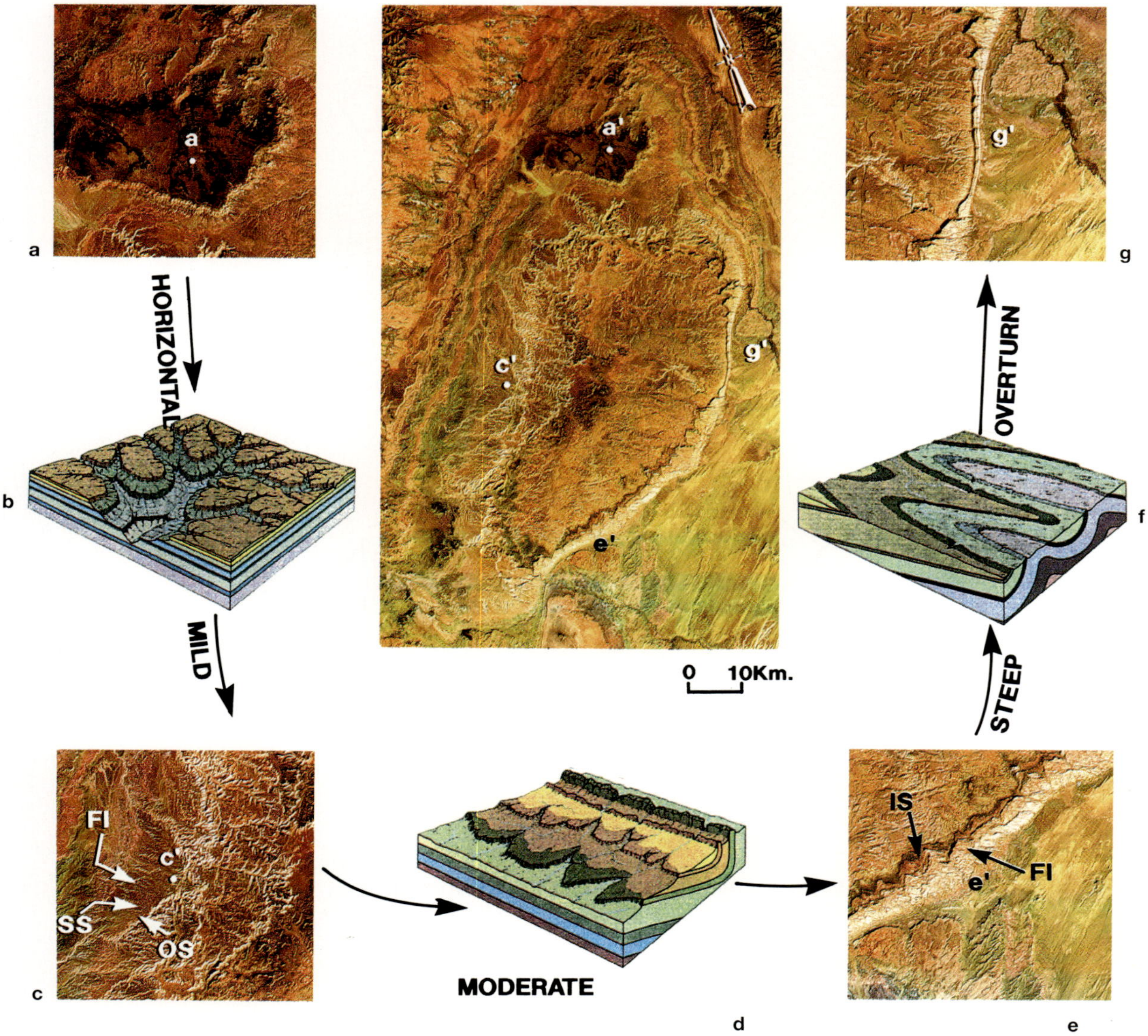

Fig. 3.3. Landsat (TM) image of San Rafael Swell, Utah
The surface expressions of exposed bedrock strata at various magnitudes of inclination are shown. The accompanying block diagrams are from Hamblin and Howard (1989). *Arrows* show direction of increasing dip magnitude. *FI* Flatirons; *SS* subsequent streams; *OS* obsequent streams; *IS* interrupted slopes. (Block diagrams published with the permission of MacMillan Publishing Company)

3.2.2 Landforms Related to Inclined Bedrock Strata

Figure 3.3 also illustrates how the magnitude of dip of inclined bedrock strata creates unique landforms visible on imagery data. A cursory analysis of these landforms, therefore, allows one to decide which type of imagery interpretation method is required for structure analysis.

The exposed rims of the San Rafael Swell in Utah can be clearly divided into several segments which exhibit different landform characteristics. Starting in the northwest corner, and proceeding counterclockwise, the exposed rims of the swell gradually increase in inclination to form an asymmetric dome with steeper slopes towards the eastern sides. Figure 3.3 a and its related block diagram (Fig. 3.3 b) show surface expressions of beds that dip at 0 to 5 °. This segment of the rim exhibits the typical landform characteristics of a structural plateau which is characterized by the presence of escarpments and gentle slopes that develop on resistant and nonresistant beds, respectively, and are dissected by a typical dendritic drainage system (i. e., “layer-cake geology”). The plateau lacks any diagnostic surface features that can be used to further determine the attitude of bedrock units in this area and, therefore, will require stereo measurements.

The southwestern and southeastern corners of the swell (Fig. 3.3 c, e) consist of beds with dips ranging from 5 ° to 10 ° and 10 ° to 60 °, respectively. In these segments, the V-shaped flatirons and stream valleys are clearly manifested and can be used to estimate the direction of dip. Note that the size of the flatirons as well as the V-shaped valleys are inversely proportional to the magnitude of dip. These types of surface folds can be easily mapped and analyzed with monoscopic data.

Finally, the eastern rim of the swell consists of dips that range from 60 ° to 90 ° as well as some overturned beds (Fig. 3.3 f, g). This segment of the rim is expressed as a profound hogback feature which becomes steeper and shorter with increase in dip magnitude. The near-vertical and overturned beds at the middle segments of the exposed limb (Fig. 3.3 g), are no longer expressed as asymmetric ridges with typical flatirons. These highly deformed features may have to be further analyzed with stereo data.

In summary, dip and strike magnitude and moderately deformed structures can be estimated from monoscopic imagery, whereas mildly and highly deformed structures must be evaluated with stereo imagery.

3.2.3 Shadowing Effects

The position of the sun during the collection of Landsat data must be considered when investigating the inclination of bedrock strata and determining related dip and strike orientations and magnitudes. As most Landsat data are acquired between 9:30 and 10:00 a. m. (local sun time), illumination is always from the southeast leading to shadows across northwest slopes. The shadowing effect can be observed nicely around the exposed rim of the San Rafael Swell (Fig. 3.3). The dip slopes on the eastern side of the swell are illuminated, whereas the scarp slopes are dark. The western side of the swell shows the opposite effect. The interpreter, therefore, must exercise caution in distinguishing dip from scarp slopes on Landsat data if shadow patterns are used. It is sometimes helpful to rotate the image during the interpretation process to better define slope with respect to sun angle.

The sun’s angle is also dependent on season, meaning that images acquired at different times of the year can display different degrees of this shadowing effect. In low-relief areas or in cases where bedrock units are covered by vegetation, the shadowing effect can provide crucial information for structural interpretation. It is often advantageous to select image data from the time of year where sun angle is low or to use multiple data sets for structural interpretation. This effect is particularly pronounced at higher latitudes as shown by the images of northern Alaska in Figs. 3.4 and 3.5. Considerably more structural details are available in this region with satellite imagery that was taken during seasons with low sun angle.

Shadowing effects can also be created on imagery by nongeological features. Most common are shadows produced by low clouds or smoke (shown in Fig. 3.5). It is not uncommon to find, on imagery, profound linear features that reflect vapor trails from passing jetliners or smoketrails from large fires. Such misleading shadow patterns can be additionally confusing when they follow topography (for example, low clouds trapped in a valley) giving the added impression that they represent geological features. Interpreters must learn through experience to differentiate between legitimate shadow features and artificial ones.

3.2.4 Geomorphic Expressions of Different Fold Types

Folds typically undergo intensive erosional processes which modify their topographic expressions. The rate of denudation (removal of the crest) and the resulting geomorphic expressions of the folds are linked to the age, prevailing climatic conditions, style, lithological composition and magnitude of deformation of the folded strata. By analyzing the surface expression of folds on satellite imagery, considerable structural information can be gained.

The erosional evolutions of exposed folds can be divided into four stages which are illustrated in Fig. 3.6 and listed in the next section.

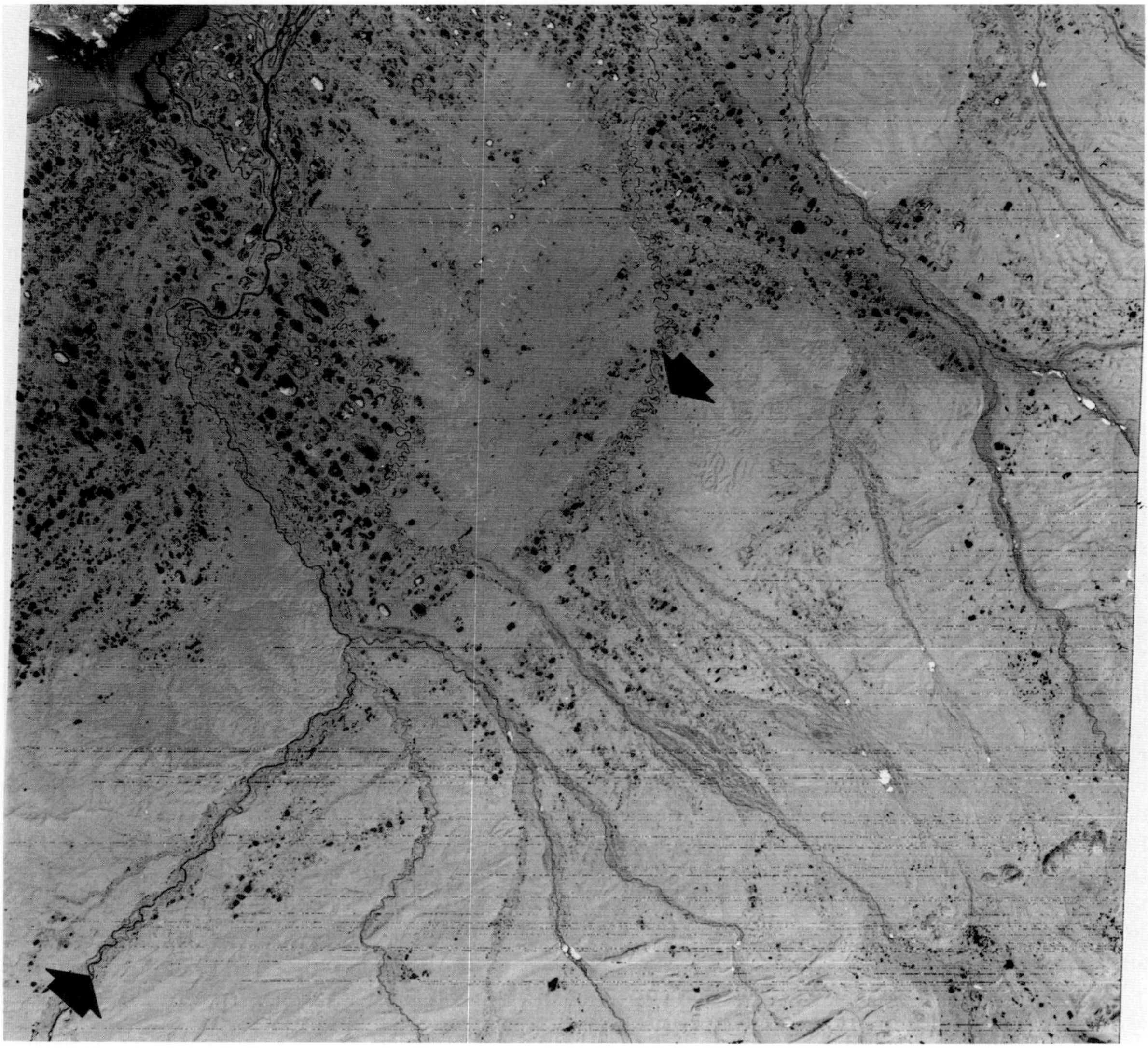

Fig. 3.4. The shadowing effect I
Landsat scene of northern Alaska taken in July with a sun angle of 41 °. It is very difficult to discern variations in topography and related structures. Note the linear features at the center of the image (marked by *solid arrows*). Scale of image is 1:1 000 000

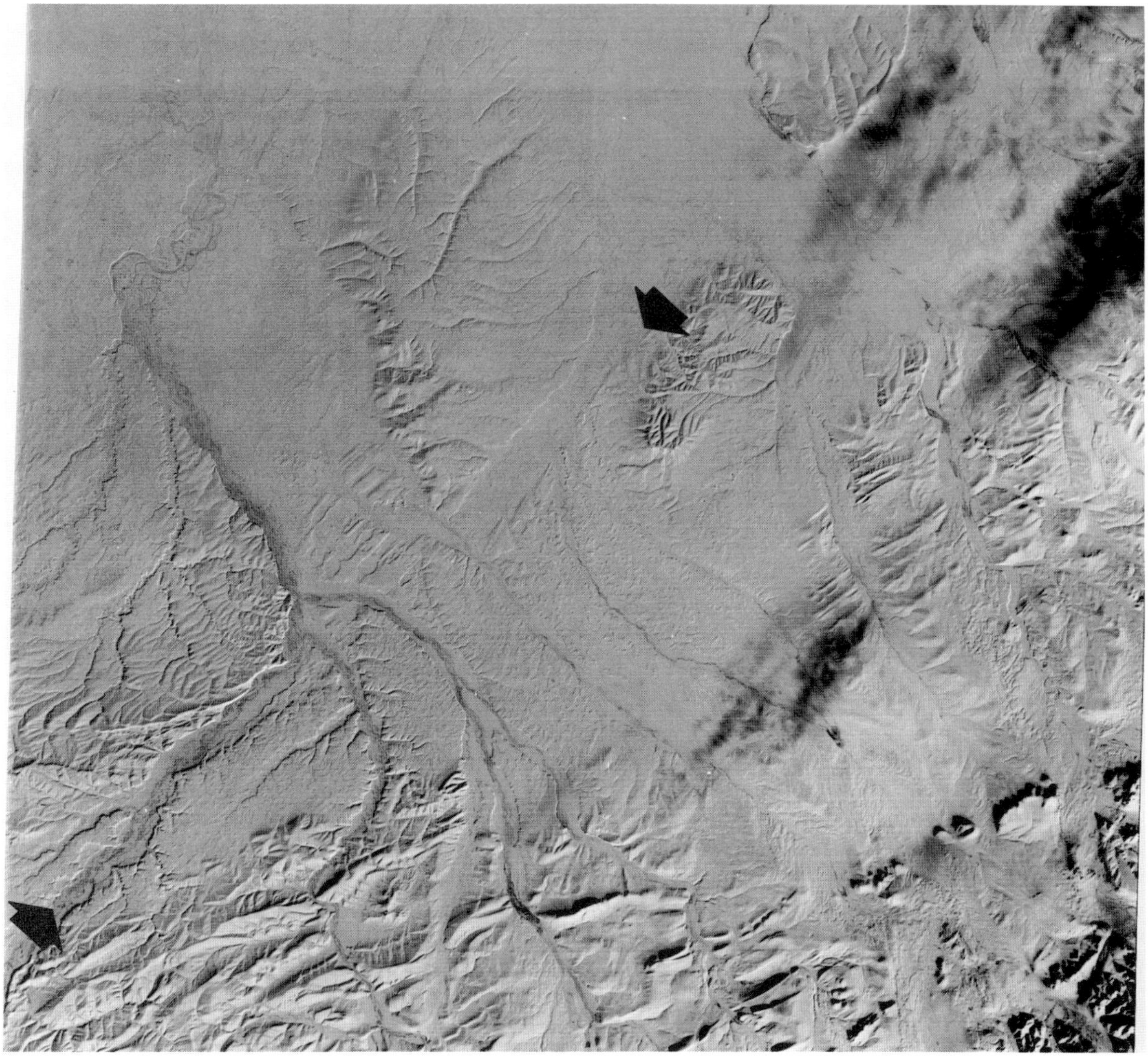

Fig. 3.5. The shadowing effect II
Landsat scene of the same area taken in February, with a sun angle of 11°. The low sun angle reveals subtle topographic features of exposed folded strata in the southern corner of the image as well as the presence of a major linear feature at the center of this scene. Scale of image is 1:1 000 000

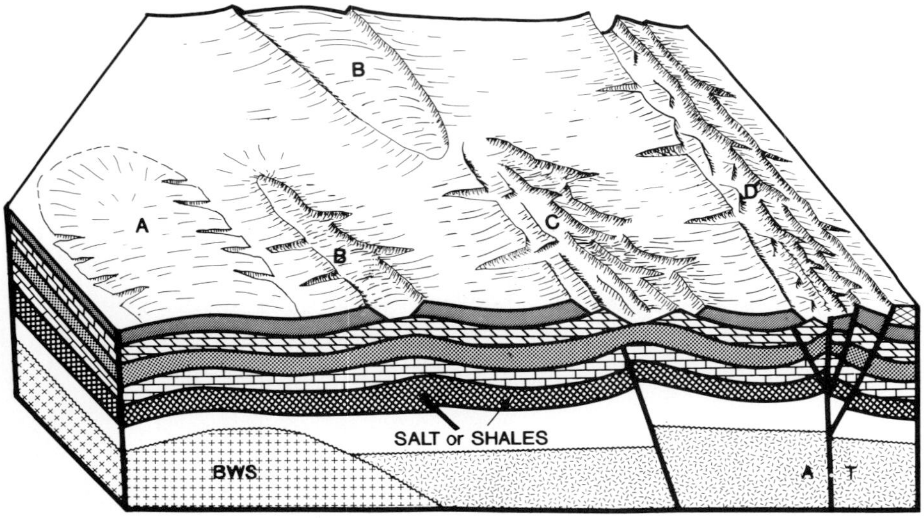

Fig. 3.6. Erosional evolution of different exposed fold types
The schematic block diagram shows four different erosional stages of exposed folds and their relationship with subsurface structures. *A* positive relief; *B* early breaching; *C* late breaching; *D* obliterative

Positive Relief Stage. A positive relief stage exists when folds and domes are at the initial stages of their erosional evolution, exhibiting a direct relationship between topography and structure (A in Fig. 3.6). Such relationships are rare in nature and usually occur in structures that are either extremely young or characterized by flat and resistant tops. This type of relationship, however, is the hallmark topographic expression of subtle and broad folds, domes that develop over basement structure and topography, flat-topped detached anticlines, extremely young folds in mountain belts, and different types of diapiric structures.

Early Breaching Stage. An early breaching stage occurs in cases where the upper layers of the fold have been removed and is manifested as a series of concentric hogbacks, whereas the crest of the fold still exhibits a positive relief topography (B in Fig. 3.6). This is the most common topographic expression of moderately deformed, detached fold belts, as well as solitary domes and anticlines that develop in front of such belts in many foreland basins.

Late Breaching Stage. A late breaching stage exists in those folds that have undergone a complete removal of their crests but have preserved limbs. In this case, there is an inverse relationship between structure and topography (C in Fig. 3.6). Highly deformed folds which developed over shallow, high-angle fault systems, particularly when the core of the exposed structure consists of nonresistant units (e.g. salt and shales), are the most common examples of this stage.

Obliterative Stage. An obliterative stage exists when erosional processes have eliminated most of the structural relief, leaving only subtle remnants which consist of concentric rims and portions of plunging noses (D in Fig. 3.6). Note that, under such conditions, there is an inverse relationship between structure and topography (i.e., anticlines appear as topographic lows). Such surface expressions typically occur in highly deformed strata involving thrusting or wrench faulting as well as in regions that were exposed to a long period of erosion and denudation.

Examples of folds which exhibit different stages of erosional evolution are given in Fig. 3.7. The subscene in the upper left shows a producing domal structure in the Central Basin Platform (West Texas) that exhibits a positive relief stage. The dome is depicted on imagery data by the unique radial drainage pattern that outlines its topographic crest (also by the drilling pads which are manifested as white dots). The surface structure is extremely subtle and thus the rock units (Cretaceous limestones) which are exposed around the dome are not expressed as flatirons but rather manifest the typical expressions of a dissected plateau (i.e., "layer-cake geology").

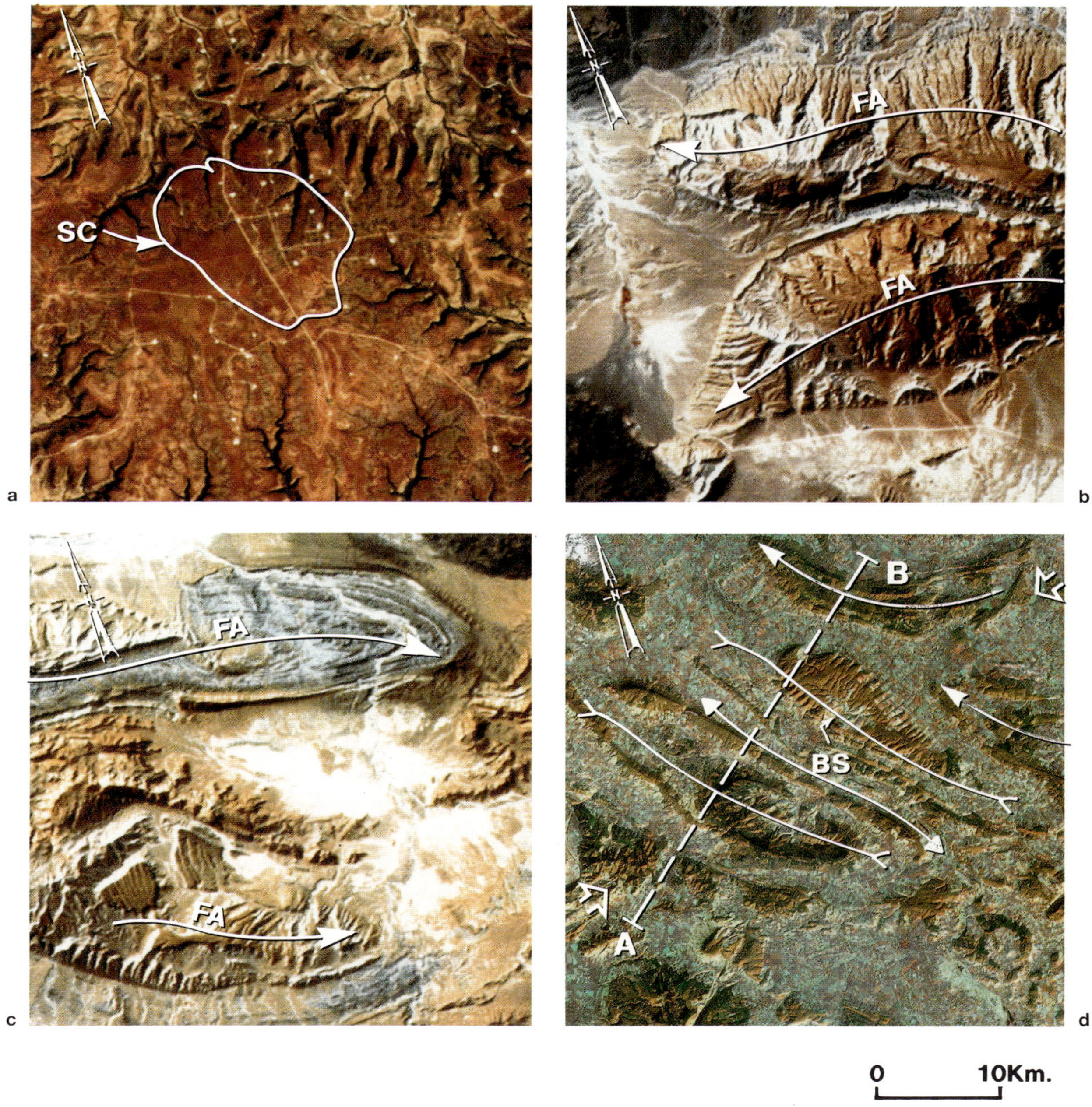

Fig. 3.7. Examples of folds exhibiting different stages of erosional evolution

a A positive relief surface expression of the Leatherwood oil fields in the Central Basin Platform of West Texas (see Fig. 3.8). **b** An early breaching stage of Mid-Pleocene folds in the Zagros fold belt, Iran. **c** Late breaching expressions of salt-cored folds in the Zagros Fold Belt, Iran. **d** Obliterated salt-cored and wrench-related anticlines in the Lower Saxony Basin, Germany. *FA* Fold axis; *SC* structural closure; *BS* breached structure. Cross section *A–B* is illustrated in Fig. 3.11

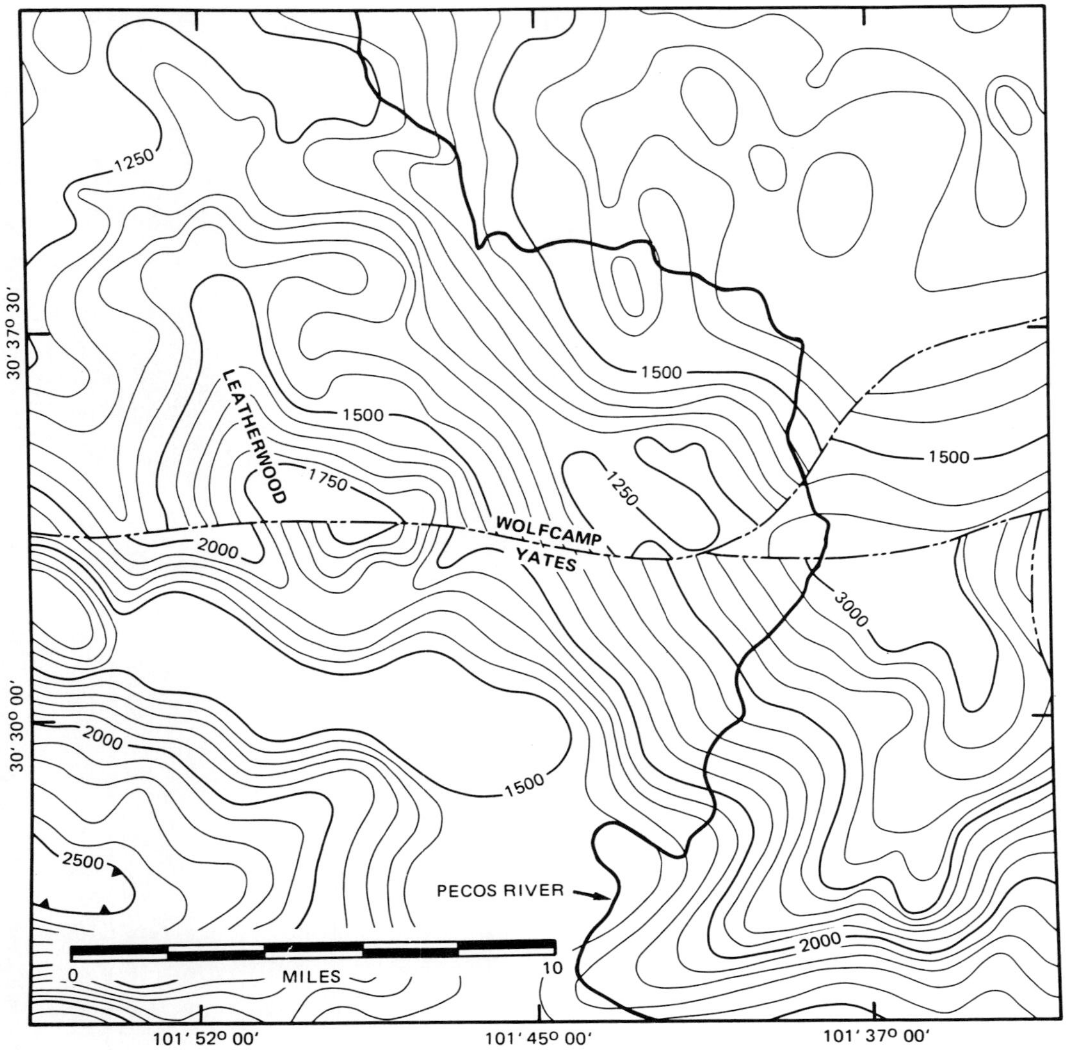

Fig. 3.8. Structure contour map of the Leatherwood oil field, West Texas
The structure closure area (top Permian) corresponds to the one shown in Fig. 3.7 a. (Courtesy of Exxon Exploration Company)

Figure 3.8 shows the subsurface expression of the producing closure (the Leatherwood Field). Although more subtle, the surface structure observed on the imagery clearly mimics the configuration of the producing Permian structure. The relationship between the surface and subsurface structures observed is attributed to the combined influence of structural reactivation and differential compaction processes. This mechanism will be discussed in more detail in Chapter 4 which covers buried and obscured structures.

The upper right and lower left subscenes of Fig. 3.7 show outstanding examples of folds from the Zagros Mountains (Iran) that exhibit early and late breaching stages of erosional evolution. The positive surface expressions of these structures are attributed to two controlling factors: (1) the extreme arid climatic conditions of this region which are characterized by a slow rate of denudation and (2) the relatively young age of this fold belt (Mid-Pliocene; Koyi 1988). The advanced breaching stage occurs in more intensely deformed folds, particularly those that include salt diapirs which are associated with the anticlines during the folding process (Fig. 3.9 and 3.10).

The bottom right subscene (d) shows examples of anticlines in the Lower Saxony Basin *(Leine Bergsland)*, Germany, that exhibit an obliterative stage of

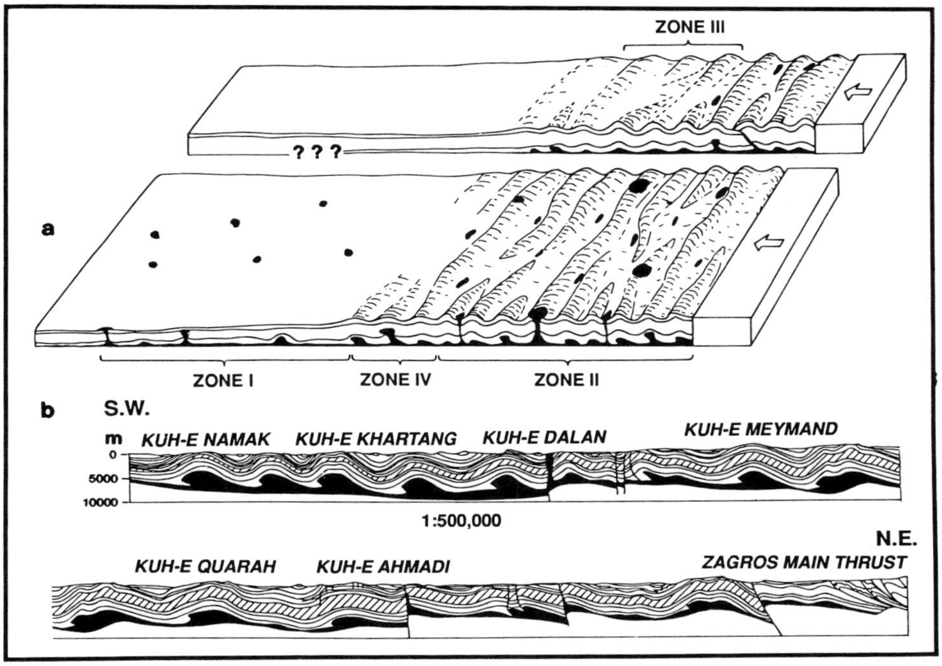

Fig. 3.9. Generalized block diagrams of the Zagros fold belt
The relationships between the folded strata, salt, and basement faults are illustrated. Note that the breaching process of folds, in most cases, occurs in folds that are cut by basement faults and are intruded by salt diapirs (generalized from Koyi 1988). Koyi proposed to divide the mountains into four zones. *Zone I* represents areas of gravity overturn alone. *Zone II* includes areas where lateral shortening has been superimposed on gravity overturns. *Zone III* represents areas where lateral shortening may precede any gravity overturn and *zone IV* represents areas where lateral shortening and accelerating gravity overturns are acting simultaneously. (Published with permission of the AAPG)

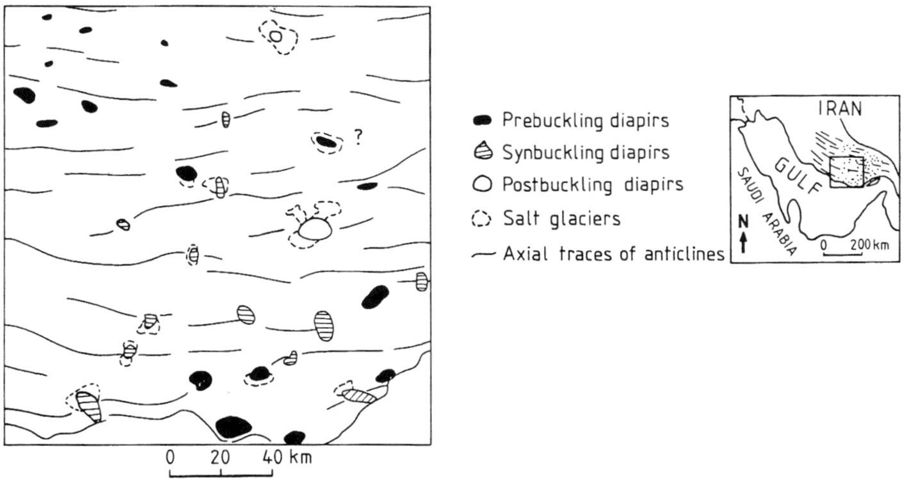

Fig. 3.10. Satellite imagery interpretation of southern Iran
Three generations of diapirs have been distinguished by Koyi (1988) on the basis of shape, size and location in folds. Intensive breaching processes in the folds of the Zagros occurred primarily in the synbuckling and postbuckling diapirs as demonstrated in the imagery data in Fig. 3.7 b, c. (Published with permission of the AAPG)

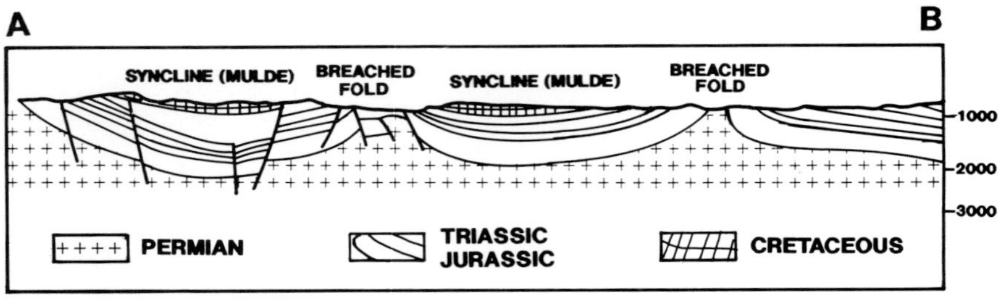

Fig. 3.11. A generalized cross section of the Lower Saxony Mountains, Germany
A complete topographic inversion of geological structures is shown. (*A* and *B* also in Fig. 3.7 d, generalized from Jordan 1979.)

topographic expressions. The intense erosional processes along the folds are attributed to their association with both wrench faults and salt tectonics (Betz et al. 1987). The inverse relationships between the structures and topography in the lower Saxony Mountains are nicely illustrated with a geological cross section of this area (Fig. 3.11).

As illustrated, the broad synclines (known in Germany as *Mulde*) are capped by thick Cretaceous carbonate rock units and are well protected from erosion. The tight, salt-cored anticlines which formed along an active wrench fault system were completely breached by erosion and form the major valleys in this area. Note, however, that the level of obliteration of the folds in the region changes abruptly along a linear topographic feature which was marked by arrows on both sides of this element. Truncated structures and offsets of local lithostratigraphic units are also observed along this element. This local complication in the topographic expression of the synclines is discussed in the next section.

The magnitude of deformation of different folds and, subsequently, their stage of erosional evolution are also manifested by their map patterns. These relationships are generalized in Fig. 3.12, but will be expanded in subsequent sections. Relatively undeformed folds which exhibit positive or early breaching stages usually produce solitary radial or random patterns. Moderately deformed folds will generally manifest parallel, multidirectional and zigzagging patterns. Highly deformed folds will exhibit more complicated patterns which may include en echelon, cross-trending and anastomosing patterns.

Fig. 3.12. Dependencies between magnitude of deformation and folding patterns
Increases in the magnitude of folding results in the development of corresponding fold patterns

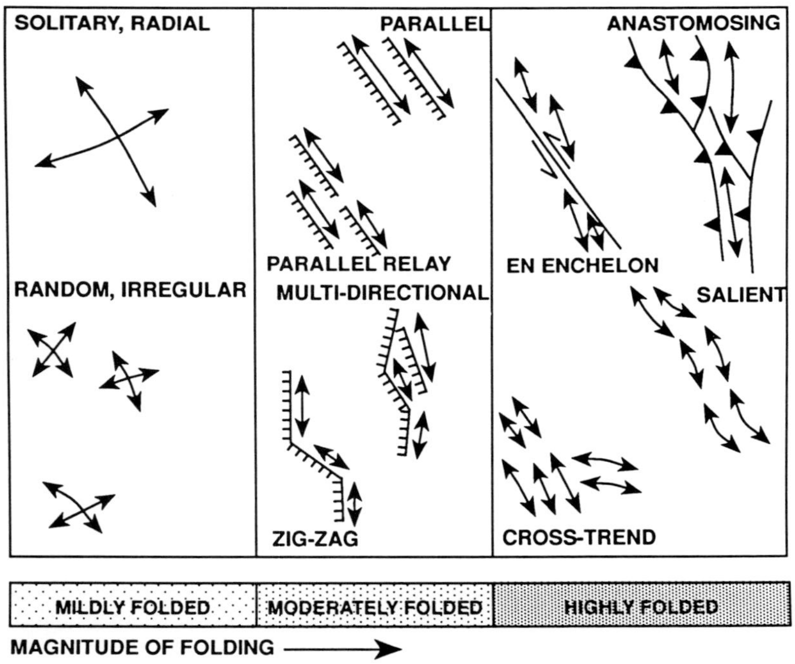

3.2.5 Complications in the Expressions of Folded Strata

In applying the topographic/structural models of folded strata, one should bear in mind that the surface expressions of an individual or a group of folds can be altered and complicated by several different factors that may reflect their unique internal stratigraphy or local variations in style and intensity of deformation. This phenomenon, on the one hand, requires that the topographic/structural models be used carefully and in conjunction with other available surface and subsurface data, but on the other hand, it can be used for cursory identification of unique geological and structural conditions in a basin or a region.

One example of variation in surface expressions of folds, which is related primarily to internal stratigraphy, was illustrated earlier with the images from the Zagros Mountains (see Fig. 3.7 b, c). As illustrated, the erosional stages of folds in this region are determined largely by whether or not the folds are associated with salt diapirs.

Fig. 3.13. Relationship between folded strata and cross-trending, reactivated faults (Leine Bergsland)
a The block diagram and **b** the generalized surface geologic map illustrate some of the surface expressions that are typically formed along such faults. Corners of **b** indicate the extent of the block diagram. *FLT* Fault-line trace; *FR* fractures with no observable offset; *BS* breached structure; *TS* truncated structure. (Geology compiled from Jordan 1979 and others)

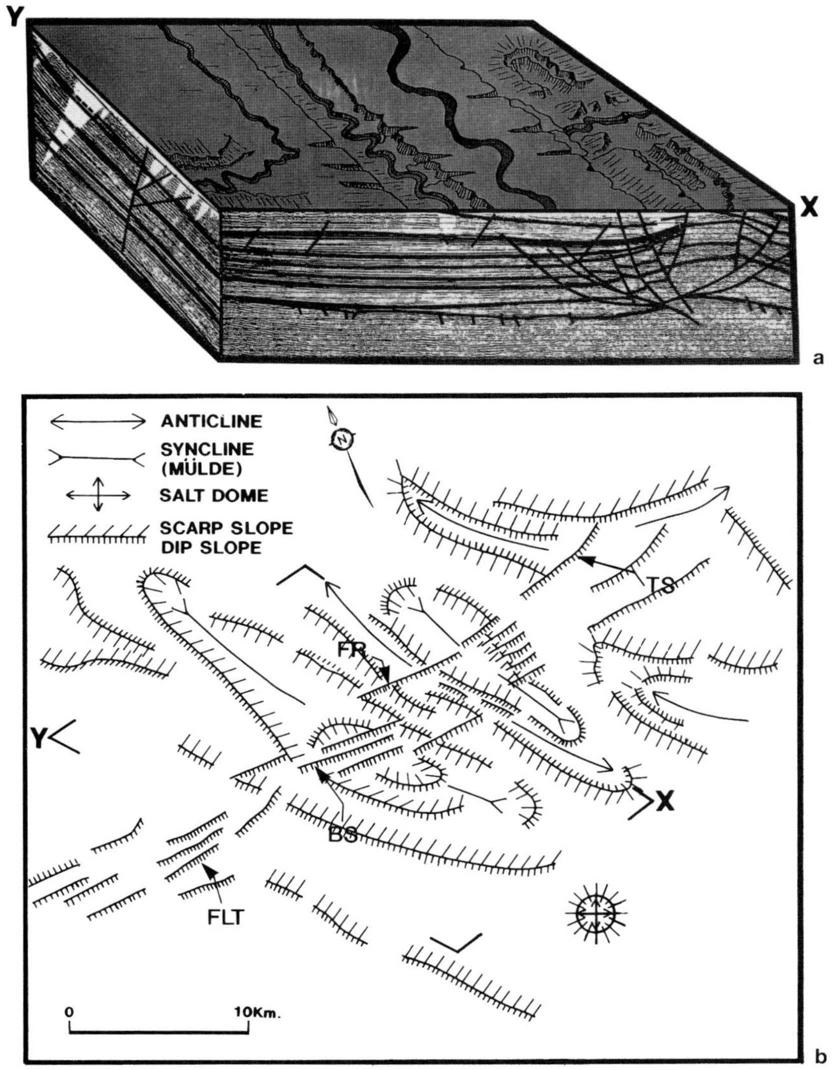

Another common complication in the topographic expressions of folded strata occurs in places where folds are cross-cut by active fault systems. This occurrence causes an increased erosional process which locally breaches the cross-trending folds. For example, the salt-related folds in the Lower Saxony Mountains (Fig. 3.7 d) were locally breached by erosion because they are cross-cut by a reactivated fault system which consists of a wide zone of individual faults and fractures. The surface and subsurface relationships between these two structural elements are illustrated by the block diagram and the interpretive map in Fig. 3.13. As shown, a through-going fault system, which is well exposed in its western segment cross-cuts and breaches the folded strata in its eastern segment, causing breaching, offsets, and truncations of structures and other linear topographic features. On the image data, the alignment of these features produces a profound linear topographic feature (often referred to as a lineament) that indirectly reveals the presence of this buried, reactivated fault.

The recognition of such linear features on imagery often leads to the identification of subtle active fault systems which had not been detected by other mapping tools. The seismic data in the block diagram of Fig. 3.13 a, which focus on the central portion of the fault zone, nicely demonstrate this phenomenon. The salt-related anticlines have a profound expression on seismic data which will receive great attention by interpreters. The cross-trending fault, however, is more subtle and its expression may be hidden by the more dominant structural trend of the salt and wrench-related anticlines. This aspect of structural identification and mapping of cross-trending fault systems and its possible implication for hydrocarbon exploration will be demonstrated later in this chapter as well as in Chapter 7 (with an example from the Jura Mountains and the Mollase Basin of Switzerland).

3.3 Analysis of Exposed Faults

3.3.1 Surface Expressions of Fault Scarps

Faults commonly produce strong linear topographic scarps that are easy to recognize and analyze on imagery. If the scarps coincide with the trace of the fault plane itself, they are referred to as fault scarps or fault lines (Bates and Jackson 1987). Such conditions, however, are very rare and require extremely recent movement along the fault. More commonly, the fault line is eroded and removed

Fig. 3.14. Surface features of fault-line traces
An idealized perspective diagram of surface features associated with rifting was modified from Morely et al. (1990) to illustrate various geomorphic features associated with fault-line traces. *AF* Alluvial fan; *AM* abandoned meander; *BS* breached structure; *DD* drainage divide; *DS* deflected stream; *GD* groundwater discharge; *HV* heavy vegetative cover; *SR* shutter ridges; *TF* triangular facets

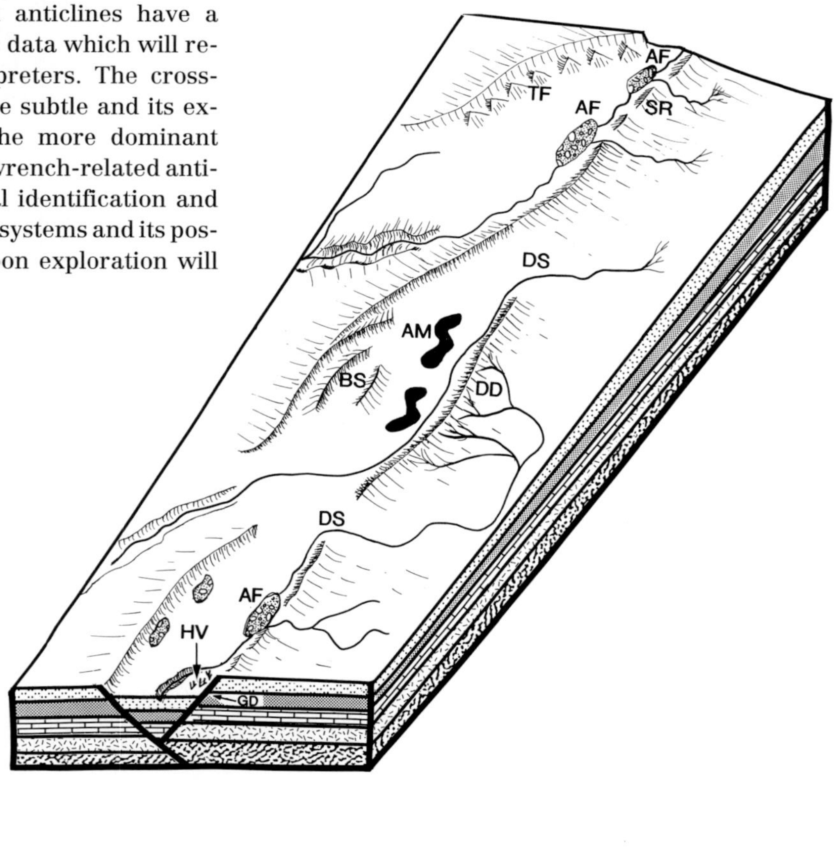

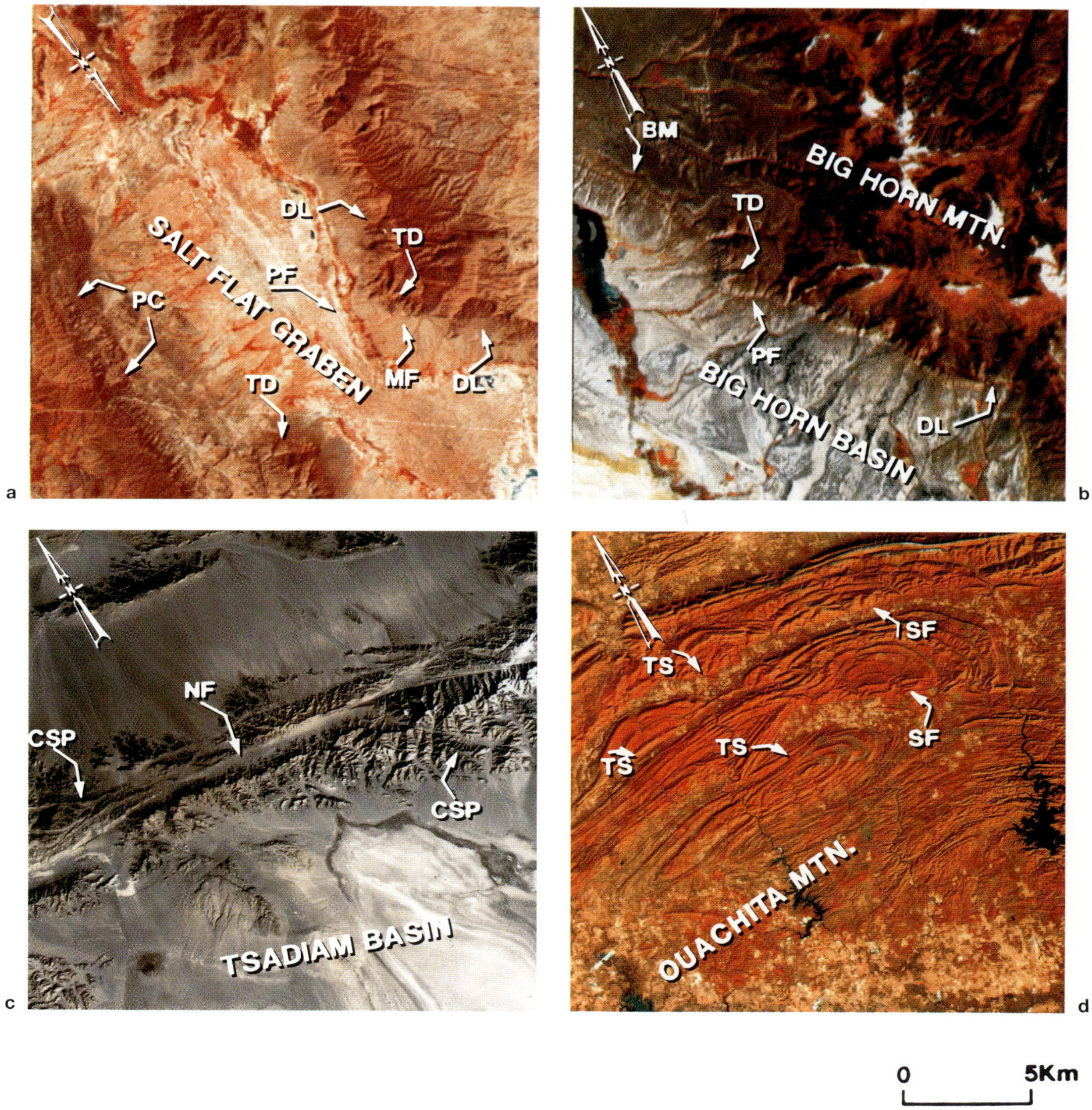

from the location of the fault plane by differential erosion to produce a series of scarps referred to as fault-line scarps or fault-line traces (e. g. Thornbury 1969). The block diagram shown in Fig. 3.14 illustrates the most common diagnostic features associated with fault-line tracess (FLTs). The observed surface features can be divided into two categories: (1) topographic features that directly reflect the FLT, such as linear scarps, triangular facets and shuttered ridges and (2) geomorphic and other fea-

Fig. 3.15. Typical expressions of different fault-line traces (FLTs)
Illustrated by satellite images of **a** Salt Flat Graben, West Texas; **b** Bighorn Mountains, Wyoming; **c** Altyn Tagh wrench fault, western China; **d** the Ouachita fold and thrust belts, Oklahoma. *DL* Dogleg feature; *TD* trap-door; *MF* multidirectional FLT; *PF* pseudo-FLT; *BM* breached monocline; *NF* negative FLT; *CSP* compressional splay; *TS* truncated structure; *SF* sinuous FLT; *PC* parallel composite FLT

tures that develop along the fault such as springs, lakes, sag ponds, linear valleys, offset drainage, alignment of alluvial fans and linear drainage divides.

Many of the features mentioned above are illustrated by the satellite imagery subscenes of different faults that are shown in Fig. 3.15. The interpretation of these subscenes will be discussed further in the next section.

3.3.2 Geomorphic Expressions of Different Fault Types

As stated earlier, fault lines are rarely found in nature because their initial surface expression is modified by erosional processes shortly after faulting has occurred. The resulting fault-line traces, however, preserve several diagnostic surface attributes which can be recognized on imagery and used to determine fault kinematics and sense of movement. On the basis of their surface expressions and aggregate map patterns, it is proposed here to divide fault-line traces into four major categories: (1) multidirectional positive, related to high-angle, dip-slip normal or reverse faults; (2) parallel composite, related to rotated fault blocks of high-angle or listric dip-slip faults; (3) linear negative, related to strike-slip faults; both basement-involved (wrench) and detached (tear) faults; and (4) sinuous positive, related to low-angle thrust faults. Table 3.1 summarizes the major characteristics of these faults and their related surface expressions and map patterns. (For terms related to fold and fault patterns, see Fig. 3.12.)

3.3.2.1 Multidirectional Positive

Multidirectional FLTs which exhibit positive relief topography are commonly associated with Cenozoic-to Mesozoic-aged faults which have not been severely eroded. Examples of this type are shown along the exposed extensional-faulted margins of the Salt Flat Graben in West Texas and the compressional faulted edge of the Bighorn Mountains in Wyoming (Fig. 3.15 a, b). As illustrated, these FLTs clearly manifest positive topographic relief that reflects the vertical movement along the fault, making it easy to map from satellite imagery. The topographically high side, which corresponds to the upthrown block, appears on imagery as a well-defined dissected plateau. In contrast, the topographically low side, which corresponds to the downthrown block, is generally characterized by the presence of unconsolidated sediments and fault-related geomorphic features (alluvial fans, ponds and related vegetation patterns). The FLT is made up of short segments of individual, multidirectional faults that intersect at oblique angles to form zigzag patterns. Distinct topographic features can be observed at the intersection of the blocks which form dogleg edge patterns and positive structural features known as trap doors. Erosional remnants of breached folds, related to block-edge flexures, can also be observed along such FLTs. In places where folding or draping over the fault was extensive, late breaching or obliterative folds can develop and their limbs form spectacular flatirons that accentuate the zigzag patterns of the faults but may confuse interpretation by introducing topographic patterns that may be mistaken for an actual FLT (see discussion of pseudo-FLTs).

Table 3.1. Major characteristics of fault-line traces

Type	Style of deformation	Main features
Multidirectional positive	Extensional and compressional fault blocks	Zigzag fault patterns, trap doors, breached folds and monoclines
Parallel composite	Extensional and compressional fault blocks	Staircase fault patterns, wide and long zones of deformation, linear appearance
Linear negative	Wrench fault assemblages and tear faults	Negative surface expressions, often occupied by valleys, terminates by compressional splays
Sinuous positive	Fold and thrust belts	Sinuous patterns, upper and lower plates manifest different topographic expressions

3.3.2.2 Parallel Composite

A series of parallel FLTs may produce a composite topographic expression in Cenozoic and Mesozoic fault systems that have not been severely eroded. Examples of this type are observed on the western rim of the Salt Flat Graben (Fig. 3.15 a). In contrast to the first type, which forms sharp, faulted boundaries, this FLT is expressed as a wide zone of parallel to subparallel faults and fractures forming a transitional region between the upthrown and downthrown blocks. This zone is heavily eroded producing a unique "staircase" topography which reflects the presence of rotated fault blocks that constitute the master fault. An important distinguishing factor between the two types of FLTs is the increased length of the parallel composite ones. Spectacular expressions of parallel, composite FLTs can also be developed in a series of normal fault systems that often splay from major wrench fault systems (see also Fig. 3.15 c).

3.3.2.3 Linear Negative

FLTs with linear and negative topographic expressions are the trademarks of strike-slip fault systems. A spectacular example of this type can be seen along the Altyn Tagh wrench fault system in western China (Fig. 3.15 c). This FLT extends several hundred kilometers and is manifested as a profound, negative topographic feature that has a linear or curvilinear appearance. (In humid areas, such FLTs are usually occupied by a major stream channel.) There is no consistent expression of high and low sides along this FLT, and, in fact, alternating high and low segments are the hallmark characteristics of wrench fault systems that exhibit different fault profiles along strike. Detached strike-slip ("tear") faults which are found in fold and thrust belts also are manifested as linear negative FLTs but they are limited in length to a single overriding thrust plate. Such features will be illustrated in the next paragraph.

3.3.2.4 Sinuous Positive

Sinuous FLTs with positive topographic expressions are the hallmark features of low-angle thrust faults. Many of the typical surface and structural characteristics of sinuous, positive FLTs related to low-angle thrust faults can be found in the Ouachita fold and thrust belt in southern Oklahoma (Fig. 3.15 d). The characteristics of these features can be summarized as follows: (1) the FLTs are expressed as positive sinuous surface features where the topographically high blocks reflect the upper, overriding plate; (2) there are observable differences in the topographic expressions of the two different plates which are related to variations in lithology and level of erosional activities; (3) the actual location of the FLT is often difficult to detect on imagery, but it may be inferred by the alignments of preexisting structures such as folds, bedding and other faults which are typically overridden and truncated along thrust faults; (4) the thrust-related FLT may be transferred to an adjacent thrust or terminated by a "tear" fault which is expressed as a profound linear negative FLT.

3.3.3 Modification of Fault-Line Traces

The surface expressions of fault-line traces can be further modified by erosion or complicated by structural reactivation to produce unique geomorphic features. As in the case of the topographic expression of folds, this phenomenon requires the use of surface and subsurface data to constrain the interpretation, but at the same time can be used for the identification of unique fault systems. Three such FLTs fall within this category: (1) inverted traces associated with advanced erosion; (2) pseudo-fault traces, related to unique geomorphic features that develop in front of the FLTs; and (3) polyphase traces that reflect repeated movement along the fault.

3.3.3.1 Inverted

This type of FLT reflects an inverse relationship between the structural movement of the fault and its present topographic expression. Such conditions occur in old fault systems which are characterized by the presence of nonresistant layers that are exposed at the upthrown block. An excellent example of an inverted FLT is illustrated in the block diagram from the Eastern Desert of Egypt (Fig. 2.14 b). In the case shown, resistant Eocene carbonate rock units were removed from the top of an uplifted horst block, causing an inversion of its topographic expression. The preservations of these rock units on the downthrown sides led to the development of a profound positive FLT which shows an inverse relationship with the fault movement.

3.3.3.2 Pseudo-Fault

These FLTs are expressed as an alignment of geomorphic features that develop in front of positive fault-line tracess, but do not reflect its actual location. For example, the front of the alluvial fans in the Salt Flat Graben (Fig. 3.15 a) is expressed as an alignment of linear vegetative areas that manifest a zigzag pattern and could be mistakenly interpreted as an FLT of high-angle dip-slip faults. The pseudo-faults are created by an apron of unconsolidated sediments that are transported by fluvial systems from the upthrown blocks. Hence, the energy available for transportation is somewhat a direct function of the slope of the uplifted blocks. It is not surprising that the pattern formed mirrors the shape of the actual faults. Another common version of pseudo-FLTs is developed by remnants of hogbacks and related flatirons of tight folds that were breached or obliterated along the faults. In this case, zigzag patterns of the hogbacks could be interpreted as actual FLTs. The faulted edges of the Bighorn Mountains are known for this particular type of geomorphic expression (Fig. 3.15 b).

3.3.3.3 Polyphase

This type of FLT exhibits a mixture of diagnostic features of different types of faults which indicate a kinematic history of several distinct tectonic events with varying stress regimens. It is extremely important for the recognition of reactivated fault systems and superimposed styles of deformations. This will be illustrated in greater detail in the next section with examples from the Rhine Valley and Jura Mountains region.

3.4 Surface and Structural Patterns of Different Structural Styles

The mode of deformation of a particular region or a basin defines its structural style which is manifested by the assemblages of individual structures, their unique geometries and aggregate map patterns. Structural style identification is, therefore, an important concept in hydrocarbon exploration and the analysis of exposed structures with satellite image data can play a significant role in this effort. For effective identification and analysis of structur styles, it is required that the interpreter be familiar not only with the surface and map pattern of individual structures as described earlier, but also with the geometries and map patterns of structural styles that are normally seen in sedimentary basins. Although the major components of different structural styles will be illustrated here, the reader is encouraged to review a more thorough treatment of this subject matter which can be found in several of the works cited at the end of this chapter, particularly those of Harding and Lowell (1979) and Lowell (1985).

Seven major types of structural styles have been recognized and are shown in Fig. 3.16. The diagram shows the hydrocarbon traps most commonly associated with these structural styles. It is important to note that the first four are basement involved, whereas the other three are detached. Whether the structure is attached or not has a great effect on the ease of recognition. There is also a significant difference in the level of structural deformation that characterizes each style. For example, arches and domes fall within the category of mildly deformed structures and, therefore, their identification with monoscopic data will be quite limited. Compressive and extensional basement-involved fault blocks are characterized typically by moderately deformed structures and are easily interpreted with monoscopic data. Wrench- and thrust-related structures fall within the category of highly deformed structures and their interpretation on imagery warrants the aid of stereo data.

Past experience shows that imagery can be used successfully to recognize all basement-involved structural styles such as extensional and compressional blocks, wrench fault systems, large arches and domes (basement warps). Dêcollement fold and thrust belts are also discernable, but successful style identification of detached normal faults or salt-cored structures has been quite limited to date. Figure 3.17 shows detailed structural interpretations of the images which were shown earlier in Fig. 3.15 for the purpose of demonstrating the unique mapping pattern of each style.

The following sections include observations regarding the identification of different structural styles which are shown in these figures. (Arches and domes related to basement warp structures are not shown here because they fall more within the category of buried and obscured structures, to be discussed in Chapter 5.) Initial observations discussed here were made by Harding and Hopkins (1977).

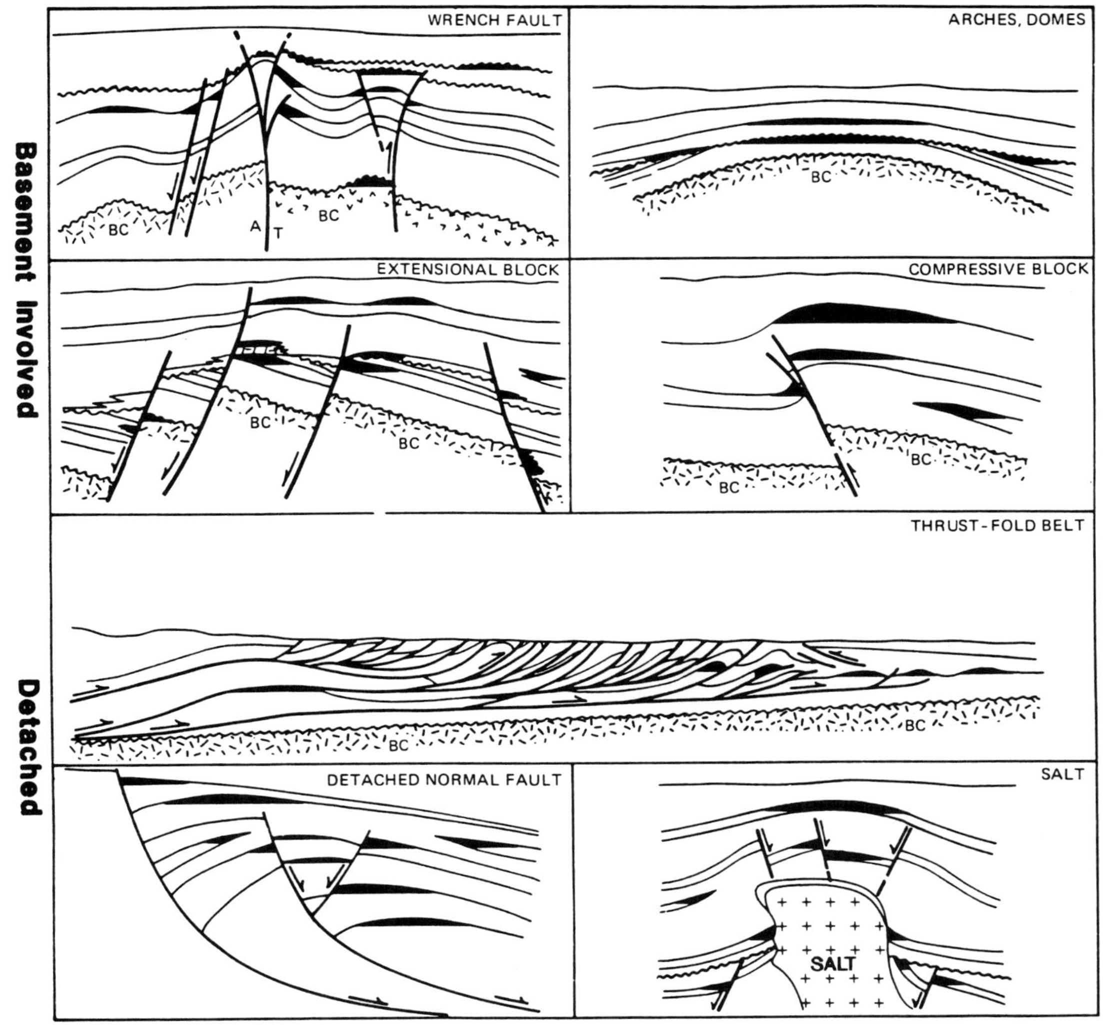

Fig. 3.16. Recognition of structural styles
Cross sections showing the major structural styles and related trap types as defined by Harding and Lowell (1979). Basement rocks are indicated by *BC*. (Published with permission of the AAPG)

3.4.1 Compressional vs. Extensional Basement-Involved Fault Block Styles

The structural styles of both compressional and extensional basement-involved fault blocks are manifested in the interpreted imagery of Fig. 3.17 a, b. Note that the image from the Salt Flat Graben was reversed in direction to point south so that the master faults of both images can be compared.

Major features of both images can be described as follows. Faults with positive topographic expressions and multiple orientations are the most diagnostic elements of both compressional and extensional styles. These faults trend oblique to the dominant grain and may intersect to form clusters of fault blocks or trapdoors. The faulted margins of the blocks are often mixed with less diagnostic sets of parallel composite FLTs that trend along the domi-

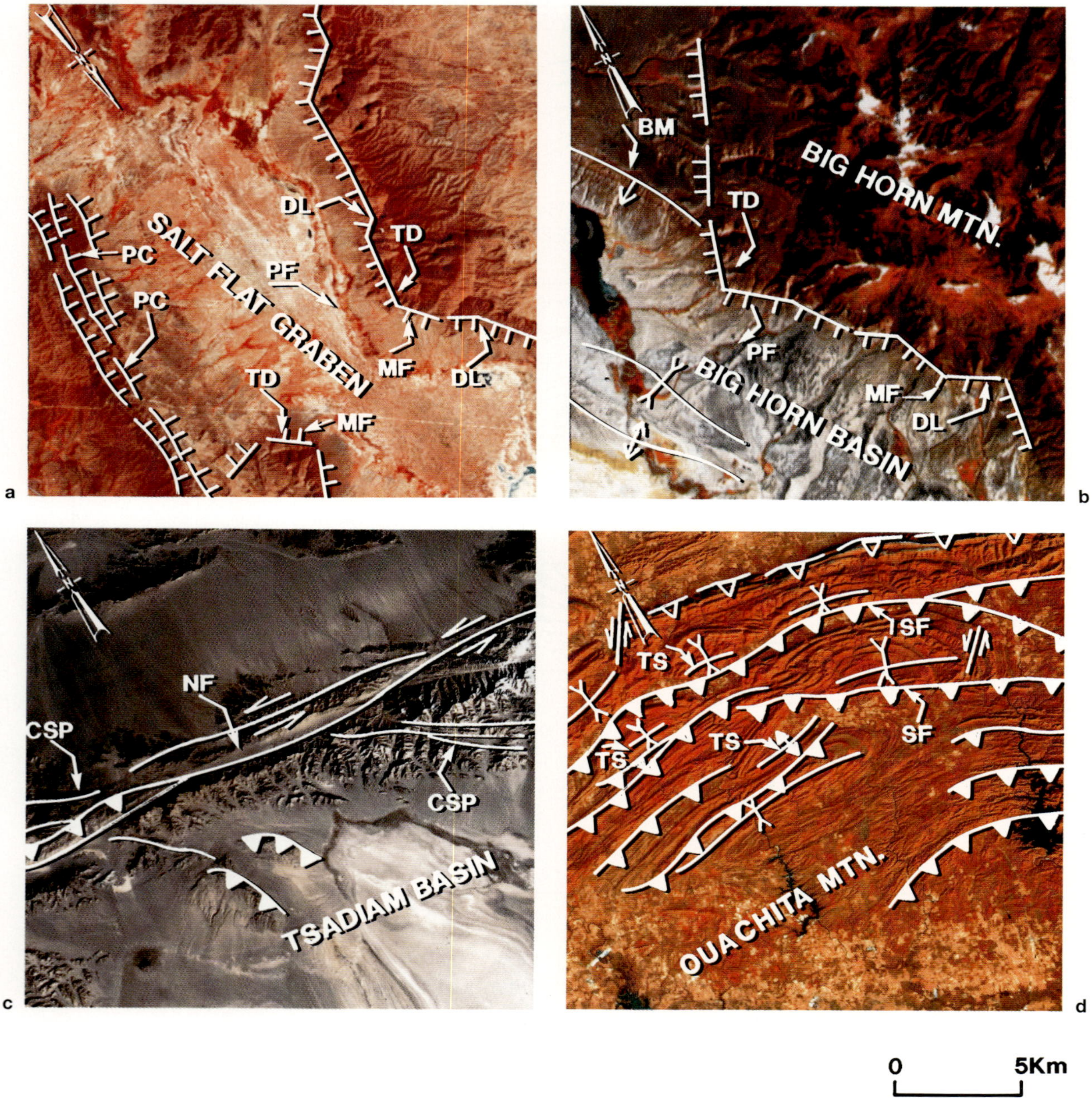

Fig. 3.17. Imagery expression of different structural styles
Detailed interpretation of faults and folds of the same images shown in Fig. 3.15. **a** Extensional fault block of the Salt Flat Graben; **b** compressional block faulting in the Bighorn Basin, Wyoming; **c** wrench fault assemblages of the Altyn Tagh fault system, western China; **d** décollement fold and thrust belt of the Ouachita Mountains, Oklahoma. Letter codes are the same as in Fig. 3.15. (After Corona and Wielchowsky 1984)

nant structural grain and reflect the branching of the single master fault into several rotated fault blocks. Drape folds with multiple orientations may be formed along the zigzagging FLT to form attractive hydrocarbon traps (i.e., "trapdoors"). These folds are usually breached and easy to recognize on imagery. Monoclinal flexures may be formed along the more linear segments of the faulted blocks, but these features are much less attractive for hydrocarbon exploration.

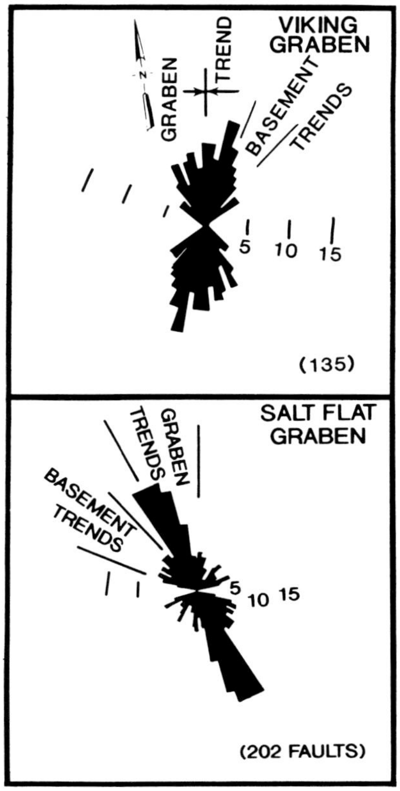

Fig. 3.18. Fault orientation of extensional fault blocks
Plots of fault orientation of two well-exposed or mapped graben features. The majority of the faults follow the major trends of the graben's axis. Other dominant trends are related to reactivation of preexisting basement structures. (Compiled from Phelps and Harding 1987)

Plots of fault orientation (Fig. 3.18) of these styles show that they fall predominantly (90%) within a 90° quadrant of the dominant structural trend (i. e., parallel to the maximum stress direction). These orientations are classified as longitudinal or oblique (depending on angle) and reflect normal results of block faulting. Other orientations (more than 45° away from the dominant trend), if present, may indicate the reactivation of preexisting zones of weakness or superimposed styles. These faults can lead to the development of the most attractive trapdoor structures in such tectonic settings. For example, the large trap-doors in the Salt Flat Graben (Fig. 3.17 a) were formed at the intersection of oblique and longitudinal graben-aged faults with transverse faults that reactivated preexisting Paleozoic structures. This phenomenon is illustrated in greater detail in Part 2.

Distinguishing between compressional and extensional block faulting is difficult unless something is known about the tectonic setting of the area. However, compressive block faulting is commonly associated with areas of extensive, well-developed folding, whereas extensional block faulting may exhibit only drape flexuring with moderately deformed folds.

3.4.2 Wrench Fault Assemblages

The typical structural style of wrench fault assemblages is illustrated in Fig. 3.17 c. To aid in the description of these features on imagery, common structural features and terminologies associated with this style are illustrated in Fig. 3.19. Wrench

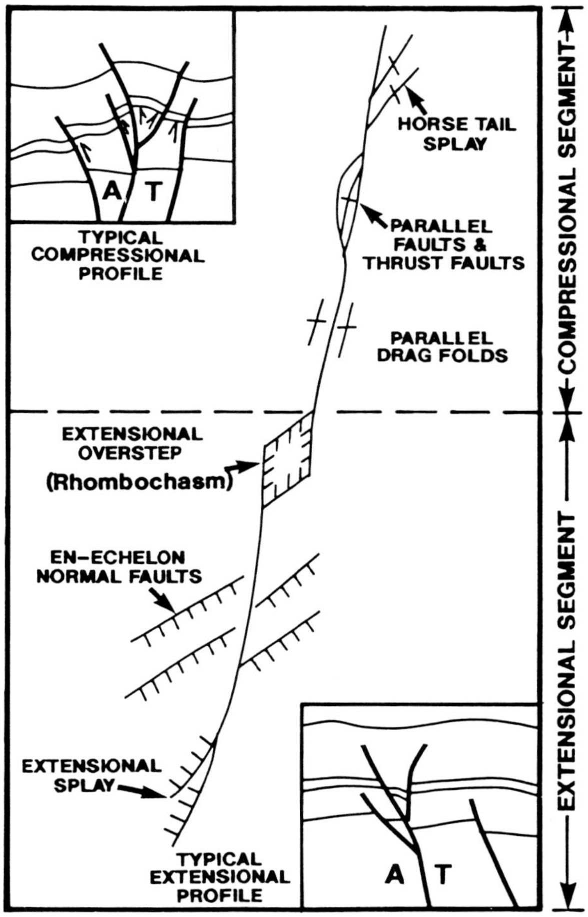

Fig. 3.19. Typical patterns of wrench fault systems
Shown are common structures that develop along right-lateral wrench fault systems and their typical seismic expression. (Compiled and modified from Harding et al. 1985)

fault assemblages can sometimes be recognized from fault zone characteristics alone, and the fault-line traces have straight to curvilinear through-going negative expressions which cannot be observed in any other type of fault system. The FLTs often splay from the master fault to form both extensional and compressional oblique features. Extensional splays ("horsetails") often appear as profound parallel composite FLTs, whereas compressional splays are manifested by thrust faults and oblique anticlines. Associated deformation is usually limited to an elongated zone which may vary considerably in width and topographic expression.

Because of the small scale of conventional displayed imagery, direct observations of strike-slip offset (offset streams, glacials or piercing points) are usually inconclusive. This is particularly true when correlation of "marker beds" is attempted as a means of identifying the horizontal movement along the fault. Optimum geological conditions would be necessary to indicate such offsets, and these must usually be accompanied by careful field mapping.

The identification of wrench fault assemblages is particularly important because these structures are extremely difficult to map and recognize on seismic data. Such identification is required because these fault systems change their profile characteristics along strike and often can be confused with other compressional or extensional fault systems.

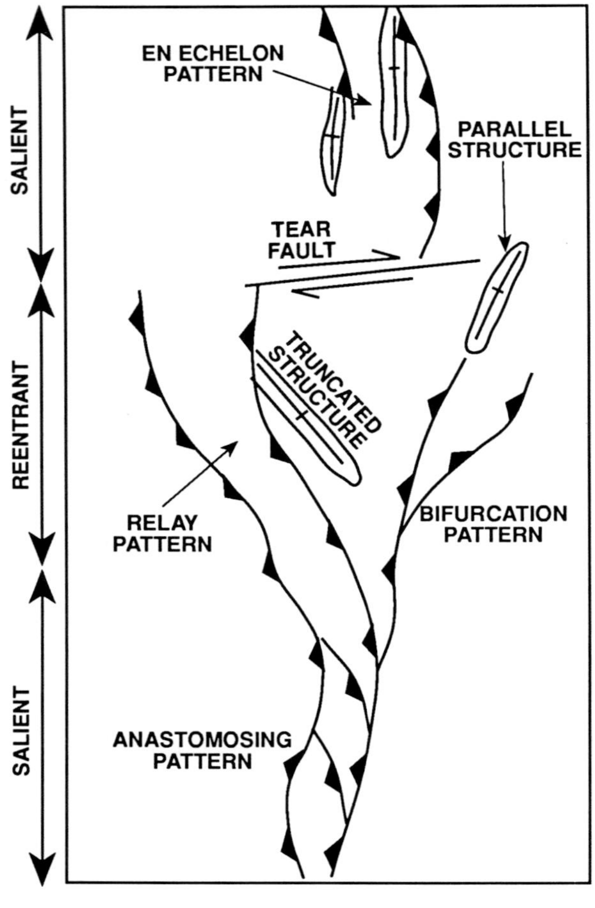

Fig. 3.20. Typical patterns of fold and thrust belts
Shown are common structures that develop in fold and thrust belts and are usually detectable on satellite data

3.4.3 Dêcollement Thrust Fault Assemblages

Most fold and thrust belts form large sinuous mountains that are easy to recognize on images by the unique appearances of positive, sinuous FLTs, abundant folds, truncated structures, various overriding plates with different physiographic appearances and other features that were described earlier and are summarized in Fig. 3.20. Structural analysis of these belts with imagery is particularly important for two reasons. First, these belts are considered to be one of the most important remaining frontiers for hydrocarbon exploration (e.g. the Canadian Rockies, the Zagros Fold Belt in Iraq and Iran and the Rif Mountains in Morocco). Second, these mountain belts are characterized by an extremely complex structural setting and typically are difficult to image with seismic data.

Because style identification is so evident on imagery, the most important contribution of the analysis lies in the recognition of local variations in structural patterns of these belts that are often compartmented into different segments with varying degrees of structural complexities and availability of large-scale structural traps. In many instances, the subdivision of fold and thrust belts into different structural compartments occurs along large-scale, cross-trending, reactivated fault systems which may appear on imagery as large "tear" faults. This phenomenon will be illustrated in the next section. Also, recognition of preexisting reactivated faults and superimposed styles can provide significant clues to structural elements that exist under the dêcollement surface. Such features might not be detectable on seismic data.

3.4.4 Regional Structural Analysis

Effective recognition of structural styles and trends in sedimentary basins often requires the establishment of the spatial relationships between the structures observed in the basin and those which are exposed in the surrounding areas. In such investigations, the interpreter must seek structural information on large areas which is not usually obtained during routine exploration activities. Regional scale satellite imagery mosaics were found to provide an effective and inexpensive tool for such analysis.

For example, three distinct styles can be distinguished on the interpreted satellite mosaic of the Rhine Valley, Jura Mountains, Black Forest Massif and Vosges Massif in Switzerland, Germany, and France (Figs. 3.21 and 3.22).

The Rhine Valley, which was formed predominantly during mid-Eocene to Early Pliocene rifting (Illies 1970), displays typical structural characteristics of an extensional fault block regime. That is, the graben's margins are made of a series of positive, multidirectional FLTs related to high-angle, dip-slip faults. The typical zigzag pattern and associated trapdoors and dogleg features are quite noticeable on this imagery mosaic. Opposing the Rhine Graben rims are two Paleozoic massifs, the Black Forest *(Schwarzwald)* and Vosges, which exhibit a wrench-related structural style. That is, the massifs are dissected by several northeast-trending faults which produce linear negative FLTs and typical splays or "horsetail" features.

The Jura Mountains, which were formed by Late Cretaceous compressional events related to the Alpine orogeny (Trümpy 1980), display typical structural characteristics of a detached fold and thrust belt. Structures in this belt consist of thrust faults that have positive sinuous FLTs, elongated folds that are almost parallel to the thrust faults, and strike-slip "tear" faults that occur obliquely to the thrust faults.

3.4.5 Reactivated Structures and Superimposed Styles

More detailed interpretation of the imagery often reveals structural and topographic features that do not fit the appropriate structural assemblages. Such anomalies usually indicate the presence of reactivated structures and related superimposed styles.

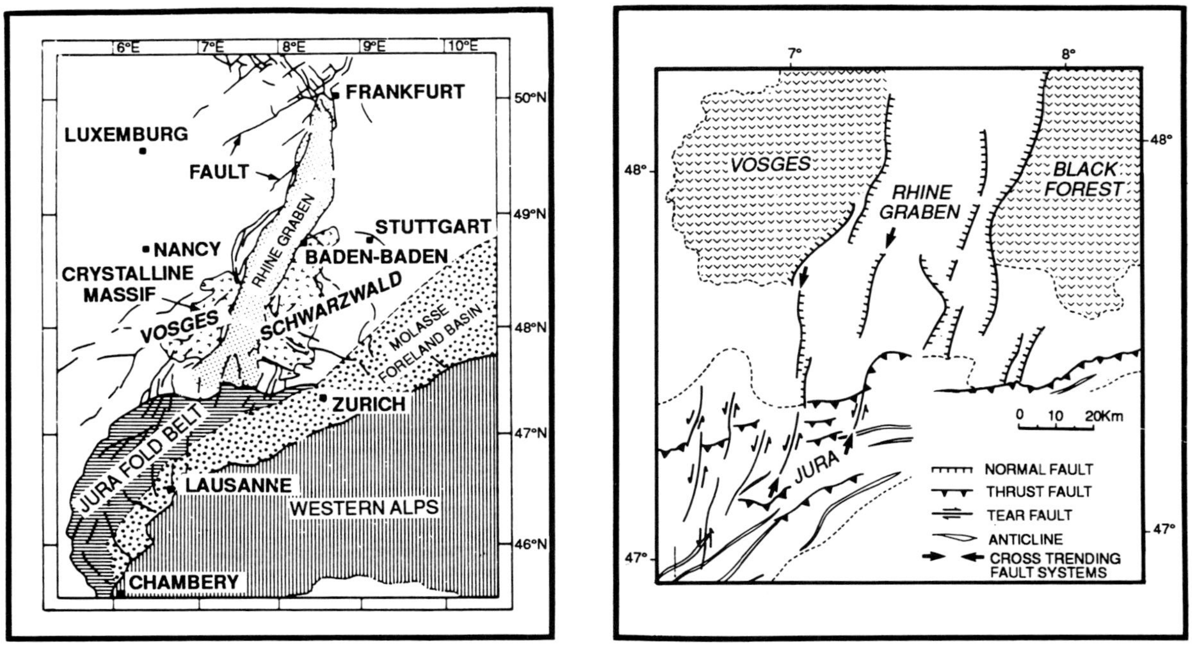

Fig. 3.21. Tectonic maps of the Rhine Graben and the Jura Mountains
Shown are the location and trends of major types of fault systems. (After Spicer 1980)

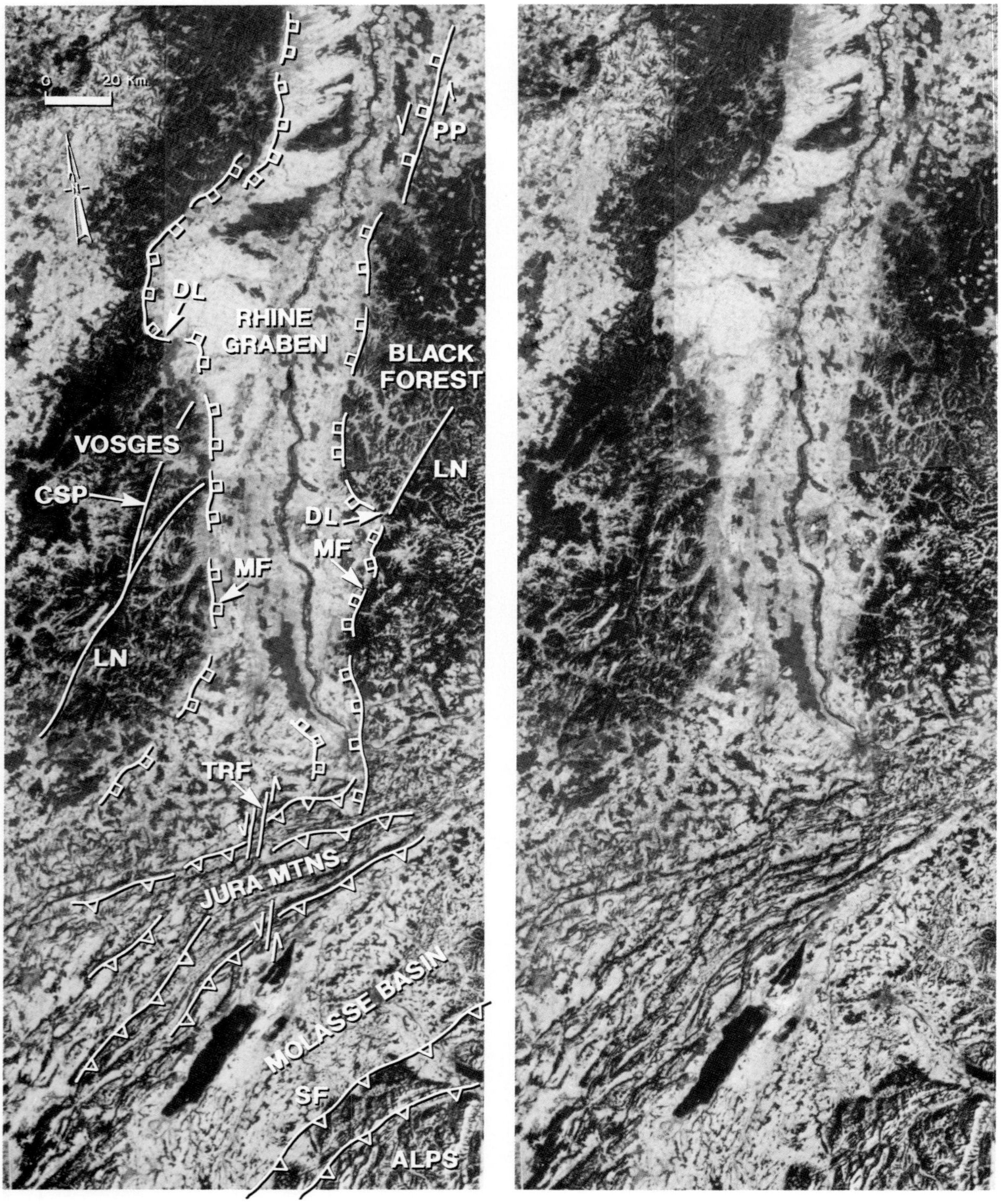

Fig. 3.22. Structural styles of the Rhine Graben and Jura Mountains
Interpreted and uninterpreted satellite imagery mosaic of the study area showing the style of deformation of different tectonic elements. *DL* Dogleg feature; *LN* linear negative FLT; *PP* polyphase fault; *TRF* tear fault; *CSP* compressional splay; *MF* multidirectional FLT; *SF* sinuous FLT

a

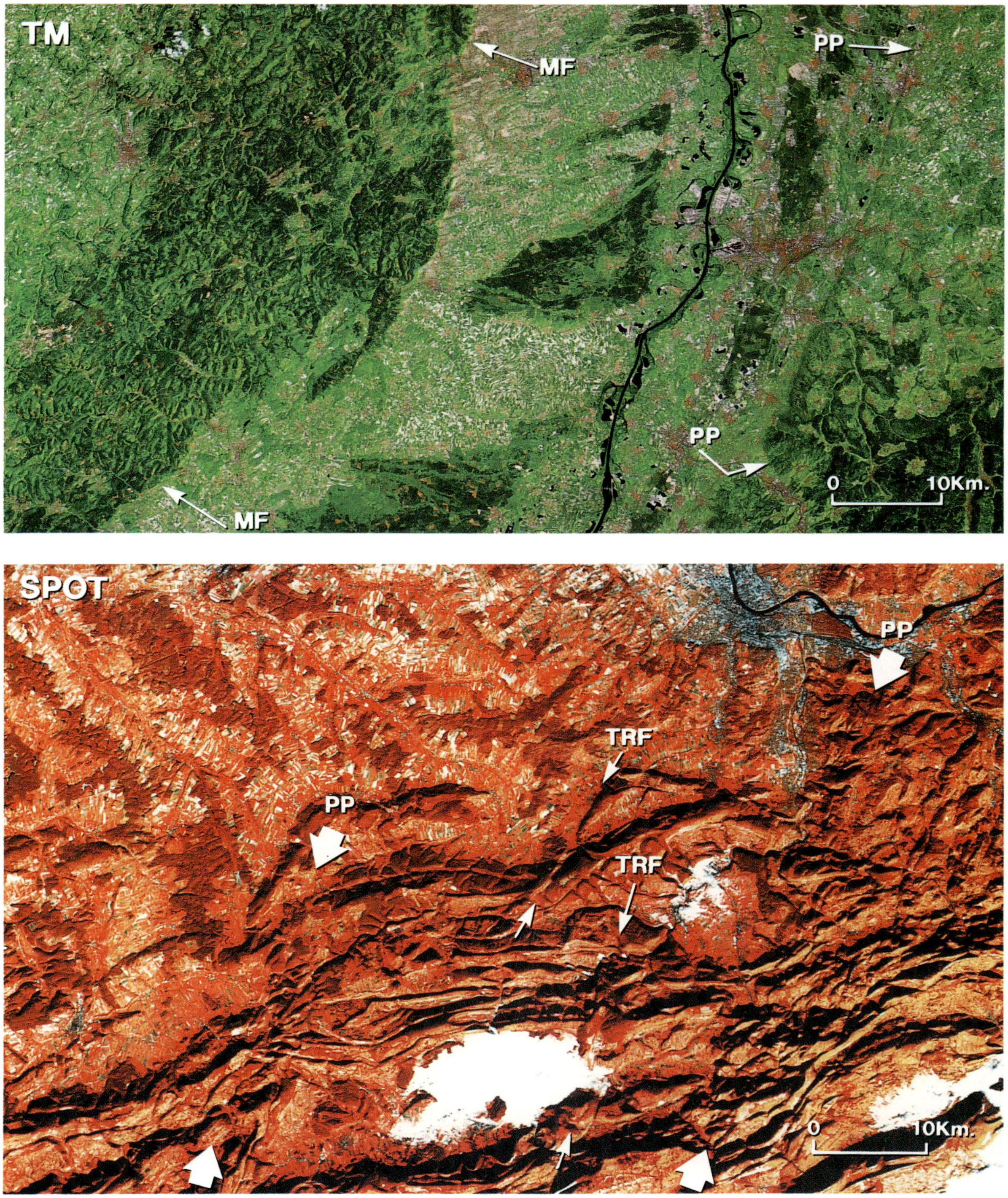

b

Fig. 3.23. Reactivation of superimposed styles
A close-up look at two cases where reactivation and superimposed styles can be observed on satellite imagery. Shown is a reactivation of Paleozoic wrench faults along the faulted edges of the Rhine Graben (**a**) and major "tear faults" in the Jura Mountains which have reactivated pre-existing graben-age faults (**b**). *MF* Multidirectional fault; *PP* polyphase fault; *TRF* tear fault

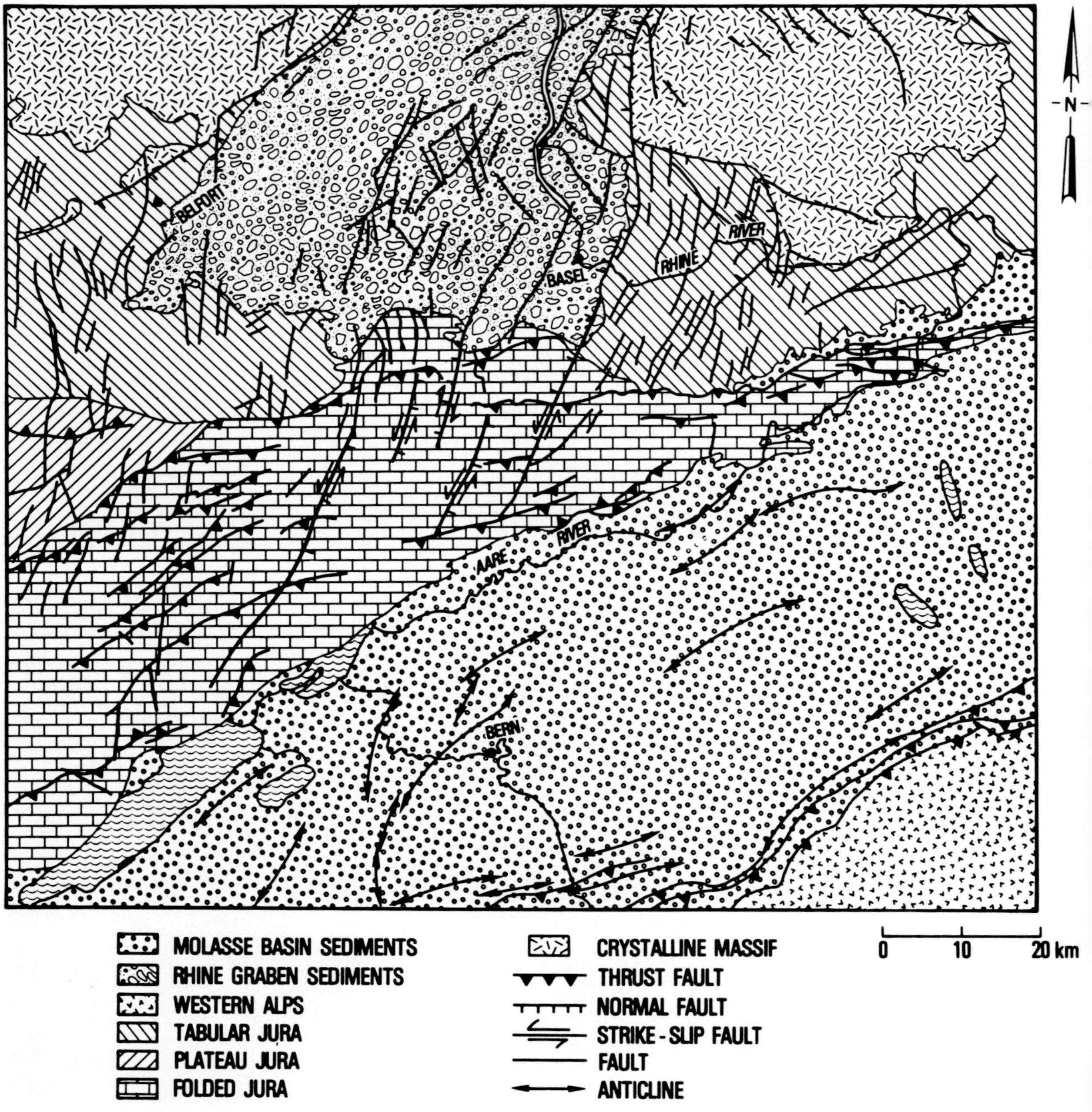

Fig. 3.24. Reactivated "tear" faults in the Jura Mountains Tectonic map of the Jura Mountains and the southern edge of the Rhine Graben showing the spatial correspondence between the faulted edge of the Rhine Graben and reactivated "tear" faults. (After Spicer 1980)

In the eastern rim of the central part of the Rhine Graben (Fig. 3.23 a), newly formed extensional faults were reactivated and followed the preexisting wrench fault systems of the Black Forest massif. Consequently, the FLT of this segment of the graben does not display the typical zigzag patterns of extensional high-angle faults, but rather exhibits the typical linear appearance of FLTs that are related to the wrench fault systems. This segment is, therefore, identified as a polyphase FLT which also trends at oblique angles to the general direction of the graben and, thus, should be regarded as a transverse fault. The profound zigzag appearance of the "normal" western rim accentuates the anomalous appearance of the eastern rim.

In the Jura Mountains region (Figs. 3.23 b and 3.24, preexisting extensional fault systems of the Rhine Graben were reactivated and subdivided (or compartmented) the Jura fold belt into several stacks of thrust sheets bounded by northeast-trending zones of faults and fractures with "tear" fault characteristics (i. e., linear negative FLTs). Consequently, each block is characterized by significantly different intensities of deformation. The interaction between the Rhine Graben faults and the tectonic evolution of the Jura has been previously suggested by several other authors (Laubscher 1977; Trümpy 1980). These relationships are also supported by other data sets such as in situ stress measurements and earthquake activities (Illies and Greiner 1978) as well as outcrop studies (Diebold et al. 1960; Laubscher 1977). The structural characteristics of these reactivated fault systems, their timing of reactivation, and their influence on hydrocarbon plays of the adjacent Molasse Basin are examined further in Chapter 7.

3.4.6 Word of Caution

In spite of the many advantages of using satellite imagery for early identification of structural styles, the reader must be aware of some serious limitations that may be encountered in relying so heavily on the interpretive surface data. First and foremost, imagery provides a summation of many structural events that may be diverse and unrelated. Tectonic overprinting may cause erroneous style patterns, and incomplete surface expressions may result in false structural appearances. Second, the style identification at the surface may not always be relevant to subsurface structure within the basin. For example, structural inversion of grabens can cause them to appear as positive topographic elements at the surface. One should always remember that structural imagery interpretation should be integrated with other available surface and subsurface data.

3.5 Summary

Exposed structures manifest clear expressions on satellite imagery that can be used to identify and analyze their geometries, style of deformation and in some cases, movement histories. Folded structures are divided into three categories: mildly deformed, moderately deformed and highly deformed. The orientations and magnitude of inclination of moderately deformed folds can be estimated with monoscopic imagery by analyzing the surface expressions of flatirons that developed along their breached limbs. The analysis of the other two types usually requires the use of stereo imagery and the application of the three-point dip measurement method.

The magnitude and style of deformation of folds, as well as their age and lithological makeup, often can be estimated by their surface expressions. The erosional expression of folds can be divided into four stages: positive relief stage, early breaching stage, late breaching stage and obliterative stage. Mildly deformed folds that develop over basement blocks usually preserve their structural expresisons to exhibit a positive-relief erosional stage. Moderately deformed folds that develop over high-angle faults exhibit various degrees of breaching of their crests. The surface expressions of highly deformed folds associated with thrusting and wrenching processes are usually obliterated by erosion, preserving only small remnants of their actual geometry. Modification in the erosional expressions of individual folds can be used to recognize local variations in their lithologies or the presence of reactivated fault systems.

Exposed faults lead to easily recognizable fault-line traces or FLTs. FLTs are divided based on their surface expressions, mapping patterns and related style of deformation into four main types: (1) multidirectional positive, related to high-angle, dip-slip normal or reverse faults; (2) parallel composite, related to rotated fault blocks of high-angle or listric normal faults; (3) sinuous positive, related to low-angle, thrust faults; and (4) linear negative, related to strike-slip, wrench or detached "tear" faults. Other, more unique types, of FLTs include (1) inverted, associated with intense erosion; (2) pseudo-fault, related to unique geomorphic features; and (3) polyphase, reflecting repeated movement along the fault.

The structural styles of particular regions or basins define their dominant mode of deformation, related tectonic habitats, and the assemblages of individual structures. Structural styles which can be detected on imagery include compressional and extensional block faulting, wrench fault assemblages, dêcollement thrust fault assemblages, and basement warp structures. The first three can be recognized on monoscopic imagery by analysis of fold and fault patterns and their surface expressions. The fourth must be analyzed with stereo data and requires considerable integration of other data sets.

Regional analysis of small-scale mosaics provides an effective tool for early identifications of different styles. More importantly, such analysis leads to the recognition of anomalous features which may indicate the presence of reactivated structures and related superimposed styles. In spite of the many advantages of using satellite imagery for structural style identification, serious pitfalls and limitations may be encountered in using this technique. These include complex tectonic overprints and surface structures that do not necessarily reflect subsurface style. Structural imagery interpretation therefore always must be integrated with other available surface and subsurface data.

References and Further Reading

Angelier J, Colletta B (1983) Tension fractures and extensional tectonics. Nature 301: 49–51

Babcock EA (1971) Detection of active faulting using oblique infrared aerial photography in the Imperial Valley, California. Geol Soc Am Bul 82: 3189–3196

Bailey GB, Anderson FD (1982) Applications of Landsat imagery to problems of petroleum exploration in Quaidam Basin, China. AAPG Bull 66: 1348–1354

Bates RL, Jackson JA (1987) Glossary of geology, 3rd edn American Geological Institute, Alexandria, Virginia

Berger Z (1988) Detection and analysis of basement structures in low relief basins using an integrated analysis of Landsat data. AAPG Bull 72 (2): 160–161

Berger Z, Corona FV (1986) Landsat structural analysis of the Rhine Valley and the Jura Mountain area, western Europe. Int Symp on Remote sensing of the environment, 5th Thematic Conf, Reno, NV. Environmental Research Institute of Michigan, Ann Arbor, pp 35–48

Betz D, Fuhrer F, Greiner G, Plein E (1987) Evolution of the Lower Saxony Basin. Tectonophysics (Amsterdam) 137: 127–170

Boyer SE, Elliott D (1982) Thrust system. AAPG Bull 66 (9): 1196–1230

Corona FV, Wielchowsky CC (1984) Landsat imagery analysis of exposed structures. Exxon Production Research Company, Internal Rep

Diebold P, Laubscher HP, Schneider A, Tschopp R (1960) Geologic atlas of Switzerland, Sheet 1085, St. Ursanne

Halbouty MT (1980) Geological significance of Landsat data for 15 giant oil and gas fields. AAPG Bull 64: 8–36

Hamblin WK, Howard JD (1989) Exercises in physical geology, 5th edn). Macmillan, Minneapolis

Harding TP (1984) Graben hydrocarbon occurrences and structural style. AAPG Bull 68: 333–362

Harding TP, Hopkins HR (1977) Identification of structural styles with Landsat imagery. Exxon Production Research Company, Internal Rep

Harding TP, Lowell JD (1979) Structural styles, their plate-tectonic habitats, and hydrocarbon traps in petroleum provinces. AAPG Bull 63 (7): 1016–1058

Harding TP, Vierbuchen RC, Christie-Blick N (1985) Structural styles, plate-tectonic settings and hydrocarbon traps of divergent (transtensional) wrench fauls. In: Biddle KT, Christie-Blick NH (eds) Strike-slip deformation, basin formation, and sedimentation. SEPM Spec Publ 37: 51–77

Hopkins HR, Navail H, Berger Z, Merembeck BF, Brovey RL, Schriver JS (1987) Structural analysis of the Jura Mountains-Rhine Graben intersection for petroleum exploration using SPOT stereoscopic data. In: SPOT 1 image utilization assessment, results. Centre National d'Etudes Spatiales, Toulouse, France, p 803–810

Illies JH (1970) Graben tectonics as related to crust-mantle interaction. In: Illies JH, Mueller ST (eds) Graben problems. E Schweizerbart'sche Verlagsbuchhandlung, Stuttgart, pp 4–27

Illies JH, Greiner G (1978) Rhinegraben and the Alpine system. Geol Soc Am Bull 89: 770–782

Illies JH, Baumann H, Hoffers B (1981) Stress pattern and strain release in the Alpine foreland. Tectonophysics 71: 157–172

Jordan H (1979) Geologische Wander-Karte, Leineberg-land, 1:100 000, Niedersächsisches Landesamt für Bodenforschung, Hannover, Germany

Kent PE (1970) The salt plugs of the Persian Gulf region. Leicester Lit Philos Soc Trans 64: 56–88

Kent PE (1979) The emergent Hormuz Salt plugs of southern Iran. J Pet Geol 2: 117–144

Krohe A, Eisbacher GH (1988) Oblique crustal detachment in the Variscan Schwarzwald, southwestern Germany. Geol Rundsch 77 (1): 25–43

Koyi H (1988) Experimental modeling of role of gravity with lateral shortening in Zagros Mountain Belt. AAPG Bull 72 (11): 1381–1394

Laubscher HP (1977) Fold development in the Jura. Tectonophysics 37: 337–362

Laubscher H, Bernoulli D (1980) Cross-section from the Rhine-Graben to the Po Plain. In: Laubscher H, Bernoulli D (eds) Geology of Switzerland, guidebook, part B. Wepf, Basel, pp 183–209

Laubscher HP, Bernoulli D (1982) History and deformation of the Alps. In: Hsü KJ (ed) Mountain building processes, Academic Press, London, pp 169–180

Lisenbee AL (1978) Laramide structure of the Black Hills uplift, South Dakota-Wyoming-Montana. In: Matthews V

III (ed) Laramide folding associated with basement block faulting in the western United States. Geol Soc Am Mem 51: 165–196

Longwell CR (1926) Structural studies in southern Nevada and western Arizona Geol Soc Am Bull 37: 551–584

Love JD, Coe Christiansen A (1985) Geologic map of Wyoming. USGS, 1:500 000 scale

Lowell JD (1985) Structural styles in petroleum exploration. OGCI Publ, Tulsa, OK

Miller VC (1961) Photogeology. McGraw-Hill, New York

Miser HD (1954) Geologic map of Oklahoma. USGS, 1:500 000 scale

Morley CK, Nelson RA, Patton TL, Munn SG (1990) Transfer zones in the east African rift system and their relevance to hydrocarbon exploration in rifts. AAPG Bull 74 (8): 1234–1253

Peltzer G, Armijo R, Tapponnier P (1987) Rate of slip on the Altin Tagh fault (North Tibet, China). In: SPOT 1 image utilization assessment, results. Centre National D'Etude Spatiales, Toulouse, France, pp 709–729

Phelps PW, Harding TP (1987) Extensional fault blocks guidebook of Salt Flat Graben, West Texas. Exxon Production Research Company, Internal Rep

Spicer A (1980) Tectonic map of Switzerland. Commission Geologue Suisse, Basel, Switzerland, scale 1:500 000

Stephenson TR, VersPloeg AJ, Chamberlain L (1984) Oil and gas map of Wyoming. Wyoming Geol Surv Map Ser 12

Thornbury WD (1969) Principles of geomorphology. John Wiley and Sons, New York

Trümpy R (1980) An outline of the geology of Switzerland. In: The geology of Switzerland, guidebook, part 2: Wepf, Basel, p 104

Wilcox RE, Harding TP, Seely DR (1973) Basic wrench tectonics. AAPG Bull 57 (1): 74–96

Chapter 4 Image Interpretation Techniques: Obscured and Buried Structures

4.1 Introduction

As illustrated in the previous chapter, satellite imagery can provide an excellent tool for detection and analysis of geological structures that are well exposed at the surface and manifest clear expressions of inclined bedrock strata and fault-line traces. However, a large percentage of the world's onshore hydrocarbon reserves is either obscured by thick cover of vegetation and soil in areas of low topographic relief or is completely buried under younger and relatively undeformed rock units. In these regions, the recognition of subtle topographic expression of structures can no longer be accomplished by measurement of exposed outcrops. Rather, the interpreter must rely on the recognition of local drainage, moisture and fracture patterns which indirectly reveal the presence of subsurface structures in the area.

In low relief areas, where such geological and surface conditions prevail, obscured and buried structures must be divided into two categories (see Fig. 3.1). The first includes shallow but well-defined subsurface structures that, once recognized on imagery, can be further constrained with conventional exploration tools such as seismic and well data. The second category includes deep-seated basement warp structures (BWS) that manifest similar surface expressions as the first group, but require tentative interpretation of gravity, magnetics and specially enhanced seismic data for subsurface constraints. Analysis of this type of basement warp structure also requires the use of various paleogeographic reconstruction techniques of erosional surfaces and drainage systems. These two groups also vary significantly in their influence on hydrocarbon plays, with the first actually forming hydrocarbon structural traps, whereas the second influences, in most cases, reservoir distributions and related stratigraphic traps. The principles of structural detection of the two groups on imagery are more or less similar and will be covered in the present chapter, whereas the unique aspects of basement warp structure analysis will be discussed in Chapter 5.

Most of the techniques for such detection of obscured and buried structures were developed in the early days of onshore exploration using conventional aerial photography. Later, these ideas were successfully adapted for use with various satellite image data. Numerous examples of such interpretation can be found in the remote sensing literature. Obscured and buried folds (Norman 1976), faults (Lattman 1954, 1959) salt domes (Berger and Aghassy 1980), uplifted basement blocks (Thomas 1974), and wrench faults (Peterson 1979) have been mapped successfully beneath a variety of cover types, including glacial soils (Berger 1988), alluvium (Voüte 1962) and eolian deposits (Mosely 1971), as well as in low relief areas of deep saprolite and thick vegetation (Berger 1982).

Proficiency in the recognition of obscured and buried structures requires comprehension of four fundamental principles. First, one must thoroughly understand the topographic expressions of exposed geological structures at various stages of their erosional evolution. This subject was covered extensively in the previous chapter. Second, the mechanisms that cause structures to continue manifesting surface expressions must be identified. Third, one must learn to recognize, on imagery, diagnostic features that indicate the presence of subtle obscured and buried structures. Finally, the first three skills should be combined to provide one with the ability to predict the shape of obscured or buried structures through the application of geomorphic models which describe their various topographic expressions. It is this step that also requires the integration of subsurface data. The goal of this chapter is the development of the latter three skills.

4.1.1 Mechanisms of Surface Expression

Because obscured structures are partially exposed at the surface, it is easy to understand how they might influence surface conditions such as topography, drainage, soil moisture content, and vegetative health and type. Buried structures, which are completely buried under soft or consolidated sediments, must influence surface conditions in a more indirect manner. Though the exact mechanisms by which this occurs are not fully understood, it is possible to postulate that the following are important:

Differential Loading. Differences in the type and thickness of sediment in the vicinity of the buried structure can result in differential vertical stresses. These stresses may cause partial reactivation of structures and lead to local increases in topographic relief or increased fracture density at the surface.

Differential Compaction. Lateral variations in the sedimentary column covering buried structures may result in differential compaction and subsidence and lead to local changes in topographic relief or surface fracturing (Fig. 4.1). This process may be repeated several times as newly deposited sediments are dewatered and differentially compacted over and adjacent to a buried structure (Nevin and Sherrill 1929; Wilson 1948).

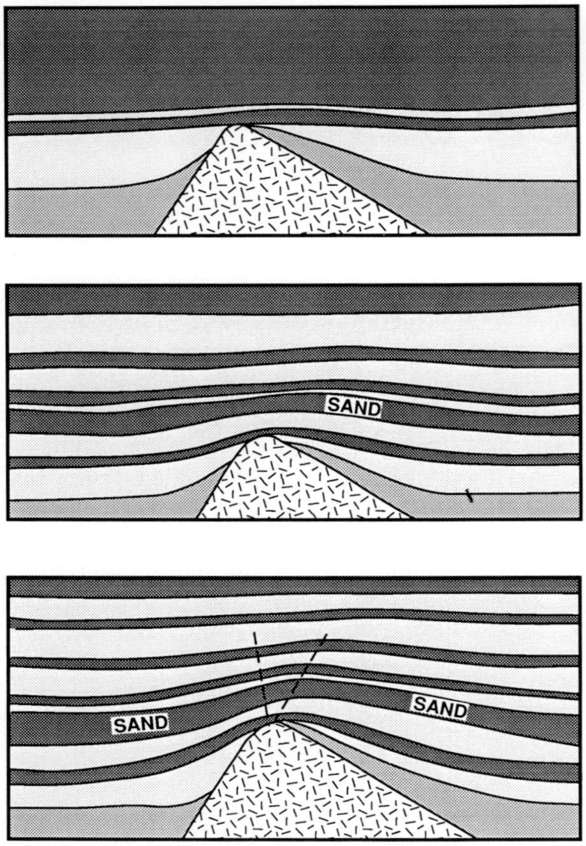

Fig. 4.1. Sand compaction
A laboratory experiment showing the different stages in compaction of sand layers over an asymmetric "hill" representing the basement. This example shows that the phenomenon has evoked curiosity in geologists for decades. (After Nevin and Sherrill 1929)

Fig. 4.2. Reactivation
Shown is a cross section of a reactivated fault from the Paris Basin, France. Two periods of movement along the fault can be seen. During Triassic time, the southwestern block was downthrown; during mid-Cretaceous time, it was upthrown. Repeated reactivations along the fault are likely to produce detectable expressions at the surface
▼

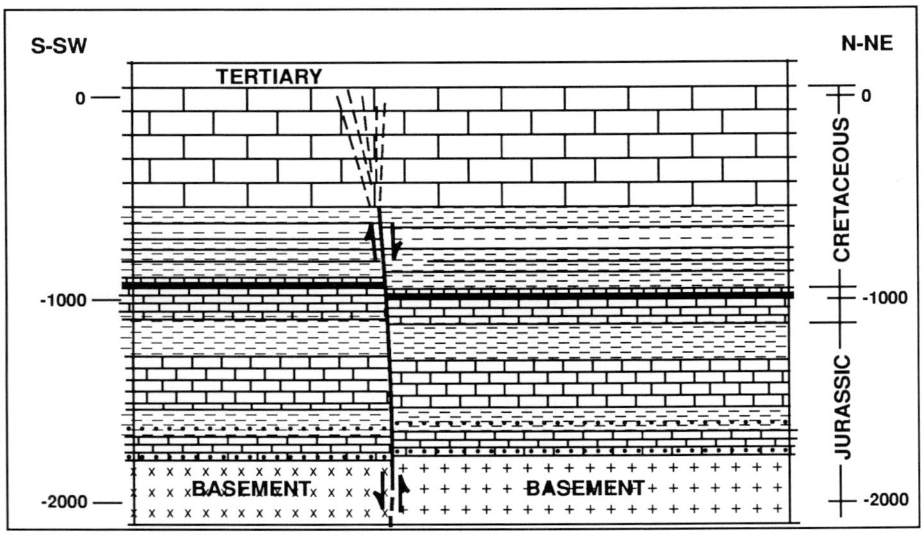

Reactivation of Geologic Structure. New, renewed, or continued stresses may cause reactivation of buried structures (Fig. 4.2). New stresses refer to changes in orientation of the principal stresses that generated the buried structure. Renewed stresses refer to an increase in stress magnitude to values that cause permanent deformation. Continued stresses refer to a stress field with similar orientation to that which generated the buried structure, but of a reduced magnitude such that the overlying cover is not significantly deformed.

Disruption of Near-Surface and Groundwater Flow. Disruption in the uniform flow of groundwater may occur above and in the vicinity of buried structures (Fig. 4.3). This disruption in flow occurs when faulting or breaching causes the leakage of confined aquifers, leading to an upward movement of groundwater (Kudryakov 1974; Töth 1980). The differences in depth of the water table over a buried structure can alter near-surface moisture content as well.

Fig. 4.3. Permeability contrasts
Schematic diagram showing the effect of structures, such as faults (A) and folds (B) and lithologies on the local groundwater surface as a result of permeability contrasts. (After Töth 1980)

Underground Water Flows and Related Erosional Features. The significant permeability contrast between the buried structure and the overlying unconsolidated sediments can trigger a unique erosional process known as piping and sapping (Fig. 4.4). Groundwater may penetrate the unconsolidated sediments via vertical fractures and pipes and continue to flow in caves that are created at the contact between the buried structure and the unconsolidated sediments. This "karst-like" process can result in the development of unique channels, pipes and collapse features over and around the buried structures.

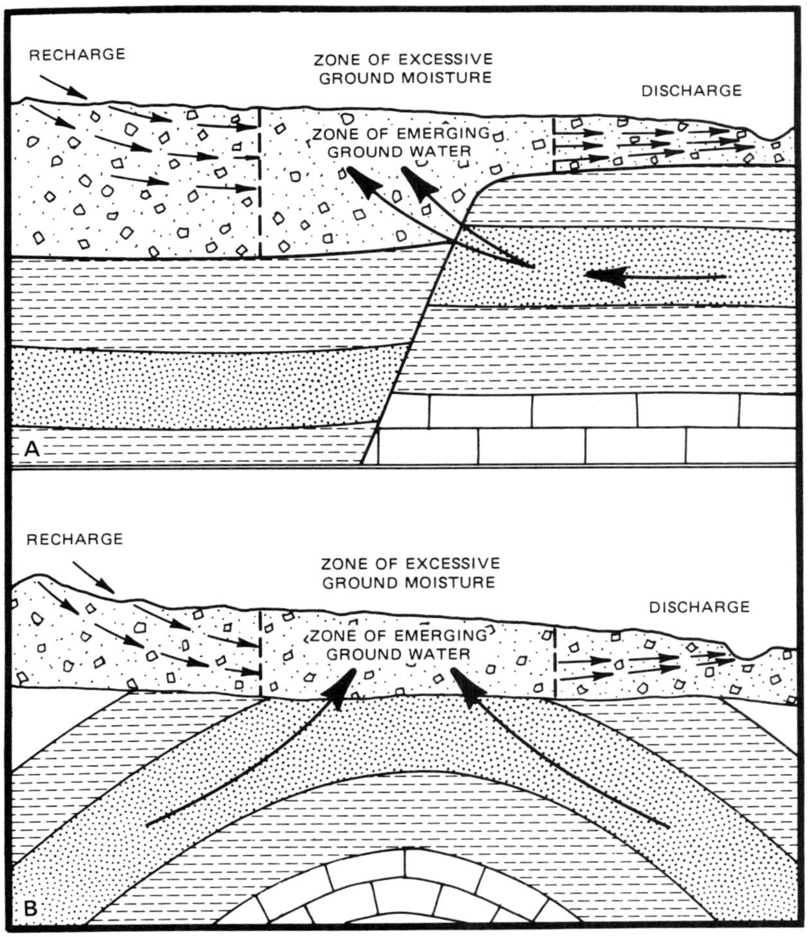

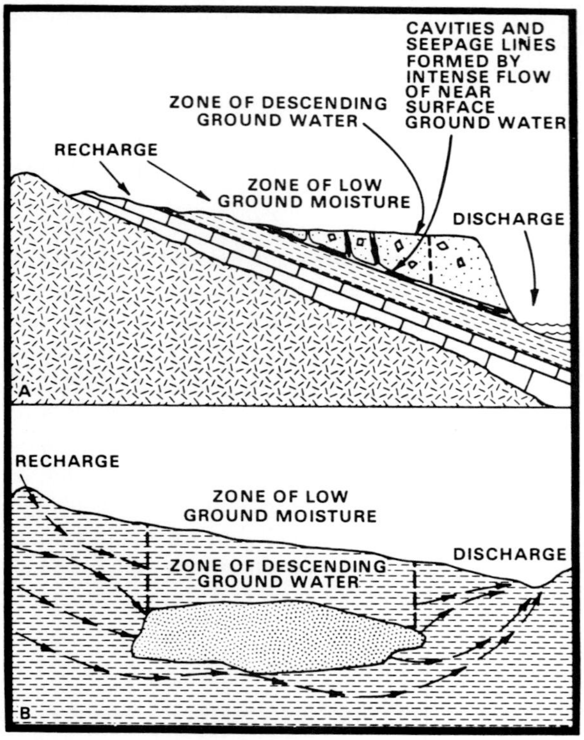

Fig. 4.4. Groundwater erosion
Schematic illustration showing common geological settings that promote a zone of descending movement of near-surface groundwater and areas of low ground moisture. **A** Shallow groundwater flows along contact between bedrock and underlying sediments and soil. **B** Uniform flow and pressure of near-surface groundwater is disrupted by a lense of highly permeable sand enclosed in matrix of low permeability. (After Töth 1980)

4.2 Analysis Criteria

4.2.1 Structurally Controlled Streams

Streams in low relief terrains usually have gentle gradients and tend to flow down the regional slope in wide, shallow valleys. If no obstacles exist, they accommodate downstream changes in slope and runoff by gradually adjusting the size and shape of their channels and valleys. Such streams usually follow the most direct route from major drainage divides towards base level (Fig. 4.5).

However, because of their low gradient, such streams are sensitive to changes in surface conditions that often occur over and in the vicinity of obscured and buried structures. In adjusting their channels to such changes, these streams develop unique drainage patterns and features that are

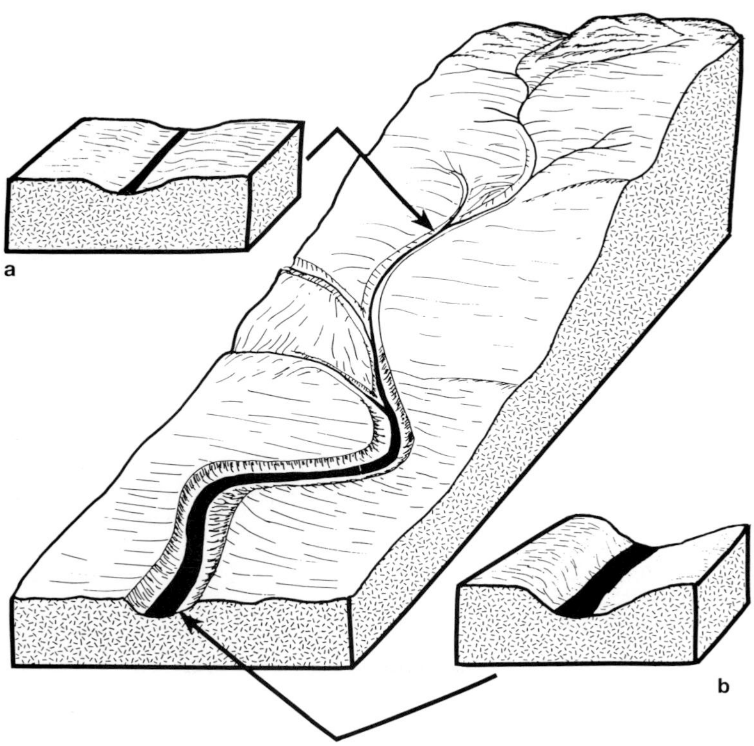

Fig. 4.5. Channel geometry
Schematic illustration showing the gradual changes in width, depth, and slope of an undisrupted stream channel in the downstream direction. (After Leopold et al. 1964)

Fig. 4.6. Channel adjustment
Plan and profile of Cottonwood Creek, Wyoming. The river changes its channel and floodplain configuration over the interval *A* to *C* because of a slight increase in slope at *B* (after Leopold and Wolman 1957). Such subtle changes in slope could reflect the presence of obscured and buried geological structures

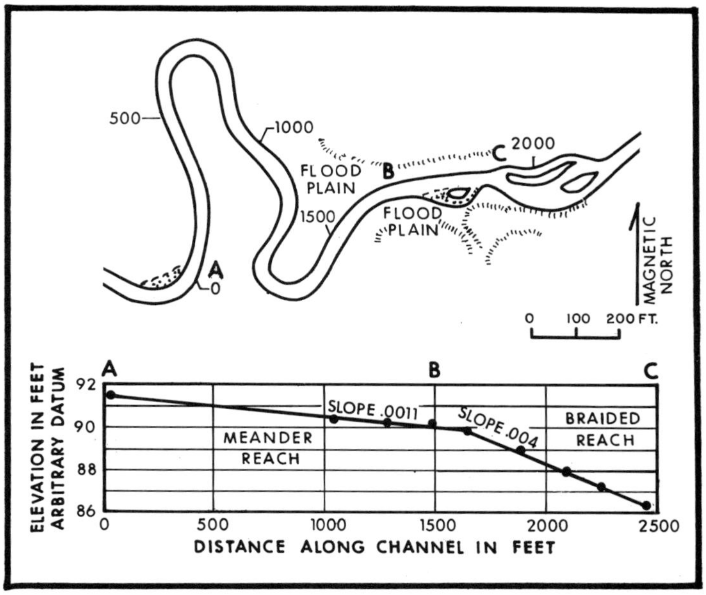

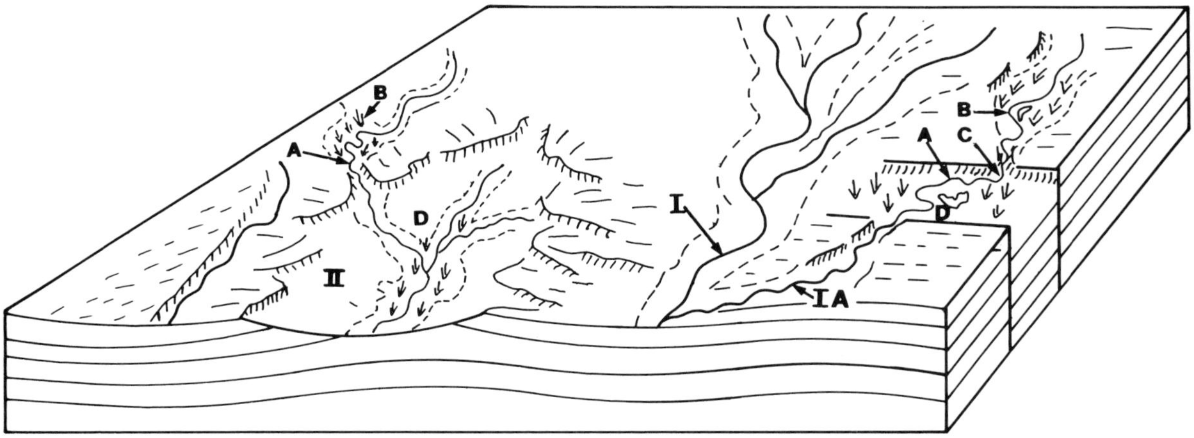

Fig. 4.7. Model of structurally controlled streams
A conceptual model showing downstream changes in valley forms and drainage patterns within a low relief region. Elements visible on Landsat imagery are highlighted. Channel *I* represents a channel that developed without the influence of structural features. Channel *IA* was disrupted by the presence of an obscured graben feature and channel *II* is cutting through a breached fold. *A* Locally pertubed drainage patterns; *B* abrupt change in channel and valley width; *C* local deflection of a stream; *D* local areas of anomalous drainage densities and topographic dissection

anomalous with respect to the regional drainage systems. These local drainage phenomena are often called "drainage anomalies" or "structurally controlled streams" (Lattman 1959; Leopold and Wolman 1957; Fig. 4.6).

One way of recognizing structurally controlled streams on satellite imagery is by applying a conceptual model such as that presented in Fig. 4.7. Based primarily on empirical data, the model illustrates the most common changes in regional streams that can occur in relation to obscured or buried structures. Four major types of structurally controlled stream patterns are large enough to be detected on satellite imagery. These are summarized as follows:

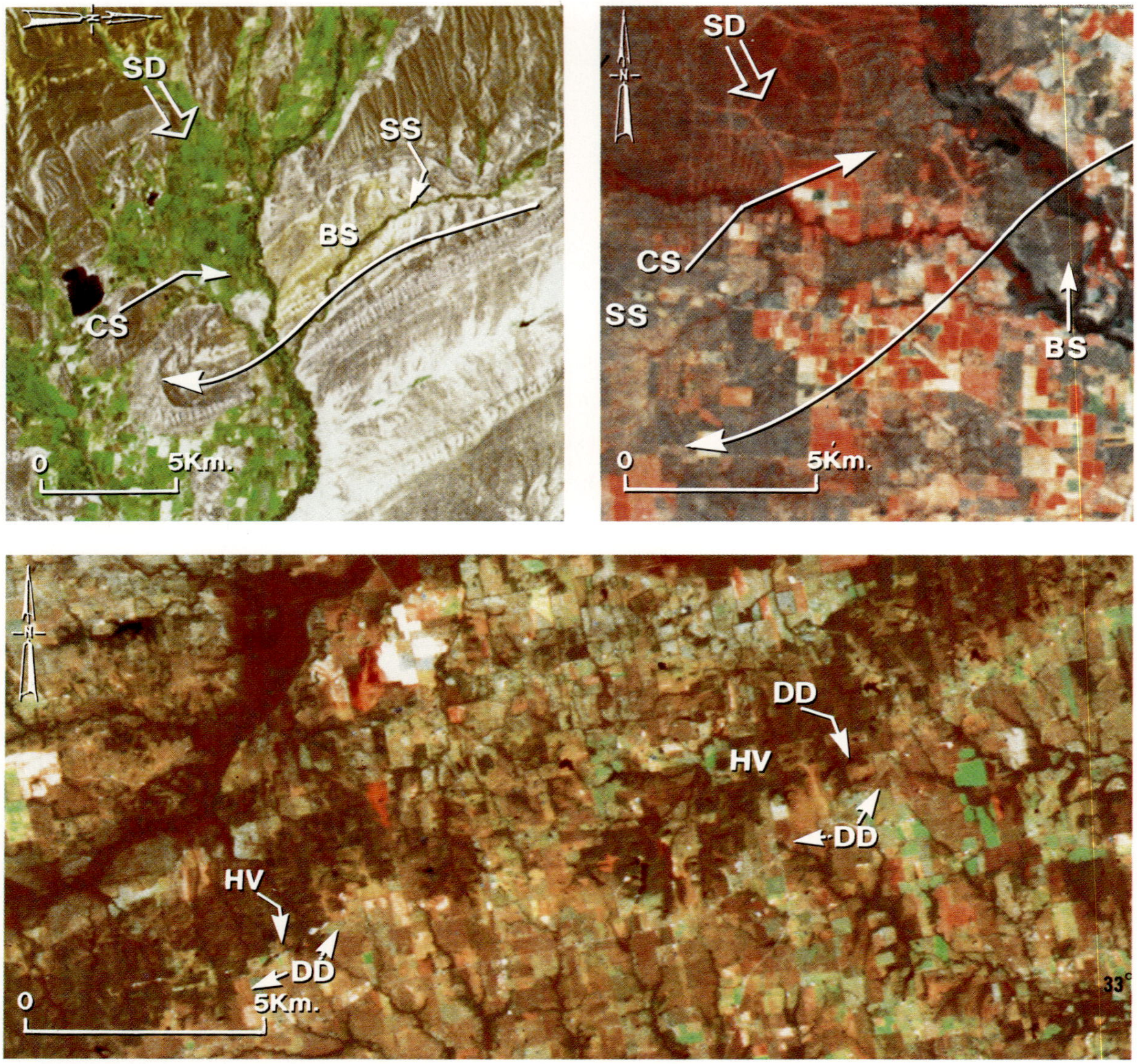

Fig. 4.8. Examples of structurally controlled streams
Landsat images showing several structurally controlled drainage features at various levels of surface-structural exposures. **A** An exposed breached fold in the Wind River Basin, Wyoming. **B** A buried fold of the Pearsall oil field in southern Texas. **C** Obscured normal faults of the Mexia-Talco fault system in eastern Texas. *CS* Cone-shaped valley representing abrupt change in width; *SS* Subsequent stream following the strike of the eroded limbs of the anticlines; *DD* structurally controlled drainage divides; *HV*-heavy vegetative cover related to increase in moisture conditions; *SD* slope direction

- Locally perturbed drainage patterns may be formed over areas that are affected by obscured and buried structures. The orientation of the local streams and drainage divides may contrast with the orientation of regional stream patterns. Obscured and buried faults usually produce angular drainage patterns. These patterns are characterized by linear tributaries joining the main trunk at high angles rather than at low angles. Obscured and buried folds may cause various types of radial patterns to develop (A in Fig. 4.7).
- Abrupt changes in channel and valley width may occur in preexisting (antecedent) streams that cross the structures. The most common feature

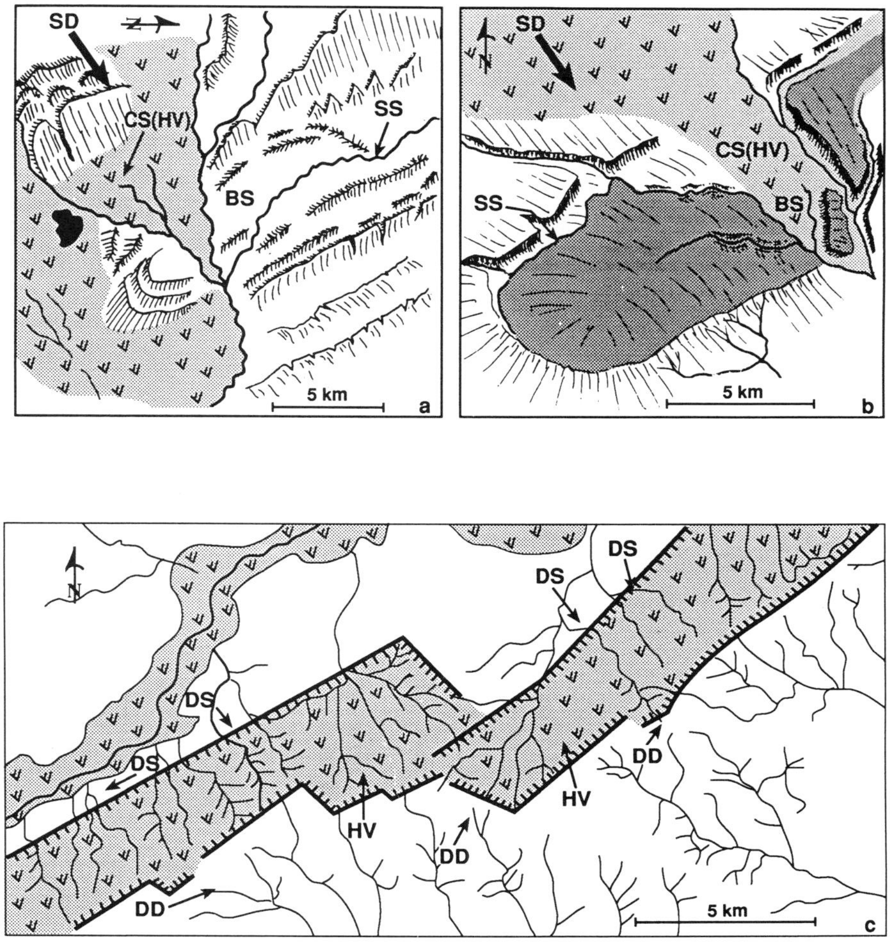

Fig. 4.9. Interpretation of structurally controlled streams
Structural geomorphic interpretation of the Landsat images shown in Fig. 4.8. *Letters* and *symbols* as in Fig. 4.8

related to this phenomenon and visible on satellite data is a "cone-shaped" stream valley that develops at the breached areas of the structure (B in Fig. 4.7).

- Local deflection of a stream's orientation may occur in preexisting streams upon crossing or circum-navigation of obscured and buried structures (C in Fig. 4.7).
- Local areas of anomalous drainage density or topographic dissection may be caused by an increase in relief and surface fracturing. These areas produce textural patterns that are visible on Landsat imagery (D in Fig. 4.7).

Figures 4.8 and 4.9 show the surface expressions and related drainage patterns of structures at different levels of bedrock exposure. The upper left image (a) shows an example of a partially exposed anticline in the Wind River Basin in the western United States. The most important drainage feature to note here is the abrupt change in the channel and valley width that occurs in the valley that crosses and breaches the structure. This surface element is depicted on the image as a profound "cone-shaped" stream valley. Also, a tributary on the right half of the image joins the main stream valley at a relatively obtuse angle, following the general strike direction of the anticline rather than the general slope of the basin. This segment of the stream should be regarded as structurally controlled, exhibiting anomalous drainage orientation for this area.

The upper right (b) image shows the surface expression of a completely buried anticline. Here, the buried Pearsall structure, which forms one of the most prolific fields in the Austin Chalk Play of southern Texas, produces structurally controlled stream valleys similar to those observed in the previous example. First, the main stream valley that crosses the crest of the anticline forms a cone-shaped valley with a wide floodplain which progressively increases in width away from the anticline. Second, a small tributary in the left portion of the scene closely follows the strike orientation of the plunging nose of the structure, forming a structurally controlled stream valley with anomalous drainage orientation.

Structurally controlled drainage divides clearly reveal the location of major FLTs of the Mexia-Talco fault systems in East Texas (c in Figs. 4.8 and 4.9). The presence of individual faults within the system is revealed by local deflections in channel orientations of tributaries that cross these faults. These structurally controlled stream valleys are well depicted on imagery data even though their expressions are severely obscured by thick vegetative cover.

4.2.2 Structurally Controlled Streams Related to Piping and Sapping

A special type of structurally related stream valley may be formed in low relief areas, where structures are partially or completely covered by unconsolidated sediments. The permeability contrast between the obscured and buried structures and the overlying unconsolidated sediments often becomes the focal point of surface and groundwater flows. Such flows can result in the initiation of these unique erosional processes and lead to the development of anomalous drainage features which can be easily recognized on imagery (e.g. Berger and Aghassy 1983). The piping and sapping phenomenon, although recently receiving great attention by the geomorphological community, has been quite ignored as a process which enhances the surface expression of geological structures, and much more research is needed in this area.

The development of unique drainage features over buried and obscured structures by the processes of piping and sapping is demonstrated by the two photographs in Fig. 4.10. The photographs show deeply incised gullies in the vicinity of the San Andreas Fault near the Mecca Hills region. These valleys, which are developing along small-scale faults and fractures of the San Andreas fault system, are much too deep and wide to reflect valley evolution by normal flow of surface water in this arid area. The formation of these valleys occurred primarily due to collapse of pipes, subsurface tunnels and other erosional features that developed along the contact between the bedrock units of the fault zone and the unconsolidated sediments that cover it. Subsequently, a badland-like landform can be formed as a "halo" around the buried and obscured structures, manifesting strong expressions on imagery. For example, the Mecca Hills (Fig. 4.11) are presently manifested by a profound expression of highly dissected mountains that have undergone a complete inversion and obliteration of their topographic/structural relief.

The process of piping is not unique to arid environments. It can also be documented in humid areas and, in many instances, in permafrost regions. The satellite image shown in Fig. 4.12 demonstrates how the process of piping and sapping can enhance the detectability of obscured and buried structures. The image shows the buried limbs of a large domal feature in the Northwest Territories, Canada. Although the limbs are completely covered by unconsolidated fluvial glacial sediments, deeply incised stream valleys are developed along the strike of the structures, exhibiting unique drainage patterns which could not have developed in this region by normal action of surface flows. (There is no large-scale drainage basin to support such channels.) The example shown here can also be used to demonstrate some of the unique characteristics of this drainage system, described as follows:

1. Drainage density: Parts of the drainage area that are affected by piping and sapping may have drainage networks of densities different from those of their surroundings, even though they are developed over relatively similar surface conditions.
2. Drainage network patterns: The dendritic drainage pattern that is common to low relief sediments and soil-covered basins is notably absent in areas affected by piping and sapping. More common are angular drainage patterns. These

Fig. 4.10. Piping and sapping ▶
Photographs from the alluvial plain of Mecca Hills, California, showing deeply entrenched stream valleys that develop by the process of piping and sapping. This phenomenon occurs along individual faults which splay from the San Andreas fault system. This process is manifested on imagery as a profound erosional "halo" illustrated in Fig. 4.11

Fig. 4.11. Piping and sapping halos
Landsat TM image of the San Andreas fault showing the highly dissected plains which form a "halo" (outlined in *dashed lines*) of badland topography in the vicinity of the fault

are characterized by numerous short tributaries, which join the main trunk at high junction angles rather than at low angles.

3. Drainage orientation: Orientations of major trunks and other tributaries are often dictated by the attitude of the bedrock and by major joint systems. Thus, drainage orientation may be in contrast to the present topography and the regional drainage network orientation.
4. Valley forms: Incised tributaries and main trunks with steep valley walls characterize areas affected by piping and sapping. Short tributaries join the major trunk in hanging valleys and terminate abruptly in amphitheater valley heads. Such valley forms are particularly developed in areas where active groundwater flow undermines the basal support and causes the valley to extend headward, owing to the collapse of the valley head.

In Jones County, Mississippi, a buried fault system was recognized by several unique features (Fig. 4.13). First, the fault is expressed as a linear topographic ridge and drainage divide that appears anomalous with respect to the topographic setting of this region. Second, unique box-shaped stream valleys and drainage patterns are observed along the fault scarp suggesting that piping and sapping processes were initiated at the contact between the fault scarp and the overlying unconsolidated sediments. In fact, field studies clearly link the presence of such anomalous drainage features with the pip-

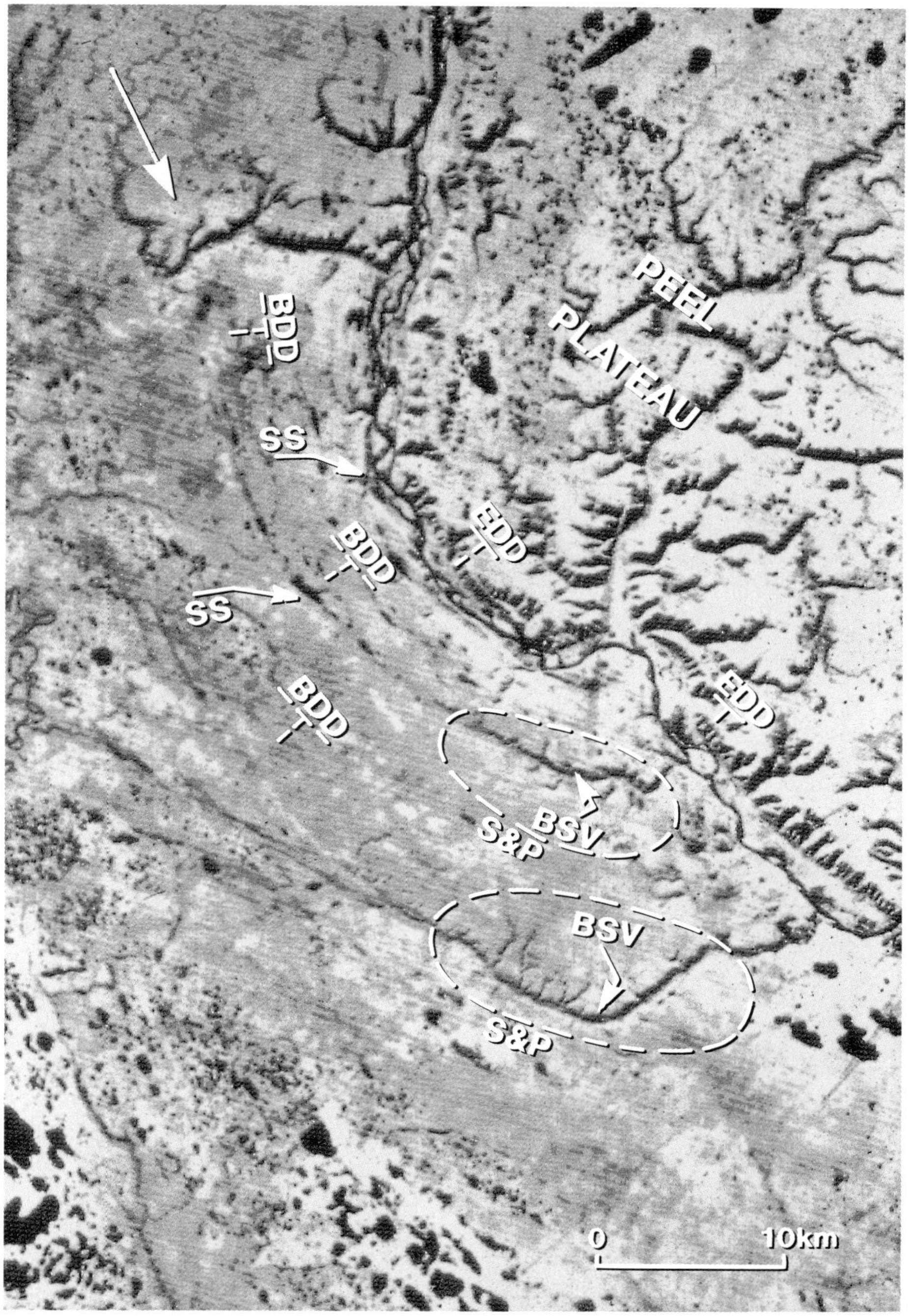

Fig. 4.12. Drainage pattern formed by piping and sapping Landsat MSS image showing the buried western limb of a large domal feature in the Peel Plateau and plain area, Northwest Territories, Canada. Deeply entrenched, box-shaped channels develop along the buried structure and highlight its geometry at the subsurface. *BSV* Box-shaped valleys; *S&P* sapping and piping; *EDD* exposed dip direction; *BDD* buried dip direction of a buried stratum; *SS* subsequent streams. The interpretation shown here has not been constrained by field observations

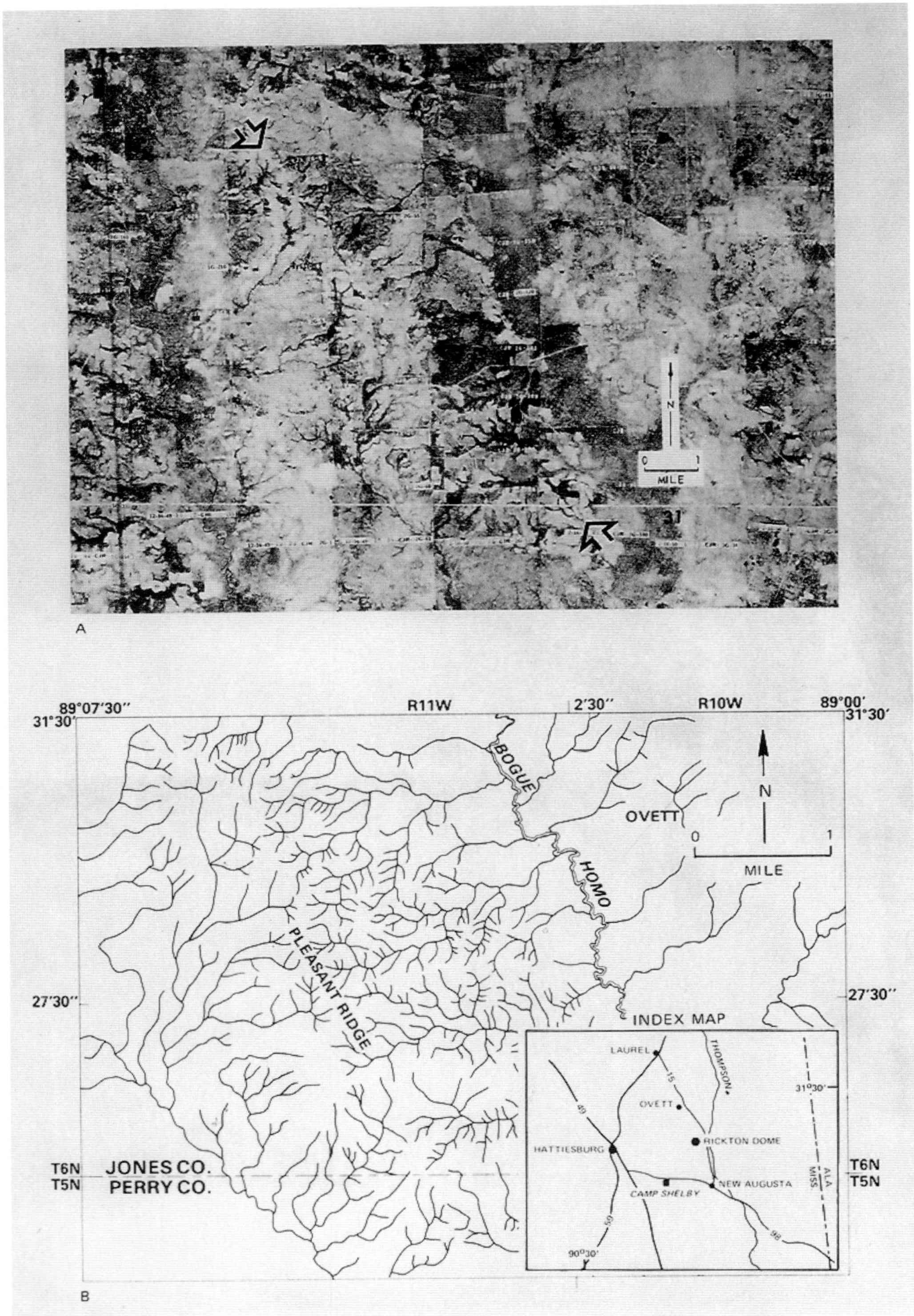

Fig. 4.13. Fault-related drainage networks
Aerial photograph and a drainage map of Pleasant Ridge in Mississippi showing the unique drainage pattern of the buried fault system. Location of the fault is highlighted with *arrows* at both ends. (After Berger and Aghassy 1983)

Fig. 4.14. Piping and sapping in humid regions
Sequential photograph taken from the drainage divide to the bottom land in tributaries that develop over a buried fault in Jones County, Mississippi. **a** Initiation and expansion of upper tributaries. This occurs primarily due to the collapse of pipes, subsurface tunnels, and other openings in the ground. **b** The water reappears at the bottom as seeps and springs. Groundwater flow triggers headward erosion of valleys that extend upstream along subsurface tunnels

a

b

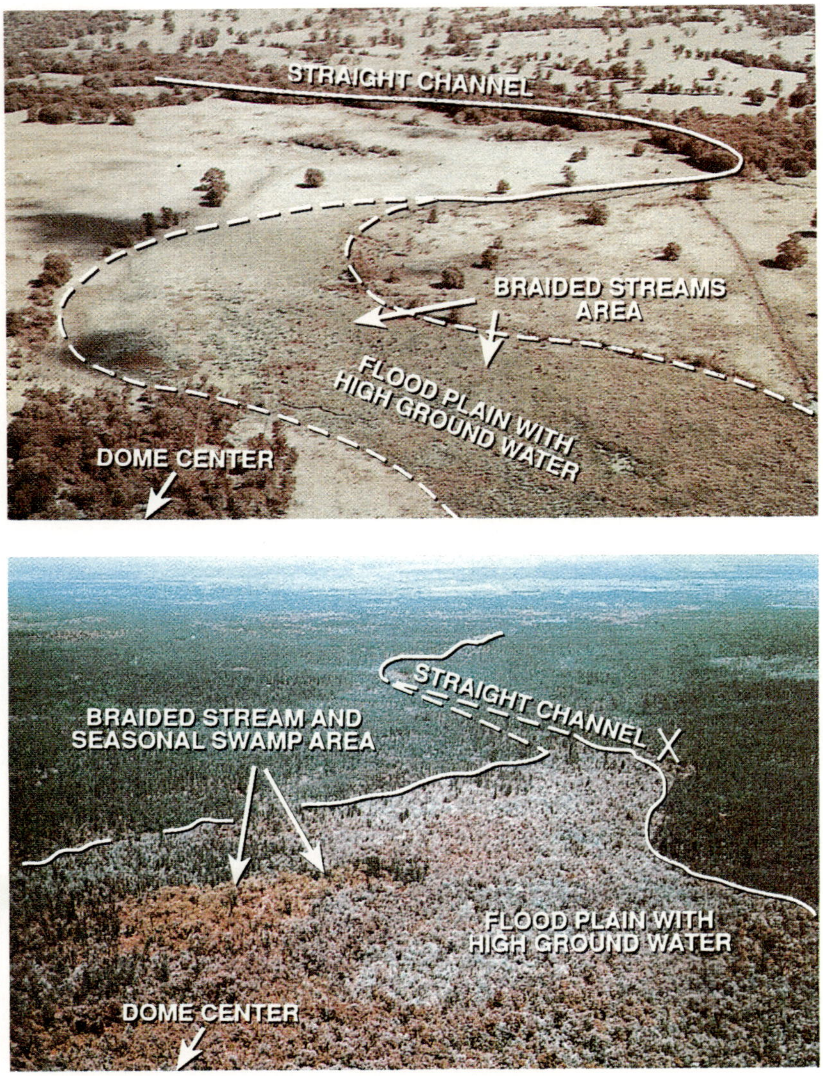

Fig. 4.15. Spectral anomalies
Low-altitude, color-infrared photographs showing the spectral expression of two breached salt domes in the Gulf Coast region, USA. (After Aghassy and Berger 1981)

ing and sapping processes (Fig. 4.14). Finally, the downthrown side of the fault appears darker than the upthrown side, indicating the presence of a high concentration of ground moisture along the fault scarp. This last phenomenon is an example of brightness and spectral variation which is discussed in the next section.

4.2.3 Brightness and Spectral Variations

Landsat spectral bands, or various composites of bands presented in black and white or color, often contain areas whose brightness and spectral reflectance characteristics differ significantly from the rest of the scene. In low relief terrains, differences in brightness may be caused by changes in surface roughness, orientation of surface features, soil moisture, and vegetation density. Differences in spectral reflectance are caused by changes in soil or rock color and the spectral character of the vegetation.

In low relief areas, subtle topographic ridges formed over structural highs are usually well drained and are characterized by reduced surface moisture conditions. These ridges appear brighter

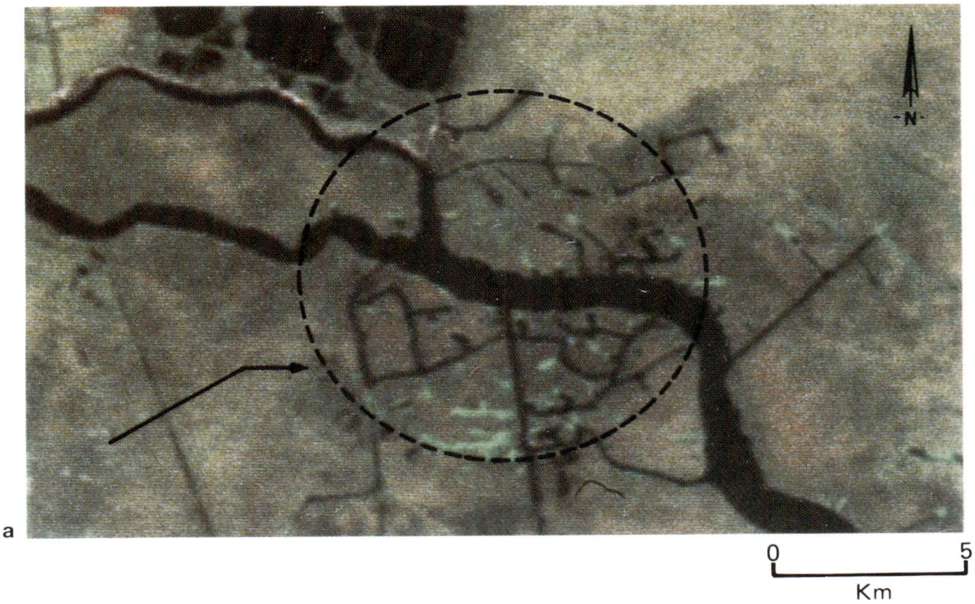

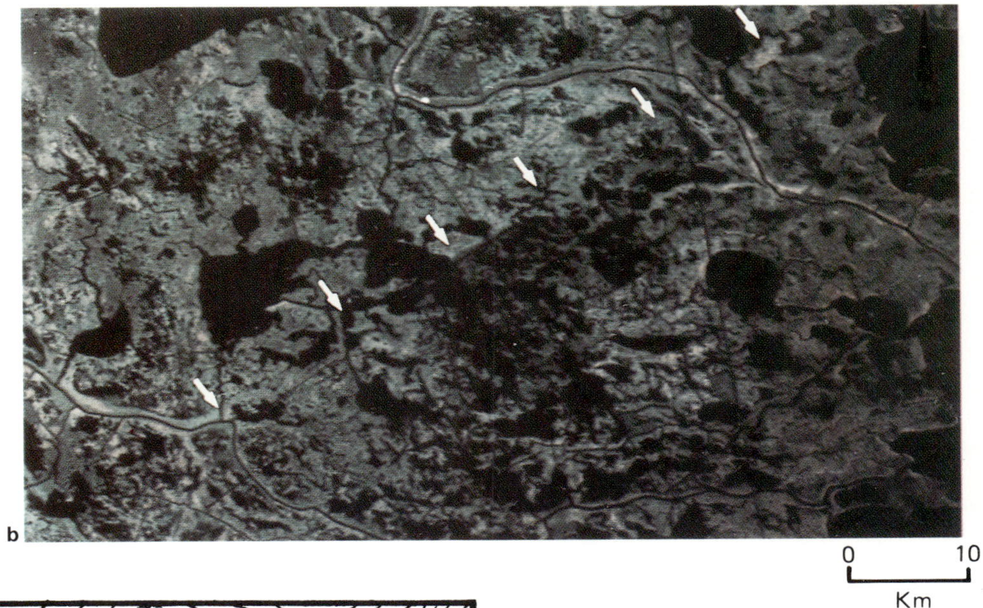

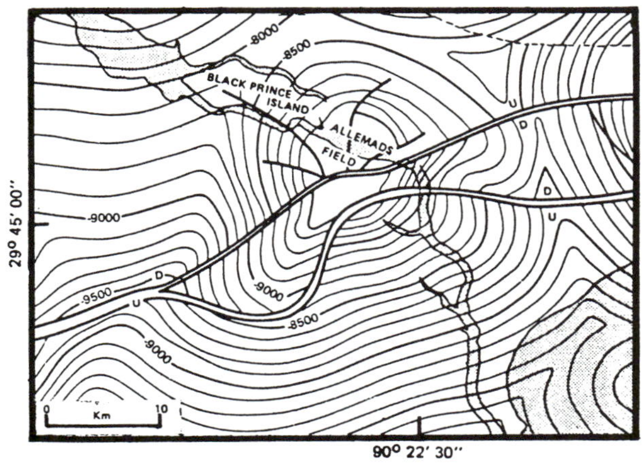

Fig. 4.16. Circular and linear tonal anomalies
Landsat TM subscenes from the Gulf Coast region near New Orleans, Louisiana, showing typical circular and linear tonal anomalies that develop in association with salt domes (**a**) and growth faults (**b**)

◀ **Fig. 4.17. Subsurface constraint**
A subsurface structure contour map (top Miocene) showing listric growth faults and salt domes in area covered in Fig. 4.16 a. (Courtesy of Exxon Exploration Company)

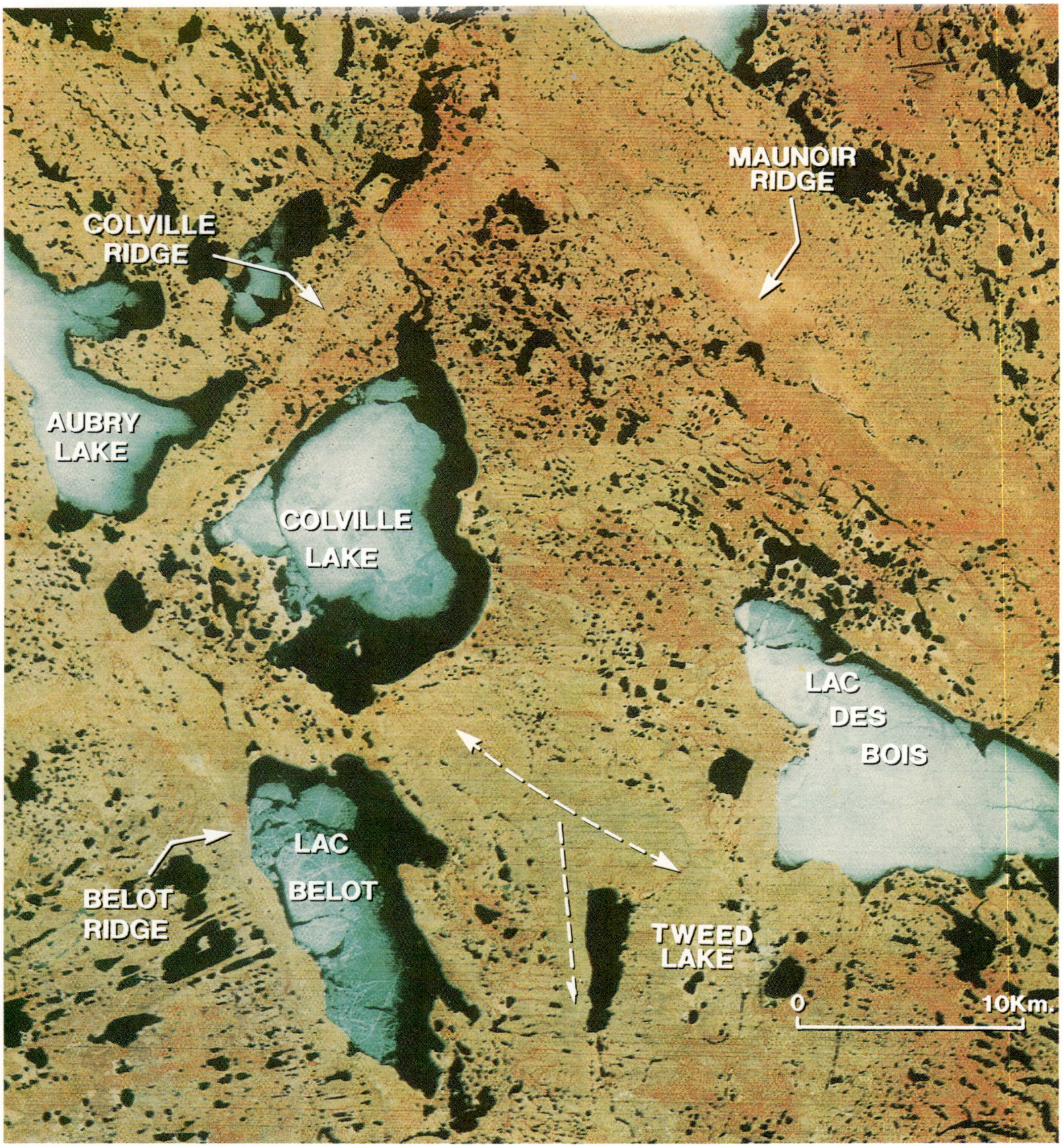

Fig. 4.18. Bright tonal anomalies
Landsat MSS image and structural interpretation of the Colville Lake area, Northwest Territories, Canada. The crests of the ridges are depicted by their bright signatures which are attributed to dry moisture conditions

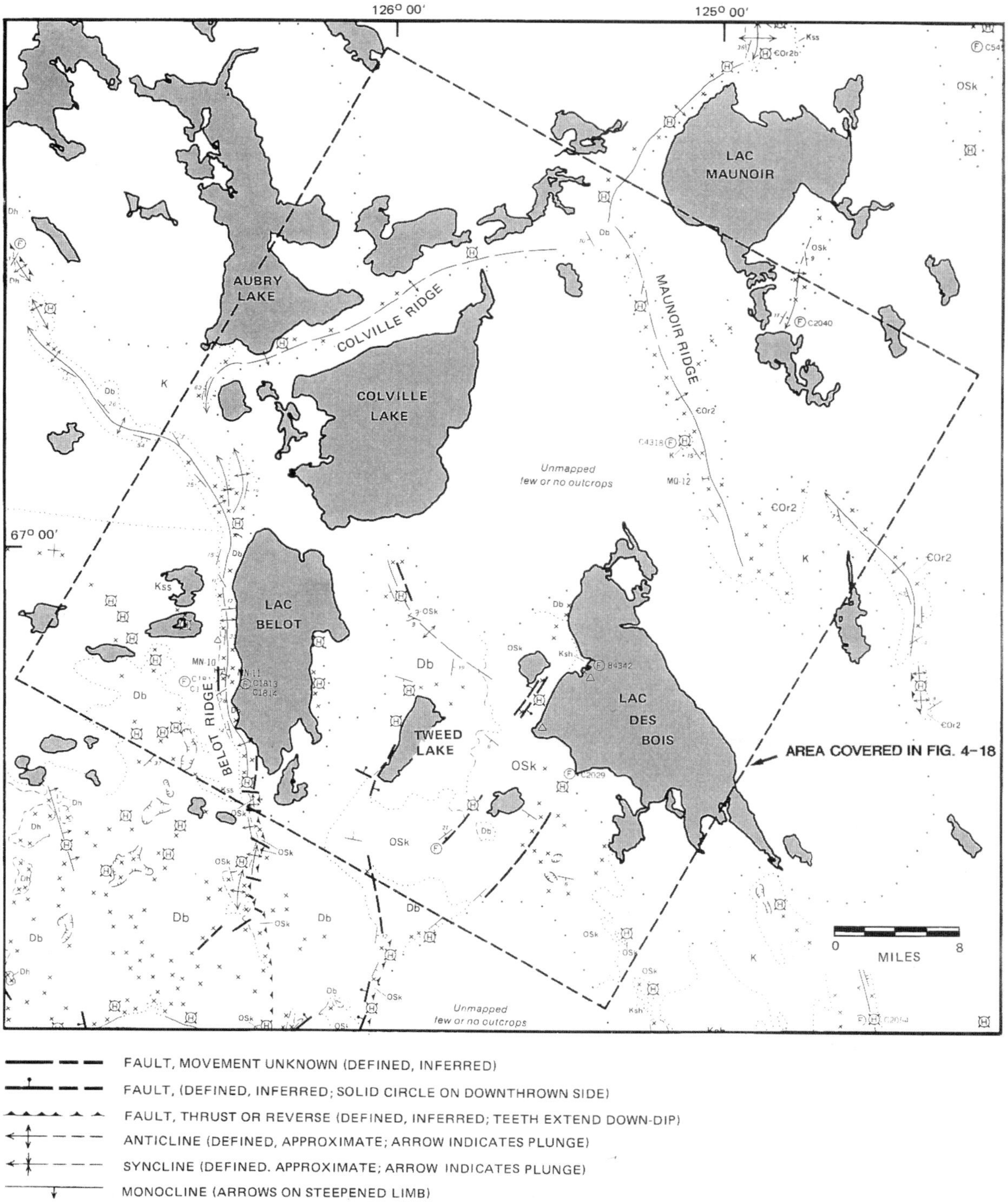

Fig. 4.19. Surface constraints
Geologic map of the Colville Lake area, Northwest Territories, Canada. The *dashed square* indicates the coverage of the satellite image in Fig. 4.18. (After Haimilia 1975; published with the permission of the Canadian Geological Survey)

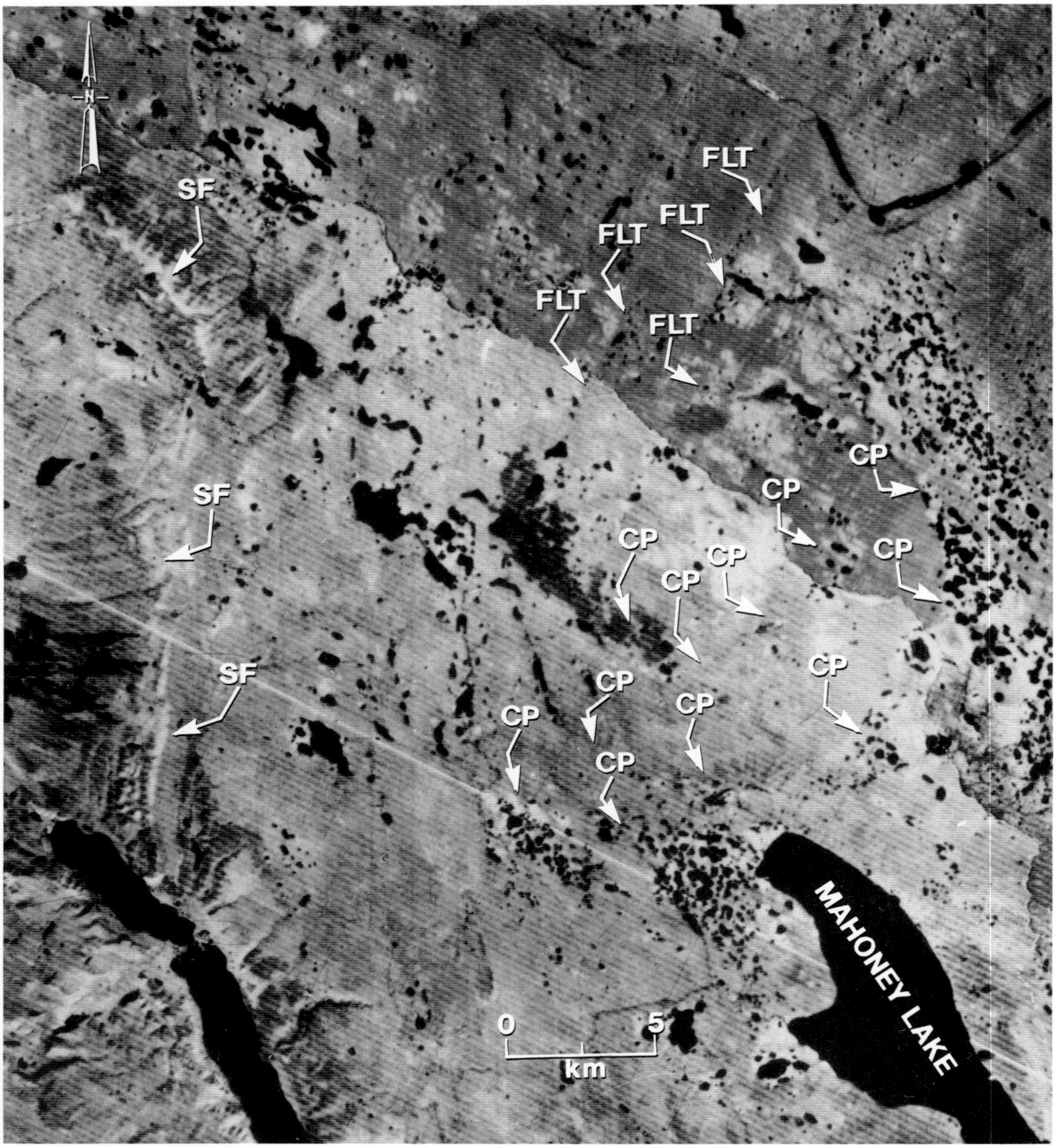

Fig. 4.20. Dark tonal anomalies

Landsat MSS image and structural interpretation of the Mahoney Lake Dome area, Northwest Territories, Canada. The breached topography of the dome is highlighted by the presence of concentric lakes and small ponds which represent areas of high-moisture conditions. *CP* Concentric patterns; *FLT* fault-line trace; *SF* sinuous FLT

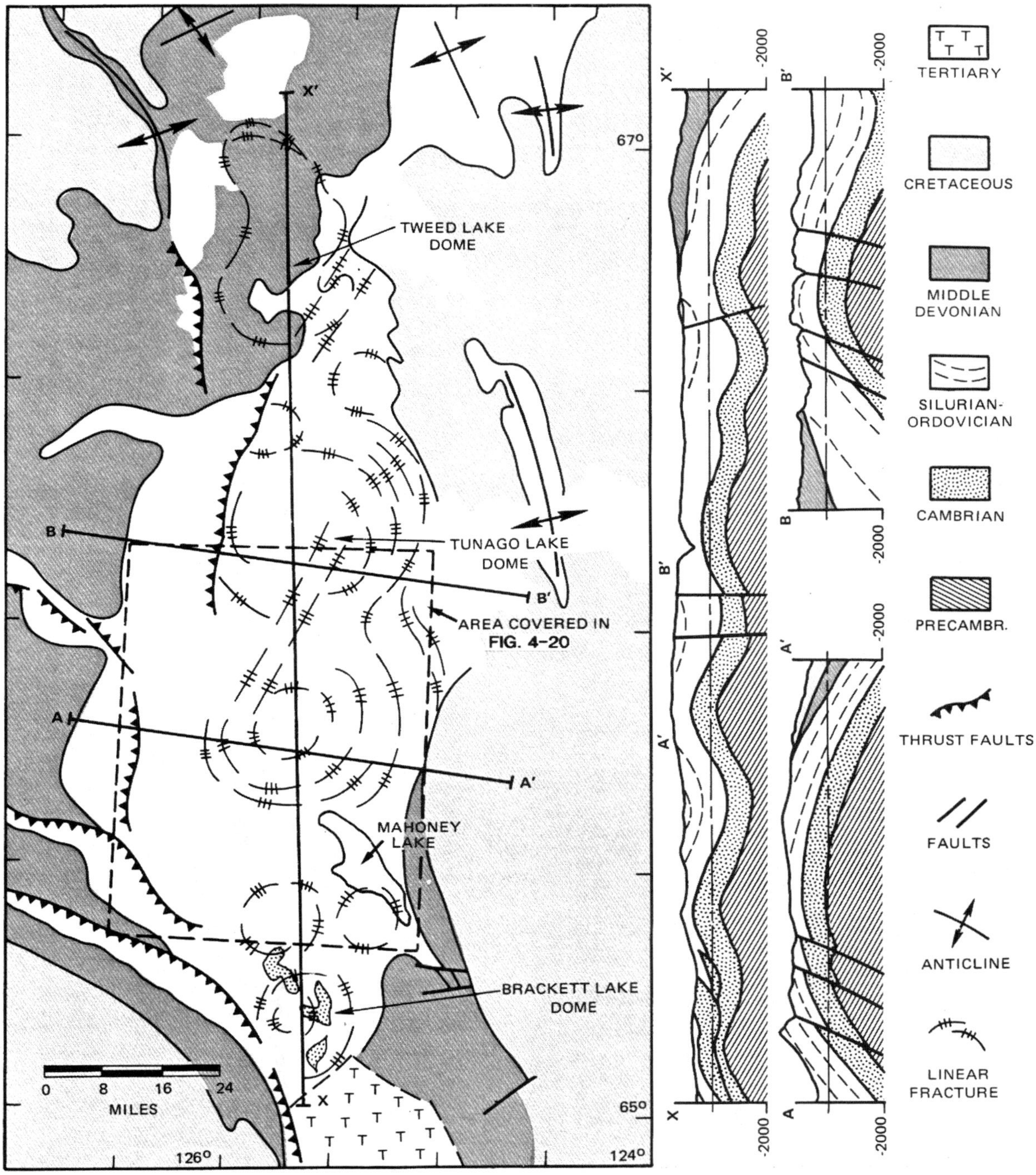

Fig. 4.21. Surface and subsurface constraints

Generalized geological map and a cross section of the Mahoney Lake area, Northwest Territories, Canada. (After Haimilia 1975) on satellite data than their surroundings. Similarly, subtle valleys and depressions which may reflect structural lows are usually poorly drained and become areas of excessive ground moisture concentrations. These low areas appear darker on satellite data than their surroundings.

A closer look at structurally related spectral and brightness variations is provided by low-altitude infrared photographs which were taken with a hand-held camera over two buried salt domes in the Gulf Coast Region, southern USA (Fig. 4.15). The upper photo shows the spectral signature of a "cone-shaped" valley that breached the crest of the Oakwood Dome, Texas. The darker appearance of the valley is clearly attributable to the excessive surface moisture and dense vegetative cover that characterizes the valley floor. The lower photo shows how the excessive moisture conditions of a stream valley that crosses the Cypress Creek Dome (Mississippi) causes an observable contrast in vegetation which is reflected on remote sensing data as a cone-shaped, spectrally unique feature.

In the passive-margin setting of the Gulf Coastal Plain of the southern USA, listric normal growth faults and salt domes create several important hydrocarbon plays (e. g. Murray 1961). Because the groundwater table is very shallow, minor fluctuations in groundwater levels associated with the presence of buried structures strongly influence surface moisture conditions, which can be detected on imagery data. The satellite images and the accompanying subsurface structure map from New Orleans, Louisiana, can be used to demonstrate this phenomenon (Figs. 4.16 and 4.17).

The upper subscene (Fig. 4.16 a) shows the presence of a circular dark feature that reflects the presence of a salt dome in the subsurface (Fig. 4.17). The darker area represents the breached and heavily faulted center of the dome, which is characterized by poor drainage conditions and high concentrations of groundwater. (Similar conditions are shown in the low-altitude photographs in Fig. 4.15.)

The bottom image shows a linear feature (marked by arrows) that forms a linear boundary between bright and dark regions. This linear feature reflects the presence of a growth fault that propagated near the surface. Note that the darker areas, which represent a zone of saturated ground, correspond to the downthrown side of the fault, whereas the upper side, which is relatively drier, appears brighter on the image. Recall that a similar relationship was observed along the fault in Mississippi shown earlier. The linear features observed here, when aligned with other structurally related elements, could form lineaments. These features will be described in the next section.

Two additional examples of this relationship between structure, topography, and spectral variations are given by satellite images and accompanying geologic maps of the Northwest Territories, Canada. In the first example (Figs. 4.18 and 4.19), elongated salt anticlines in the frontier region of the Colville Hills are depicted on the image by their uniquely bright signature. In the second example, subtle, breached domal features near Mahoney Lake are depicted by the presence of concentric dark anomalies which reflect vegetation patterns and glacial lakes that are formed along the exposed hogbacks of these structures (Figs. 4.20 and 4.21).

4.2.4 Classification and Analysis of Lineaments and Linear Features

A lineament is formed on image data by the alignment of small-scale linear surface features such as stream segments or changes in vegetation and tonal patterns. Such a feature is believed to reflect the surface expression of major obscured and buried faults or zones of weakness in the sedimentary section.

There is great confusion in the remote sensing literature between the terms "linear features" and "lineaments" which must be clarified here. Linear features are relatively short and believed to reflect individual faults and other fracture systems (for example, basement layering and grain). Lineaments are longer in general and consist of a wide zone of faults and fractures which may be manifested by the composite alignment of several different linear features. Therefore, an interpreted satellite image may include numerous linear features, but few or possibly no lineaments at all. Criteria must therefore be established to allow the interpreter to identify those surface elements that truly represent major fault systems or zones of weaknesses in the subsurface. The next section will deal with this issue.

4.2.4.1 Lineament Analysis

It is proposed here that the term "lineament" be used by interpreters conservatively and only when a linear surface element exhibits one or a combination of the following conditions: (1) it must extend on the order of 100 km in length, possibly crossing

a b d c

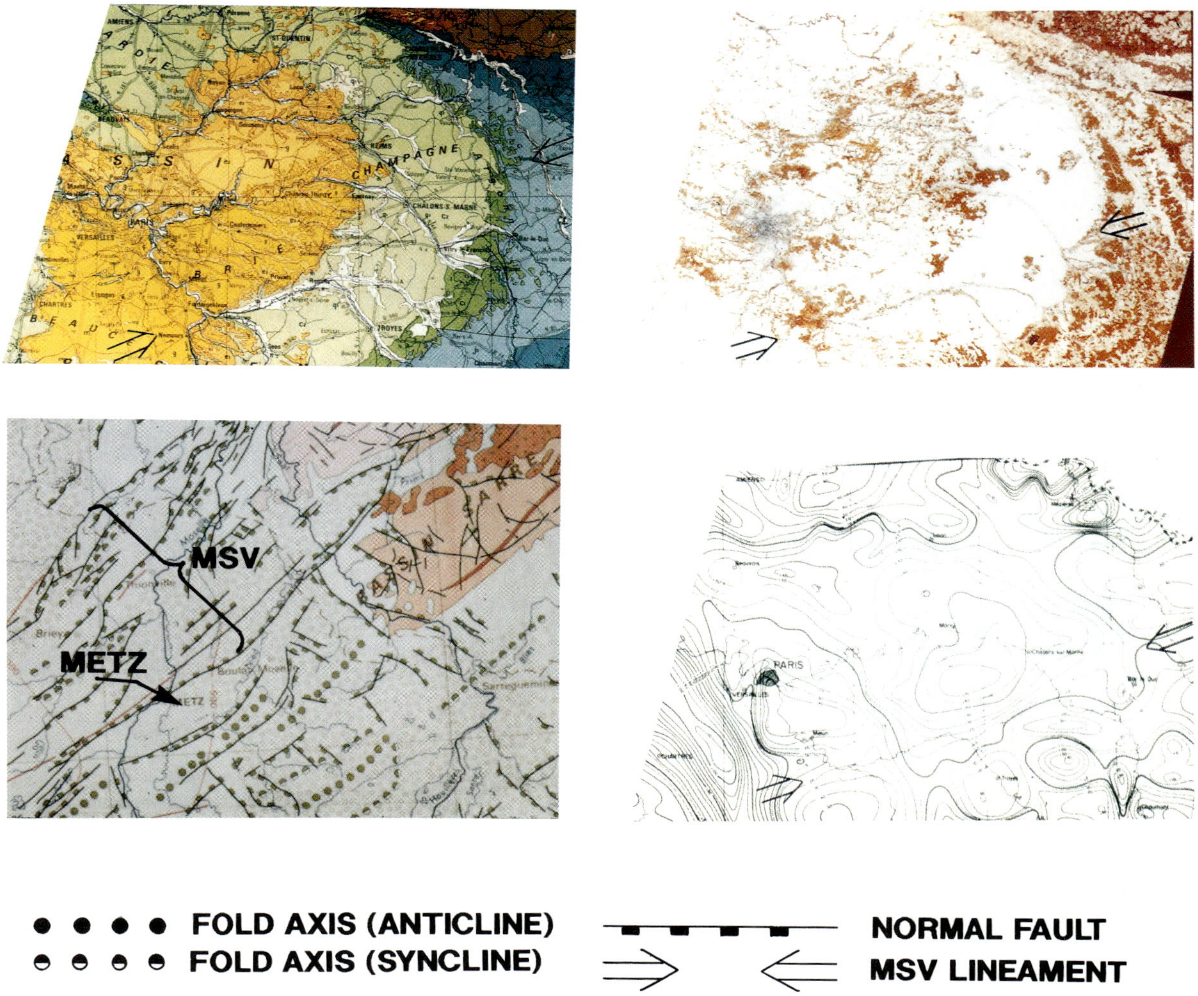

Fig. 4.22. First-order lineaments
Surface and subsurface data of the Paris Basin, showing the expression of the Metz-Seine Valley Lineament. **a** Geological map; **b** satellite imagery; **c** total intensity magnetics; **d** tectonic map of the exposed segment near the city of Metz. Note that the MSV lineament is highlighted by *arrows* at both ends in **a, b** and **c**. (Compiled from various maps produced by the French Geological Survey; published with their permission)

several different tectonic units; (2) it should be aligned with fault systems or lithological boundaries which are exposed in the adjacent tectonic units; (3) it may be spatially coincident with linear edges of gravity and magnetic anomalies and subtle deflections in regional structure and isopach maps; and (4) exposed structures and outcropping units in the basin area which are cross-cut by this surface feature should show evidence of truncation and breaching.

An excellent example of a lineament that displays many of the characteristics mentioned above is the Metz-Seine Valley (MSV) lineament in the Paris Basin, France (Fig. 4.22). On surface geologic maps as well as satellite imagery, the lineament appears as a through-going linear surface feature that causes the breaching of the elliptical outcropping pattern of the basin and the development of a linear cuesta along the valley of the Seine River. The magnetic map of the basin shows that the MSV lineament developed along the edges of two regions that exhibit different magnetic signatures that most likely reflect the faulted boundaries between two basement units with different lithologies. North of the MSV lineament, the magnetic basement is manifested by broad and low-frequency anomalies whereas south of the lineament, the magnetic basement is characterized by small and high-frequency

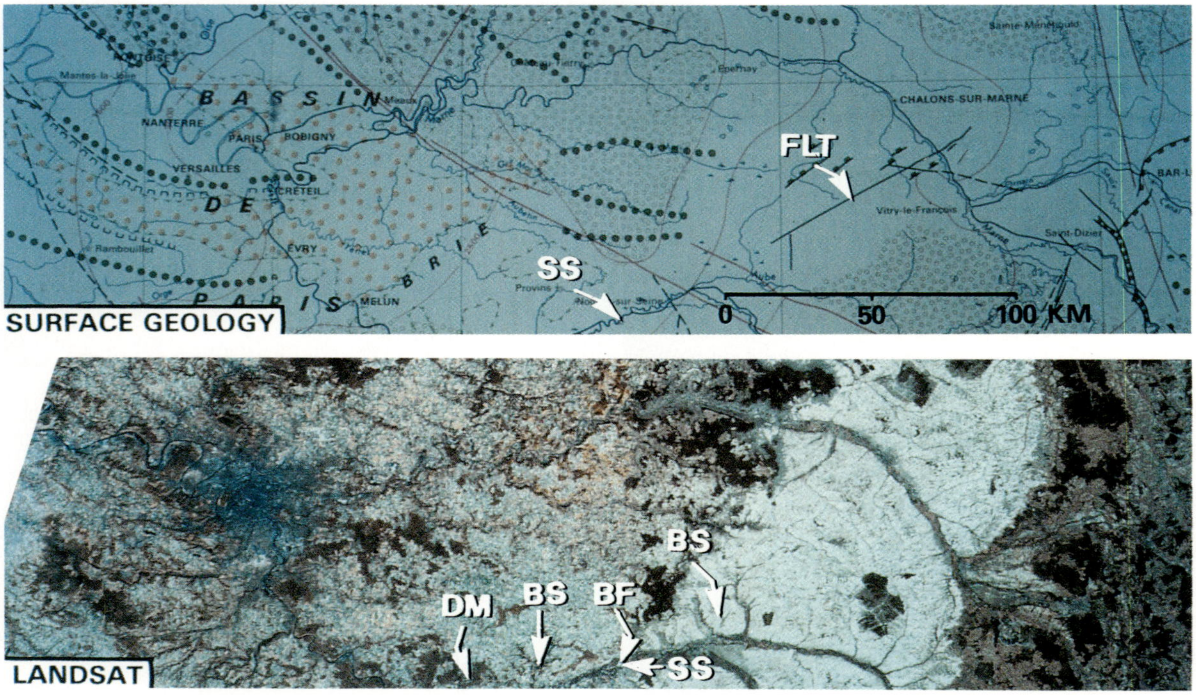

Fig. 4.23. Metz-Seine Valley lineament
The surface geological map and satellite image of the central segment of the MSV lineament. Circular features identified on the image reflect the presence of individual obscured folds that develop along the MSV lineament. Also illustrated are the exposed and buried expressions of the faulted segments of these lineaments. *DM* The breached fold of the Donne Marie Field which is further illustrated in Fig. 4.25; *BS* breached structure; *BF* buried fault; *FLT*-fault-line trace; *SS* subsequent stream

Fig. 4.24. Geological model of the MSV lineament
The block diagram shows the vertical relationship among basement features, subsurface, and surface structures along the MSV lineament. Data were compiled from potential field data, satellite image interpretation and available seismic and subsurface structure maps. *BS* Breached structure; *BF* buried fault; *SS* subsequent stream; *BWS* basement warp structure
▼

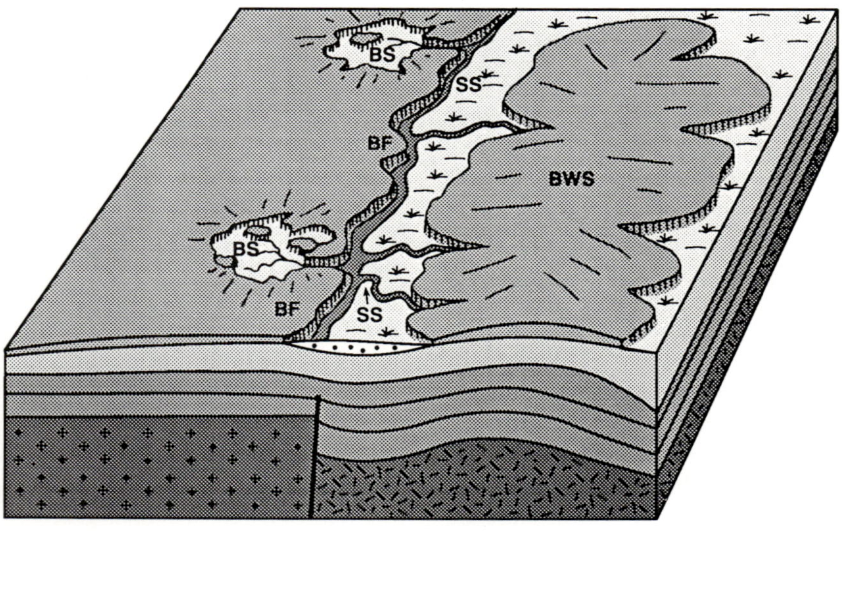

magnetic anomalies. Finally, the lineament is spatially coincident with a wide zone of faults and fractures that are well exposed at the eastern margin of the basin (shown with the enlarged tectonic map in Fig. 4.22 d).

More detailed information on the surface characteristics of the central segment of the MSV lineament is provided in Fig. 4.23. On the geologic map of the basin, the western segment of the lineament is manifested by a single exposed normal fault that dies out where it is cross-cut and offset by the Bray Fault Zone. On satellite imagery data, the eastern, buried segment of the lineament is manifested by the presence of several circular drainage and topographic features that are located along or parallel to the linear segment of the Seine Valley and the cuesta that was formed along its northern valley wall. These features are interpreted to reflect the surface expressions of individual folds which developed over a deep-seated basement fault. A conceptual block diagram illustrating the relationships between surface and subsurface structures observed along the MSV lineament is shown in Fig. 4.24.

4.2.4.2 Analysis of Linear Features

Linear features are defined as individual, small-scale topographic drainage- or moisture-related features which are believed to reflect the presence of individual faults and fractures in the sedimentary section. Excluded, whenever possible, are geological phenomena such as glacial, marine, eolian, and fluvial features as well as compositional layering. Also excluded are nongeological alignments such as fire burns, tornado tracks, clear-cutting, highways, man-made canals, etc. The analysis of linear features is probably the most common form of satellite image interpretation for exploration geology, to the point that "satellite imagery studies" and "linear feature studies" have become almost synonymous. Such studies offer the analysis of surface elements

Fig. 4.25. Surface expression of the Donne Marie Field
Satellite image of the Donne Marie Field showing (1) its breached crest and other circular, topographic expressions of its eroded limbs; (2) individual faults and fractures; and (3) their density contours (length of linear features per unit area)

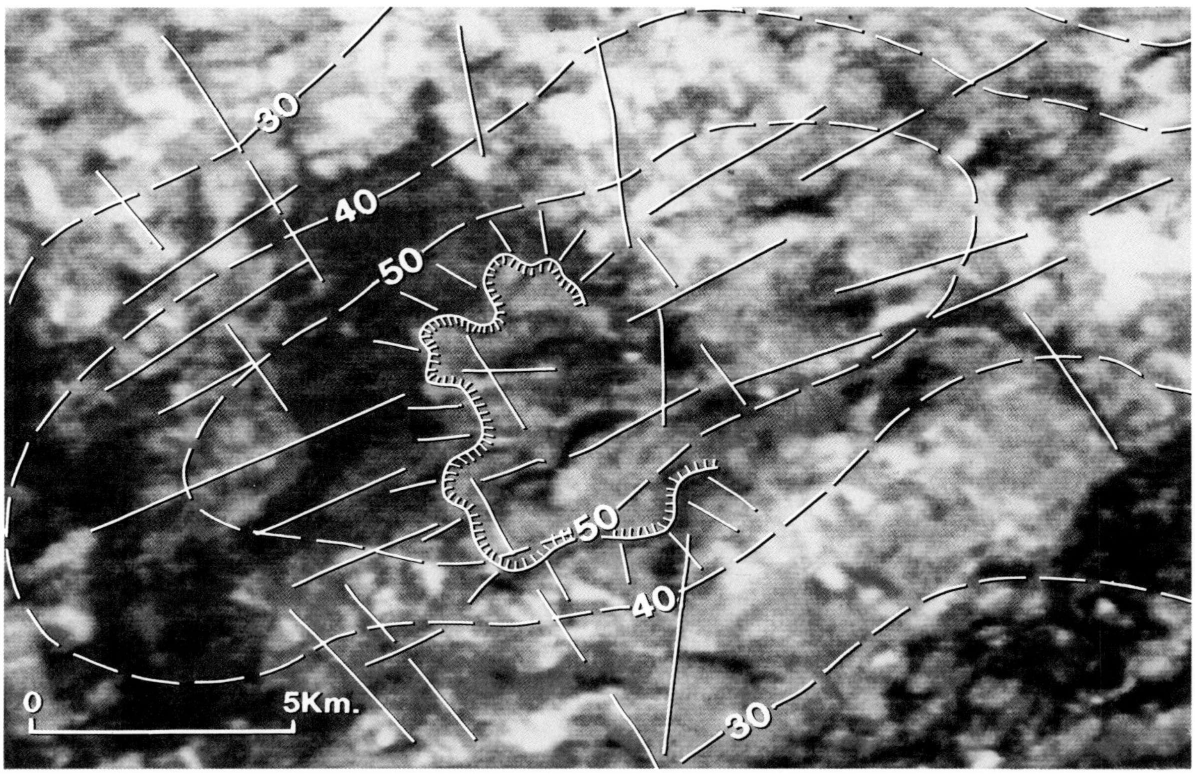

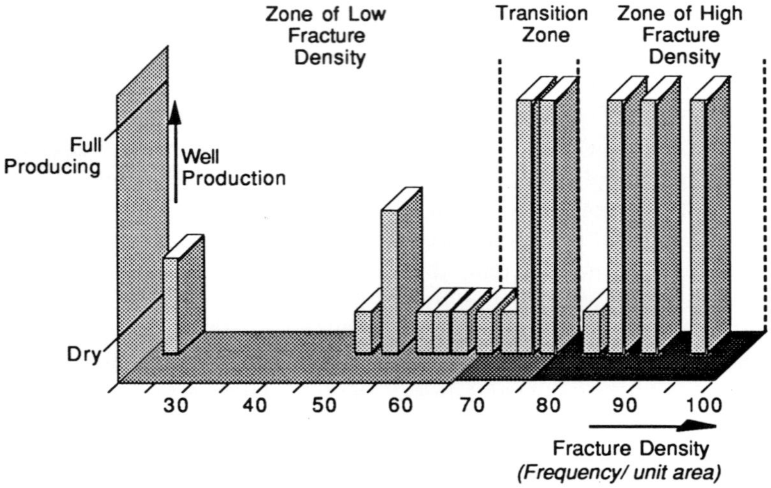

Fig. 4.26. Linear feature density analysis
A plot of fracture (linear feature) densities observed from Landsat data and producing fields in the Paris Basin. The general correspondence suggests that production in this basin may be related to increases in fracture densities

that are not usually visible with other exploration tools, providing, perhaps, new clues to the distribution of hydrocarbon plays or trends, particularly those which are related to fractured reservoirs. At the same time, this approach contains some serious limitations and pitfalls because a satellite image usually contains thousands of linear features which represent many different types of fractures. Their tectonic origin, timing of deformation and possible link to subsurface structures and related reservoirs cannot be reliably identified.

Linear feature studies usually follow three main steps. First, all linear features are identified on the imagery. Second, the distribution of the linear features is treated statistically in terms of density, orientation and length. Finally, attempts are made to relate the statistical attributes of the lineaments to known producing trends in a basin. A typical example of the first step is shown in Fig. 4.25 which captures the surface expression of an individual fold that was formed along the MSV lineament. Several surface features were interpreted on the imagery: (1) the circular breached area of the surface fold, in this case the producing Donne Marie Field; (2) a wide zone of individual linear features that trend parallel to the MSV lineament and, most probably, reflect surface fractures associated with this deep-seated basement block boundary; and (3) individual linear features that cross-cut the MSV lineament in various directions. These features most likely reflect the presence of individual faults and fractures of different ages and origins. In the second step, the linear features were analyzed in terms of densities, which are represented in Fig. 4.25 by contour lines. These contour lines, which represent the total length of a linear feature per unit area, are usually mapped with the aid of specially designed computer programs.

The close correlation between the densities of linear features and the location of the producing field encouraged the interpreters, in the third step, to continue this analysis over a wide area and test this relationship over all the producing fields in the basin. The results, shown in Fig. 4.26, suggest that production in this basin is associated with anomalous concentrations of linear features and, therefore, linear feature studies could be used as a tool for early identification of leads in this area.

In spite of the encouraging results of this study, more careful analysis of the data has shown some serious flaws in the assumptions and interpretations that were made. First, it was found that not all the oil fields in the area produce from fracture-related reservoirs. Second, it was found that the density of fractures identified on imagery is a direct function of the level of bedrock exposure in the area. This means that the high degree of correlation between structural closures and linear densities was not related to increases in fracture densities but simply reflected better bedrock exposure conditions. The author's experience to date has been quite disappointing regarding the effectiveness of linear feature studies for hydrocarbon exploration, and it is suggested that such endeavors should take a secondary role in satellite imagery studies.

4.3 Reconstruction Techniques

4.3.1 Topographic Models of Obscured and Buried Folded Strata

Obscured and buried domes or other folded structures exhibit similar relationships between topography and structures as their exposed counterparts, except that their topographic expressions are more subtle and require a careful analysis of topographic slopes and drainage elements. Berger and Aghassy (1980), using examples of buried salt domes in the Gulf Coast region of the USA, proposed a conceptual framework for a systematic analysis of domal topography. They first established a basic terminology for slope and drainage patterns that outline a hypothetical dome and then used these elements to describe the topographic expressions of domes and folds at various erosional stages. The basic elements of the model are shown in Fig. 4.27 and can be described as follows:

1. Marginal subsequent (strike) streams. These form as preexisting major streams and stream segments and tend to adjust their flow around the newly formed round obstacle. They form a general circular pattern which outlines the outer perimeter of the dome.
2. Outbound consequent (dip) streams. These form in a radiating pattern from the dome center. They may be either collected by the marginal subsequent components or flow directly outward.
3. Other (strike) consequent streams. These streams develop along the eroded consequent cuestas of the inner portion of the dome to follow a concentric drainage pattern.
4. Inbound obsequent (anti-dip) streams. These streams flow toward the dome center. They are usually collected by the central breaching stream following inversion of topography.
5. Isoclinal slopes. These are the long and gentle slopes or slope segments conforming with the arched sediments. These slopes may parallel the

Fig. 4.27. Superdome topography
Slope and drainage basic components over an idealized dome. (After Berger and Aghassy 1980)

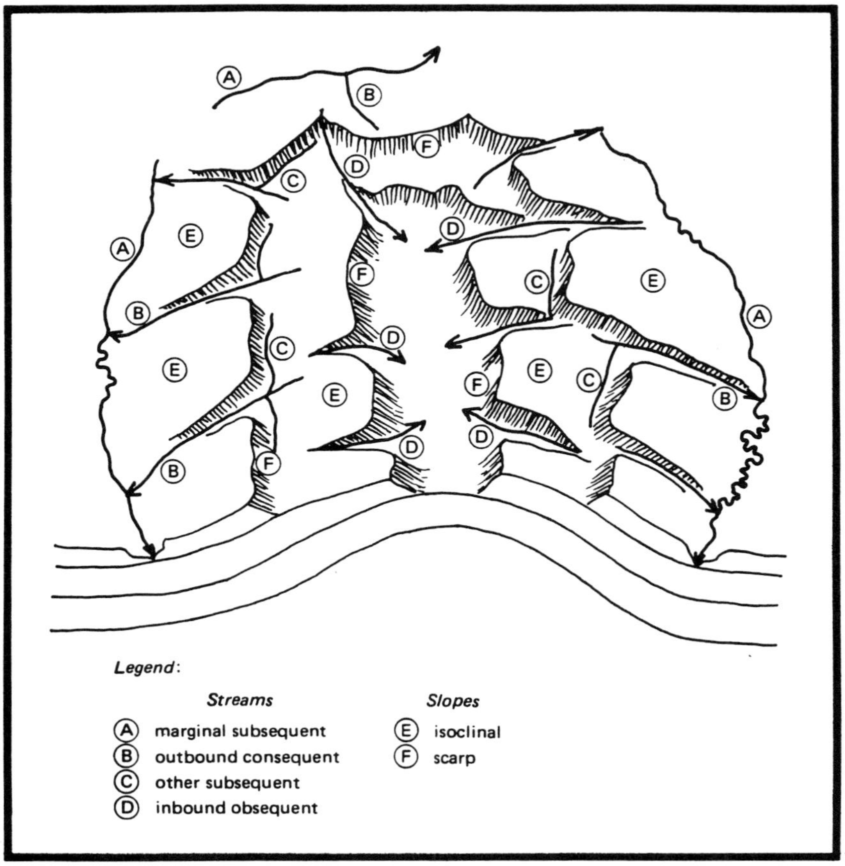

gently dipping bedding planes. However, they may also evolve into other inclined erosional surfaces.

6. Scarp slopes. These are short and steep slopes marking the lithological contact between different eroded layers. They always dip towards the breached crest of the dome.

The erosional evolution model of obscured and buried domal structures follows, more or less, the same pattern as exposed structures. However, because of the subtle topographic expressions of such structures, more tentative evaluations of their geomorphic components must be described. Also, because of their subtleties, only three stages (as opposed to four for exposed structures) can usually be ascertained. These stages are shown in Fig. 4.28 and are covered in the next few sections in their evolutionary order.

Stage 1: Positive Relief Stage

The surface configuration of the dome area exhibits a central topographic high at this stage. From the dome center, long and gentle isoclinical slopes extend down to the marginal subsequent streams with little or no interruptions. Radial drainage by outbound consequent streams is the most dominant element over the major dome body and causes dissection of the long isoclinal slopes. The major marginal subsequent streams have adjusted to the newly formed environment. Therefore, they exhibit gentle gradients and an abundance of depositional features along their courses. Toward the end of this stage, a new stream category begins to appear in the form of tributaries to the outbound consequent streams, often meeting them at right angles, but mainly following the local strike orientation. They

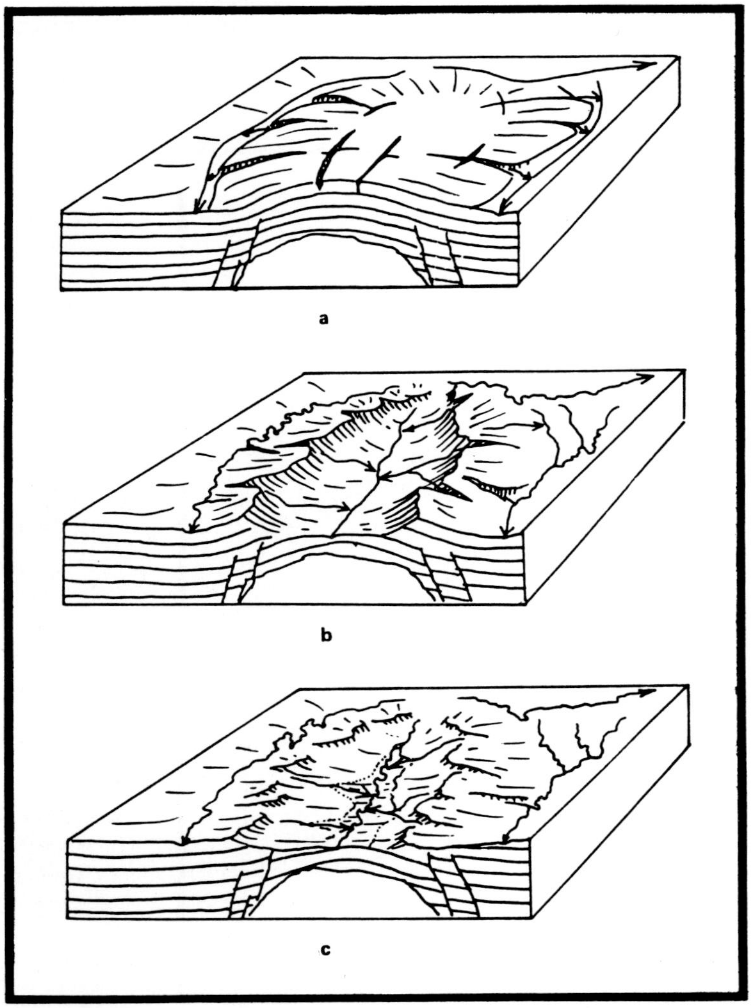

Fig. 4.28. Erosional models for buried and obscured structures

Typical surface expression of obscured and buried structures as defind by slope and drainage components. **a** Positive relief stage; **b** breached stage; **c** obliterative stage. (After Berger and Aghassy 1980)

are referred to as other subsequent streams and are not numerous at this stage.

Stage 2: Breached stage

The increased erosional activity that occurs at the dome's center results in a gradual lowering of the initial topographic high that existed there. This is done mainly by the headwaters of one or two dominant outbound streams which take over the drainage of the inner parts of the dome. An inversion of topography begins to take shape, and a major depression develops at the dome's center. Slopes facing the central depression become numerous. They will be referred to as scarp slopes. These are shorter and steeper than the isoclinal slopes and face the opposite direction. Once these slopes exist, a new category of streams appears which follows the direction of the scarp slopes toward the central depression. These are referred to as inbound obsequent streams and are destined to grow at the expense of the outbound consequent streams as they gradually capture increasing areas of their drainage.

Stage 3: Obliterative Stage

As erosion continues, the inbound obsequent streams expand headward beyond the rims and capture the better parts of the consequent outbound basins. The upper parts of the central breaching streams are no exception to this rule and, in most cases, end up dividing the dome into two halves as it widens its breaching central stream. This process results in substantial lowering of relief and rounding of forms leading to the oblitertion of their original distinct outline. Extensive sediment yield of the inbound streams, coupled with a low gradient, results in widening of the floodplain. At times, marshy areas form, especially when the outlet is restricted by rim relics. At this stage a dominance of inbound obsequent streams rather than outbound consequent streams is quite outstanding.

4.3.2 Geomorphic Investigation of Domal Topography

Drainage and slope elements that are detected on imagery can then be used to establish the erosional stages of buried and obscured domes and folds and to relate their topographic expressions to the underlying subsurface structures. For example, in the Gulf Coast Region, salt-related domal structures manifest different topographic expressions that are related, in most cases, to the depth of the salt stock and the degree of faulting associated with the salt diapir (Berger and Aghassy 1983). In the example shown in Fig. 4.29, the first three salt domes exhibit early to late stages of breaching (a–c), whereas the fourth one (d) shows an advanced stage of obliteration with a typical marshy area at the core of the dome. In the four cases shown, the topographic expressions of the salt-related domes are directly related to the proximity of the cap rock to the surface. In the first three cases, the crest of the dome maintains some positive topographic relief because its cap rock has not been exposed to the process of dissolution by groundwater and surface water. In the fourth stage, the cap rock was subjected to erosion and dissolution processes, allowing major streams to completely breach its crest.

4.3.3 Subsurface Reconstruction of Domal Topography

Once the erosional stage of buried and obscured structures has been identified and their drainage and slope patterns are delineated, it is possible to use this information as a guide for improved subsurface mapping of prospective structures. The steps involved in this technique are illustrated in Fig. 4.30.

First, imagery are used to trace structurally controlled stream patterns (Fig. 4.30 a). Second, stream segments that appear to follow the strike or dip of surface structures are identified as structurally controlled stream valleys (Fig. 4.30 b). Finally, the established patterns of the streams are integrated with well or seismic data to guide the mapping of subsurface structures (Fig. 4.30 c).

The use of drainage features as a guide to subsurface mapping naturally is most warranted in low relief areas that lack surface and subsurface controls. In fact, many of the early oil discoveries in the Central Basin Platform of West Texas and the Gulf Coast regions of the southern USA were attributed to the detectable surface expressions and drainage patterns that outline these structures (e. g. Hennen and Metcalf 1929; Christner 1940). The Soso Field in Jones and Jasper counties, Mississippi, is one of the best-known structures that is marked by drainage patterns that nicely follow the shape of the buried

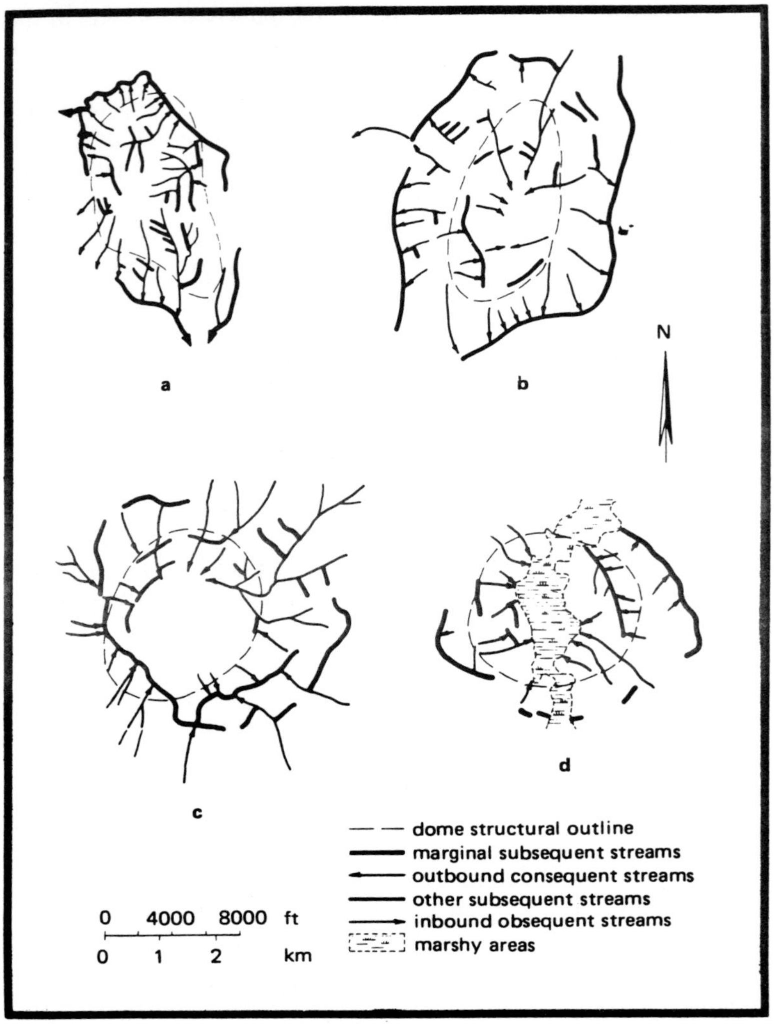

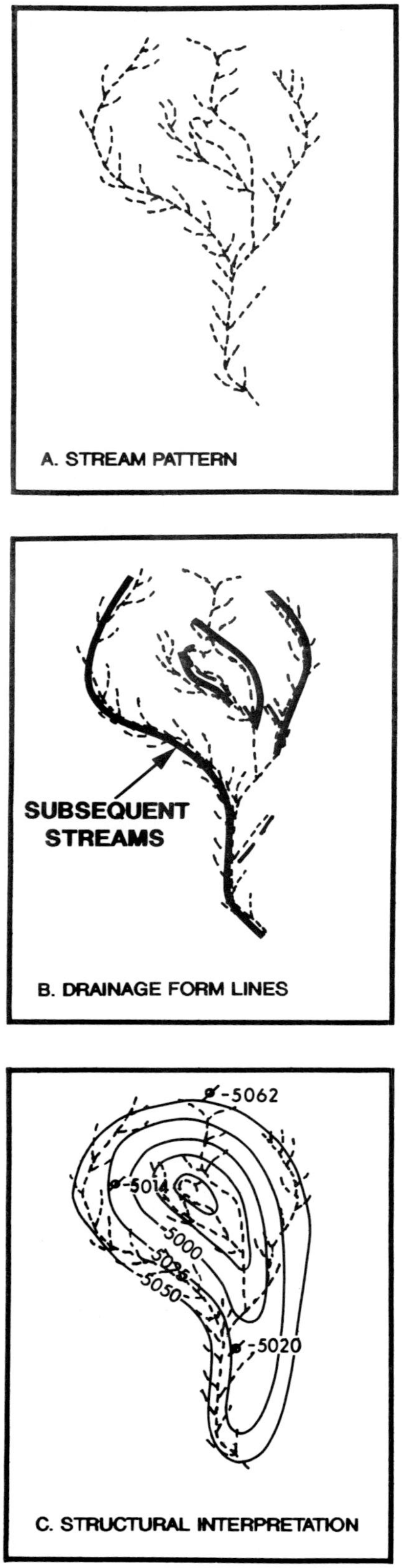

Fig. 4.29. Geomorphic investigation of domal topography
Drainage component sketches of four salt domes which exhibit variations in their topographic evolution. **a** Lampton Dome; **b** Keechi Dome; **c** Palestine Dome; **d** Steen Dome. (After Berger and Aghassy 1980)

Fig. 4.30. Drainage reconstruction techniques ▶
The three-step procedure for using drainage pattern analysis as a guide for subsurface mapping. (After Kite 1959)

Fig. 4.31. Structural reconstruction of the Soso Field ▶▶
An outstanding example of the surface manifestation of the Soso Field, Mississippi, and the potential use of drainage pattern analysis to guide the interpretation of subsurface structures. In **a**, structurally controlled stream valleys were used independently from subsurface control to outline the surface manifestation of a buried structure. **b** Shows the outline of the producing closure after the development has been completed. (After Kite 1959)

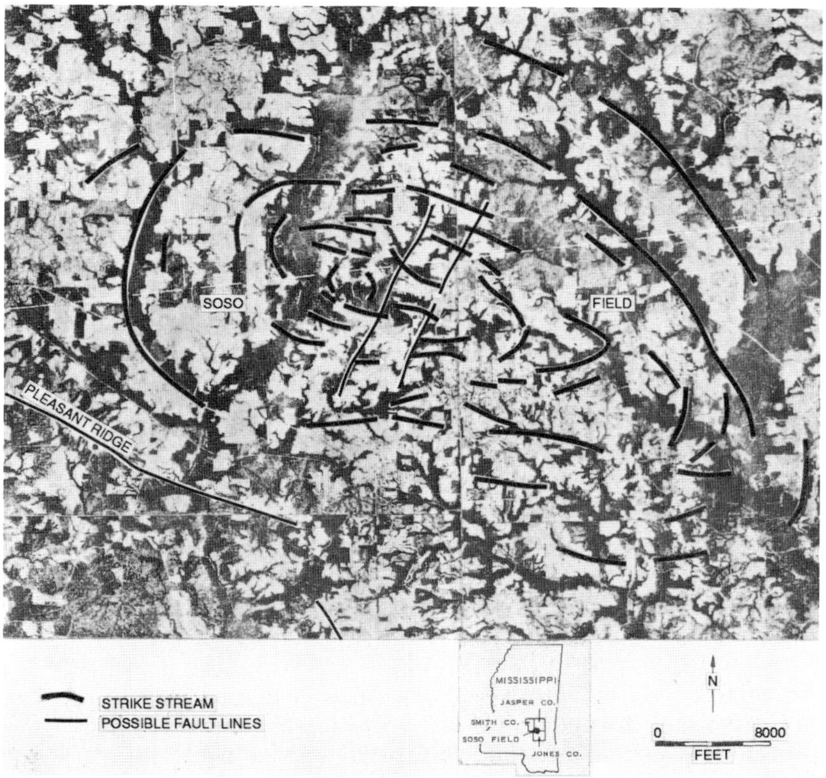
SOSO
FIELD
PLEASANT RIDGE
STRIKE STREAM
POSSIBLE FAULT LINES
MISSISSIPPI
JASPER CO.
SMITH CO.
SOSO FIELD
JONES CO.
N
0
8000
FEET

a

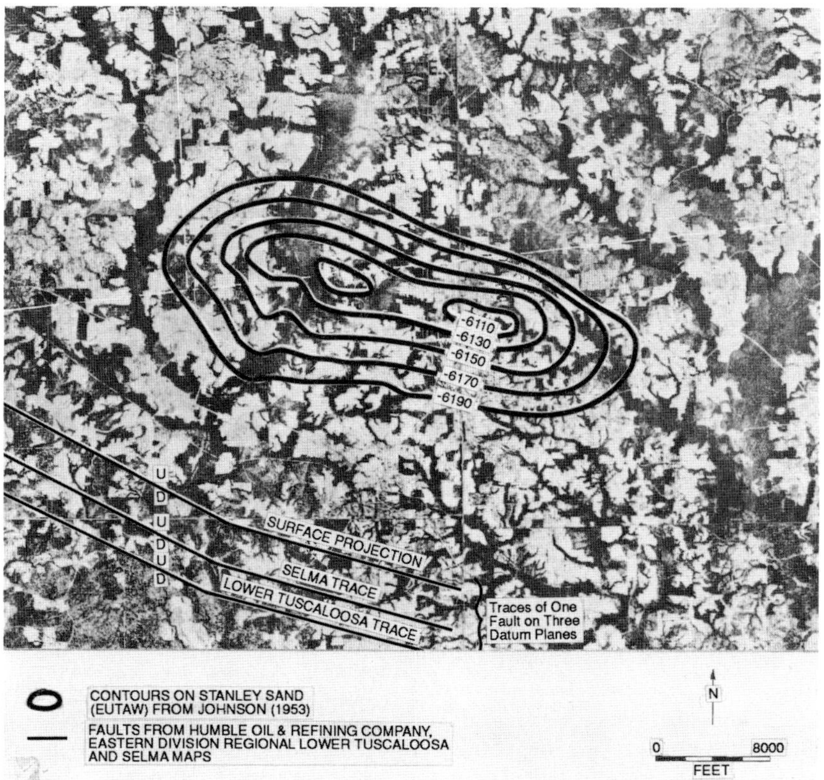
-6110
-6130
-6150
-6170
-6190
U
D
U
D
U
D
U
D
SURFACE PROJECTION
SELMA TRACE
LOWER TUSCALOOSA TRACE
Traces of One Fault on Three Datum Planes
CONTOURS ON STANLEY SAND (EUTAW) FROM JOHNSON (1953)
FAULTS FROM HUMBLE OIL & REFINING COMPANY, EASTERN DIVISION REGIONAL LOWER TUSCALOOSA AND SELMA MAPS
N
0
8000
FEET

b

structure. The analytical steps that were used to guide the subsurface mapping of this structure are illustrated in Fig. 4.31.

4.4 Summary

Obscured and buried structures are those which do not manifest clear expressions of bedrock strata, and their detection is accomplished through the analysis of their geomorphic expressions. From a remote sensing point of view, such structures can be divided into two categories, shallow obscured and buried structures, and basement warp structures. Both categories employ the same surface interpretation techniques but the latter requires the use of additional analytical tools, as discussed in the next chapter.

Although the exact development of surface expressions of buried structures is not well understood, several mechanisms may be postulated. These include differential loading, differential compaction, structural reactivation, disruption of near-surface and groundwater flows, underground water flows, and related erosional features. All of these lead to the development of surface structures that, although more subtle, mimic the configuration of the underlying features.

The analysis of obscured and buried structures is based on recognition of four surface and geomorphic features:

- Structurally controlled streams, which are detected by their unique patterns, include locally perturbed drainage, abrupt changes in channel and valley width, local deflection of stream orientation, and local areas of anomalous drainage density.
- Structurally related drainage network patterns and valley forms related to piping and sapping also help to detect obscured and buried structures by forming "halos" around them.
- Brightness and spectral variations on imagery reflect variations in moisture and soil content which can indicate the presence of topographically high or low areas. Bright areas usually reflect the crest of topographic features, whereas dark areas characterize low areas with poor drainage conditions.
- Lineaments and linear features are surface indicators of faults and fractures. The term lineaments refers to regional scale faults and zones of weaknesses in the basement. Linear features are individual faults and fractures of diverse origins.

Various types of techniques can be used to reconstruct the topography of buried and obscured structures and their possible geometries in the subsurface. One method is to use a model of a typical domal feature describing its basic slope and drainage component. One can then see the evolutionary changes as the dome undergoes the erosional processes which modified its original topographic expressions. Three basic erosional stages are recognized and include the positive relief stage, breached stage and obliterative stage. The recognition of the erosional stages of structures can be used to make predictions about their subsurface conditions as well as to guide the interpretations of their subsurface geometries.

References and Further Reading

Aghassy J, Berger Z (1981) An application of side-looking color infrared (photography) for structure detection in subtle topography. Proc 15th Int Symp on Remote sensing of the environment. Environmental Research Institute of Michigan, Ann Arbor, pp 491–498

Babcock EA (1971) Detection of active faulting using oblique infrared aerial photography in the Imperial Valley, California. Geol Soc Am Bull 82: 3189–3196

Bailey GB, Anderson FD (1982) Applications of Landsat imagery to problems of petroleum exploration in Quaidam Basin, China. AAPG Bull 66: 1348–1354

Baker VR (1980) Some terrestrial analogs to dry valley systems on Mars. NASA Tech Mem TM 81 776: 286–288

Berger Z (1982) The use of Landsat data for detection of buried and obscured geological structures in the East Texas Basin, U. S. A. Proc 2nd Thematic Conference Remote sensing explor Geol, Dallas, TX. Environmental Research Institute of Michigan, Ann Arbor, pp 577–589

Berger Z (1984) Structural analysis of low-relief basins using Landsat data. Proc Int Symp on Remote sensing of the environment, 3rd Thematic Conf, vol 1, Colorado Springs, CD. Environmental Research Institute of Michigan, Ann Arbor, pp 251–273

Berger Z (1988) Detection and analysis of basement structures in low relief basins using an integrated analysis of Landsat data. AAPG Bull 72 (2): 160–161

Berger Z, Aghassy J (1980) Geomorphic manifestations of salt dome stability. In: Craig RG, Craft JL (eds) Applied geomorphology. Allen and Unwin, Winchester, MA, pp 72–84

Berger Z, Aghassy J (1983) Near-surface moisture and evolution of structurally controlled drainage in soft sediments. In: LeFleur RG (ed) Ground-water as a geomorphic agent. 13th Annu Binghampton Geomorphology Symp, Allen and Unwin, Winchester, MA, pp 59–77

Berger Z, Corona FV (1986) Landsat structural analysis of the Rhine Valley and the Jura Mountain Area, western Europe. Int Symp on Remote sensing of the environment 5th Thematic Conf, Reno, NV. Environmental Research Institute of Michigan, Ann Arbor, pp 35–48
Berger Z, Joyce C, Simcox A, Sullivan J (1980) The influence of differential erosion on the stratigraphic sequence near salt domes in northern Louisiana. Geol Soc Am Abstr Prog 12 (7): 386
Blom RG, Crippen RE, Elachi C (1984) Detection of subsurface features in Seasat radar images of Means Valley, Mojave Desert, California. Geology 12: 346–349
Bunting BT (1961) The role of seepage moisture in soil formation, slope development, and stream initiation. Am J Sci 259: 503–518
Bureau de Recherches Geologiques et Minieres (1980 a) Carte geologique de la France: scale 1 : 1,500,000
Bureau de Recherches Geologiques et Minieres (1980 b) Carte tectonique de la France: scale 1 : 1,000,000
Bureau de Recherches Geologiques et Minieres (1981) Carte seismotectonique de la France: scale 1 : 1,000,000
Christian LB (1979) Cretaceous paleo-geology of the Paris Basin. Oil Gas J 77 (14): 172–176
Christner DG (1940) Todd Ranch (Oil) discovery, Crockett Country, Texas. AAPG Bull 24 (6): 1126–1127
Cook DG, Aitken JE (1970) Geology of Colville Lake map-area and part of Coppermine map-area, Northwest Territories. Geol Surv, Can Pap 70–12
Dikkers AJ (1977) Sketch of a possible lineament pattern in northwest Europe. Geol Mijnbouw 56: 275–285
Dunne T (1980) Formation and controls of channel networks. Prog Phys Geogr 4: 211–39
Flawn PF (1968) Palestine sheet, Geologic atlas of Texas: Bureau of Economic Geology, Univ Texas, Austin, Texas, scale 1 : 250,000
Fogg GE (1980) Aquifer modeling at Oakwood Dome. In: Geology and geohydrology of the East Texas Basin. Bureau of Economic Geology, Geol Circ 80–12: 33–39
Gary M, McAffee R Jr, Wolf CL (eds) (1974) Glossary of geology. Am Geol Inst, Washington, DC
Haimilia NE (1975) Possible large domal structures along a regional arch in the Northern Interior Plains. Geol Surv Can Pap 75-1: 63–67
Halbouty MT (1980) Geological significance of Landsat data for 15 giant oil and gas fields. AAPG Bull 64: 8–36
Hennen RV, Metcalf RJ (1929) Yates Oil Pool, Pecos County, Texas. AAPG Bull 13 (12): 1509–1556
Higgins CG (1982) Drainage systems developed by sapping on Earth and Mars. Geology 10 (3): 147–152
Higgins CG (1984) Piping and sapping: development of landforms by ground-water outflow. In: LaFleur RG (ed) Ground-water as a geomorphic agent. Allen and Unwin, Boston, pp 18–58
Higgins CG, Coates DR (1990) Ground-water geomorphology: the role of subsurface water in earth-surface processes and landforms. Geol Soc Am Spec Pap 252
Hobbs WH (1904) Lineaments of the Atlantic border region. Geol Soc Am Bull 15: 483–506
Horton RE (1945) Erosional development of streams and their drainage basins; hydrophysical approach to quantitative geomorphology. Geol Soc Am Bull 56: 275–376
Keefer WR (1970) Structural geology of the Wind River Basin, Wyoming. USGS Prof Pap 495 D: 35 pp
Kirkby MJ, Chorley RJ (1967) Throughflow, overland flow and erosion. Bull Int Assoc Sci Hydrol 12: 5–21
Kite RL (1959) Stream patterns – a guide in structural contouring. Exxon Production Research Company, Int Rep
Kudryakov VA (1974) Piezometric minima and their role in the formation and distribution of hydrocarbon accumultions. Dokladv Acad Sci, USSR 207: 240–242
Lattman LH (1954) The one-sided development of tributaries in tilted sedimentary rocks in eastern Allegheny plateau of West Virginia. Michi Acad Sci Arts Lett Pap 39: 361–365
Lattman LH (1959) Geomorphology applied to oil exploration. Min Indus 28 (6): 1–4
Lattman LH, Matzke RH (1961) Geological significance of fracture traces. Photogramm Eng 27: 435–438
Leopold LB, Wolman NG (1957) River channel patterns, braided, meandering, and straight. USGS Prof Pap 282-B: 85
Leopold LB, Wolman NG, Miller JP (1964) Fluvial processes in geomorphology. Freeman, San Francisco
Löffler E (1974) Piping and pseudokarst features in the tropical lowlands of New Guinea. Erdkunde 28: 13–18
Love JD, Coe Christiansen A (1985) Geologic map of Wyoming. USGS, scale 1 : 500,000
Miller VC (1961) Photogeology. McGraw-Hill, New York
Mosely FA (1971) A reconnaissance of the Wadi Beihan, South Yemen, with notes on basement control of gully alignment in superficial deposits. Proc Geol Assoc 82: 61–69
Murray GE (1961) Geology of the Atlantic and Gulf Coastal Province of North America. Harper and Brothers, New York
Nevin CM, Sherrill RE (1929) Studies in differential compaction. AAPG Bull 13: 1–30
Norman JW (1976) Photogeological fracture trace analysis as a subsurface exploration technique. Inst Mining Metallur Trans Sect B, 85: B52–B61
Parvis M (1950) Drainage pattern significance in air photo identification of soils and bedrocks. Photogramm Eng 16 (3): 387–409
Peterson RM (1979) Oil and gas exploration by pattern recognition. In: Shahrokhi F, Paludan T (eds) Remote sensing of earth resources. University of Tennessee Space Institute, Tullahoma, Tenn, vol 8, pp 28–60
Rowan LC, Wetlaufer PH (1981) Relation between regional lineament systems and structural zones in Nevada. AAPG Bull 65: 1414–1432
Small RJ (1964) The escarpment dry valleys of the Wiltshire Chalk. Trans Pap Inst Br Geo Publ M34: 33–52
Small RJ, Lewin J (1965) The role of spring sapping in the formation of chalk escarpment valleys. Southampton Res Ser Geogr 1: 3–29
Stearns RG (1967) Warping of the western Highland Rim Peneplain in Tennessee by ground-water sapping. Geol Soc Am Bull 78: 1111–1124

Sterns DW, Berger Z, Hopkins HR, Nelson RA (1988) The contribution of remote sensing data to exploration of fractured reservoirs. 6th Thematic Con on Remote sensing for exploration geology, Houston, TX. Environmental Research Institute of Michigan, Ann Arbor, pp 357–371
Stewart JH, Walker GW, Kleinhampl FJ (1975) Oregon-Nevada lineament. Geology 3: 256–268
Tator BA (1958) The aerial photograph and applied geomorphology. Photogramm Eng 24 (4): 549–561
Terzahgi K (1931) Earth slopes and subsidence from underground erosion. Eng News-Rec: 90–92
Thomas GE (1974) Lineament-block tectonics: Williston-Blood Creek Basin. AAPG Bull 58: 1305–1322
Tóth J (1980) Cross-formational gravity-flow of groundwater: a mechanism of the transport and accumulation of petroleum. In: Roberts WH, Cordell RJ (eds) Problems of petroleum migration. AAPG Stud Geol 10: 121–167
Voüte C (1962) Contributions of photo-interpretation to engineering projects in various stages of execution. Photogrammetria 19: 179–191
Wilson IF (1948) Buried topography, initial structures and sedimentation in Santa Rosalie area, Baja California, Mexico. AAPG Bull 32 (9): 1762–2807
Wise DU (1982) Linesmanship and the practice of linear geo-art. Geol Soc Am Bull 93: 886–888

Chapter 5 Interpretation Techniques: Detection and Analysis of Basement Warp Structures

5.1 Introduction

This chapter explores the second main branch of the obscured and buried structures category, basement warp structures (BWSs). The recognition of these structures on imagery is quite similar to the first group and employs the same image interpretation techniques. However, because these structures are more subtle and originate in the deep basement, they differ from the first group in two significant aspects. First, their subsurface constraints are often difficult to define with conventional interpretation techniques. Second, they usually do not form pure structural traps, but rather exert significant control on the development of reservoir rocks in their vicinities, particularly the development of structurally controlled incised valleys.

Detection and evaluation of BWSs therefore require the use of several analytical techniques which have not yet been illustrated. First, and foremost, the recognition of these structures on imagery must be accompanied by the identification of corresponding basement signatures on gravity and magnetic data. (Satellite images along with gravity and magnetic data are often referred to collectively as reconnaissance tools.) Second, the expressions of these features in the sedimentary cover must be ascertained through the implementation of subsurface mapping techniques that are specially designed to capture such subtle subsurface structures. For example, second-derivative structural isopach and isochron maps, as well as highly squeezed and enhanced seismic lines are often used in these circumstances. Finally, the reconstruction of the paleogeomorphic surfaces of these structures and their related incised valley systems requires the use of structural and geomorphic principles gained from the analysis of surface structures on imagery.

The objective of this chapter is to illustrate some of the additional analytical steps mentioned above. The chapter is divided into four parts:

- A method is established for early detection of BWSs through the integrated analysis of satellite imagery, gravity, magnetic and other subsurface data.
- A classification scheme for BWSs, based on their sizes, and means of detection are presented.
- The most common spatial relationships between BWSs and incised valley systems are demonstrated.
- The relationships between BWSs and known production are demonstrated with the well-documented case study of the Belle Fourche Arch in the Powder River Basin, in the western United States.

A summary of previous work on this subject is provided in the next section to illustrate the significance of BWSs in hydrocarbon exploration in general and for the development of prolific incised valley systems in particular.

5.2 Previous Studies

Several authors have illustrated the presence and influence of BWSs on hydrocarbon plays. Harding and Lowell (1979) show how such structures dominate the style of many low-deformed basins and illustrate their structural geometry and patterns. They state that these structures, which are the hallmark features of many foreland basins, intracratonic sags, and stable platforms, often become major focal points of hydrocarbon migration and accumulations. Gay (1989) provides a comprehensive review of this subject matter with special emphasis on the (gravitational) compaction phenomenon. His paper also describes some early classical work by Mehl (1920), Blackwelder (1920), Teas (1923), Nevin and Sherrill (1929), and others. Hennen and Metcalf (1929) and Christner (1940) show examples of major producing closures in the Central Basin Platform of West Texas that developed over deep-seated basement structures and were discovered by their surface expression (see also Berger et al. 1992). Halbouty and Halbouty (1982) present a reconstruction of the paleogeomorphic expressions of the deep-seated Sabine Uplift and demonstrate its role in the development of the East Texas Giant Oil Field. The development of porous dolomite in the Scipio-Albion Trend in the Michigan Basin has been attributed to the presence of fracture systems that developed along a deep-seated, reactivated basement (possibly wrench) fault system (Harding and Lowell 1979). An excellent summary of different types of paleogeomorphic traps related to deep basement structures is also provided by Martin (1966).

The development of prolific Cretaceous incised valleys related to BWSs is demonstrated in the Western Canada Basin by Jackson (1984) and Master (1984; see Fig. 5.1) and for the foreland basins of the western US, particularly the Powder River Basin, by Weimer (1980), Slack (1981), and Dolson et al. (1991). These studies, however, usually demonstrate the influence of basement structure and topography on the development of known pools, but do not offer tools and techniques for predicting the presence of such channels in the less explored parts of the basin where seismic and well data are quite sparse.

The integration of satellite imagery, gravity and magnetic data, which will be discussed in this chapter, attempts to fill this gap. It is assumed that the reader is familiar with the principles of gravity and magnetic surveys and their use in geological exploration. There are several good textbooks providing background information on these topics including, Grant (1972), Nettleton (1976), Steenland (1965), and Telford et al. (1990). Also, the reader must remember that the identification of BWSs as presented here must be followed by more thorough mapping of incised valleys with well and core data and the application of sequence stratigraphic concepts. Comprehensive reviews of this subject matter can be found in Dolson et al. (1991), Vail et al. (1977), Van Wagoner et al. (1990) and Wiemer (1992).

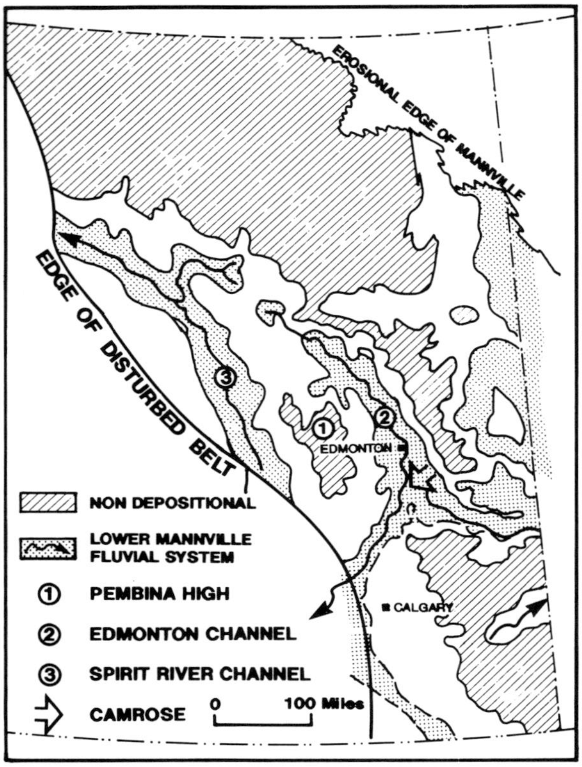

Fig. 5.1. The Pembina High
The Lower Mannville isopach is shown, illustrating the presence of a large incised valley system that developed around the Pembina High. The location of the Camrose Dome is shown by an *open arrow*. (After Master 1984. Published with permission of the AAPG)

5.3 Principles of Integrating Reconnaissance Data Sets

Basement warp structures are usually extremely subtle and, as stated earlier, their presence often cannot be conclusively demonstrated using conventional interpretation techniques of seismic and well data. In the process of detecting these structures, the explorationist must rely on the interpretation and integration of reconnaissance data sets. Also of use are seismic lines that are extremely squeezed, thus enhancing subtle features which are usually ignored during routine mapping of subsurface structures.

Reconnaissance tools, however, provide non-unique solutions and, thus, cannot be used as a stand-alone method for interpretation of BWSs. For example, magnetic features can be interpreted as reflecting variations in the magnetic susceptibilities of the basement, or they may indeed reflect the presence of basement structures with considerable topographic relief. Similarly, surface structures identified on images may or may not originate in the basement. (They could be related to detached thrust structures, for example.)

By the same token, identification of subtle features on seismic data is also problematic. Such fea-

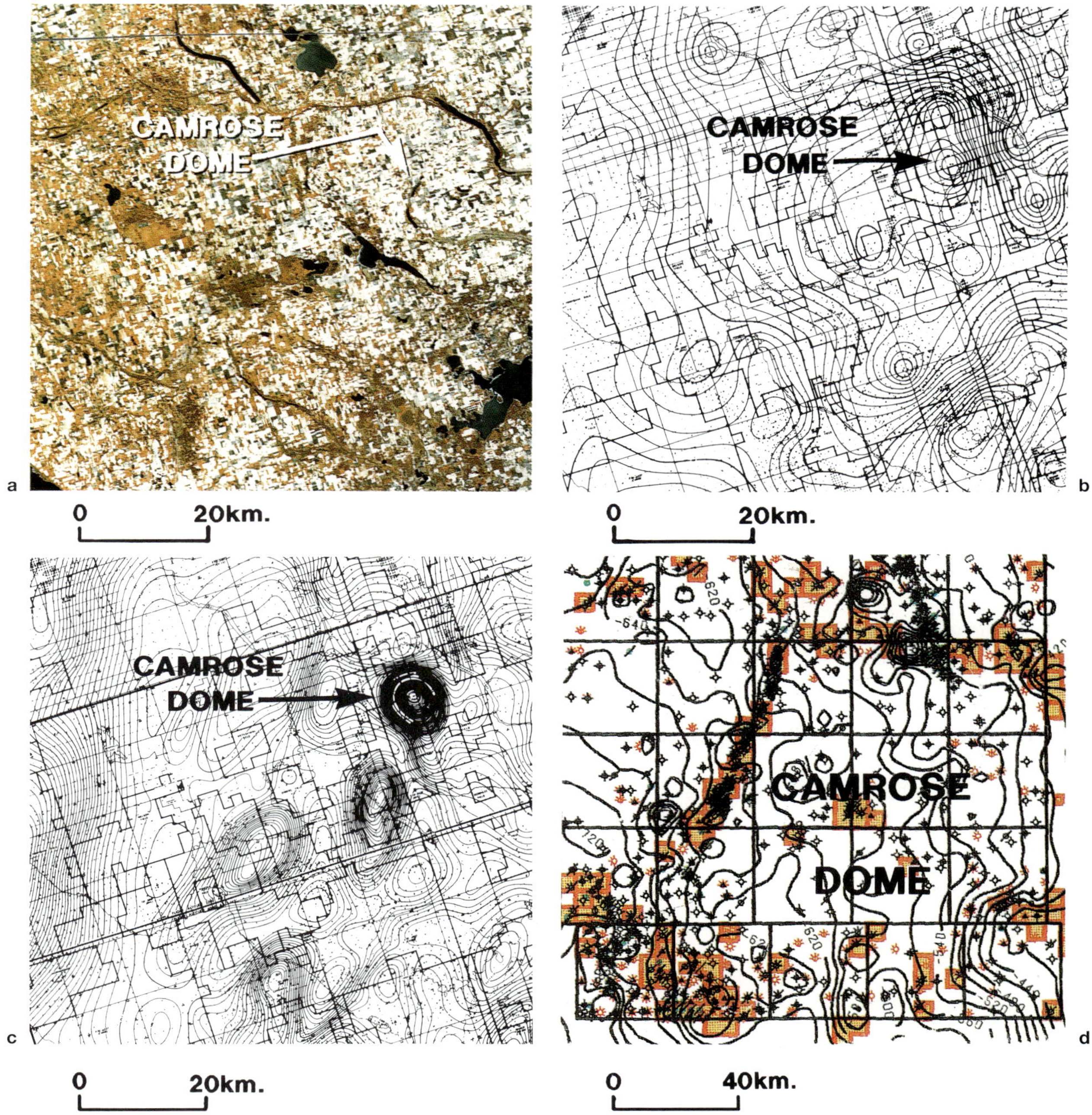

Fig. 5.2. The Camrose Dome
The expression of this basement warp structure is illustrated by **a** satellite imagery; **b** gravity; **c** magnetics; **d** production data around the Camrose Dome. (Potential field data courtesty of Aquaterra, Inc.)

tures could be related to the presence of subtle structures but could also be caused by static problems (near-surface noise). In fact, it is not uncommon to find that rigorous processing of noisy seismic data may lead to the loss of such subtle features. (See discussion on the interpretation of subtle folds in the Western Canada Basin in Figs. 6.24 and 6.25.)

The key to reliable detection of BWSs, therefore, is dependent on the use of converging lines of evidence which are extracted from all of these data sets and on a series of assumptions which are deduced from their observed spatial relationships.

The proposed integrated process for detecting BWSs with reconnaissance tools, the related assumptions which are used during this process, and the potential contributions to a hydrocarbon exploration program are illustrated in Figs. 5.1 and 5.2 with an example from the Western Canada Basin. The area shown is located along segments of a large Lower Cretaceous (Mannville) valley system that incised into the erosional surface of the Western Canada Foredeep Basin. This valley system consists of the Edmonton and Spirit River channels which were incised around a northwesterly trending topographic ridge of Paleozoic carbonates, known as the Pembina High (Fig. 5.1). This channel system produced a prolific hydrocarbon play which includes several thousand oil and gas fields (Jackson 1984).

A detailed examination of a small segment of the Edmonton Channel, near the town of Camrose, is provided in Fig. 5.2. It shows satellite imagery, potential field, and production data around a profound domal feature (referred to as the Camrose Dome). Strong spatial correlations are observed among all four data sets: the topography and present stream valleys, positive gravity and magnetic features observed on the potential field maps, and production data for Lower Mannville reservoirs which outline the periphery of the dome. These relationships allow the interpreter to make the following assumptions:

- The present topography is influenced by and mimics the configuration of a positive, dome-shaped basement feature.
- This basement feature exerted significant control on the pattern and level of incision of the Mannville channels.
- The sediments that filled the incised valleys constitute the prolific portion of the Lower Mannville reservoir in this area.

The integrated analysis illustrated here can be used in two stages of exploration. In frontier areas, the analysis may lead to the recognition of new play concepts and leads and can be used to guide data collections and interpretations. In more mature areas, it may improve the exploitation of existing pools as well as lead to the delineation of subtle additional features which have been overlooked by routine mapping.

5.4 Classification of Basement Structures

5.4.1 Exposed Basement Structures

Prior to the analysis of BWSs, it is important to examine exposed basement massifs and establish their common surface and structural expressions. The examination of many exposed massifs with satellite images suggests a classification scheme based on size and magnitude of surface expression. This classification can be nicely demonstrated with imagery and photos of the well-exposed Arabian Shield (in the Sinai Desert, Egypt) shown in Figs. 5.3, 5.4 and 5.5.

5.4.1.1 First-Order Basement Structures

The segment of the shield which is shown on the image is a first-order basement feature. It is a regional scale element (on the order of hundreds of square kilometers) that is characterized by considerable topographic relief and well-defined topographic boundaries. To the west, the massif is bounded by a series of normal fault systems, whereas its eastern margins are defined by a series of profound, parallel cuestas that were formed along the eroded and tilted sedimentary units that cover its margins.

5.4.1.2 Second-Order Basement Structures

The massif can be further subdivided into several second-order features (on the order of 10 km^2 in size) that are characterized by different color, texture, and topographic relief. The color differences

Fig. 5.3. Geomorphic expression of basement structures ▶
a An overview of the topographic expression of basement structures. Two second-order BWSs can be observed. Immediately behind the vehicle is a highly dissected terrain made of metamorphic rocks whereas, in the far background, is the highest point in the Sinai Desert, Mt. St. Catherine, which is made of granitic rocks. **b** Dike swarms which are resistive to erosion forming linear topographic ridges that may be depicted on satellite imagery as third-order features. **c** A close-up look at the third-order, positive, linear features formed by a resistant dike. In the background is the Gulf of Eilat showing the contact between the massif and the overlying Mesozoic strata

a

b

c

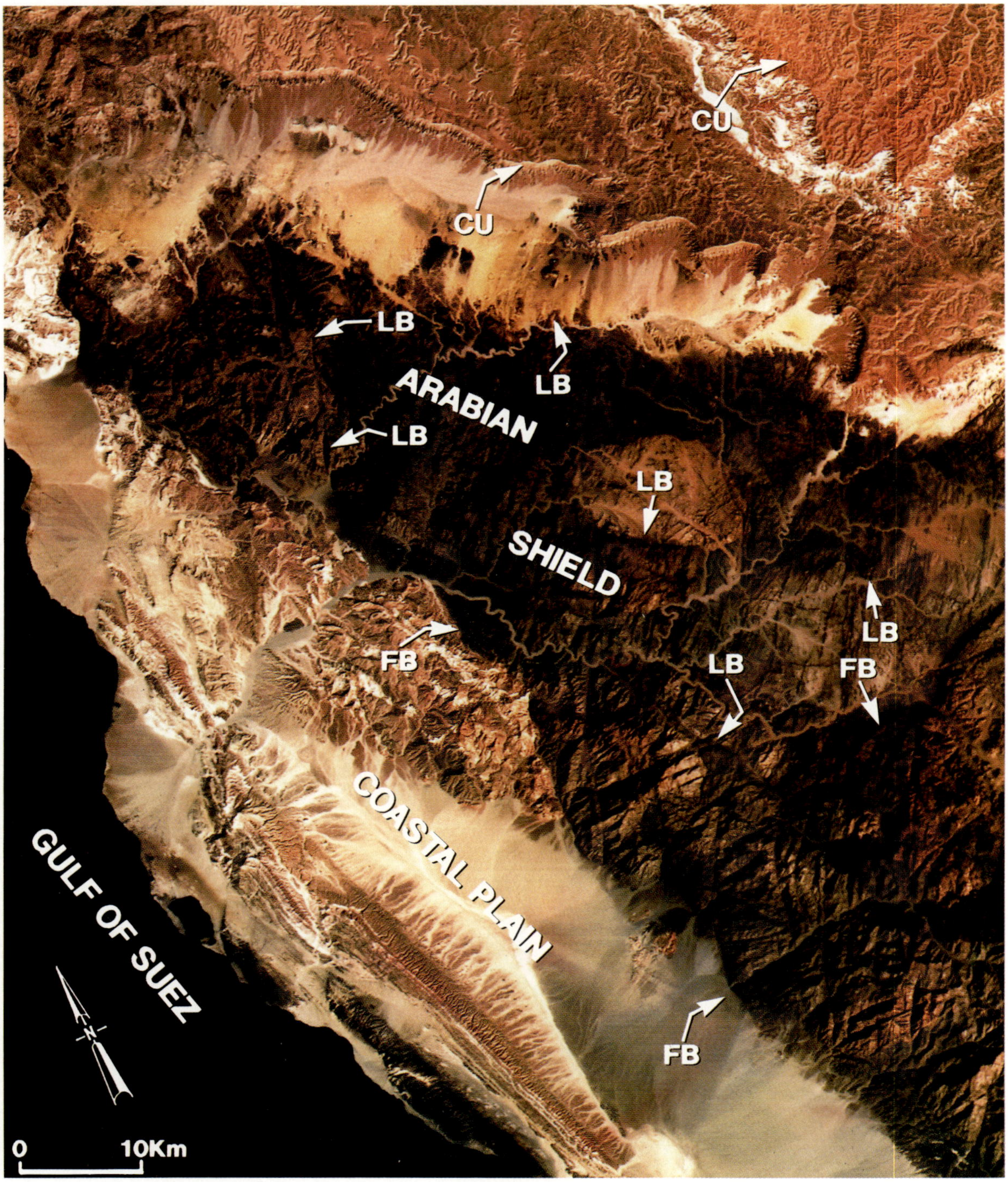

Fig. 5.4. Surface expression of basement structures
Satellite imagery of the exposed Arabian Shield in the Sinai Peninsula, Egypt, showing the expression of various types of basement structures. *FB* Fault-related boundaries; *LB* lithostratigraphic boundaries; *CU* cuesta

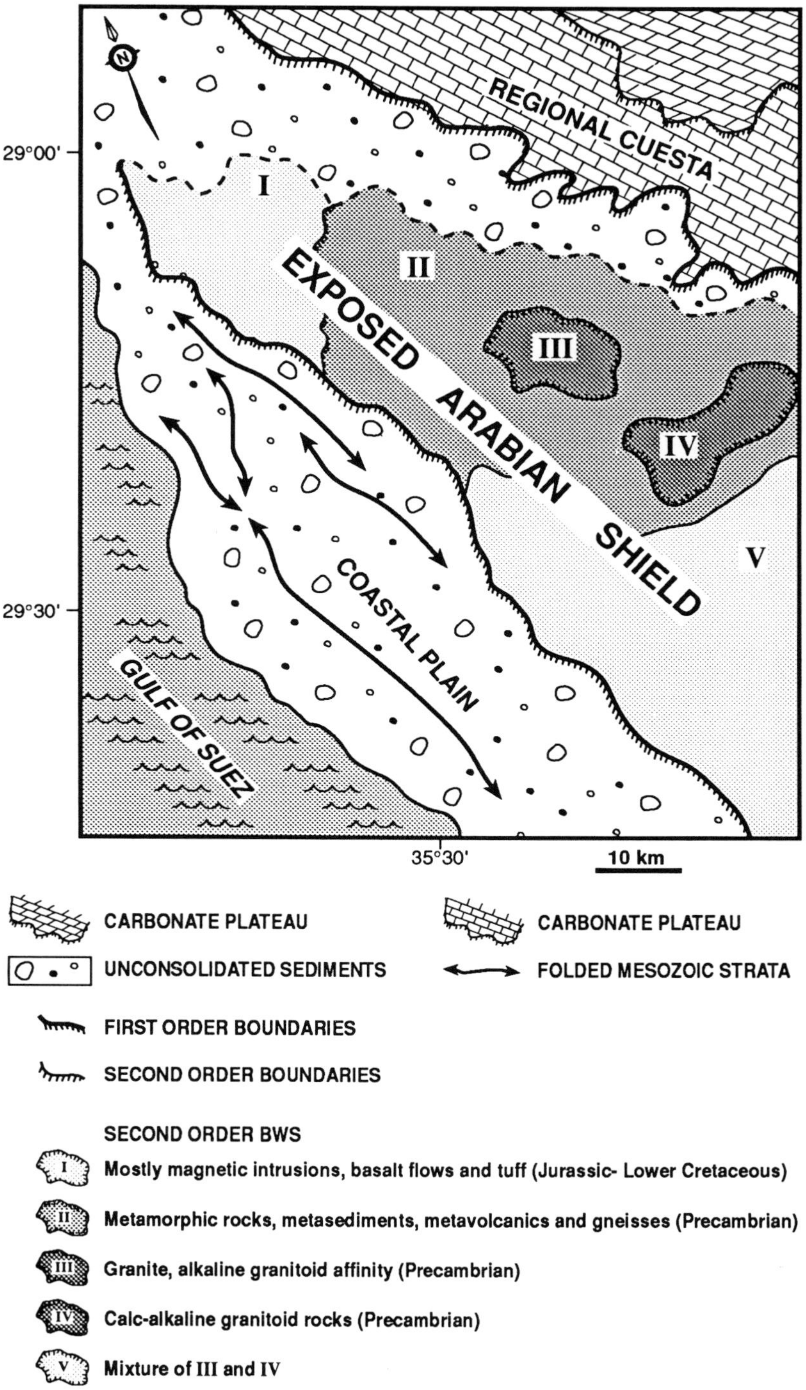

Fig. 5.5. Classification of exposed basement structures
Integrated interpretation of first- and second-order basement structures and related topographic features of the exposed Arabian Shield shown on the satellite image (Fig. 5.4). (Geology generalized from Eyal et al. 1980)

are attributed to variations in the lithological composition of each feature. The texture characteristics are related to differences in faults and fracture density. The significant variation in topography is attributed to different rates of erosion of each second-order feature. Note that the overall topographic expression of the massif is defined by these second-order basement features which are bounded, along some segments, by large-scale faults.

5.4.1.3 Third-Order Basement Structures

Third-order elements consist of individual faults and fractures that are limited to several kilometers in length and make up the "structural grain" of the massif. The individual faults and fractures are manifested on imagery as linear topographic depressions and stream valleys. The overall characteristic of the third-order features is a complex pattern of linear features that trend in several different directions and exhibit significant variations in densities. Basement layering of highly deformed metamorphic rocks as well as dikes could also produce third-order basement features, but these are not evident in this portion of the massif. Referring back to the discusion in Chapter 4, these features should be regarded also as third-order linear features.

5.4.2 Basement Warp Structures (Buried and Obscured)

BWSs can be classified in the same manner as the underlying basement features. This classification determines the type of data that can be used to constrain the structures as well as to determine the possible influences that they may exert on hydrocarbon plays.

5.4.2.1 First-Order Basement Warp Structures

First-order BWSs are regional-scale uplifts, domes and other positive features on the order of hundreds of square kilometers. They manifest strong expressions on potential field data and exert significant control on the paleogeomorphic setting of a basin or a region. The Pembina High, shown earlier, is a good example of such first-order BWSs. Other examples of this kind include the Belle Fourche Arch, shown later in this chapter and the Sabine Uplift, known for its role in the development of the East Texas Oil Field, and illustrated in Part 2 of this book. Large-scale fault systems that mark boundaries between different basement lithologies, such as the MSV lineament in the Paris Basin (shown earlier in Chap. 4), may also be considered first-order BWSs.

Table 5.1. Classification of basement warp structures

Classification	Size	Description of underlying basement features	Comments	Examples cited in the book
First order	Regional scale (hundreds of kilometers)	Domes, arches, major intrusive bodies, major zones of weaknesses in the basement	Unique regions of magnetic signature; large-scale gravity anomalies, control environmental deposition; form major erosional surfaces and related incised valleys; could form major hydrocarbon traps	• Sabine uplift (Part 2) • Pembina High (Chap. 5) • Belle Fourche Arch (Chap. 5) • Metz-Seine Valley Lineament (Chap. 4) • Bray Fault (Chap. 4)
Second order	Prospective scale (tens of kilometers)	Local domes, horst blocks, local intrusive bodies, basement steps	Positive or negative gravity and magnetic anomalies; form local stratigraphic traps	• Camrose Dome (Chap. 5) • Monias, east and west (Chap. 5)
Third order	Local phenomenon (<tens of kilometers)	Linear features such at individual faults and fractures, basement layering and dikes.	Produce complex patterns of linear surface features; usually hard to relate to gravity and magnetic data	• Linear features in the Paris Basin (Chap. 4)

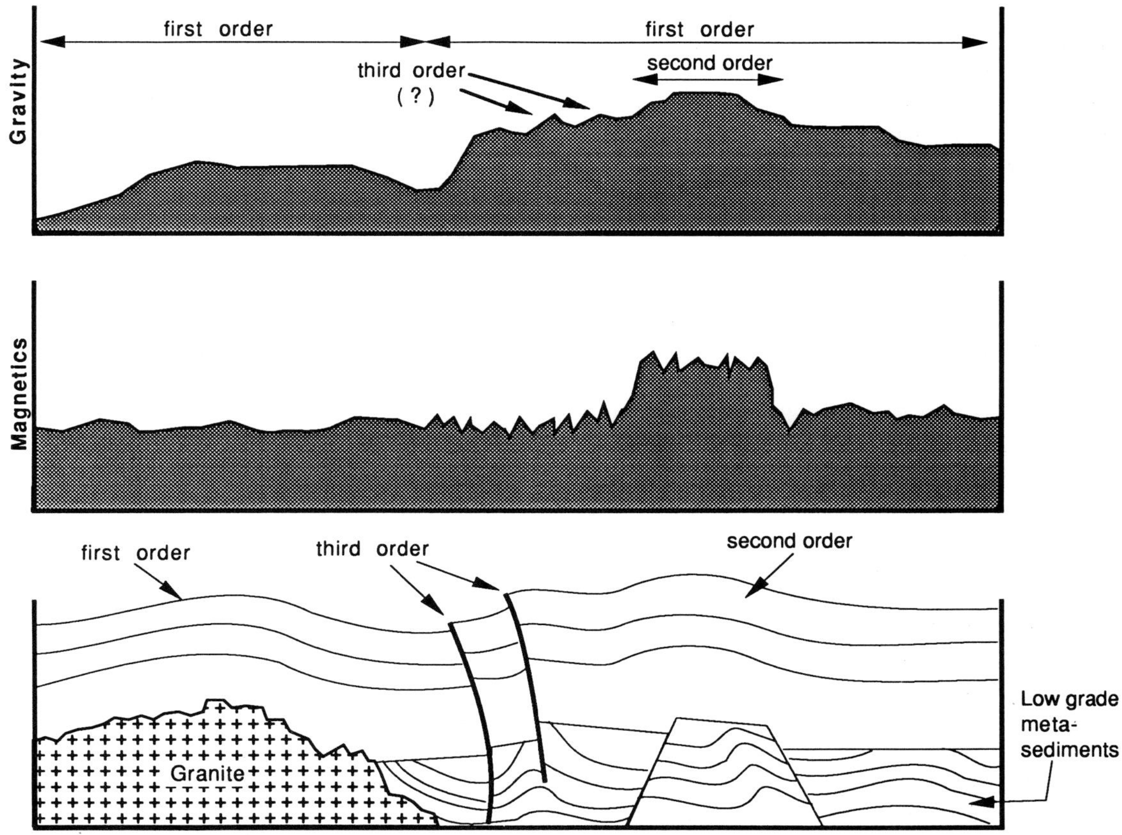

Fig. 5.6. **Classification of basement structures**
Idealized cross sections showing the stacked vertical relationship between basement and subsurface structures and their expression on potential field data. The cross sections are modeled after data from the Piedmont Region, US eastern coastal plain which is shown in Fig. 7.18

5.4.2.2 Second-Order Basement Warp Structures

Second-order BWSs consist of prospect-scale features such as local horst blocks, domal structures and monoclinal flexures that develop over basement faults. They exert local control on the distribution of potential reservoir rocks. These features are commonly on the order of 10 km^2. The Camrose Dome from the Western Canada Basin (shown earlier on magnetic, gravity and satellite images, in Fig. 5.2) provides an excellent example of such a feature. Another example of this kind, the Hilight Dome, will be shown later in this chapter together with the example from the Powder River Basin.

5.4.2.3 Third-Order Basement Warp Structures

Third-order BWSs consist of small-scale faults and fracture systems in the sedimentary section that reflect the influence of the "basement grain". These features are extremely small and hard to constrain with gravity or magnetic data, and their influence on hydrocarbon accumulations is quite limited. (See the discussion of the Paris Basin in Chapter 4.)

Table 5.1 and Fig. 5.6 summarize and illustrate the classification of the basement-related structures discussed above.

5.5 Diagnostic Surface Features of Basement Warp Structures

Structurally controlled incised valley systems that develop in the vicinity of BWSs manifest drainage patterns that are similar to those observed over other buried and obscured structures. In the interpretation of these structures, two different methods can be used.

The first method utilizes the basic slope and drainage components of domal topography which were illustrated earlier in Chapter 4 (Fig. 4.8). This method, however, requires detailed knowledge of the paleogeomorphic surface of the structure. Therefore, it is limited mostly to mature areas and well-developed and mapped producing fields.

The second method, which is more suitable for investigation of newly recognized basement warp structures, groups incised stream valleys into two categories, subsequent and transverse. The first group consists of incised valleys that were formed by streams that could not cut through the positive structure and were forced to follow its strike direction, developing a typical pattern of marginal streams. The second group consists of incised valleys that were formed by streams that cut through the positive structure and breached its crest forming typical cone-shaped valleys (illustrated earlier in Figs. 4.8 and 4.9). Figure 5.7 schematically illustrates these two types. Note that, in comparison to the first drainage classification scheme, this one is extremely generalized and focuses only on major tributaries that can be constrained with limited subsurface information.

Fig. 5.7. Structurally controlled incised valleys
Generalized classification of structurally controlled incised valleys that develop over two basement warp structures

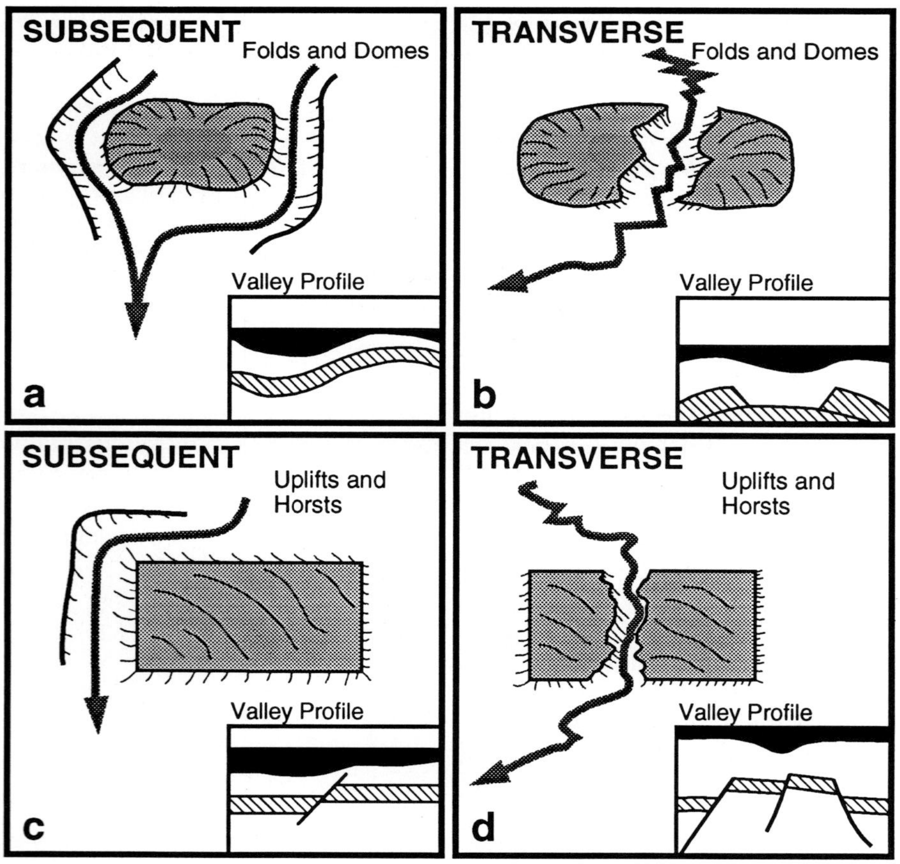

5.5.1 Subsequent Incised Valleys

Subsequent incised valleys are the dominant features in regions where the rate of uplift of the structures is greater than the rate of incision. The development of this type is nicely illustrated by the geomorphology experiment in Fig. 5.8, which demonstrates the interaction between the development of surface structures over reactivated basement structures and the evolution of incised valley systems in soft sediment. The experiment shows the development of radial subsequent stream valleys that outline the periphery of a reactivated basement dome.

Fig. 5.8. Subsequent incised stream valleys
A geomorphic experiment simulating the evolution of drainage systems in the vicinity of an uplifted dome. In this case, rate of uplift is greater than rate of downcutting, forcing streams to circum navigate the structure. (Courtesy of S. Schaum)

Two outstanding examples of subsequent incised valley systems, which outline BWSs, are shown in Fig. 5.9. The first case shows the development of a radial drainage pattern over an uplifted basement block in the Fort St. John Graben in British Columbia, Canada. The second example is of the Leatherwood Field in West Texas. In both cases, the producing closures are related to deep-seated basement structures which are difficult to observe on seismic data but manifest strong expressions on potential field as well as satellite imagery. Note the similarity between the drainage patterns that develop over this field and those which are shown in the geomorphology experiment in Fig. 5.8. Note also that the segments of the valleys that are controlled by the surface structure are substantially more incised than any other segments of the streams, both in the upstream and downstream direction.

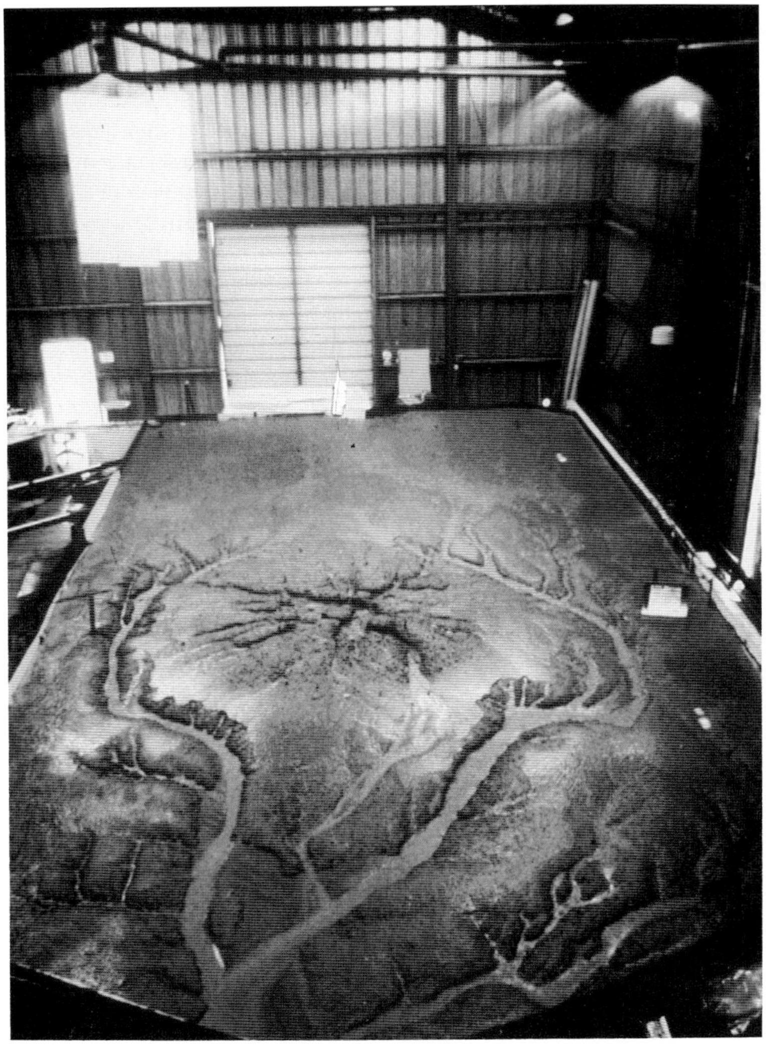

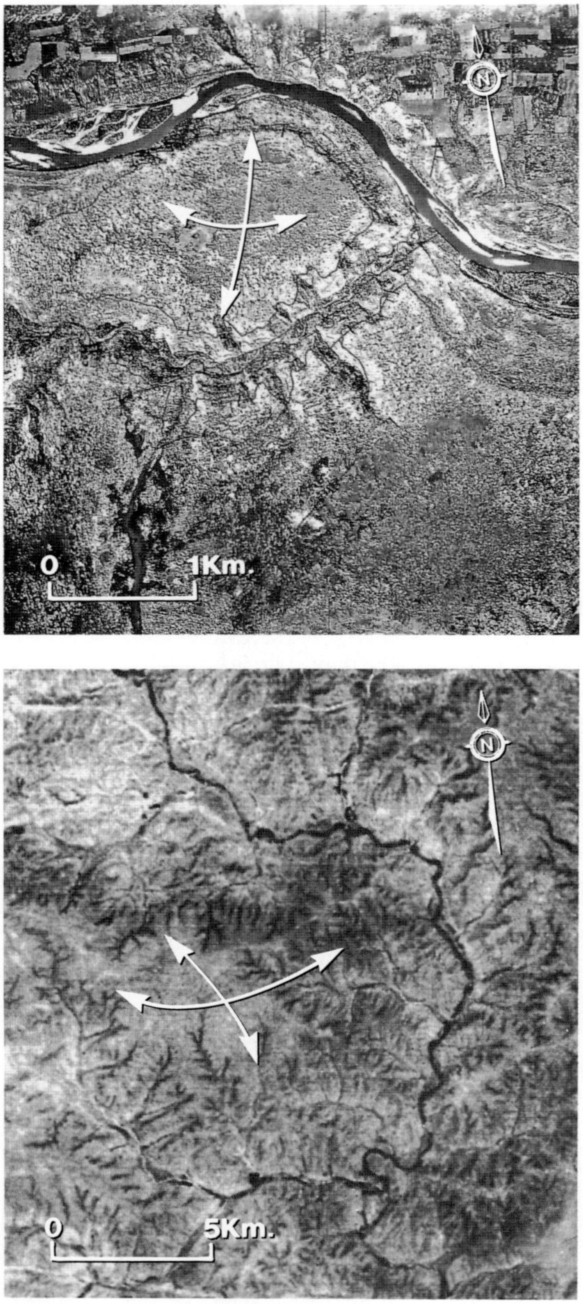

Fig. 5.9. Subsequent incised stream valleys
Shown are outstanding subsequent valleys that outline uplifted basement blocks in **a** the Wilder Field in the Fort St. John Graben and **b** the Leatherwood Field in West Texas

5.5.2 Transverse Incised Valleys

A similar geomorphic experiment using the same buried domal feature, but with different rates of uplift, simulated the development of a transverse incised stream valley (Fig. 5.10). In this case, the incision process was greater than the rate of uplift, causing noticeable differences between the patterns of incision observed here and those observed for the first type. In this case, a relatively short and deeply incised valley developed across the crest of the structure. Note the widening of this valley on both sides of the structure which no longer exhibits well-defined valley walls. Again, the stream valleys on both sides of the structures form a cone-shaped feature which was illustrated earlier in Chapter 4.

Excellent examples of this type of incised valley were demonstrated with imagery data from the Wind River Basin and the Pearsall Dome in southern Texas shown in Fig. 4.8 a and b, respectively. The Wind River Basin example is further illustrated with the surface geological map shown in Fig. 5.11. On this geological map, the Quaternary deposits in the valleys were highlighted in white patterns to illustrate the dramatic narrowing of transverse valleys which cross a series of elongated anticlines in the Wind River Basin. The cone-shaped valleys are well illustrated on both sides of the anticlines. The incised valley system observed over the Camrose Dome area in Fig. 5.2 reflects the characteristics of transverse incised valleys. Most of them are short valley systems that are incised across a northeast-trending topographic ridge that follows the crest of a basement high which is characterized by high-frequency magnetic and gravity anomalies.

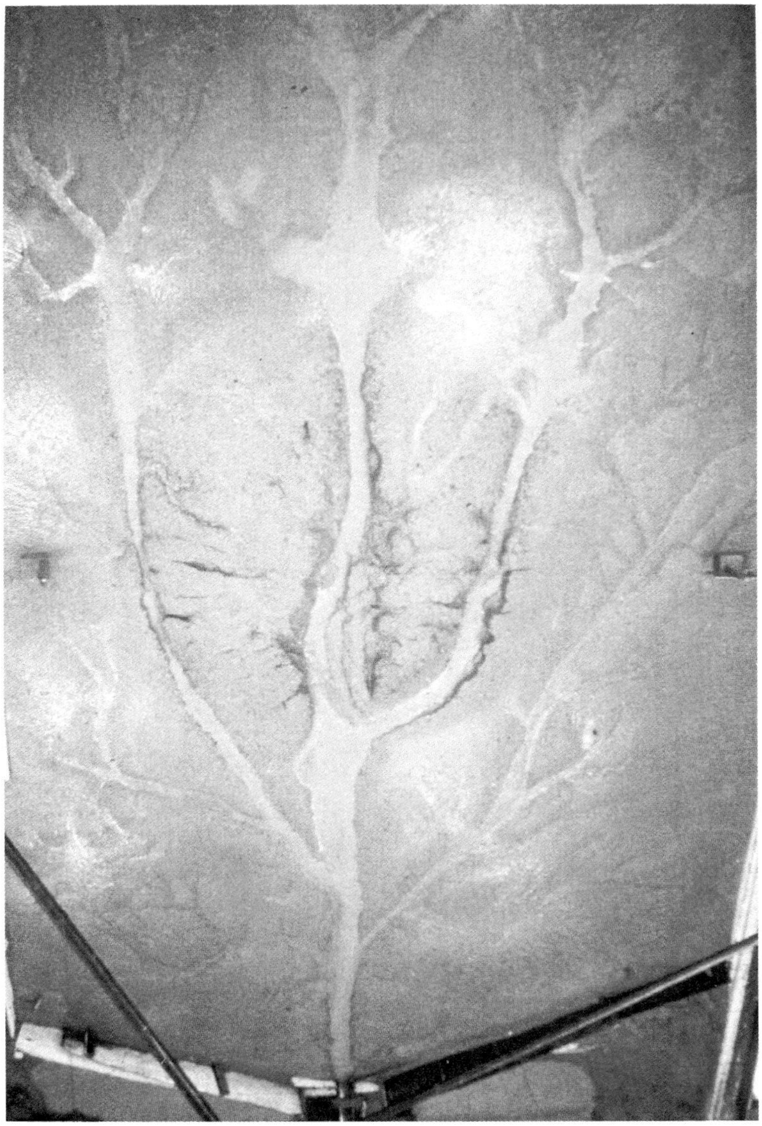

Fig. 5.10. Antecedent incised stream valleys
An experiment similar to the one in Fig. 5.6 is shown. In this case, rate of uplift was lower than rate of incision leading to the development of antecedent incised valleys

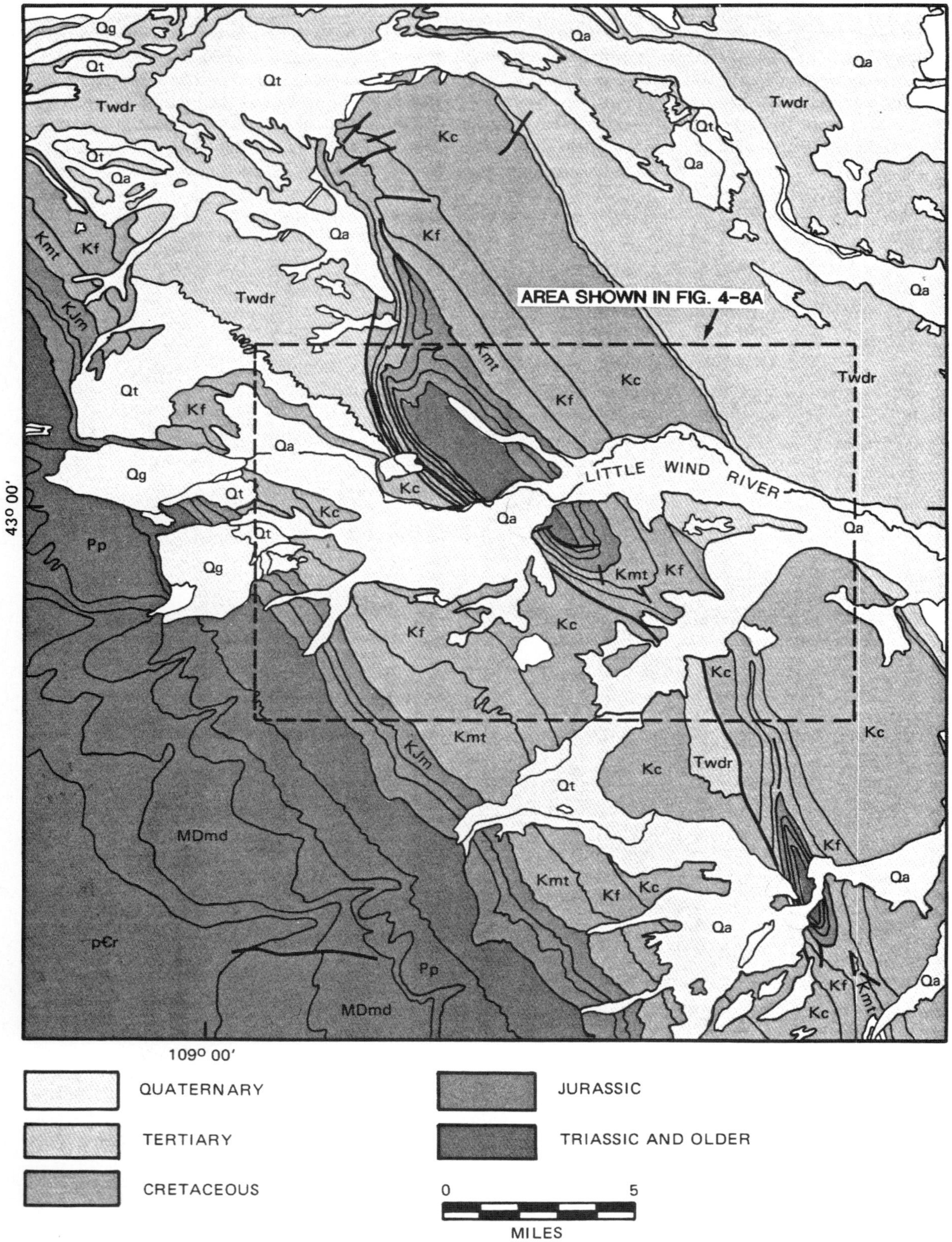
AREA SHOWN IN FIG. 4-8A
LITTLE WIND RIVER
43° 00′
109° 00′
Qa
Qt
Qg
Twdr
Kc
Kf
Kmt
KJm
Pp
MDmd
p€r
QUATERNARY
TERTIARY
CRETACEOUS
JURASSIC
TRIASSIC AND OLDER
0
5
MILES

◀ **Fig. 5.11. Surface expression of transverse incised valleys**
Surface geological map of the Wind River Basin, illustrating the development of transverse incised valleys across the anticlinal ridges of this foreland basin. The *dashed box* indicates area covered by the image in Fig. 4.8 a. (Geology after Keefer 1970)

5.5.3 Cuesta Topography

Cuesta topography is another important diagnostic feature that develops over basement warp structures. Many positive basement warp structures are outlined by a series of concentric asymmetric ridges, similar to those observed around the exposed Arabian Shield, except that these features are much more subtle. Cuesta topography may persist throughout the geological history of a basin or a region leading to the development of important hydrocarbon plays in the valleys that develop parallel to these cuestas. This phenomenon is illustrated later in this chapter with a case study from the Powder River Basin of the western United States. Another case of cuesta topography related to major BWSs occurs in the Sabine Uplift, which is examined in Part 2 of this book.

5.6 Specialized Interpretation Tools

As stated in the introduction to this chapter, successful identification of basement warp structures requires the use of tools and techniques which are not routinely used in exploration. First and foremost, potential field data must be processed to enhance the expression of basement structures and topographies. These processes may involve several filtering techniques which are designed to remove the expression of deep crustal features and enhance the expression of basement structures and topography right above the sedimentary fill. For example, the second vertical derivative filtering method enhances near-surface effects at the expense of deeper anomalies (e. g., Telford et al. 1990).

An image processing system can be effectively used to filter, merge and enhance the spatial relationships between features observed on potential field data and satellite images. This process is particularly effective when the filter parameters can be adjusted in "real time" to achieve optimum enhancement of the relationships. An example of this process is illustrated with data from the Pecos Arch in the Central Basin Platform in West Texas which is shown in Fig. 5.12. In Fig. 5.12 a, a second vertical derivative of the magnetic data set is displayed as a three-dimensional model, illuminated from the southeast. The data show that the northeast-trending magnetic anomaly of the Pecos Arch is divided into several sub-blocks by north-trending linear features (basement faulted blocks?). The spatial relationship between these basement blocks and drainage features can be ascertained when a satellite image is superimposed on this data set (Fig. 5.12 b, c). For example, the southernmost basement block is the Leatherwood Oil Field which was shown earlier in this chapter.

Detecting the presence of BWSs on seismic data is made extremely difficult by several factors: (1)-these structures are very subtle; (2) they can extend over large areas and, thus, cannot be fully captured by any single seismic line; (3) the processing of the seismic data is usually designed towards capturing shallow features; (4) the typical layering of the sedimentary section is usually absent, giving the basement a "uniform" seismic signature. One way to overcome these difficulties is to use advanced seismic display techniques (i. e., seismic workstations). One can then view a large number of seismic lines displayed on the screen in an extremely condensed format. Subtle basement structures and related features in the sedimentary cover can then be identified (Fig. 5.12 d).

The availability of seismic workstations allows the interpreter to assemble large-scale regional lines and further investigate the tectonic history of these basement features and their influence on the distribution of potential reservoir rocks (Fig. 5.12 e). A similar procedure can be followed using regional-scale cross sections which are constructed across these basement features. This procedure was illustrated by Halbouty and Halbouty (1982) who demonstrated the tectonic history of the Sabine Uplift in West Texas and its influence on the development of the giant East Texas Oil Field.

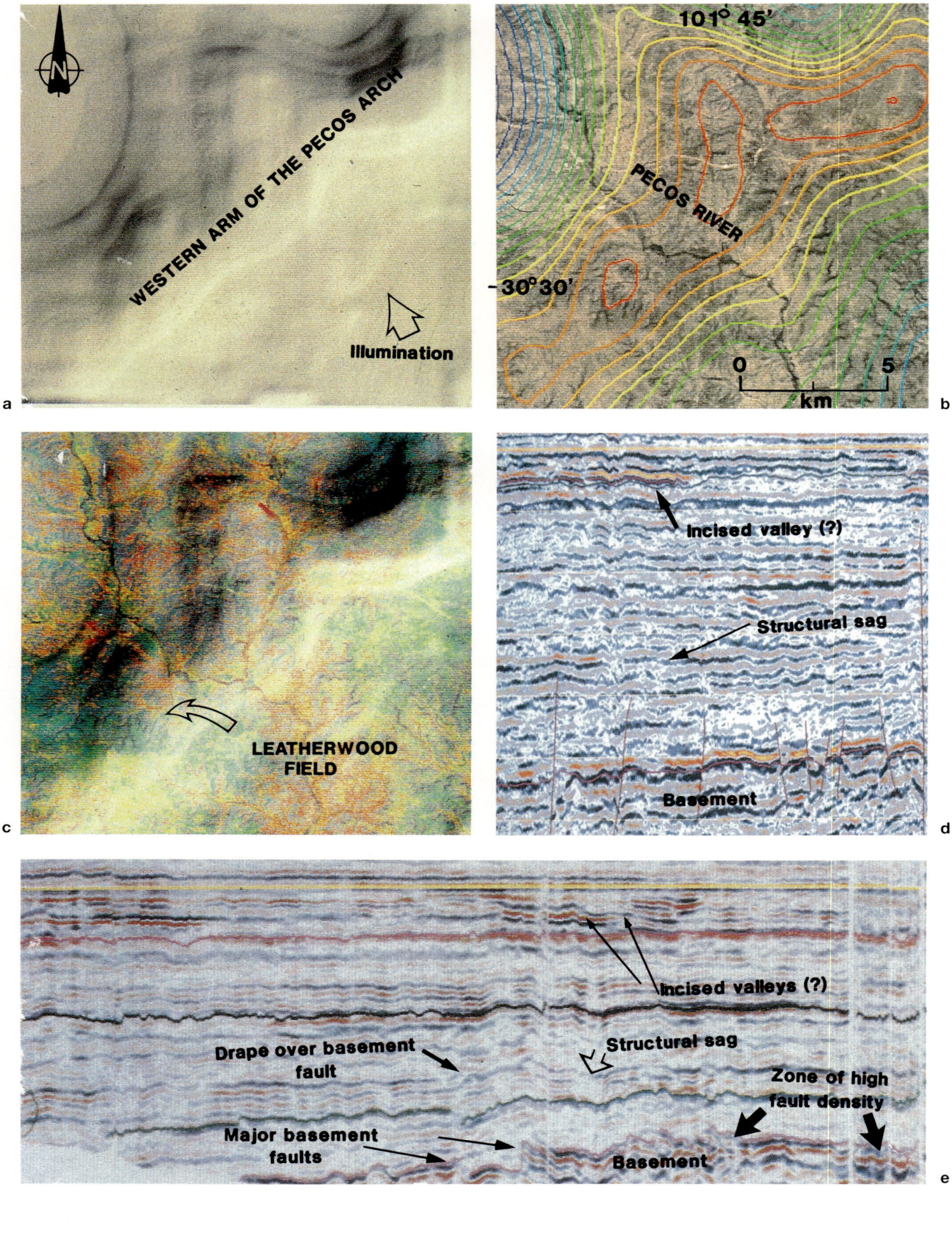
a
N
WESTERN ARM OF THE PECOS ARCH
Illumination
b
101° 45'
PECOS RIVER
30°30'
0
5
km
c
LEATHERWOOD FIELD
d
Incised valley (?)
Structural sag
Basement
e
Incised valleys (?)
Drape over basement fault
Structural sag
Zone of high fault density
Major basement faults
Basement

◀ **Fig. 5.12. Geophysical expression of basement warp structures**
a Illuminated three-dimensional representation of magnetic data of the Pecos Arch in West Texas. **b, c** Satellite imagery merged with data shown in **a**. **d** A condensed seismic section of a basement structure and related incised valley system. **e** Regional assemblages of seismic data across major basement features. (Location of **d** and **e** not released for proprietary reasons)

5.6.1 Structurally Controlled Incised Valleys in the Belle Fourche Arch Area, Powder River Basin

5.6.1.1 Introduction

An extensive network of incised valley systems dissected the erosional surfaces of the western US foreland basins during Lower Cretaceous time. This drainage system produced a prolific hydrocarbon play that so far has yielded more then 1.5 billion barrels of oil-equivalent hydrocarbons (Dolson et al. 1991; Fig. 5.13). The prime reservoir rocks in these valley systems are members of the Muddy and Dakota Formations consisting of fluvial and transgressive marine sands. The evolution of this unique stratigraphic setting occurred in three phases beginning with a widespread deposition of marine sand during a sea-level high stand, followed by a down-cutting process during a sea-level low stand and culminated by a process of backfilling of the incised valleys with fluvial and marine strata during a sea-level rise (Weimer 1983, 1992; Fig. 5.14).

The development of the incised valley systems in the western USA, in general, and in the Powder River Basin, in particular, was strongly influenced by two types of structural/topographic features illustrated in Figs. 5.13 and 5.14. The first type consists of large-scale, positive topographic features (such as the Belle Fourche Arch) which formed the drainage divide in the region. The second type consists of deep-seated fault systems and related drape structures which locally control the orientation and level of incision of the valleys in some areas. Both of these elements and their control on the process of incision have been described previously in several basins of the western United States (e. g. Weimer 1980; Stone 1972). Slack (1981, p. 730), for example, concluded his analysis of the Powder River Basin (and the Belle Fourche Arch in particular) by stating that "Virtually all stratigraphic production from the diverse assemblages of depositional environments, representative of seven formations can be directly related to subtle, repeated movement along newly identified structural trends".

The spatial relationships between these structural/topographic features, their association with deep-seated basement structures and their control on the producing incised valley systems will be reexamined in the next section in light of the proposed integrated approach and related nomenclature of BWSs. The study area is outlined in Fig. 5.13. It includes the northwestern half of the Belle Fourche Arch and the two prolific channel systems that outline its opposing limbs. Note that (1) a producing incised valley of the Dakota formation, which was independently mapped by Exxon explorationists, was also included in this map; and (2) the location of four BWSs are highlighted: the Belle Fourche Arch, the Hilight Dome, the Little Missouri Lineament and the Clareton Lineament.

5.6.1.2 Surface and Subsurface Expressions of the Belle Fourche Arch

A surface geological map of the entire basin and a Dakota structure contour map of the study area can be used to illustrate the general characteristics of the Powder River Basin (Fig. 5.15 a, b). The surface geological map shows a basin which is outlined by a typical elliptical outcrop pattern of rock units that dips towards the center of the basin where younger rock units outcrop. The outcrop pattern reveals (1) a strong asymmetry to the east where the outcropping units and the related cuestas are closely spaced and follow the faulted edge of the Bighorn Mountains; and (2) a noticeable difference in the spacing and orientation of the outcropping units of the western side of the basin. These changes occur along the Belle Fourche River, suggesting some type of local variation in the structural setting of the basin in this area. The structure contour map of the study area, however, shows no apparent influence of deep-seated structures on the producing horizon, suggesting that any such features must be extremely subtle.

The possible relationships between BWSs and production for the study area is illustrated in Fig. 5.16. The Belle Fourche Arch is expressed as a profound, north-trending positive gravity and magnetic feature that bifurcates to the south into north- and northeast-trending segments. It is clearly separated from the positive gravity and magnetic expression of the Black Hills by a profound, elongated

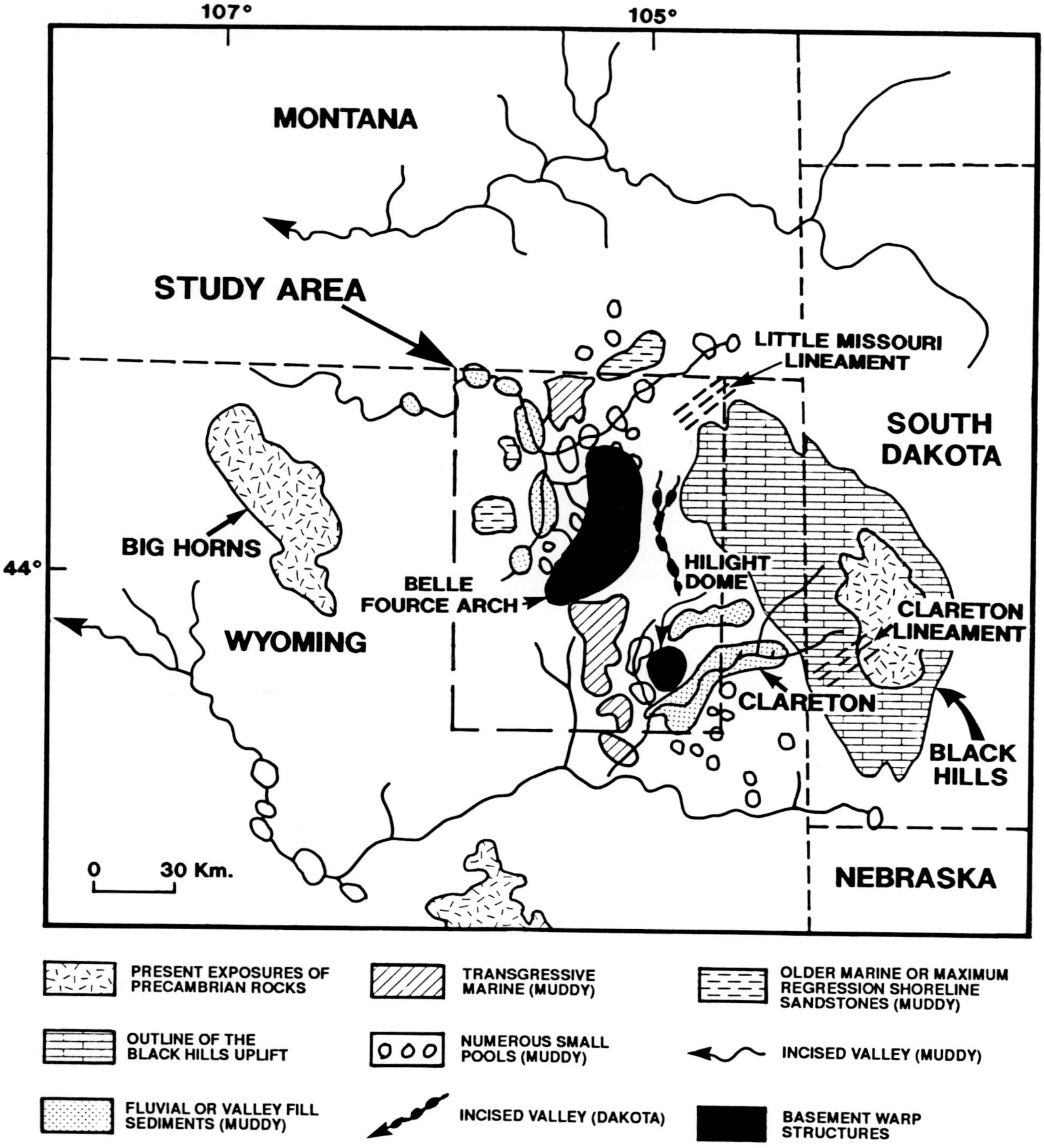

Fig. 5.13. Belle Fourche incised valley system

A paleogeographic reconstruction of the Belle Fourche Arch, Muddy and Dakota incised valley systems and related production (compiled and modified from Dolson et al. 1991). The *box* shows the location of the study area. Muddy incised valley channels were added from Exxon's internal database

Fig. 5.14. Model of incised valleys
A three-stage evolutionary model of incised valleys applicable to the Muddy Formation. *T1* and *T2* represent high stand, *T3*, low stand and *T4*, sea-level rise. (Weimer 1983; published with permission of the AAPG)

Fig. 5.15. Structural characteristics of the Powder River Basin
Left A surface geological map of the Powder River Basin. *Right* A structure contour map of the top of the Dakota formation in the study ara. The surface geological map shows the asymmetry of this basin and the influence of the Belle Fourche Arch on its outcropping pattern. The structure contour map shows a relatively unstructured basin and no apparent expression of the Belle Fourche Arch. (Courtesy of Exxon Exploration Co.) ▼

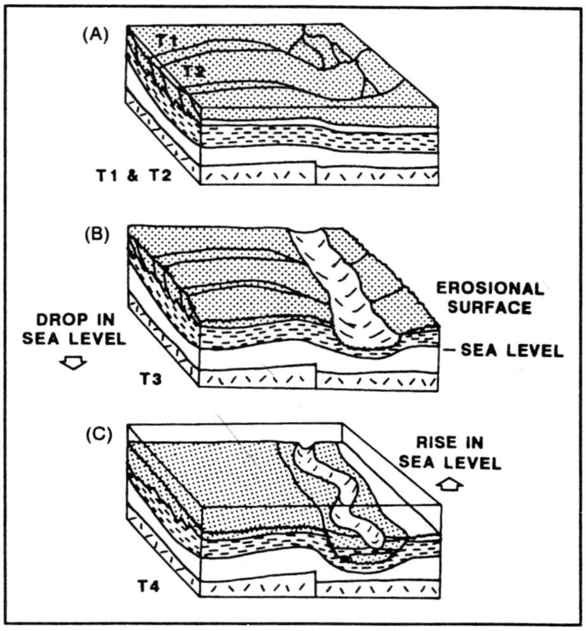

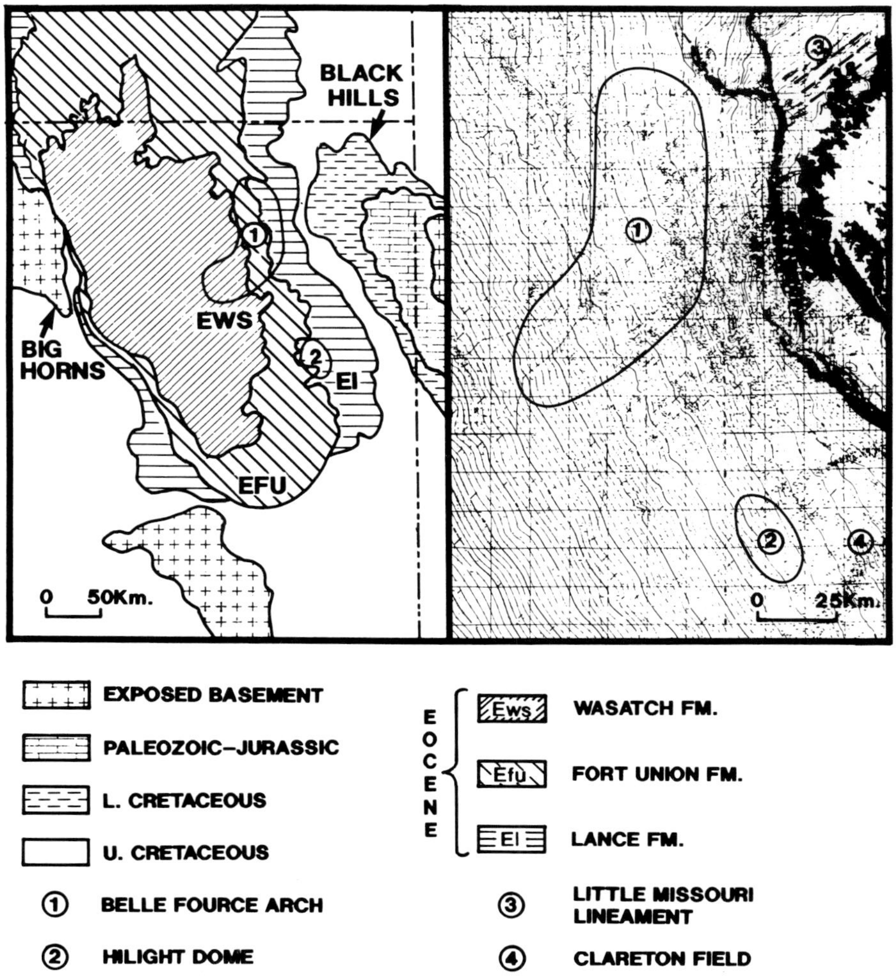

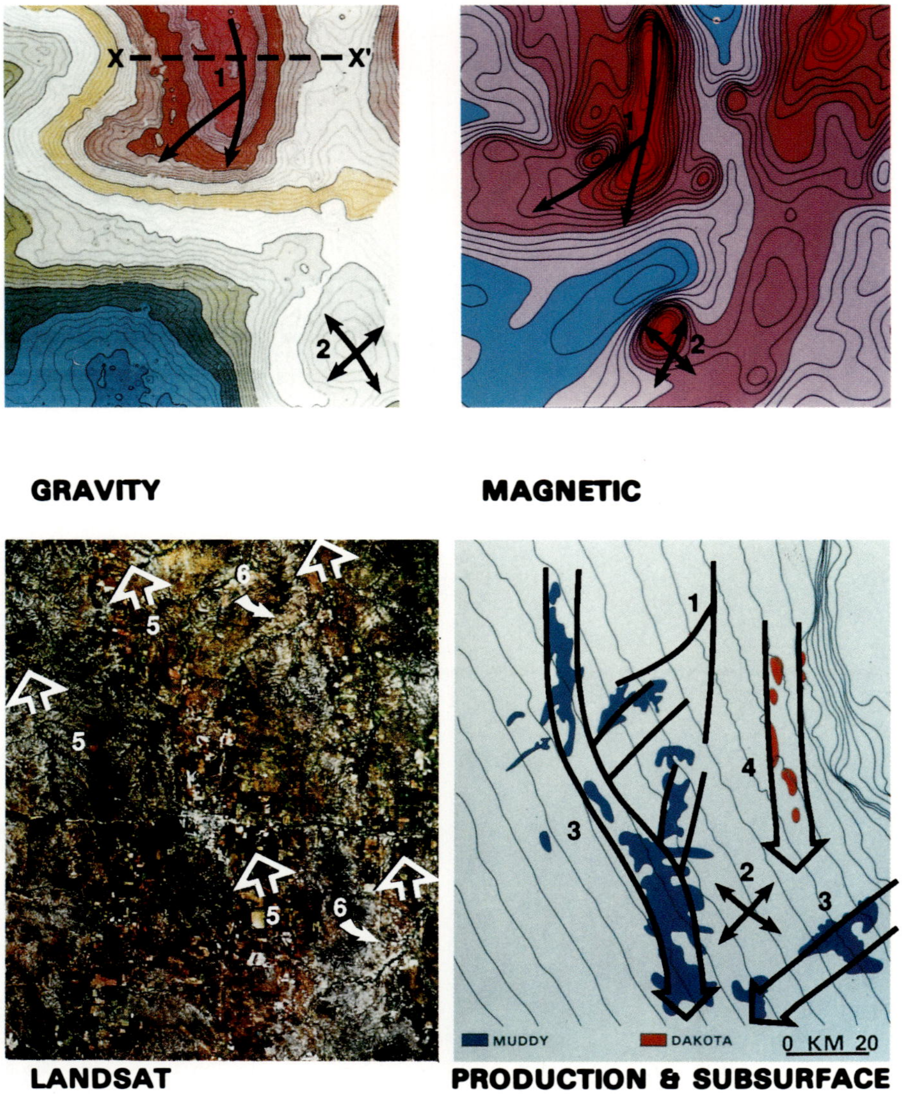

Fig. 5.16. Expression of the Belle Fourche Arch
Gravity, magnetic, satellite imagery and production data from the Belle Fourche Arch study area. The relationship between various gravity and magnetic features and Muddy and Dakota incised valley systems are shown. X-X' shows the location of the seismic line in Fig. 5.17. *1* Axis of the Belle Fourche Arch; *2* axis of the Hilight Dome; *3* Muddy Channel; *4* Dakota Channel; *5* topographic cuesta and subsequent valleys; *6* antecedent stream valley. (Gravity and magnetic data courtesy of Orxy Exploration Co. and Newmag Inc., respectively)

gravity and magnetic negative feature. The southernmost part of the study area is characterized by the presence of a circular, domal gravity and magnetic feature which has been named the Hilight Dome. The satellite image of the study area shows the presence of profound cuesta topography. Asymmetric topographic ridges and related subsequent stream valleys seem to develop along the boundaries of the Belle Fourche Arch. These ridges are dissected by transverse stream valleys with profound cone-shaped valley in two areas that spatially coincide with the location of the Little Missouri and the Clareton Lineaments. Producing Muddy and Dakota Valleys seem to develop as subsequent valleys flowing in a southerly direction around the Belle-

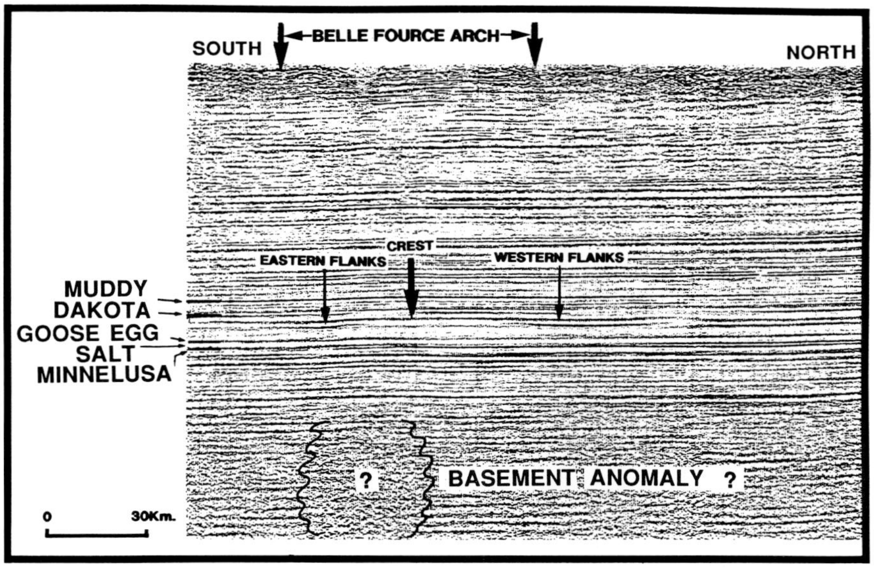

Fig. 5.17. Seismic expression of the Belle Fourche Arch A squeezed seismic line across the Belle Fourche Arch shows the presence of a subtle basement feature that may reflect the roots of this arch. Also shown are subtle topographic drape features that develop over the basement feature. (Seismic data courtesy of Exxon Exploration Co.)

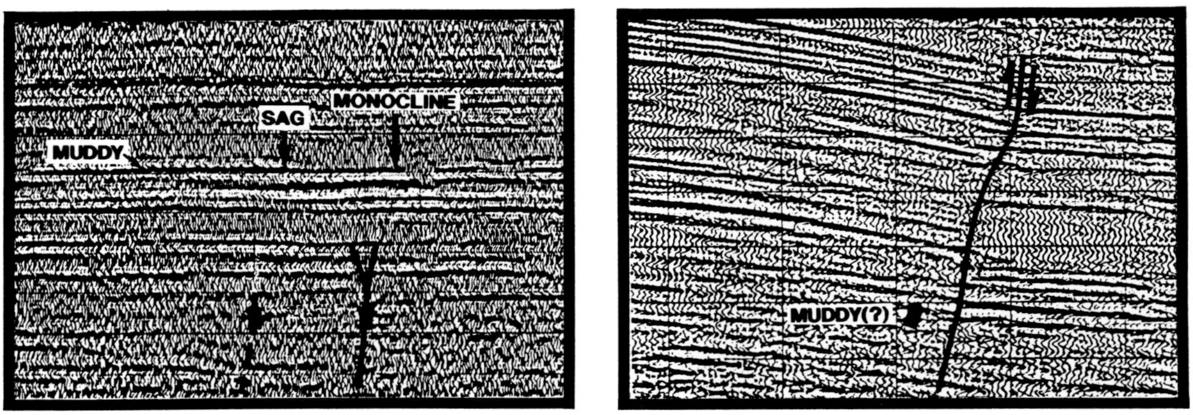

Fig. 5.18. Seismic expression of the Little Missouri Lineament Two squeezed seismic lines across the Little Missouri Lineament illustrate its profound structural expression **a** at the margin of the basin in contrast to its subtle expression in **b** the basin's interior. (Seismic data courtesy of Exxon Exploration Co.)

Fourche Arch and the Hilight Dome. Two smaller channels cut across the arch and the domal feature following the location of the Little Missouri and Clareton lineaments.

The seismic line across the Belle Fourche Arch shows the presence of a basement anomaly (a chaotic zone of seismic reflections) and a subtle drape feature of the overlying strata just above the basement feature (Fig. 5.17; see location of this seismic line in Fig. 5.16). The Little Missouri Lineament is expressed on seismic data as a profound, high-angle reverse fault (Laramide reactivation?) with a considerable vertical throw (Fig. 5.18 a). The ex-

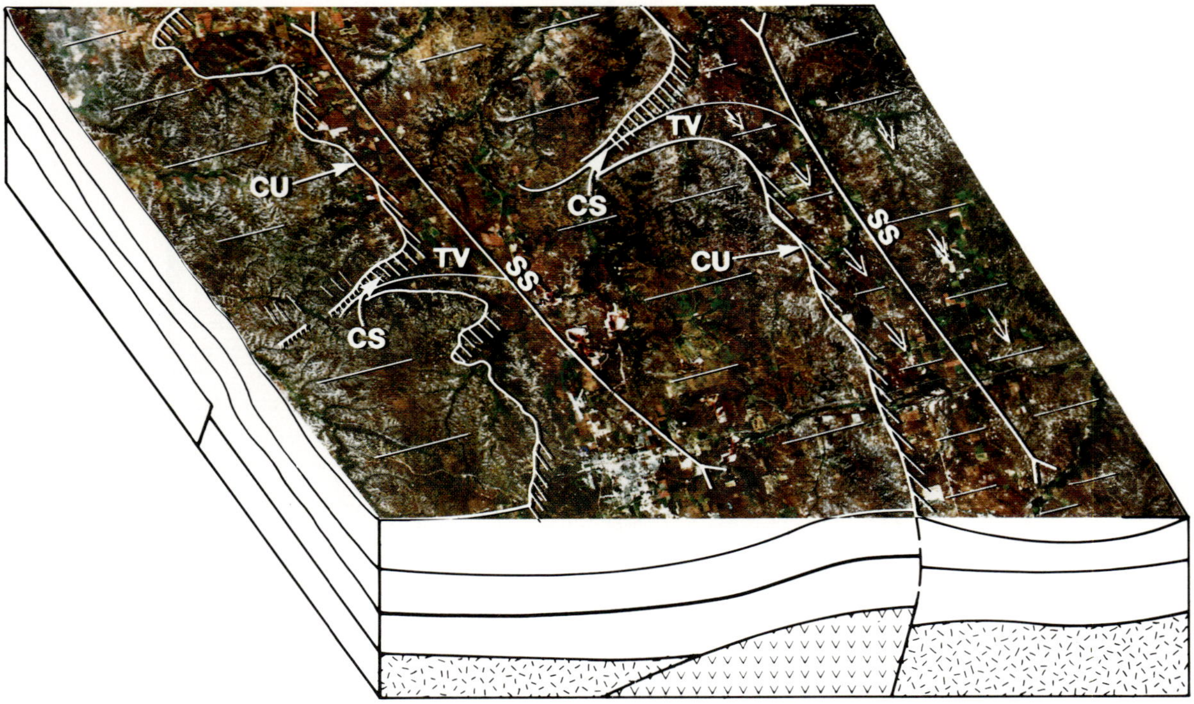

Fig. 5.19. Surface expression of the Belle Fourche Arch
The block diagram shows the development of cuesta topography over the central part of the Belle Fourche Arch. This relationship must have persisted from Lower Cretaceous time to the present day. *TV* transverse valley; *CS* cone-shaped valley; *SS* subsequent stream; *CU* cuesta

pression of this fault, however, diminishes rapidly towards the basin's interior where it is expressed as an extremely subtle, high-angle basement fault and a related flexure of the shallower section (Fig. 5.18 b). The Clareton lineament exhibits similar seismic expressions but is not illustrated here.

The present geomorphic expression of the Belle Fourche Arch and its relationship to the deep basement structures and overlying strata are illustrated by the block diagram in Fig. 5.19 which is centered on the crest of the Belle Fourche Arch and the cross-cutting Little Missouri Lineament. The arch is expressed at the surface as a subtle monoclinal flexure that developed over a tilted basement fault block. To the west, it is bounded by a profound linear topographic cuesta that developed over the faulted boundary of the arch.

A profound incised valley developed west of the cuesta following, more or less, the same orientation and location as the Dakota Valley shown in Fig. 5.13 and 5.16. A second, eastern cuesta that develops along the eroded limbs of the arch is presently occupied by incised valleys that follow, more or less, the location of the Muddy channel. Both cuestas are breached by erosion in areas where they are cross-cut by the Little Missouri Lineament showing the typical cone shape of a transverse incised valley.

5.6.1.3 Geological Model of the Belle Fourche Arch

Using the surface and subsurface information described above, it is possible to construct a geological model that illustrates, in greater detail than the available literature, the structural and geomorphic expression of the Belle Fourche Arch, its possible origin in the basement and its paleogeomorphic expression during Early Cretaceous time (Fig. 5.20).

Within the study area, the Belle Fourche Arch is interpreted as a north- and northeast-trending, first-order magnitude, basement warp structure that, most likely, reflects an uplifted and tilted basement fault block (similar to those which are exposed in the western United States; e. g. the Bighorn and Black Hills uplifts). South of the arch, a solitary, second-order, basement warp, domal feature is observed and is referred to as the Hilight Dome. The through-going, northeast-trending Little Missouri and Clareton lineaments are thought to reflect a

Fig. 5.20. Geological model of the Belle Fourche Arch area
Shown are major basement warp structures, lineaments and related incised valley systems which were recognized through the integrated analysis of satellite imagery, gravity, magnetic and other subsurface information

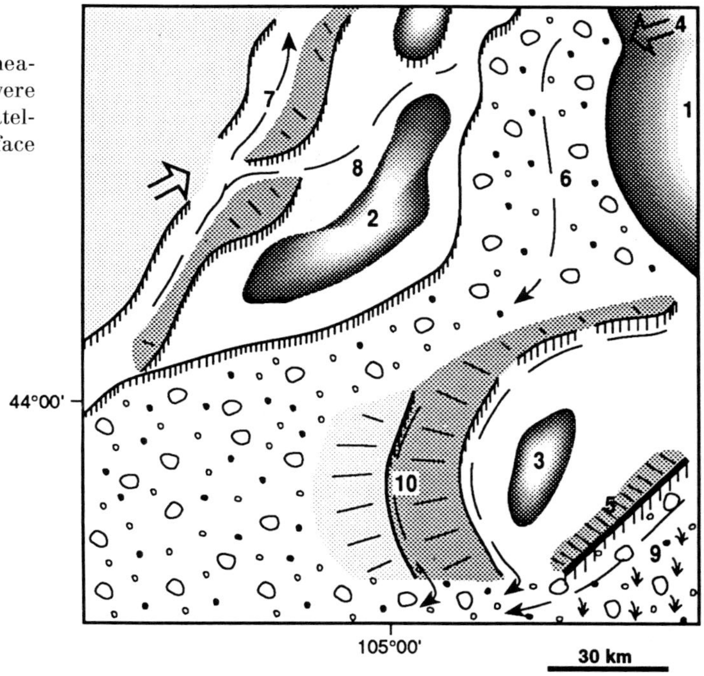

1	BLACK HILLS	6	DAKOTA SUBSEQUENT VALLEY
2	BELLE FOURCHE ARCH	7	MUDDY SUBSEQUENT VALLEY
3	HILIGHT DONE	8	L. MISSOURI VALLEY
4	LITTLE MISSOURI LINEAMENT	9	CLARETON VALLEY
5	CLARETON LINEAMENT	10	HILIGHT SUBSEQUENT VALLEY

Precambrian zone of weakness in the crystalline basement, showing evidence of minor structural reactivation.

During Early Cretaceous time, the arch was manifested as a positive topographic ridge showing an early stage of breaching with well-developed cuesta topography. Between the western faulted flanks of the arch and the Black Hills Uplift, a well-developed incised subsequent valley existed and was filled with sediments during Dakota time. This topographic feature is referred to here as the Dakota Sag. The cuesta topography of the eastern flank of the arch led to the development of subsequent valleys which were filled with sediments during Muddy time. There is evidence that small-scale, transverse valleys may have cut the arch along the Little Missouri Lineament and were later filled by Muddy sediments.

The Hilight Dome exhibits typical characteristics of a positive relief feature which was surrounded by radial subsequent valleys. Its eastern flanks became the focal point for the accumulation of the transgressive marine sands of the Hilight Field. The southern flanks of the dome may have been influenced by the presence of the cross-cutting Clareton lineament leading to the development of a profound transverse valley which was filled later by prolific Muddy sediments.

5.6.1.4 Exploration Application

The Belle Fourche Arch area has been densely drilled, and it is doubtful that the analysis presented here can lead to the recognition of new major incised valleys. A limited remaining potential in the arch area exists in small tansverse valleys that may have developed along smaller-scale lineaments which have not been recognized in this study or are too small to be detected by the reconnaissance tools used. In the less mature parts of the Powder River Basin, however, additional basement warp features can be recognized and used for further investigations. For example, Figs. 5.21 and 5.22 show the

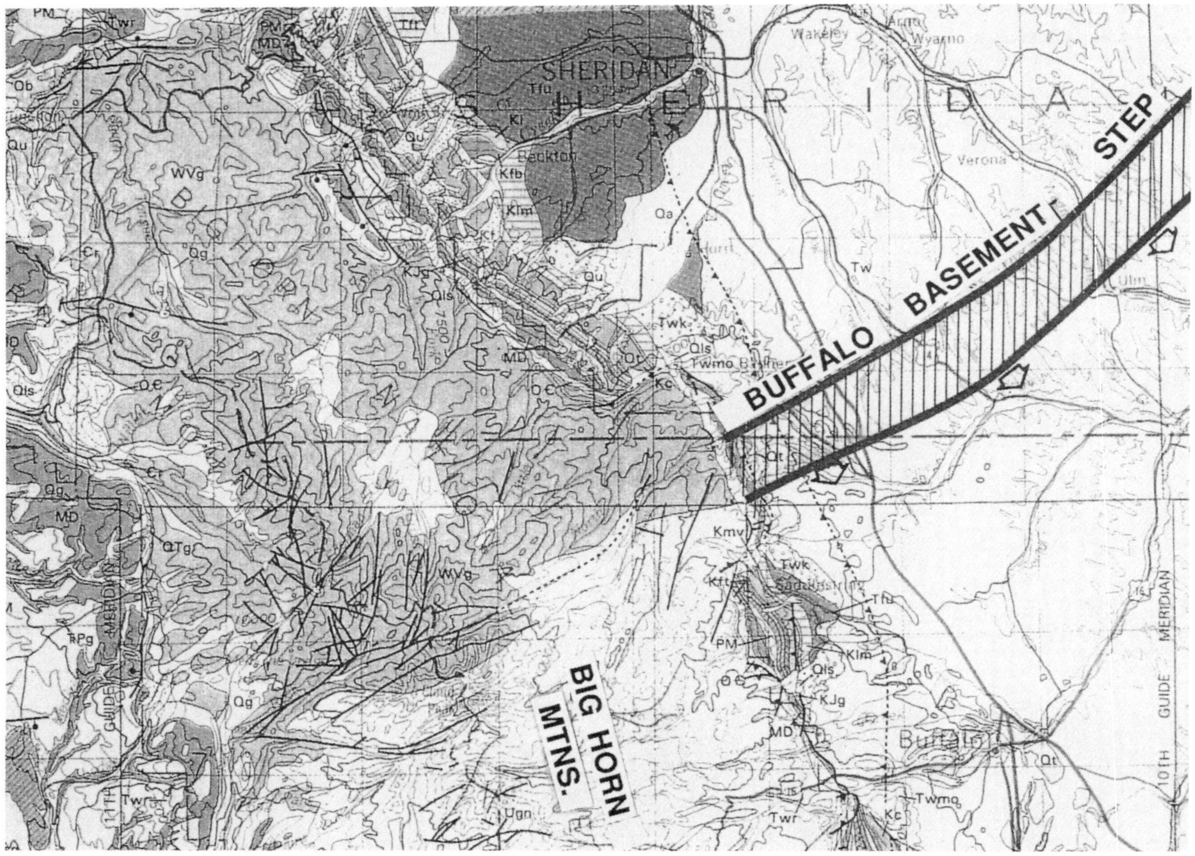

Fig. 5.21. **Geological expression of the Buffalo Basement Step** A surface geological map of the Bighorn Mountains showing the location of Buffalo Basement Step which is expressed as a major basement boundary bounded in part by a fault zone. The extension of the Buffalo Basement Step into the basin area, which is illustrated with satellite imagery in Fig. 5.22, is also shown. (Geological map from Love and Christiansen 1985, published with the permission of the US Geological Survey)

surface expression of a first-order northeast-trending lineament that developed along a major basement block boundary and is referred to as the Buffalo Basement Step. This lineament is well exposed at the Bighorn Mountains and manifests a strong topographic expression in the basin area, exerting significant control on the present drainage patterns and orientations. The present Powder River, for example, follows this lineament for over 200 km. Analysis of this lineament with well and seismic data could lead to the recognition of new incised valley features for further evaluation.

5.7 Summary

Basement warp structures constitute a unique branch of buried and obscured structures in that (1) they are more subtle and originate in the deep basement; (2) their detection requires the use of tools and techniques that are not routinely used in exploration; and (3) they usually are too subtle to form pure structural traps but, rather, exert significant influence on reservoir distribution and hydrocarbon accumulations in their vicinities. BWSs are the trademarks of many low relief basins and their influence on hydrocarbon plays is manifested in many ways, ranging from the development of fracture porous dolomite trends along active faulted boundaries to reefal growth along their fringes to the development of various clastic stratigraphic traps. The most important, and common, plays associated with these features are structurally controlled incised valley systems.

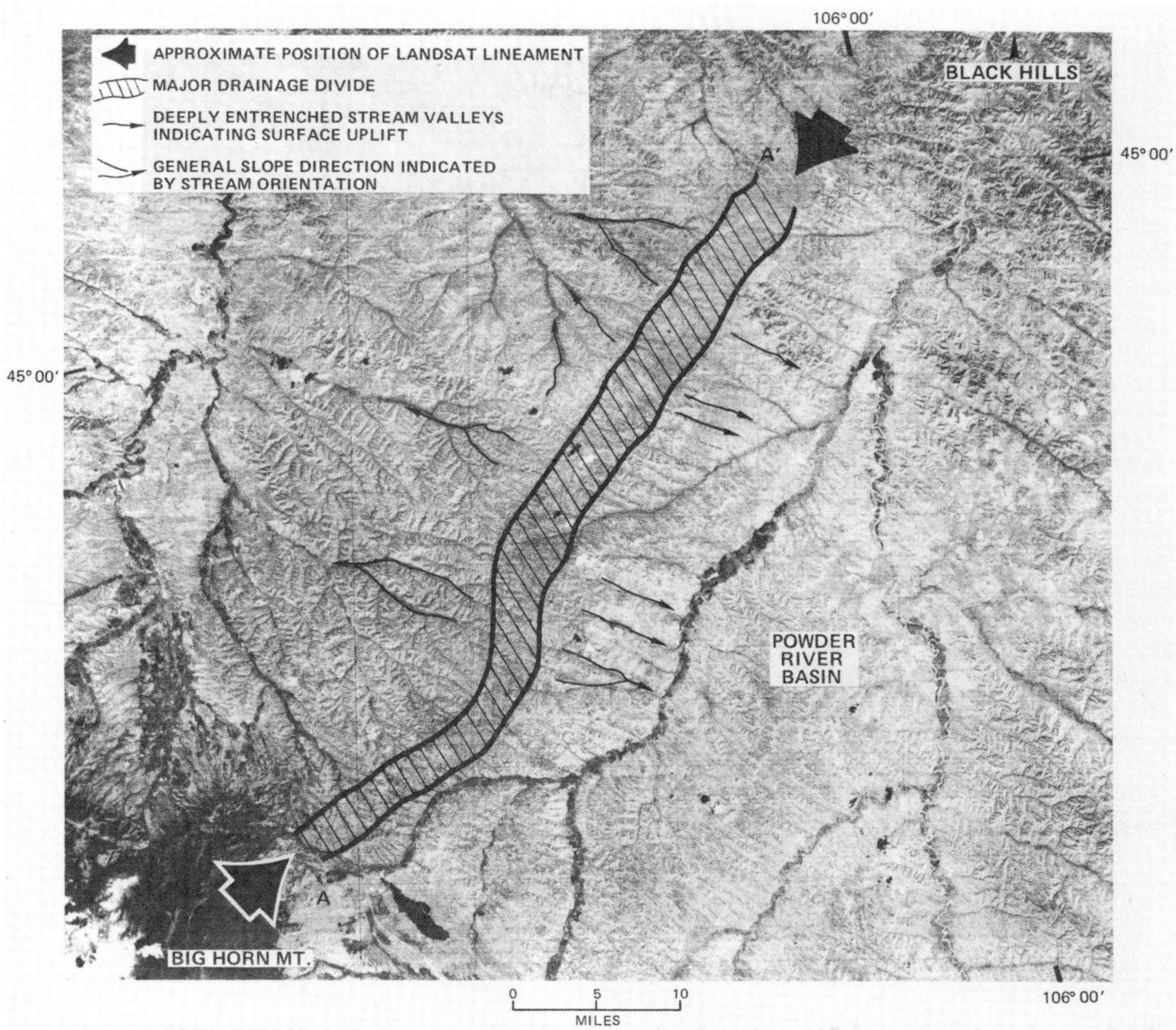

Fig. 5.22. Surface expression of the Buffalo Basement Step Landsat image showing the surface expression of the Buffalo Basement Step. This feature is expressed as a regional-scale topographic ridge which forms a major drainage divide in this basin. It is spatially coincident with the basement block boundaries shown in Fig. 5.19. (Note that the image is slightly tilted with respect to the map)

Detection and analysis of BWSs usually involve the integration of gravity, magnetic, and satellite imagery, known collectively as reconnaissance tools. Also frequently used are specially enhanced seismic lines and various paleogeomorphic reconstruction techniques. Because these tools each provide a nonunique solution, the interpretation relies on converging lines of evidence which are developed through the integration of all available data and the application of modern geomorphic analogs.

Exposed basement structures can be divided according to their size into three categories: first, second and third order. BWSs may be similarly divided, and their classification also indicates their detectability and their potential influence on hydrocarbon plays. The first two categories include domal features, arches and other positive elements which are usually large enough to be detected by available exploration tools. Their influence on hydrocarbon traps is quite noticeable. The third category consists of small-scale faults and fractures within the basement grain. Their detectability is more problematic, and their influence on hydrocarbon accumulation is quite limited.

The persistance of positive topographic expression of BWSs throughout the geological history of the basin leads to the development of a unique stacking of structurally controlled, incised valley systems. Two types of valleys can be recognized with subsurface investigation. Subsequent valleys are those which circum navigate the structures and transverse valleys are those which cut across their positive expressions. The development of these two types is largely dependent on the rate of uplift but they can both exist within the same region or within individual structures.

Integrated analysis of basement warp structures can be used in two stages of exploration. In frontier areas, such analysis provides new play concepts and leads and can be used to guide data collections and interpretations, particularly seismic surveys. In mature regions, the analysis can be used to expand existing plays in areas where subtle features have been overlooked by routine mapping. In either case, however, such analysis should be viewed as a reconnaissance mapping phase which should be followed by thorough investigation of well and seismic data and the implementation of sequence stratigraphic concepts.

References and Further Reading

Berger Z (1982) The use of Landsat data for detection of buried and obscured geological structures in the East Texas Basin, U.S.A. Proc 2nd Thematic Conf Explor Geol, Dallas, TX. Environmental Research Institute of Michigan, Ann Arbor, pp 453–464

Berger Z (1984) Structural analysis of low-relief basins using Landsat data. Proc Int Symposium on Remote sensing of the environment, 3rd Thematic Conf, Colorado Springs, CO, vol 1. Environmental Research Institute of Michigan, Ann Arbor, pp 251–273

Berger Z (1988) Detection and analysis of basement structures in low relief basins using an integrated analysis of Landsat data. AAPG Bull 72 (2): 160–161

Berger Z, Williams TH, Anderson DW (1991) Geologic stereo mapping of geological structures with SPOT satellite data. AAPG Bull 76 (1): 101–120

Bhattacharyya BK, Leu L (1975) Spectral analysis of gravity and magnetic anomalies due to two-dimensional structures. Geophysics 40: 993–1013

Blackwelder E (1920) The origin of the central Kansas oil domes. AAPG Bull 4: 89–94

Christner DG (1940) Todd Ranch (Oil) discovery, Crockett Country, Texas. AAPG Bull 24 (6): 1126–1127

Dolson J, Muller D, Evetts MJ, Stein JA (1991) Regional paleo-topographic trends and production, Muddy sandstone (Lower Cretaceous), central and northern Rocky Mountains. AAPG Bull 75 (3): 409–435

Eyal M, Bartov Y, Shimron AE, Bentor YK (1980) Geological map of the Sinai peninsula, Geol Surv Israel, Jerusalem, 1 : 500,000 scale

Gay SP Jr (1989) Gravitational compaction, a neglected mechanism in structural and stratigraphic studies: new evidence from mid-continent, U.S.A. AAPG Bull 73 (5): 641–657

Grant FS (1972) Review of data processing and interpretation methods in gravity and magnetics. 1964–71. Geophysics 37: 647–661

Halbouty MT (1983) The time is now for all explorationists to purposefully search for subtle traps. In: Halbouty MT (ed) The deliberate search for the subtle trap. AAPG Mem 32: 1–10

Halbouty MT, Halbouty JJ (1982) Relationship between the East Texas Field and Sabine Uplift in Texas. AAPG Bul 66 (8): 1042–1054

Hamblin WK, Howard JD (1980) Exercises in physical geology. Burgess, Minneapolis

Harding TP, Lowell JD (1979) Structural styles, their plate-tectonic habitats, and hydrocarbon traps in petroleum provinces. AAPG Bull 63: 1016–1058

Hennen RV, Metcalf RJ (1929) Yates Oil Pool, Pecos Country, Texas. AAPG Bull 13 (12): 1509–1556

Hobbs WH (1904) Lineaments of the Atlantic border region. Geol Soc Am Bull 15: 483–506

Jackson PC (1984) Paleography of the Lower Cretaceous Mannville Group of Western Canada, in Masters JA (ed), Elmworth – Case study of a Deep Basin Gas Field: AAPG Memoir no. 38

Keefer WR (1970) Structural geology of the Wind River Basin, Wyoming. US Geol Surv Prof Pap 495-D, p 35

Krebs W (1971) Devonian reef limestones in the eastern Rhenish Schiefergebirge. In: Miller G (ed) Sedimentology of parts of central Europe, guidebook. 8th Int Sed Congr Verlag Waldemar Kramer, Frankfurt a. Main, pp 45–81

Lisenbee AL (1978) Laramide structure of the Black Hills uplift, South Dakota-Wyoming-Montana. In: Matthews V III (ed) Laramide folding associated with basement block faulting in the western United States. Geol Soc Am Mem 51: 165–196

Love JD, Coe Christiansen A (1985) Geologic map of Wyoming. USGS, scale 1 : 500,000

Marrs RW, Raines GL (1984) Tectonic framework of Powder River Basin, Wyoming and Montana, interpreted from Landsat imagery. AAPG Bul 68: 1718–1731

Martin R (1966) Paleo-geomorphology and its application to exploration for oil and gas (with examples from western Canada). AAPG Bull 50: 2277–2311

Master JA (ed) (1984) Elmworth – case study of a deep basin gas field. AAPG Mem 38

Mehl MG (1920) The influence of the differential compression of sediments on the attitude of bedded rocks. Science, New Ser 51: 520 (Abstr)

Nettleton LL (1942) Gravity and magnetic calculations. Geophysics 7: 293–310

Nettleton LL (1976) Gravity and magnetics in oil prospecting. McGraw-Hill, New York

Nevin CM, Sherrill RE (1929) Studies in differential compaction. AAPG Bull 13: 1–22

Nichols PH (1964) The remaining frontiers for exploration in northwest Texas. In: Rhodes ML (ed) Trans Gulf Coast Assoc Geol Soc, 14th Convention, Corpus Christi, TX, pp 7–21

Slack PB (1981) Paleotectonics and hydrocarbon accumulation, Powder River Basin, Wyoming. AAPG Bull 65: 730–743

Steenland NC (1965) Oil fields and aeromagnetic anomalies: Geophysics 30: 706–739

Stone WD (1972) Stratigraphy and exploration of the Lower Cretaceous Muddy formation, northern Powder River basin, Wyoming and Montana. Mountain Geol 9: 355–378

Teas LP (1923) Differential compaction, the cause of certain Claiborne dips. AAPG Bull 17: 370–377

Telford WM, Geldart LP, Sheriff RE (1990) Applied geophysics, 2nd edn. Cambridge University Press, Cambridge, NY

Thomas GE (1974) Lineament-block tectonics: Williston-Blood Creek Basin. AAPG Bull 58: 1305–1322

Vail PR, Mitchum RM, Thompson S (1977) Seismic stratigraphy and global changes of sea level, part 3: relative changes of sea level from coastal onlap. In: Paytton CE (ed) Seismic stratigraphy – applications to hydrocarbon exploration. AAPG Mem 26: 63–97

Van Wagoner JC, Mitchum RM, Campion KM, Rahamanian VD (1990) Siliciclastic sequence stratigraphy in well logs, cores, and outcrops: concepts for high resolution correlation of time and facies. AAPG Methods Explor 7

Weimer RJ (1980) Recurrent movement on basement faults, a tectonic style for Colorado and adjacent areas. In: Kent HD, Porter KW (eds) Colorado geology. Rocky Mountain Assoc Geologists, pp 23–35

Weimer RJ (1983) Relation of unconformities, tectonics and sea level changes, Cretaceous of the Denver Basin and adjacent areas. In: Reynolds MW, Dolly ED (eds) Mesozoic paleography of west-central United States. SEPM Rocky Mountain section, Rocky Mountain Paleography Symp 2, pp 359–376

Weimer RJ (1992) Development in sequenced stratigraphy: foreland and cratonic basins, presidential address. AAPG Bull 67 (7): 965–982

Weimer RJ, Emme JJ, Farmer CL, Anna LO, Davis TL, Kidney RL (1982) Tectonic influence on sedimentation, Early Cretaceous east flank, Powder River Basin, Wyoming and South Dakota. Colorado School Mines Q 77 (4): 61 pp

Chapter 6 Interpretation Techniques: Structural Mapping with Stereo Data

6.1 Introduction

All of the structures encountered so far in this book have been analyzed using monoscopic imagery data. It was noted, however, that the accuracy of reconstructing exposed geological structures diminishes as they approach the extremes of being highly or mildly deformed. In these cases, the variable relationships between geological structures and their topographic expressions as well as the lack of diagnostic features from inclined bedrock strata create a need for some sort of stereo mapping. Such efforts can be done by using satellite stereo data directly or by supplementing monoscopic interpretation of satellite imagery data with local mapping of structures with stereo aerial photography or radar. This chapter provides a brief introduction to the principles involved in creating stereo imagery data, specific applications to structural mapping and the equipment used for this analysis. Further discussions related to stereo photography equipment and general applications in geology may be found in Ray (1960), Wolf (1974), Slama (1980) and Petrie (1992).

6.2 Principles of Stereo

6.2.1 Concepts of Visual Depth Perception

Human vision has the ability to discern three dimensions because of the principle of retinal disparity (Fig. 6.1). The separation between the two eyes is commonly called the eye base *(B_e)* and is, on average, 6 to 7 cm. As the figure shows, the rays from two points with a depth separation enter through the pupil and strike the retina. The eyes tilt to place one of the points in focus so that it strikes the center of the retina in both eyes. Because the second point is further away from the eye, its image appears displaced from the center of the retina a certain distance, in opposite directions in each eye. It is this displacement that allows the brain to perceive a

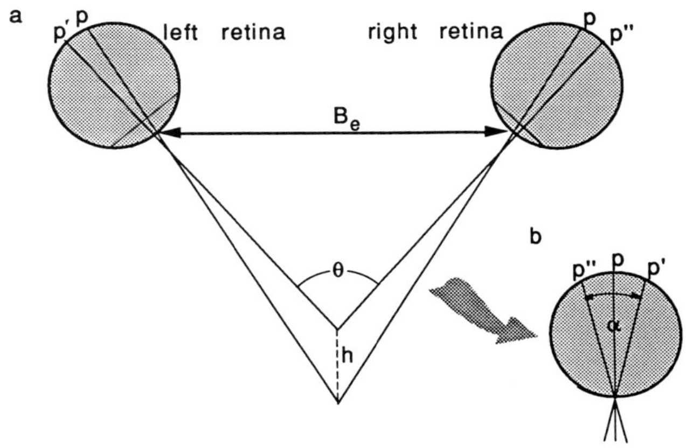

Fig. 6.1. The principle of retinal disparity
When the eye is focused on one point, a point at a different distance from the eye base is projected onto the retina offset from the first point. (After LaPrade 1980)

third dimension.[1] Figure 6.1 b shows the two eyeballs superimposed. The angle indicated as α is called the angle of disparity (or the parallatic angle) and is used by the brain to determine the distance from the object to the observer.

In the brain, the two retinal images are fused to form one image with a third dimension. If the disparity angle is too large (i. e., the object is very close to the observer) the brain is not able to fuse the two images and the object will appear to be double. (Try holding a pencil at arm's length and then focusing past it on a distant object.) Likewise, there is a maximum limit to human depth perception. The smallest discernable disparity angle is about three seconds of arc giving most people a range of 400 m

[1] The lines from the object to the center of the retina subtend the angle of convergence, θ. Detecting the muscular tension required to maintain the angle of convergence in the eyes provides another cue to sensing depth.

within which depth can be perceived. Beyond the maximum range of 400 m or without a dual image, the mind uses other methods to determine depth. Such monocular cues include overlapping, familiarity with object sizes, atmospheric fading of distant objects and perspective effects.

The only monocular cue that applies to simulated stereoscopic viewing is the focusing accomodation effect of the eye. When fixed on a point, the lens of the eye adjusts focal length so that a clear image is projected onto the retina. Objects at different distances will appear out of focus or require adjustments in the lens to be discerned. Sensing the different amount of tension required to focus the lens gives an indication of distance.

6.2.2 Concepts of Simulated Depth Perception

The most common way to simulate depth perception and measure altitude differences on an image is to compare two photographs taken of the same site but from different orientations so as to produce a parallax. Figure 6.2 shows common terminology associated with stereo pair production. The configuration shown in Figure 6.3 is the one used for satellite stereo.

The air base *(B_a)* is the distance between the two positions of the camera and determines the amount of parallax *(p)* created. Each image is then presented to its respective eye (a later section covers methods used to achieve this). yielding a perceived three-dimensional object. Shown is the ideal case where the stereo pair yields a model of the exact same proportions. When viewed, the object appears to "float" in space between the two photos and at a certain distance from the eye base called the apparent object distance, *D*. The apparent distance can be greater than, equal to or less than the distance from the eyes to the actual photos.

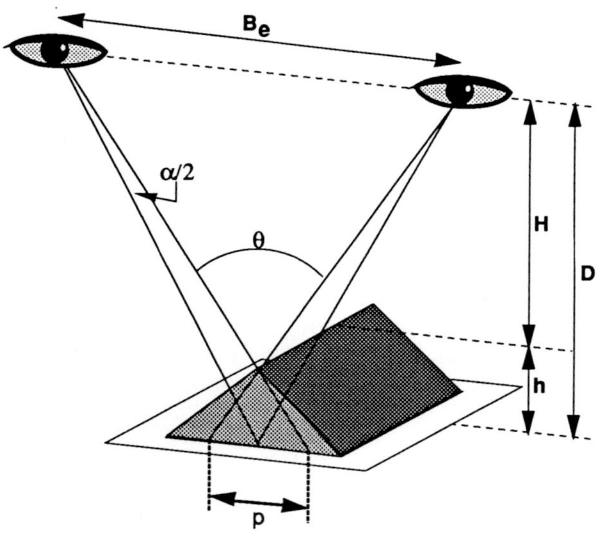

Fig. 6.2. Stereo pair viewing
When an object is viewed from two separated locations, a parallax, *p*, is produced. θ is the angle of convergence and α is the angle of disparity

6.2.3 Common Stereoscopic Distortions

6.2.3.1 Vertical Exaggeration

Figure 6.3 shows the simplest situation where the resulting image has the exact same proportions as the original. This is often not the case. Maintaining proper proportions requires that the ratios of altitude to air base and apparent distance to eye base be equal:

$$\frac{D}{B_e} = \frac{A}{B_a} \tag{6.1}$$

For most geological purposes, it is not advantageous to have correct height-to-base relationships because common reliefs would appear very flat for the scale of the map. To aid in calculating slopes and relative heights, stereo images are often created with vertical exaggeration distorting an object to appear higher than it actually is. Vertical exaggeration, *E*, is defined as follows:

$$E = \frac{\text{perceived height-to-base ratio}}{\text{actual height-to-base ratio}} \tag{6.2}$$

and

$$E = \frac{D}{B_e} \cdot \frac{B_a}{A} \tag{6.3}$$

These relationships can easily be obtained by comparing similar triangles in the geometry. A complete derivation can be found in Wolf (1974).

Because the perceived height of an object depends directly on the parallax, it is clear that increasing the air base without changing altitude will cause a vertical exaggeration. Of course, adjusting the left side of Eq. (6.1) can also cause vertical exaggeration. However, studies have shown (e. g. LaPrade, 1972, 1973) that the best perception of depth occurs when the apparent object distance *(D)* is approximately five times the eye base. (Unfortunately,

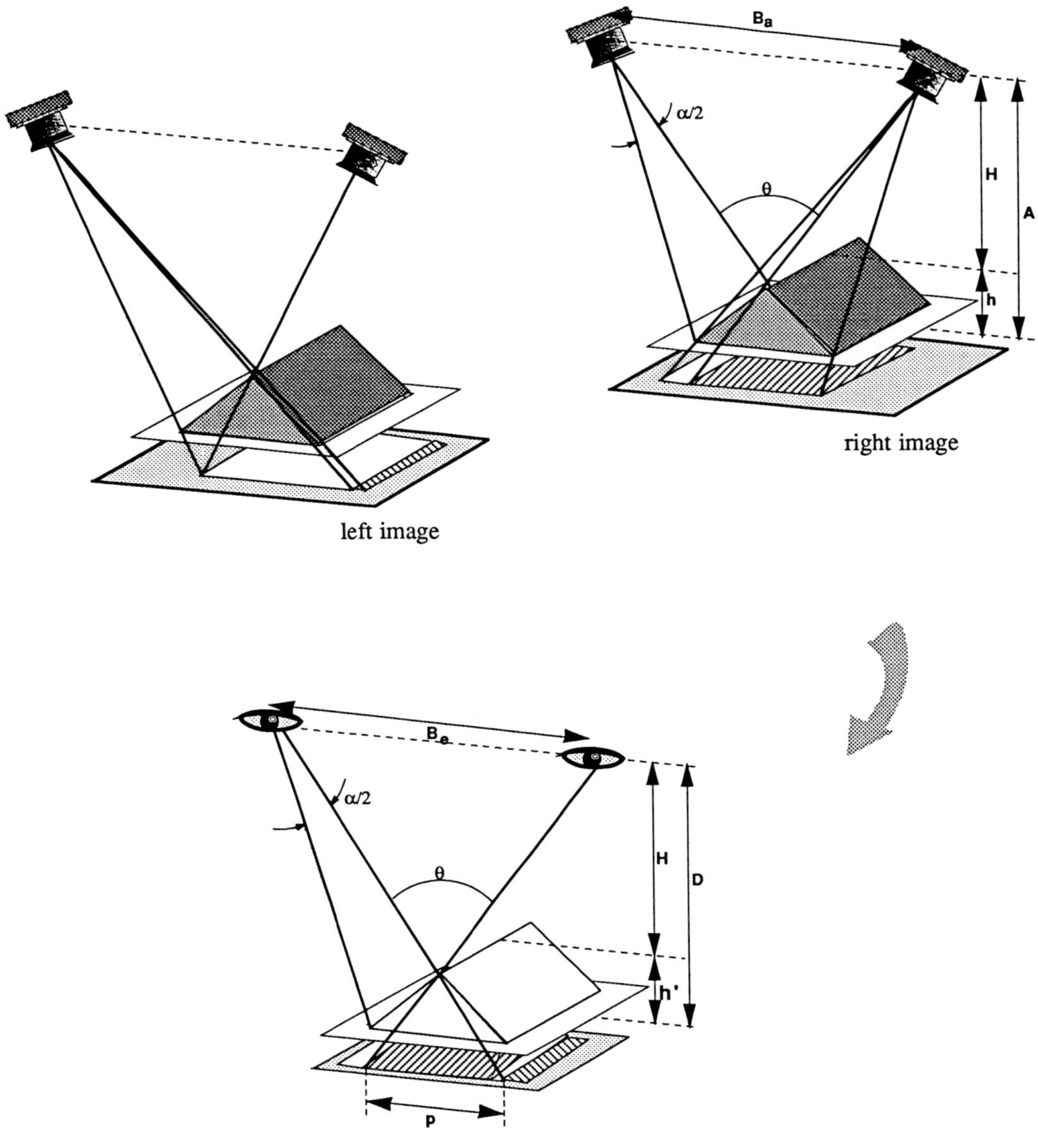

Fig. 6.3. Fundamentals of stereo pair production
Each image of the symmetric mountain (height h) is distorted due to the parallax. When the images are presented to their respective eye, the brain reproduces the three-dimensional, symmetric mountain (height h'). Cameras are shown tilted to illustrate the production of stereo with SPOT's off-nadir capability. This is not the case for traditional photography

we cannot adjust our eye base.) Introducing this constant term in Eq. (6.4),

$$E \approx 5\frac{Ba}{A} \tag{6.4}$$

6.2.3.2 Pseudoscopic Illusions

The common phenomenon of pseudoscopic illusion occurs as a result of two situations. In the first case, left and right images are presented to the opposite eye. This yields the perception of depth as inverted making positive relief appear negative – mountains

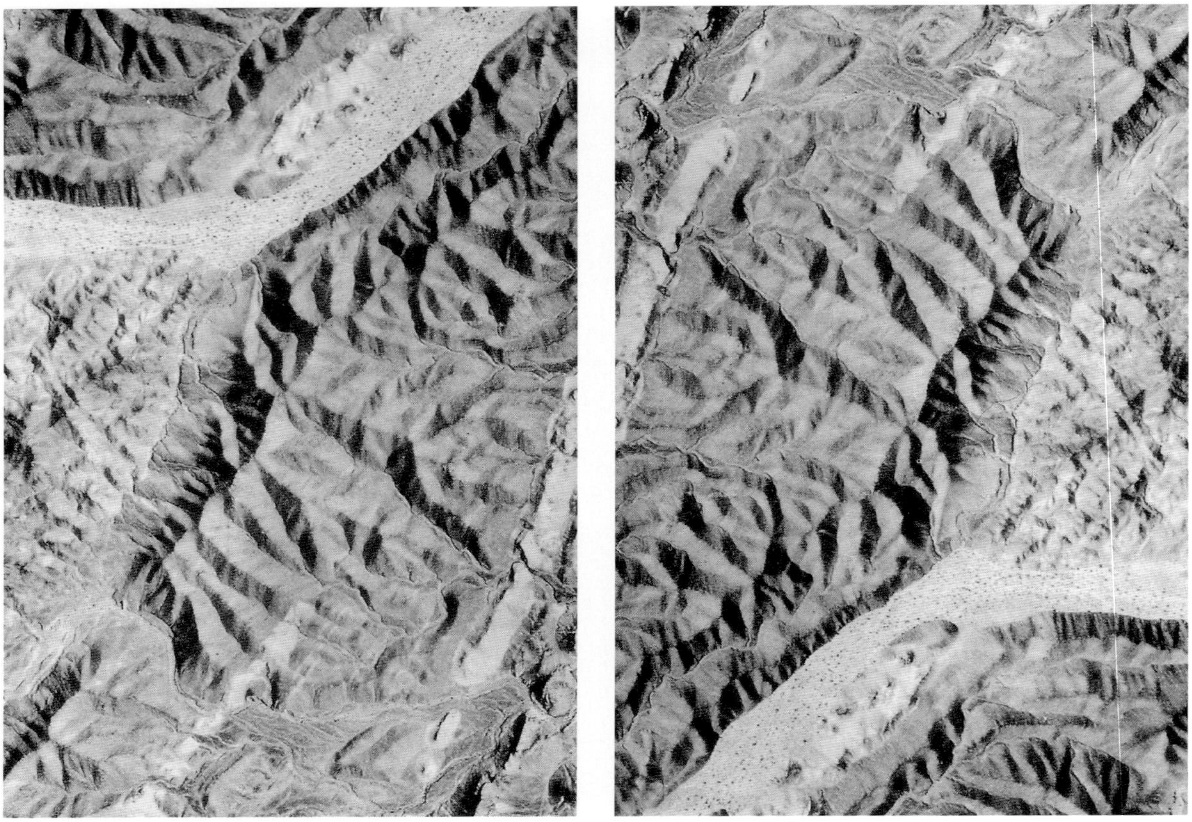

Fig. 6.4. The pseudoscopic effect
Because of the brain's predisposition to an overhead light source, images with shadows coming towards the viewer appear inverted. The *right* image shows the pseudoscopic view where the uplifted blocks appear as topographic depressions. The correct view of uplifted tilted fault blocks can be seen in the *left* image

become valleys. This problem is readily avoided by aligning pictures correctly (as described presently). The second manifestation of the illusion occurs when pictures are aligned so that shadows fall away from, instead of towards, the viewer (Fig. 6.4). Because illumination is encountered most often from above in the everyday world, an image with illumination from underneath causes havoc with interpretation. (Approximately three out of ten people are able to view pseudoscopic images correctly.) This problem is also easily rectified by rotating the image until the correct perspective view is seen.

6.3 Principles for Geological Mapping with Stereo Data

6.3.1 The Floating Dot Principle and Parallax Bars

Several methods for making measurements from stereo pairs are commonly used. Most are based on creating a "floating dot" that can be moved in three dimensions over the apparent image. Using the notation from Fig. 6.3, the perceived height difference of points is expressed in Fig. 6.5 as,

$$h = \frac{A\Delta p}{B_a} \tag{6.5}$$

where Δp is the difference in parallax and h is the vertical separation. A simple tool called a parallax bar is used to generate the floating dot. It consists of two, connected transparent plates each of which has a dot (or cross hairs) engraved on it. Each plate is placed on one photo and a finely adjustable dial allows adjustment of their separation, thus giving a measure of parallax.

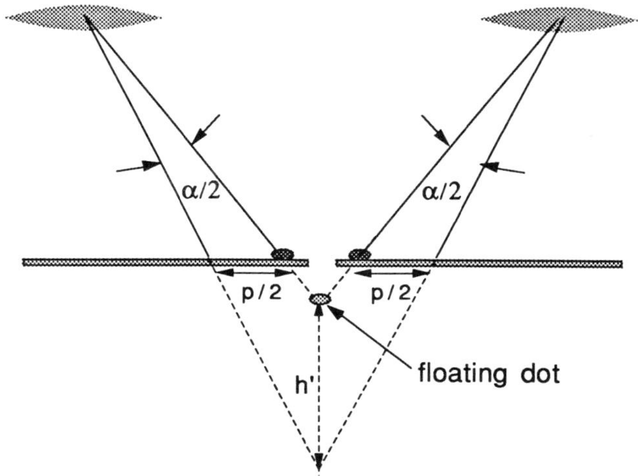

Fig. 6.5. Geometry of a floating dot
Adjusting the space between the dots makes the apparent dot rise and fall with respect to the apparent stereo image

6.3.2 Quantitative Measurements of Dips and Strikes

Dips and strikes can be measured from any image that has elevation data and a well-defined lithostratigraphic unit (often referred to as a marker bed). Using the three-point system illustrated in Fig. 6.6, the interpreter uses a floating dot to identify two points with the same elevation that will constitute the strike line. A third point is chosen on a line (the dip line) bisecting the strike line. The dip is then found as follows:

Fig. 6.6. The three-point system
Points A and B are at the same elevation, and the line connecting them forms the strike line. A line perpendicular to the strike then indicates the dip direction

$$\text{dip} = \arctan \frac{\text{elevation change along dip line}}{\text{length of dip line}} \tag{6.6}$$

The accuracy of dip and strike measurements from the three-point system depends on two main factors: the precision of the elevation measurements themselves (which are dependent on the remote sensing data and equipment used), and the accuracy of determining the bedrock surface (dependent on the level of bedrock exposure). Both of these elements will be examined later in the section dealing with evaluation of various stereo data.

6.3.3 Creating a Form-Line Surface Structure (FSS) Map

One of the prime goals behind stereo mapping of geological structures is the construction of their precise geometries at the surface. Such measurements are particularly warranted over and around structures that do not manifest clear expressions of inclined bedrock strata or are characterized by extremely complex geometries. In these cases, the structures have shapes, sizes and geometries that cannot be fully evaluated. A form-line surface structure (FSS) map describes the shape of the structure as it is constructed on top of a key stratigraphic unit (which constitutes the marker bed). The process of constructing an FSS map consists of four steps which are illustrated in Fig. 6.7. The block diagram in the center of the figure should be considered as the three-dimensional view of the stereo data, whereas the other maps show the information derived at different stages of the process.

During the first step, the interpreter identifies and maps major lithological contacts and relates them to the known geological units (if available). During the second step, dip and strike measure-

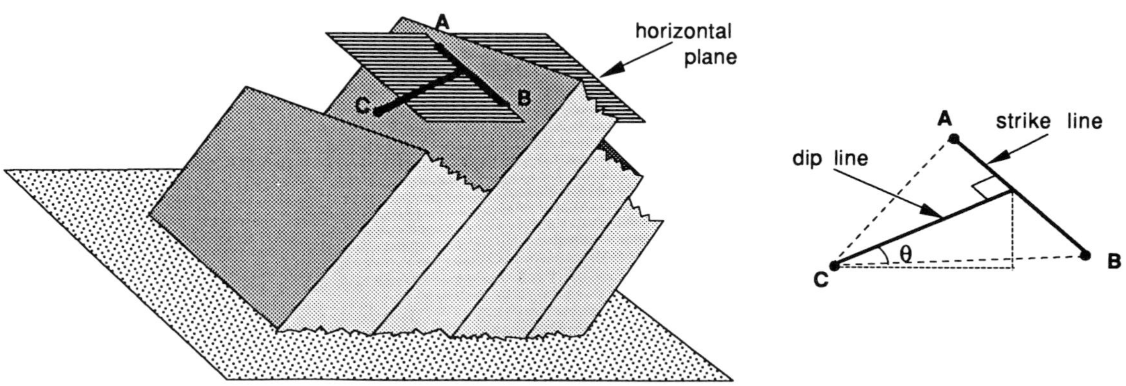

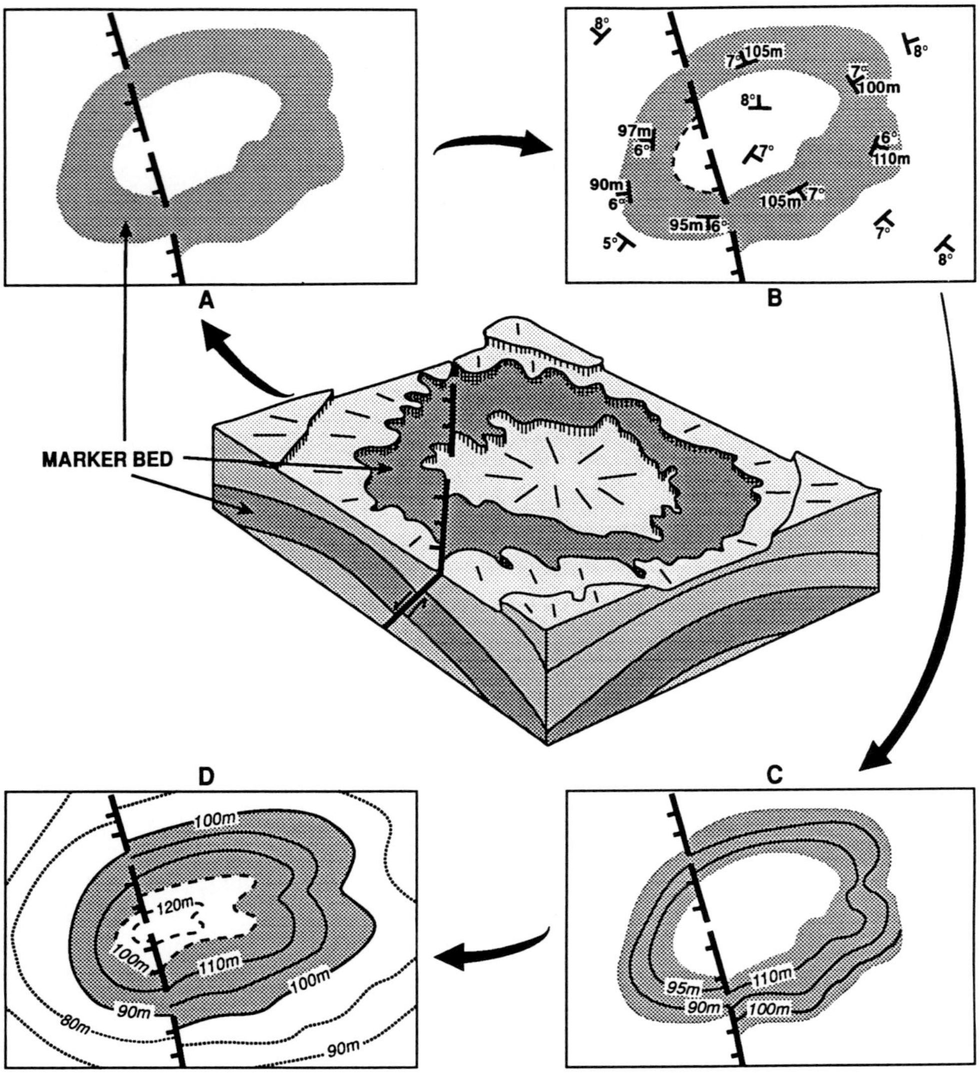

Fig. 6.7. Creating a form-line surface structure (FSS) map
A block diagram of a breached dome is used to demonstrate the four steps involved in the reconstruction of FSS maps. **A** Lithological units are identified on the map; **B** dip and strike measurements are obtained; **C** the geometry of a key exposed marker bed is constructed; and **D** the remaining eroded segments *(dotted contours)* and buried segments (in *dashed contours*) are interpolated

ments are obtained over and around the structure particularly in places where well-defined bedrock units can be identified. In the third step, structure contour lines are drawn along a well-exposed marker bed. The shape, spacing and elevation of the contour lines are guided by the dip and strike measurements (i.e., orientation, magnitude, and elevation) which were obtained in the previous step. During the final step, the buried and eroded segment of the structure is reconstructed using dip and strike measurements of bedrock strata above and below the marker bed. Faults with detectable offsets can also be displayed on this map and their exact vertical displacement may be further refined by measuring elevation differences along their FLTs. The accuracy of the reconstructed maps and the level of structural details obtainable depends on (1) the degree of exposure of the marker bed and (2) the accuracy of the mapping procedure used.

FSS maps are presented in the same format as conventional subsurface contour maps which are constructed from well data and, more or less, are obtained by similar interpolation techniques. A comparision between these two maps can be used for two purposes: first, to establish the relationships between surface and subsurface structures in the study area, and second, to guide and improve the mapping of subsurface structures in problematic areas. The first step is particularly critical because marker beds are not always directly correlated with subsurface structures at the reservoir level. For example, thickness variations, local salt dissolution phenomena and disharmonic folding can create shallow structures which do not exist in a deeper section. These two steps will be demonstrated later in this chapter.

6.4 Tools for Geological Mapping with Stereo Data

Many different types of stereo equipment and related methods, from simple optical devices to complicated computer systems, can be used to view and measure stereo data. The more commonly used systems for stereo display are discussed below.

6.4.1 Stereo Pair Alignment

As an initial step, stereo pairs must be aligned correctly. The procedure is relatively simple and illustrated in Fig. 6.8. The first step is to find and mark the center of each image. Next, the center of the left image must be found on the right image (it will be displaced by parallax) and vice versa for the left image. The center is found by recognizing patterns on the photo. Next, the two images must be arranged so that all four points fall on the same line. The displaced centers should be closer together than the actual centers. (The opposite arrangement yields the pseudoscopic effect.)

6.4.2 Simple Optical Stereoscopes

A typical lens stereoscope is presented in Fig. 6.9. As mentioned before, the distance *D* is set to be five times the eye base for viewing. Note that in the lens stereoscope, the angle of disparity, α, as well as the convergence angle γ, are the same as under normal viewing conditions. The final depth cue, the focusing accommodation, is determined by the focal

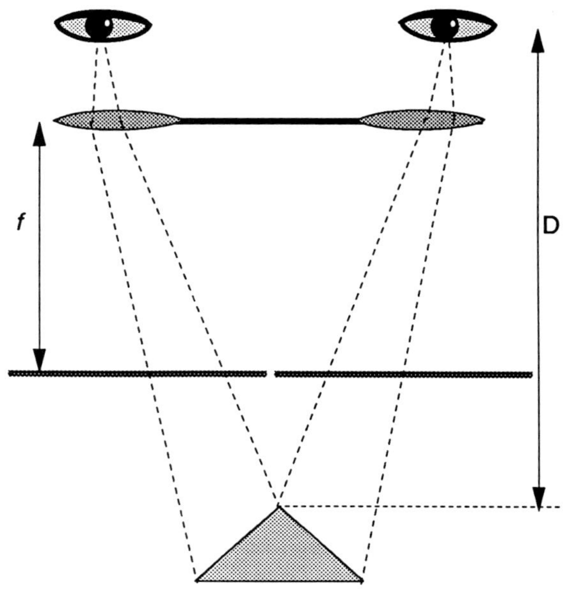

Fig. 6.9. A lens stereoscope
Shown is a schematic diagram of the simplest stereoscopic viewing instrument. The lens refracts the light from the photos into parallel rays

Fig. 6.8. Stereo pair alignment
After the two centers and the two transferred centers are found, the photos must be arranged so that all four points are colinear ▼

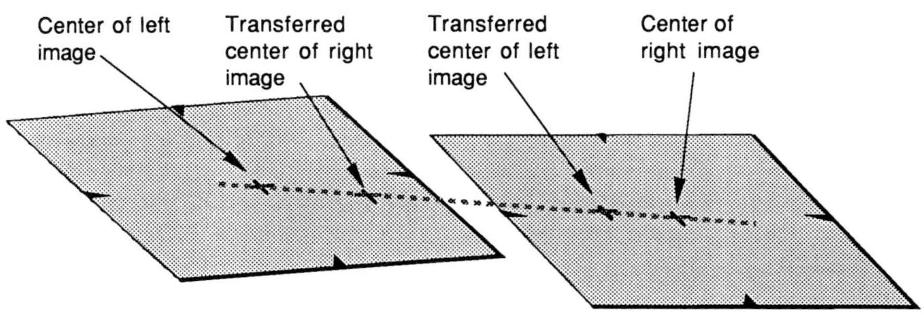

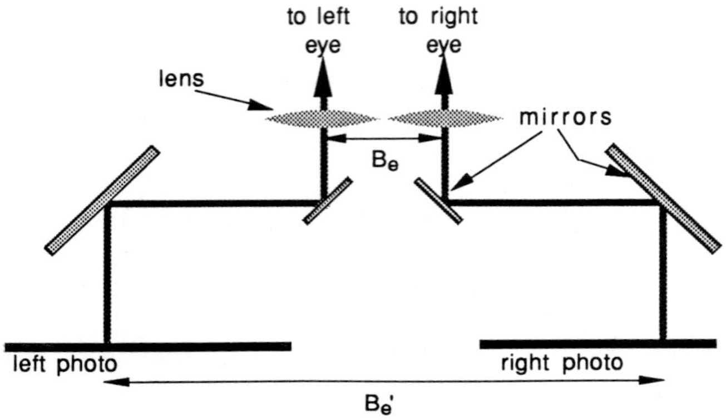

Fig. 6.10. A lens-mirror stereoscope

Fig. 6.11. Stereoplotting equipment
a The Kelsh plotter, an analog stereoplotter; b The Leica digital video plotter

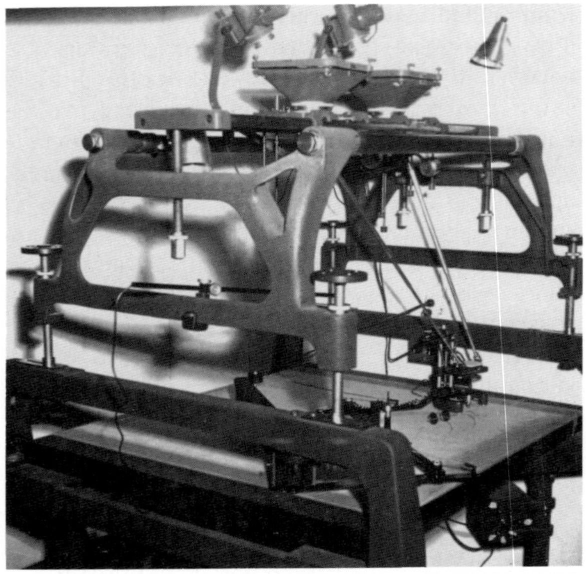

a

b

length of the lenses, *f*. The focal length could be set to correspond to the perceived distance to the object *(D)*. In this case, focusing accomodation and convergence angles would be correct but disparity angles would not. There is some evidence to suggest that such a situation would improve discrimination of small distances. Common procedure, however, is to set the focal length to infinity to decrease eye strain. (Thus, to see the stereo image, the viewer looks "through" the page and then focuses on the object.)

The lens stereoscope is extremely popular because it is inexpensive, portable, and easy to use. A variety of models, including ones for fieldwork, are available.

The simple lens-mirror stereoscope is a slight improvement on the lens stereoscope and is illustrated in Fig. 6.10. The principles involved are much the same. The focal length of the lens must be increased to accommodate the longer light path leading to a reduction in model size. (Vertical exaggeration is not affected.) The main advantage of the lens-mirror stereoscope is the effective widening of the eye base by mirrors allowing larger photos to be viewed without overlapping at the center.

6.4.3 Complex and Analytical Stereoscopes

More complicated versions of the stereoscope are available for use in comprehensive mapping. A stylus can be connected through armatures to the parallax bar allowing one to create a map by plotting various points and their elevations. Most commonly, stereoscopes will employ a double-projection system using glass-plate diapositives. While more expensive to produce, these plates allow the left and right images to be projected onto a table, called the platen or stage, and then viewed in stereo using several methods. The height of the platen can be adjusted to align the topography image with the floating dot (Slama 1980). Some plotters have a mechanism to convert platen position directly into elevation readings. (The Kelsh plotter, a typical early model, is shown in Fig. 6.11 a.) This type of arrangement is an example of an analog stereoplotting device and was the mainstay of photogrammetric mapping from the 1920s until recently (Petrie 1992).

Recent innovations in computer technology have allowed the industry to introduce more and more computer control to the stereoplotting process. The first stage of computer integration replaces the mechanical analog computers in stereoplotters with encoders that send information about the position of the cross hairs to a computer. The user then positions the cross hairs on the stereopair while viewing the perceived floating dot through a lens and mirror arrangement. When the correct orientation is achieved, the computer performs the parallax calculation and records a set of x, y, z coordinates. This type of arrangement is called the semi-analytical stereo plotter (e. g. the Zeiss Stereocord G3-PC).

The next degree of complexity is the analytical stereoplotter which gives control of the cross hairs to the computer via motors and feedback signals (e. g., the Zeiss Jena Dicomat). The photographs are mounted on stages whose movement is controlled by the computer. The user views the pair through high quality optics and directs the motion of the image through a track ball, mouse or other input device. The high precision which is achievable by these machines spelled the demise of the manually controlled stereocomparator. In recent years, video displays, which are becoming increasingly less expensive, have been added to analytical stereoplotters to allow the user to view, simultaneously, the data set that is being constructed such as contour lines, feature outlines, etc. (Petrie 1992).

As a low-cost alternative to complex stereoplotters, the digital video plotter (DVP) has appeared as a system suitable for use when less accuracy is acceptable (e. g., when feature boundaries will be subject to interpretation anyway). The two stereo images are displayed side-by-side, and a device similar to a mirror stereoscope is placed in front of the screen as in Fig. 6.11 b (Klaver et al. 1992).

Another common method for displaying a stereo pair may be referred to as the flicker system (shown in Fig. 6.12). A special monitor is used that alternately displays the left and right images in rapid succession. At the same frequency, a polarizing screen placed in front of the monitor alternates between two polarized states. The user wears glasses with each lens polarized to match the appropriate state on the monitor. The sequence of images shown on the screen is coordinated with the reversal of polarity on the modulator, so that the user's left eye sees only the left image and the right eye sees only the right image, giving the effect of a clear, three-dimensional, stereoscopic image. A mouse or similar pointing device can be used to manipulate a floating dot (generated by the same principle as before) around the screen. Thus, elevations can be recorded directly in the computer to be superimposed

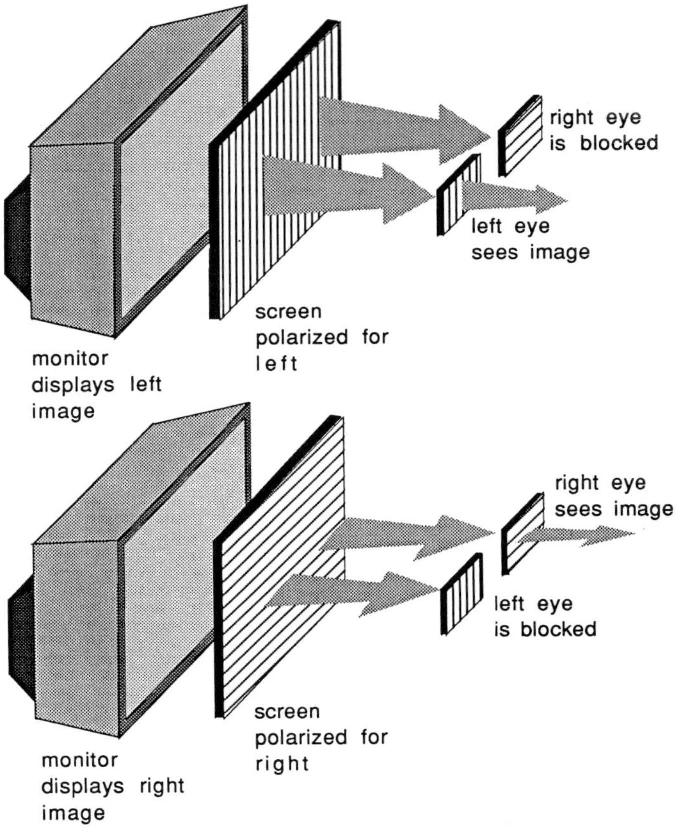

Fig. 6.12. The flicker system
By alternating the polarization state of the screen with the image displayed on the monitor, the viewer successively sees the left and right images. If the frequency is high enough, the illusion of a three-dimensional image is created

on other data sets later. One variation of this technique can be used on a standard PC equipped with a special display board to drive the monitor and liquid crystal modulator (Welch 1989). Another approach entails connecting the stereo monitor to an existing image-processing system (Williams and Thompson 1988). Both systems are extremely easy to operate, relatively inexpensive to install, and useful for many types of stereo data.

6.4.4 Digital Image Correlation

With the increase in the availability of digital data, attention focused on a fully digital implementation of the stereoscope. That is, one which goes from initial image data to final output without the production of any hardcopy images. Geographical data

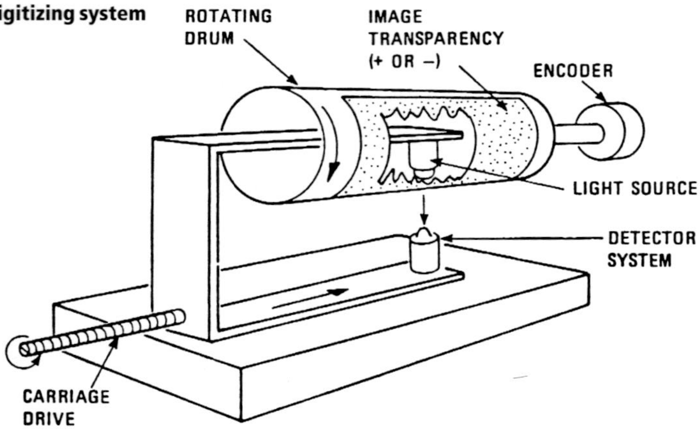

Fig. 6.13. Schematic of a digitizing system (Bryant 1974)

can be input into a computer through the use of digitizing systems (e.g. Fig. 6.13) or directly, if the data are already in a computercompatible format (as is the case with most satellite data).

The final goal in such systems is to create a digital elevation model (DEM digital terrain model is a synonymous term). A DEM is essentially a database of x, y, z coordinates corresponding to the area of interest. Once obtained, the DEM can be displayed a variety of ways (as a contour map, a shaded relief map, a 3-D block model, etc.) or merged with other data.

Software is now readily available to create a DEM directly from a satellite stereo pair. The computer automatically correlates points on the stereo pair by using pattern-matching algorithms and subsequently performs the necessary parallax calculations. The two images are combined into one which is orthographically correct (i.e., the relative distances between points on the image are the same as the relative distances between the points on the ground) and which contains accurate elevation data. In order to constrain the data, these systems need, as additional input, ground control points (GCPs) which are points whose elevation is known (from ground surveys, etc.). The sensors have six degrees of freedom (tilting along three axes and translation in three directions), all of which must be determined for the entire time during which an image was captured. Brockelbank and Tam (1991) report that the SPOT satellite requires 9 GCPs to capture an image and that at least eight GCPs are required along with satellite orbit information to estimate all necessary parameters.

Following this procedure, two methods exist for computerized correlation of points: area-based and feature-based. Each can be used separately or in combination. The feature-based system first performs a second-derivative edge identification to produce outlines of prominent features. After this is done on both images, pattern recognition routines are used to match a pattern on an image with its stereo counterpart. The difference in position between the two images, which is due to parallax, is used to derive elevation data.

The area-based approach consists of three steps. First, a regular grid of dots – test points – is superimposed on one of the images. Second, the pattern of pixels around each test point is compared with the second image until a match is found. A corresponding test point is then placed on the second image. Finally, after all test points are identified on both images, parallax measurements are used to correct the points to their correct orthographic positions. Figure 6.14 shows the steps in an area-based DEM production.

It is not practical or required to find the elevation of each pixel in an image. Rather, a suitably dense grid for the application at hand is selected and the elevations of all other points are interpolated. Thus, DEMs can lead to contour maps of any vertical *resolution*. This specification, however, should not be confused with the contour map's vertical *accuracy* which can never exceed the distance between test points (which is ultimately limited by the spatial resolution of the image).

DEM systems have been available from several companies since the 1970s. The task is usually performed on a mainframe or minicomputer, but PC-based systems are developing at a rapid pace (see, for example, Ostrowski et al. 1989; Jachinski and Zielinski 1992). Also available are systems capable of processing stereo radar images. Presently, some of these systems allow partial correction of the

Fig. 6.14. Area-based procedure for creating a digital elevation model (DEM)
a Test points are projected onto the left image from the stereo pair. By pattern matching, the corresponding points are mapped on the right image. **b** An orthographic image is created and contoured. **c** and **d** The orthographic image can be viewed in three dimensions. (Courtesy of Erdas)

many geometrical distortions that plagued radar imagery in the past and their proficiency in this matter is constantly improving (e. g., Toutin et al. 1992; Tilley 1992).

6.5 Evaluation of Different Stereo Data

Stereo imagery is available from several sources such as SPOT images, high- and low-altitude aerial photography as well as radar data. Successful use of stereo data in an exploration venture requires familiarity with the advantages and disadvantages of each source. To gain some appreciation of the use of different stereo data sets in different geological environments, and their accuracy as compared to field measurements, examples are given here from two areas which represent the typical geological conditions which require stereo measurements.

6.5.1 SPOT vs. High-Altitude Photography in Low Relief Areas

The Pecos Arch of the Central Basin Platform in West Texas provides a good example for demonstrating the use of stereo measurements for structural interpretation in subtle low relief terrain (Fig. 6.15). The area is characterized by large-scale, basement-involved structures buried under a thick section of very gently deformed Cretaceous rock units. Production is from multiple strata ranging in age from Ordovician to Permian (Galley 1958; see also Fig. 6.16).

Because of the combined influence of differential compaction and loading, as well as structural reactivation, many uplifted basement blocks and associated drape folds are manifested at the surface as low-amplitude structures that, although subtle, closely mimic the configuration of the deeper traps (Fig. 6.17). The Yates and Todd oil fields are only two of the many early oil discoveries attributed to workers recognizing such surface structures with field mapping techniques (Hennen and Metcalf 1929; Christner, 1940).

Panchromatic SPOT imagery of the Yates Field area (Fig. 6.18) provides an excellent view of the regional surface geology. The area coverage of a SPOT image is about 25 to 75 times larger than the area covered by high- and low-altitude aerial photographs; however, the resolution of the SPOT image is still sufficient to detect exposed bedrock needed

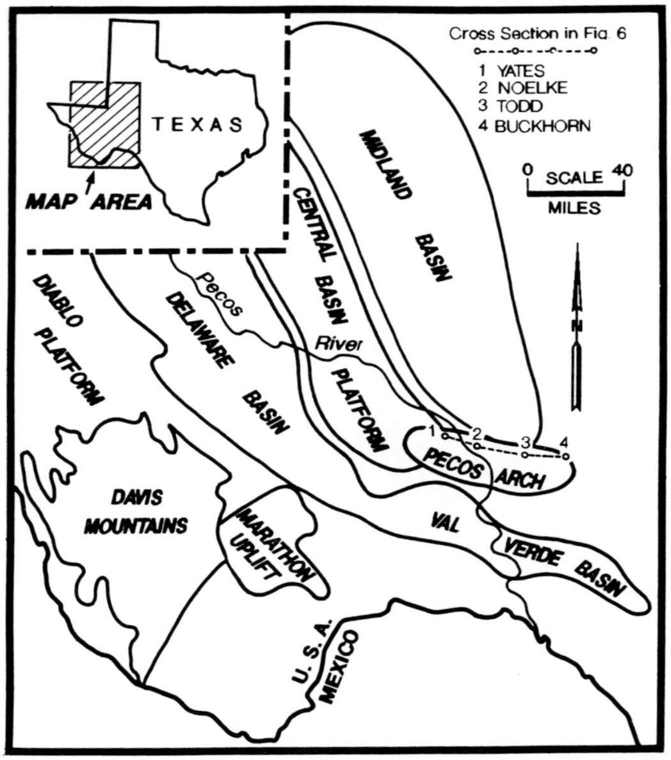

Fig. 6.15. Tectonic map of West Texas
The location of the Pecos Arch study area, Central Basin Platform is shown. (After Renfro et al. 1973) See Fig. 6.16 for cross section

Fig. 6.16. Generalized cross section of the study area
(Courtesy of Exxon Exploration Company)

Fig. 6.17. Relationship between surface and subsurface structures ▶
Comparison between **a** subsurface structure maps and **b** surface structure maps of the Yates Field area (generalized from Hennen and Metcalf 1929). The surface structure map, done by plane table mapping, is about five times more subtle than the producing trap. (The surface structure map is done on a Cretaceous unit just above the top of the basement sandstone)

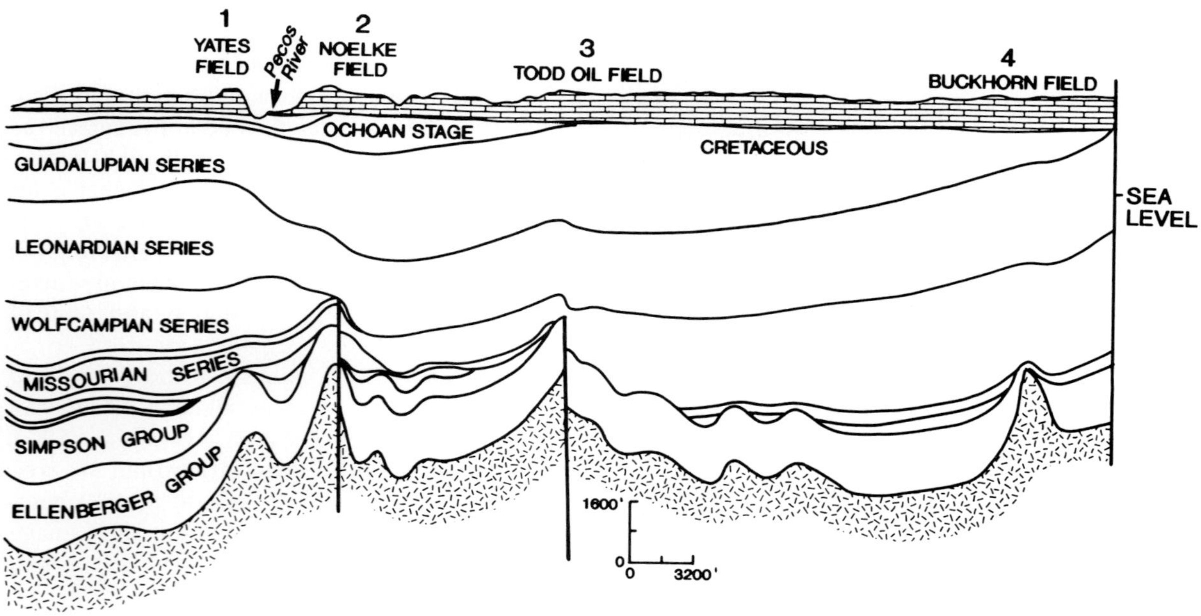

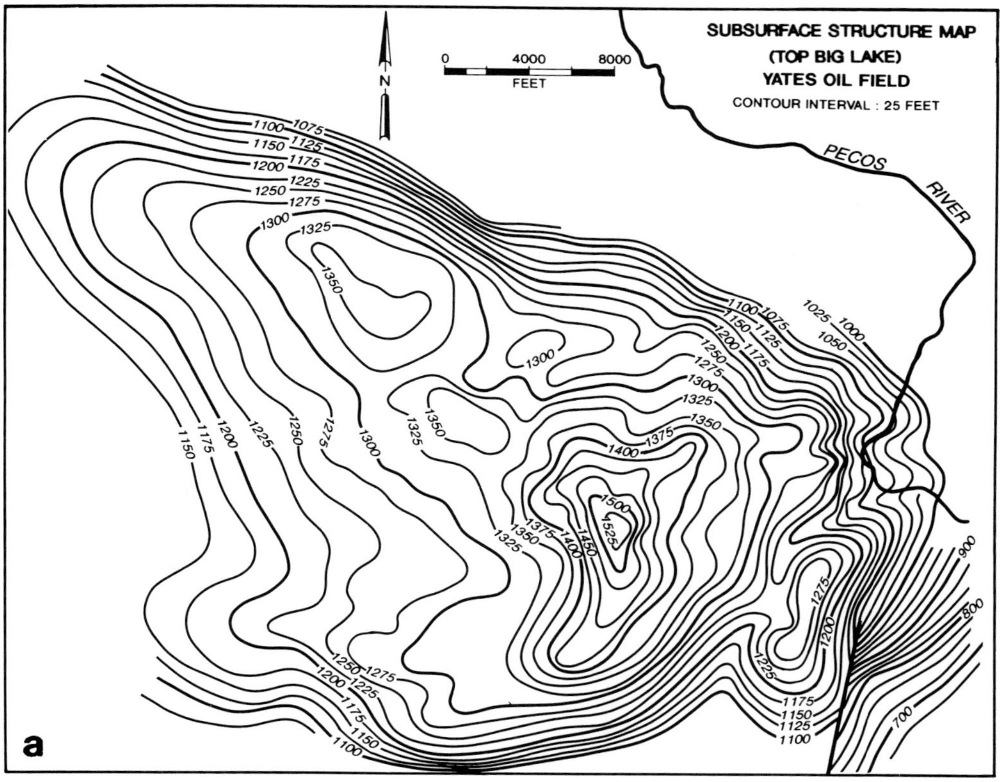
SUBSURFACE STRUCTURE MAP
(TOP BIG LAKE)
YATES OIL FIELD
CONTOUR INTERVAL : 25 FEET
0 4000 8000
FEET
N
PECOS RIVER
a

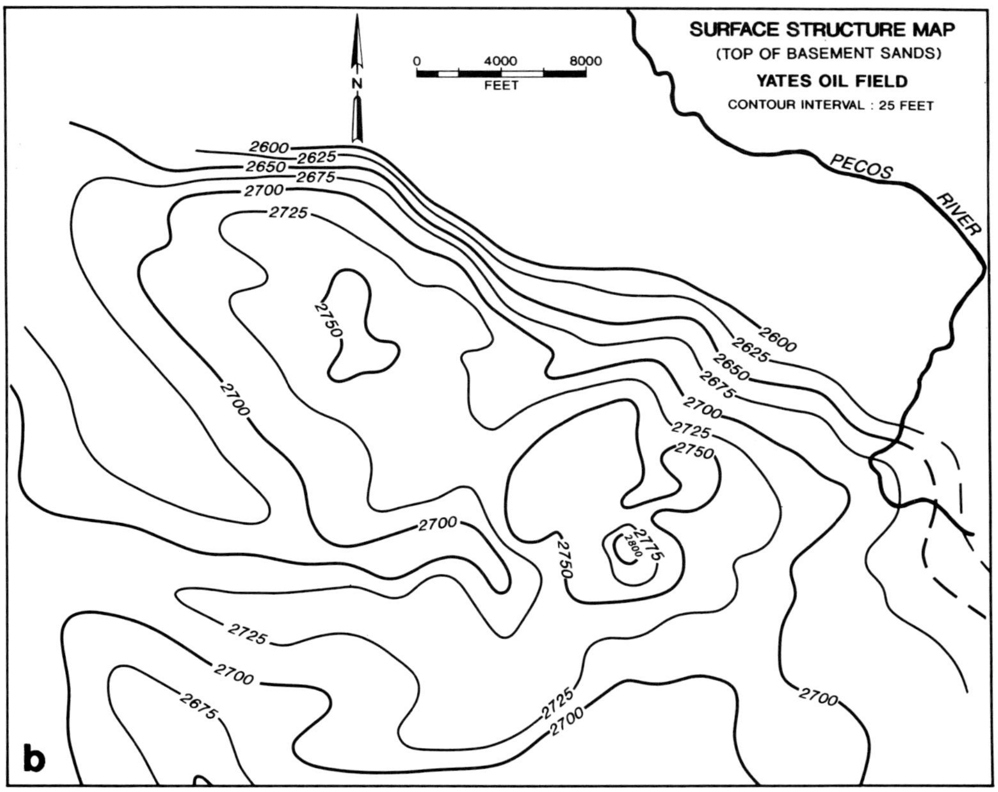
SURFACE STRUCTURE MAP
(TOP OF BASEMENT SANDS)
YATES OIL FIELD
CONTOUR INTERVAL : 25 FEET
0 4000 8000
FEET
N
PECOS RIVER
b

Fig. 6.18. Panchromatic subscene
About a quarter of the entire SPOT image of the Yates oil field area. Note approximate relative coverage of low- and high-altitude photographs in the *lower left corner*

Fig. 6.19. Comparison between SPOT and high-altitude aerial photography
Stereo SPOT (**a** and **b**) and high-altitude aerial photography (**c** and **d**) of a breached fold north of the Yates Field, showing the expression of major marker beds. Also shown are dip and strike measurements which were obtained for each data set. Stereo can be viewed with a pocket stereoscope

for elevation measurements. Note the lack of flatiron exposure in this terrain, which instead exhibits "layer-cake" topography defined by flat to gently dipping bedrock.

Figure 6.19 shows both the stereo mapping capabilities of SPOT and aerial photography for a breached fold north of Yates Field. Lithostratigraphic units can be mapped on both data sets, but the aerial photography provides sharper definition of bedrock than the SPOT data. Consequently, dip measurements made with SPOT stereo have a 2° resolution as compared to the 1° or less resolution using aerial photography. Differences in the level of structural detail that can be extracted from SPOT as compared to aerial photography determine the type of structural mapping that can be accomplished using these data sets (Fig. 6.19). SPOT data can sup-

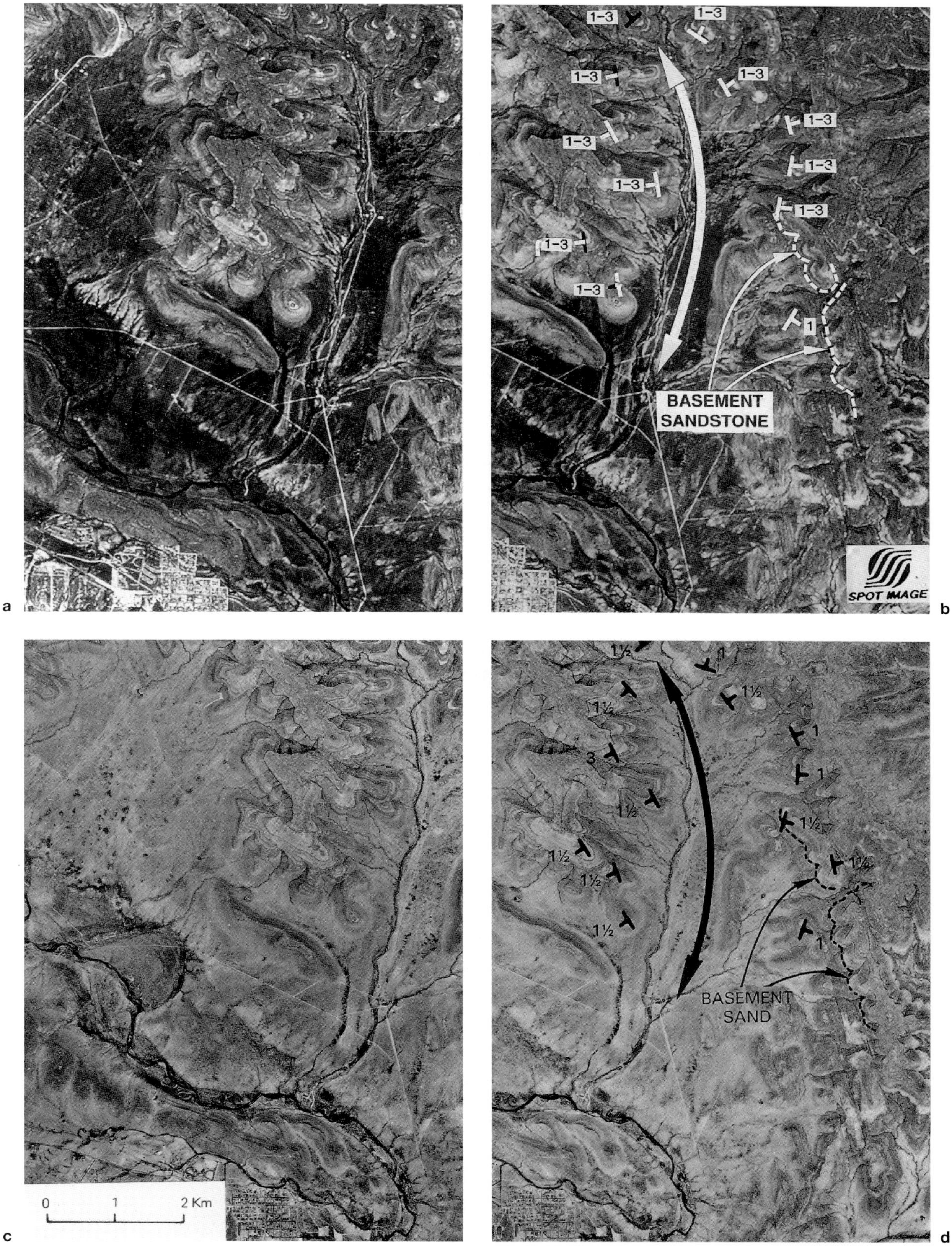
a
b
c
d
1-3
1
BASEMENT
SANDSTONE
SPOT IMAGE
0 1 2 Km
1½
3
BASEMENT
SAND

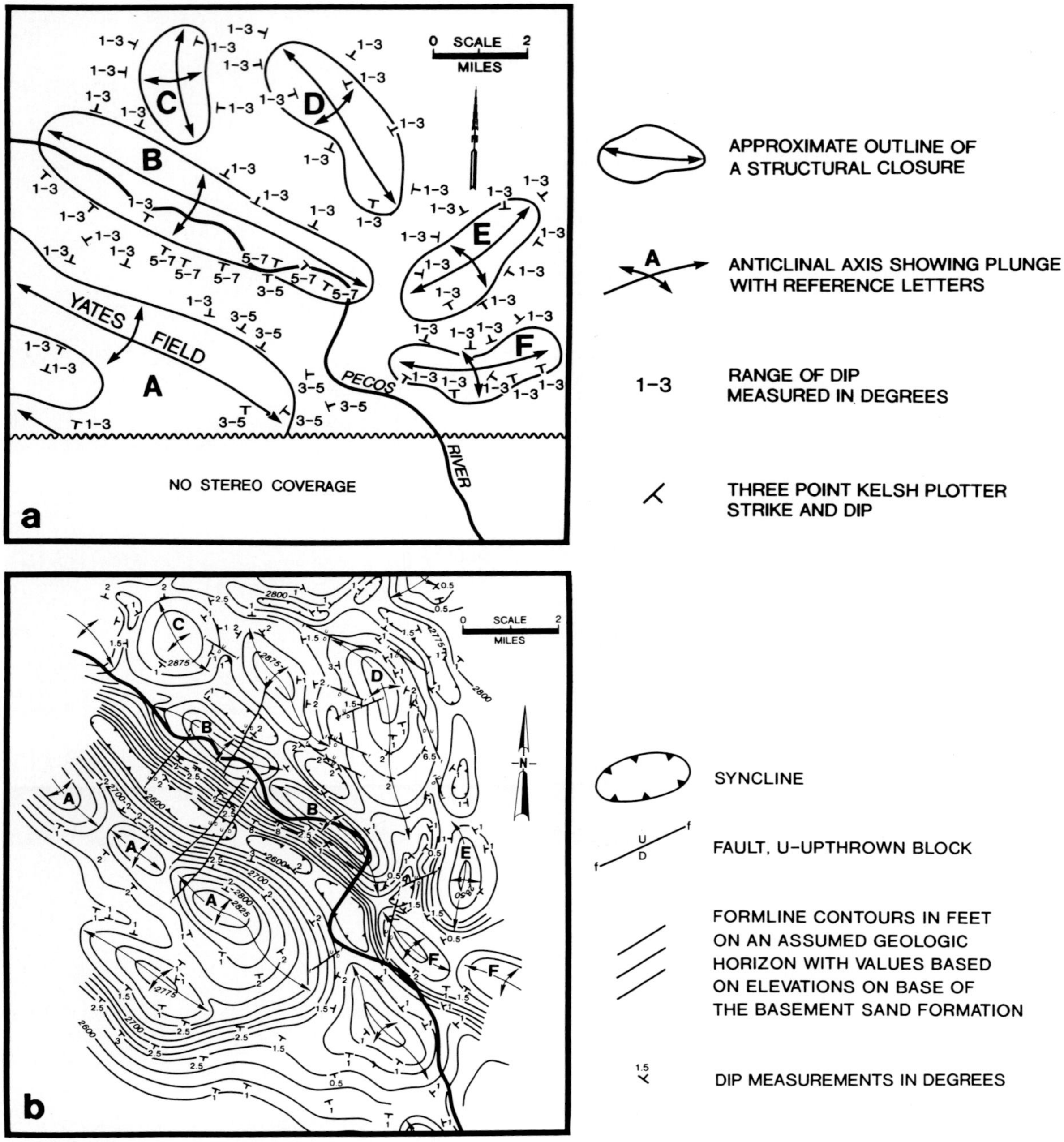

port semiquantitative mapping of structural closures in the region (Fig. 6.20 a), whereas data obtained from aerial photography can be used to contour individual structures on a key marker bed at a 25-m interval (Fig. 6.20 b). Both maps, however, provide compatible information on structural closures and compare well with data obtained by field mapping techniques (e. g. Fig. 6.17).

Fig. 6.20. Comparison of FSS maps
Stereo mapping of the breached fold shown in Fig. 6.19. Shown are FSS maps derived from SPOT data (**a**) and aerial photography of the same area (**b**). Strike and dip measurements were obtained by a Kelsh plotter. (Stereo work by D. Anderson)

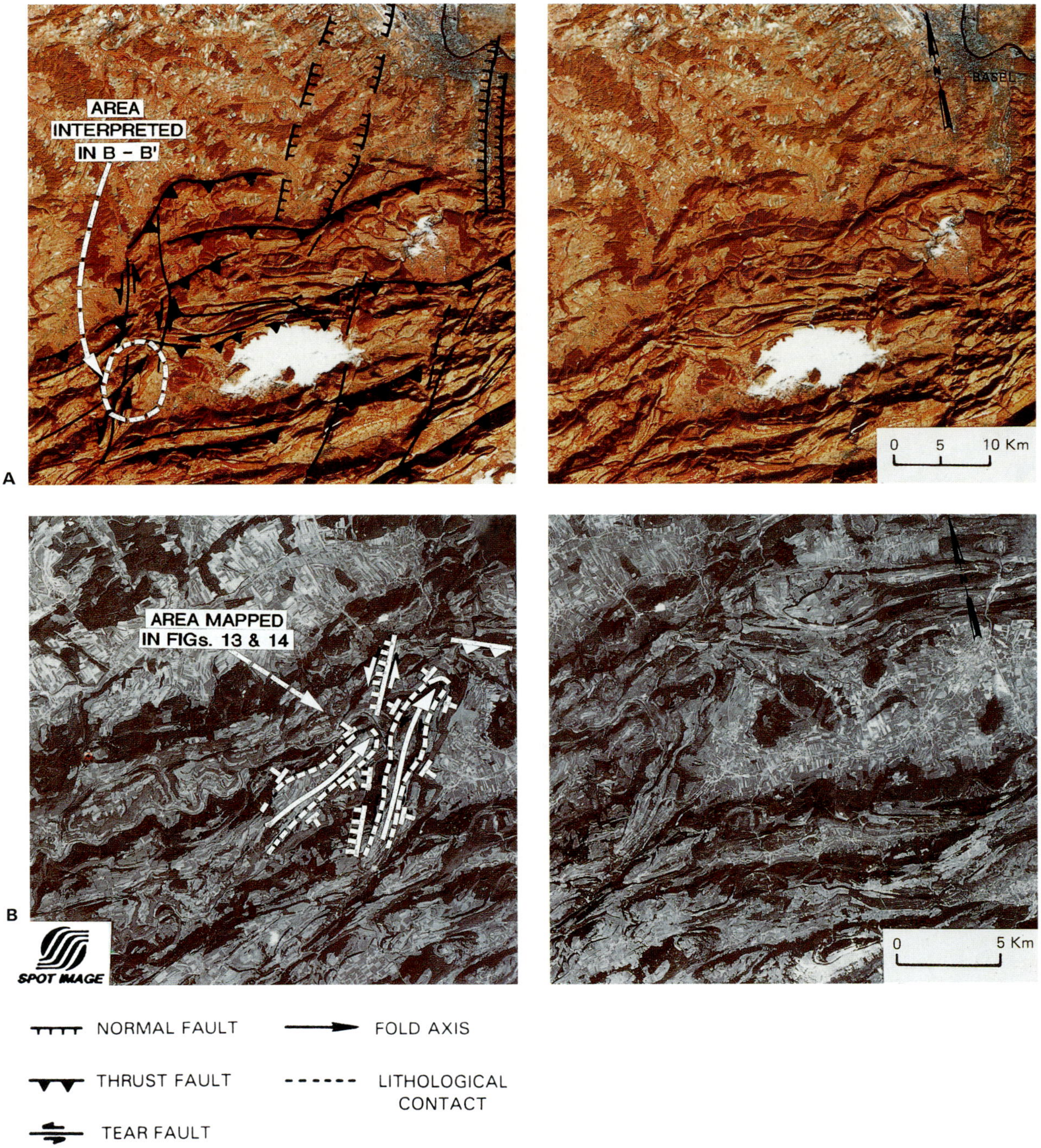

Fig. 6.21. SPOT imagery data of the central part of the Jura region
Multispecral imagery (**A, A'**) showing the surface expression of major structural elements (not in stereo). Panchromatic stereo image (**B, B'**) of the western margin of the Delémont basin with interpretation of major structures. (Stereo can be viewed with a pocket stereoscope)

6.5.2 SPOT vs. Field Data in Highly Deformed Areas

The Jura Fold Belt in Switzerland provides an example of a highly deformed area where the exposed structures are covered by thick vegetation. The dominant structural style of the Jura is disharmonic folding above Triassic evaporite detachment, ac-

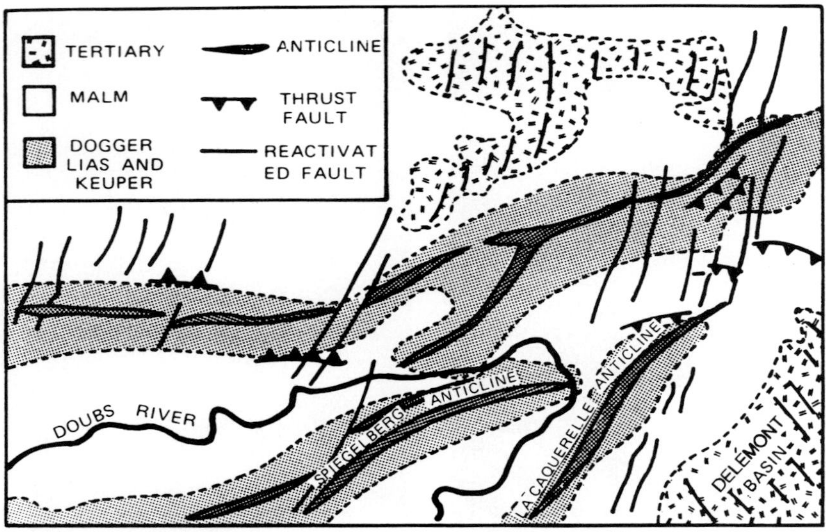

Fig. 6.22. Tectonic map of western margin of the Delémont Basin
(After Diebold et al. 1960)

companied by minor thrust faults partly exposed at the surface. These large structures are cross-cut by complex fault systems that extend into the Jura from the adjacent Rhine Graben (Laubscher 1977; Trümpy 1980; see tectonic map in Fig. 3.21). A generalized structural interpretation of this region (Fig. 6.21, A and A') is possible with monoscopic imagery (Berger and Corona 1986). Individual plunging folds of the Jura form elongated ridges with well-defined limbs. Thrust faults are manifested as irregular to sinuous FLTs parallel to the fold axes. Major cross faults appear as an alignment of topographic ridges and valleys that cut and breach many of the folds.

Due to the complex outcrop patterns and dense vegetation of the Jura, many structures must be analyzed with stereo imagery to obtain details and resolve ambiguities in the monoscopic interpretation. An example of stereo analysis with SPOT panchromatic imagery is illustrated by a closer examination of the western margin of the Delémont Basin (Fig. 6.21, B and B'). Dip and strike measurements of key marker beds, which were obtained with a Kelsh plotter, allowed the recognition of two folds truncated by a reactivated normal fault which pres-

Fig. 6.23. Comparison between dip and strike measurements in the Delémont Basin
Obtained in the field by Diebold et al. (1960) with SPOT stereo and a Kelsh plotter at increments of 5°

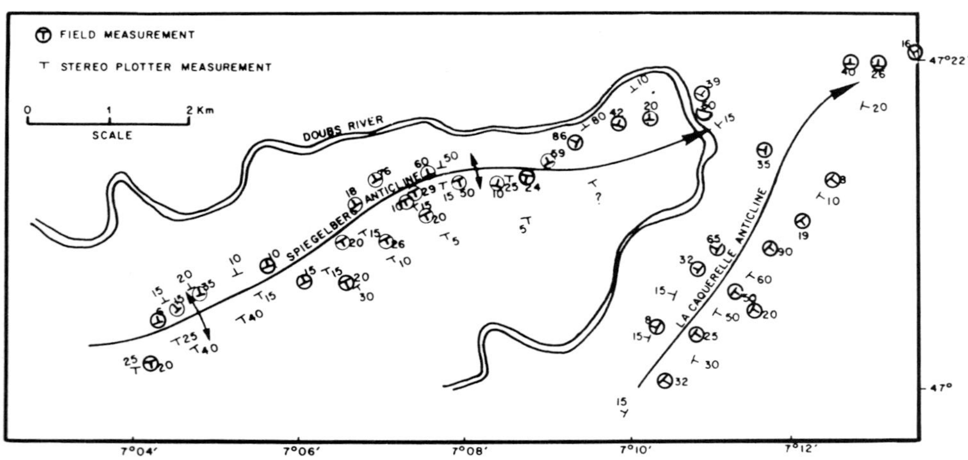

ently exhibits oblique offset. This detailed interpretation of structures not discernible on monoscopic imagery is generally compatible with interpretation obtained by field mapping (Fig. 6.22; see also Laubscher and Bernoulli 1980). It was found that dips from SPOT stereo compare well with those obtained in the field, indicating that imagery-derived dips can be used in quantitative mapping of structures in the region (Fig. 6.23).

6.6 Exploration Application

Surface structure maps generated from stereo imagery can be used at two different stages of exploration. During early phases, regional maps, such as those shown for the Yates Field in Fig. 6.20, can be used to establish the style, trend, size, and spatial distribution of structures in a basin or region. This information, in turn, can be used to assess the remaining hydrocarbon potential in the basin, to guide the orientation and spacing of seismic programs, and to upgrade areas for future exploration. At this stage, detailed mapping of key structures, such as those shown for the Jura region in Fig. 6.21, can be used to investigate the structural evolution of the basin as well as to establish an inventory of different types of potential structural traps.

During the advanced prospecting stage of exploration, detailed mapping of surface structures is usually aimed at improving subsurface mapping of prospective structures. Such effort is particularly useful in areas where the collection and interpretation of seismic data are problematic. This type of mapping may be done with SPOT stereo, but often requires data with higher resolution. (See the comparison of the Yates Field map in Fig. 6.19 and the discussion in the "Selecting Stereo Data for Structural Mapping" section.)

In low relief areas, stereo imagery mapping is focused on the more rugged portions of the basin, particularly along structurally controlled stream valleys which exhibit anomalous levels of downcutting and unique drainage patterns (e. g., Lattman 1959; Berger and Aghassy 1983). The entrenchment of stream valleys over and around deep-seated structures provides bedrock exposure and is an excellent aid to the explorationist who has elected to use surface structure maps. Also, these areas often constitute the relatively unexplored parts of the basin for two reasons: (1) collecting seismic data in these parts of the basin is usually difficult, costly or both, and (2) seismic data quality is usually poor due to a low signal-to-noise ratio.

A typical application of stereo mapping of subtle structures is illustrated in Figs. 6.24 and 6.25. A deeply entrenched stream valley in the Western Canada Basin caused the exposure and breaching of a subtle surface fold, which was identified and mapped with stereo imagery data (Fig. 6.24). Seismic data collected along the river are of poor quality, but show the presence of a deep-seated reactivated fault and a related inverted fold (Fig. 6.25). The close spatial relationships between the surface and subsurface structure allow the explorationist to (1) interpret the structure through noisy data; (2) estimate potential trap size; (3) predict a four-way closure; (4) assume that the subsurface structure is real and not related to velocity pull-up; and (5) justify the cost of collecting expensive seismic data across these valleys to complete the mapping of this prospect.

Surface structure maps generated from stereo imagery play an even greater role in mapping subsurface structures in fold and thrust belts. Complex thrust-related structures and the generally poor-quality seismic data common in these mountain belts usually present the explorationist with several structural interpretation options. Figure 6.26 illustrates this situation and shows a relatively good set of seismic lines used to identify and map two prospective anticlines within the fold and thrust belt region of the Western Canada Basin (structures A and B in Fig. 6.26 a–c). The explorationists' attempt to extend the mapping of these prospects to the south was hampered by poor seismic data, allowing only an extremely generalized interpretation of the two prospective structures (structures A and B in Fig. 6.26 d–e).

A surface structure map of the area of poor seismic quality was generated from stereo aerial photographs (Fig. 6.26 f). The surface mapping was done independently of subsurface control and confirmed the presence of the two prospective anticlines separated by a late-stage normal fault, which is slightly offset due to surface projection. The availability of the surface map allowed the interpreter to (1) constrain the structural interpretation of the seismic data; (2) evaluate the size of the prospect; (3) determine the presence of southward-plunging closures; and (4) reduce the number of additional seismic lines that may be required to complete the mapping of this prospect.

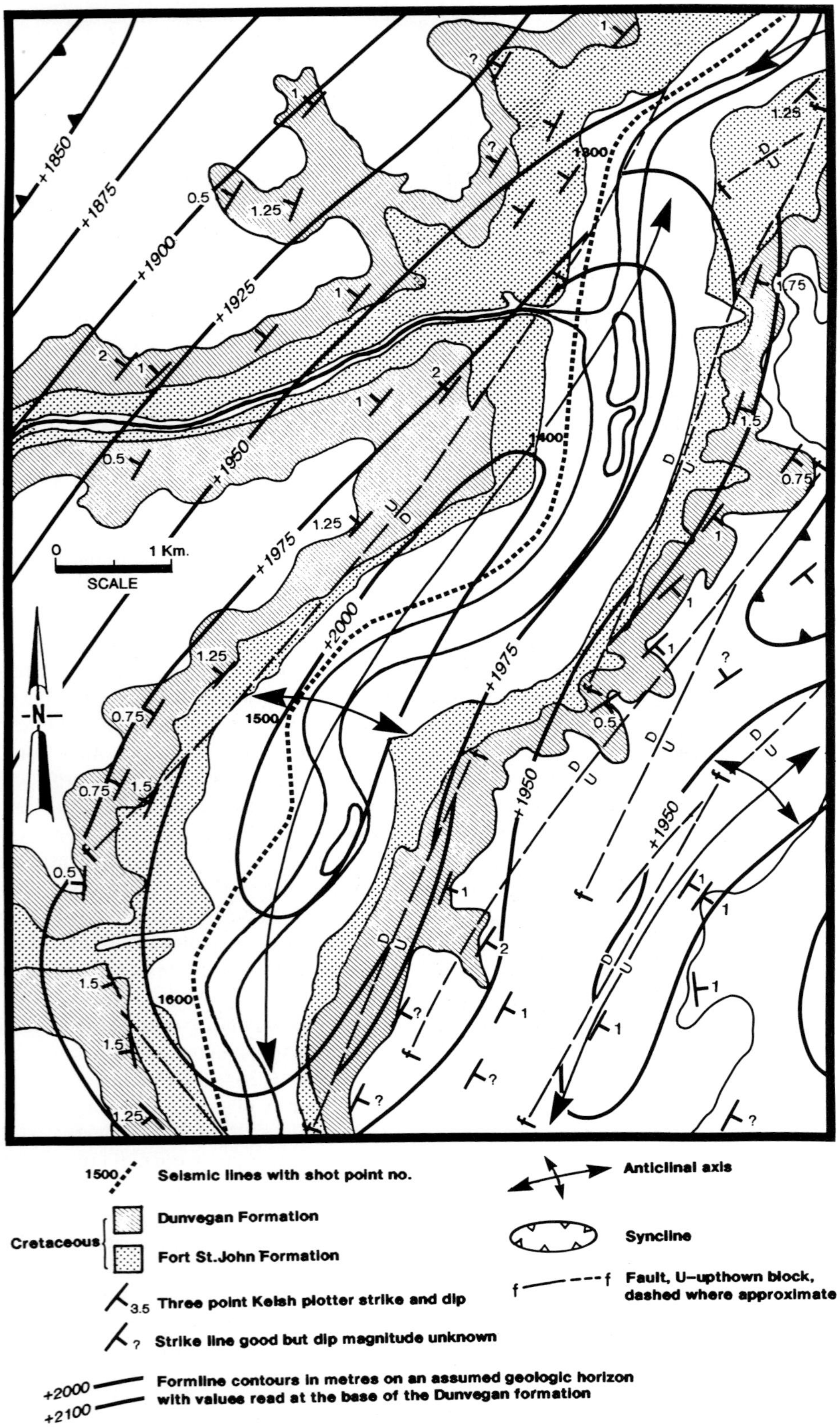

+1850
+1875
+1900
+1925
+1950
+1975
+2000
+1975
+1950
+1950
1800
1400
1500
1600
0 1 Km.
SCALE
N
1500 Seismic lines with shot point no.
Cretaceous
Dunvegan Formation
Fort St.John Formation
3.5 Three point Keish plotter strike and dip
? Strike line good but dip magnitude unknown
+2000
+2100
Formline contours in metres on an assumed geologic horizon with values read at the base of the Dunvegan formation
Anticlinal axis
Syncline
f f Fault, U–upthown block, dashed where approximate

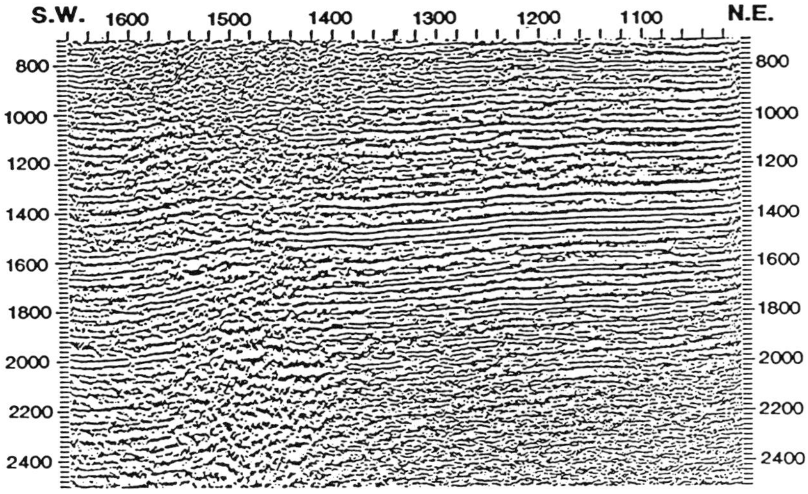

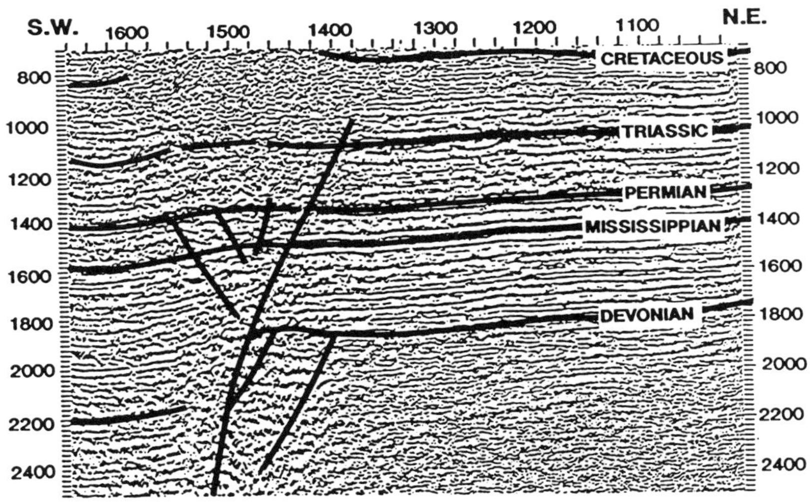

◀ **Fig. 6.24. Form-line surface structure (FSS) map of a breached fold in the Western Canada Basin**
The map was done with stereo aerial photography and a Kelsh plotter. Surface structure mapping is done at the base of a well-defined Cretaceous unit. Note the location of seismic data collected along the floodplain of the river. (Stereo mapping by D. Anderson)

Fig. 6.25. Typical noisy seismic data
Collected along the strike of the breached fold shown in Fig. 6.24

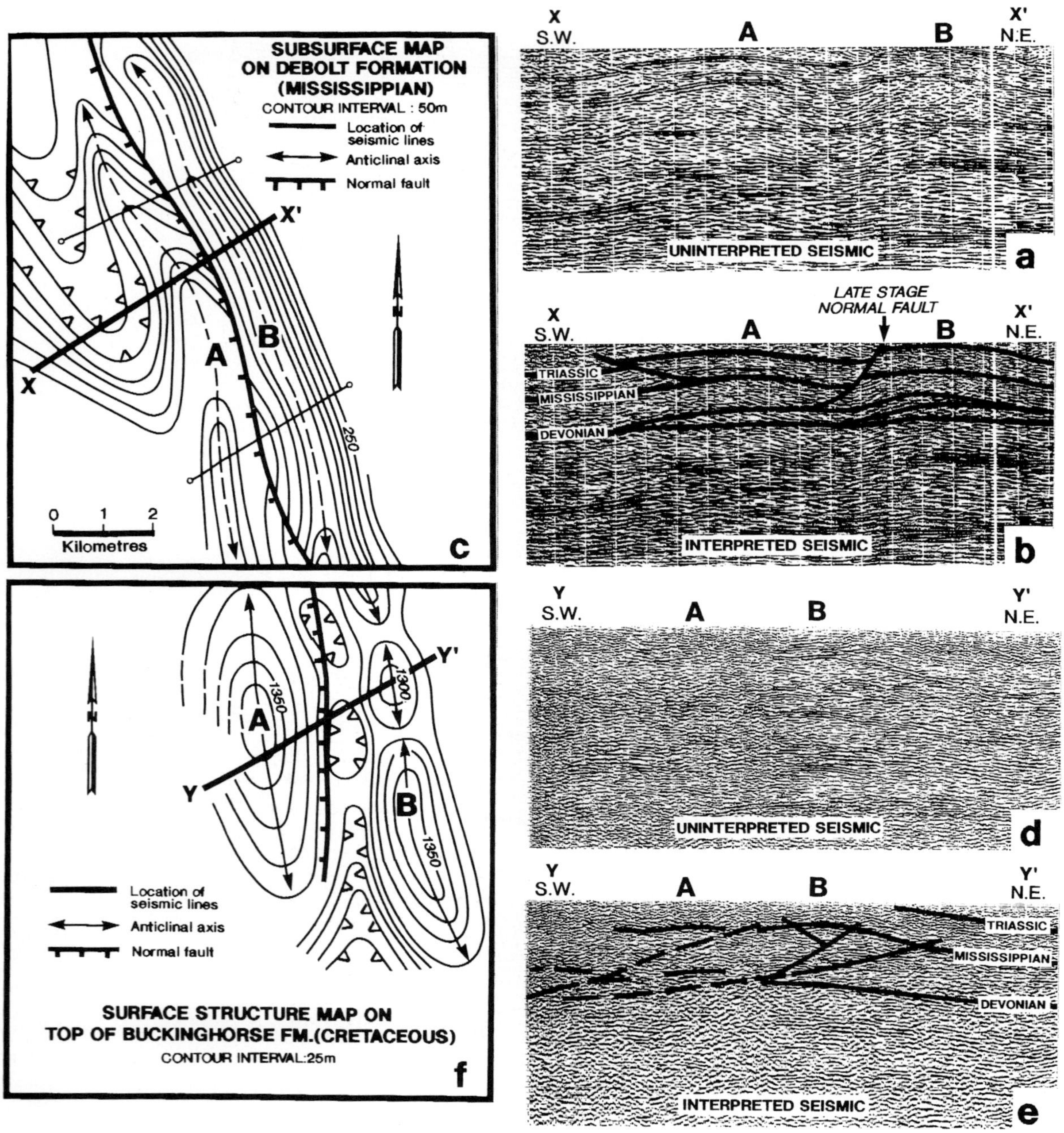

Fig. 6.26. Surface and subsurface data in the Canadian Rockies fold and thrust belt

Surface and subsurface information of two prospective structures with variable seismic resolution are shown. Subsurface interpretation by D. B. Halwas and J. D. Thomson, Imperial Oil Resource Ltd.

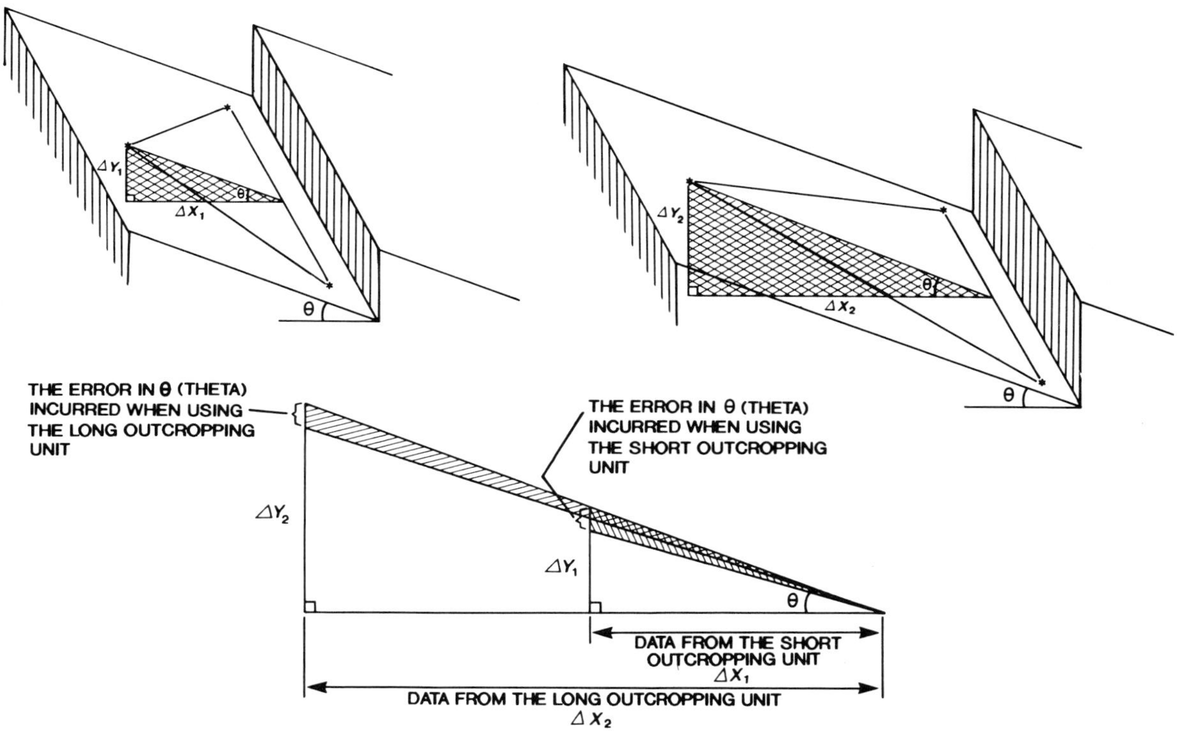

Fig. 6.27. Accuracy of three-point measurements
A simplification of the error in dip calculations using the three-point system is shown. Two outcrops are shown with the same angle of dip, but with different lengths. The *triangles* showing the dip angle from each outcrop have been superimposed to show that applying the same error in altitude (Δy) to each yields a smaller error in the angle (θ) in the longer outcrop

6.7 Selecting Stereo Data for Structural Mapping

SPOT stereo imagery is advantageous for structure mapping because it has wide-area coverage compared to stereo arial photography. Prior to selecting remote sensing data for stereo mapping, however, the interpreter must determine the level of structural detail required to perform the task at hand. For example, a study aimed toward early identification of potential traps would require less detailed information than that needed for detailed mapping of prospective structures. The data from the Central Basin Platform (Fig. 6.20) illustrate how one determines the required level of detail. The identification of potential traps around the Yates Field would have required dip measurements accurate to 2°, whereas a detailed mapping of the Yates or other individual structures would have required dip measurements accurate to 1° or less.

Prior to commitment to a particular type of stereo imagery data, the interpreter must examine samples of stereo imagery or existing geological maps and determine whether the expected accuracy of dip measurements will be met. Two independent factors influence the level of accuracy in dip measurements obtained from remote sensing data. The first factor is related to the accuracy of the instrument, which in turn, is related (in part) to the altitude of the sensor. Sensors at a higher altitude incur a larger error in elevation measurements, leading to larger errors in the dip angle. The second factor arises from the nature of the three-point measurements system used and depends on the length of the slope being measured. Figure 6.27 illustrates that in measuring longer slopes, the percent error on dip measurement is reduced.

One must remember that the length of the outcrop often is related to the inclination of the strata. That is, near-horizontal strata commonly have longer outcrops than do tilted strata (see the progression of exposed slope length on the San Rafael Swell in Fig. 3.3 and differences in average slopes of the Jura and West Texas in Fig. 6.28). This is a fortui-

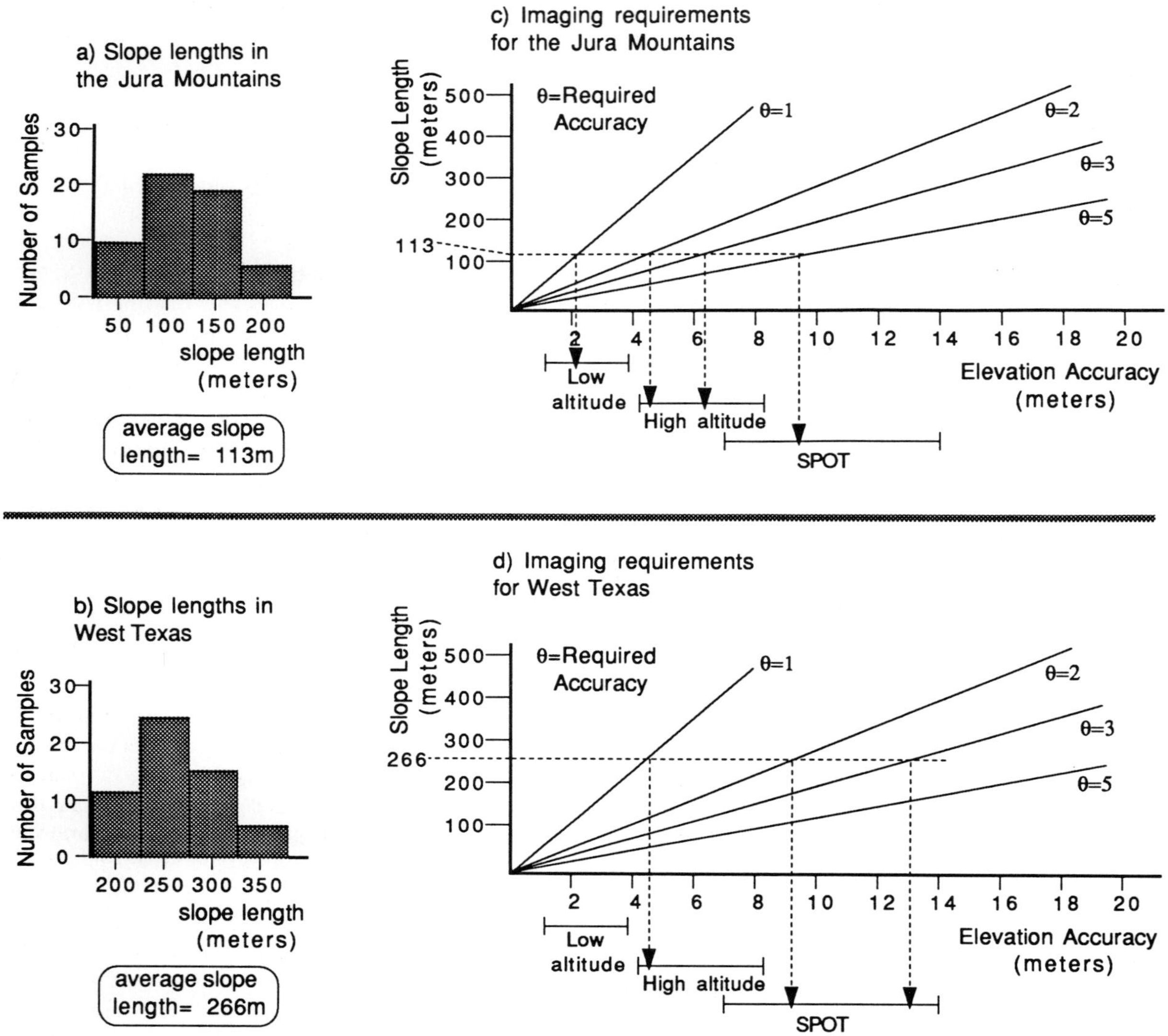

Fig. 6.28. Evaluating accuracy of dip measurements
The graph illustrates a two-step procedure used to evaluate the accuracy in dip measurements obtained from stereo data. **a** and **c** represent data from the Delémont Basin in the Swiss Jura Mountains area. **b** and **d** represent data for the Yates Field area in West Texas

tous relationship, because low relief areas require the greatest accuracy of measurement.

Figure 6.28 illustrates a simple, two-step procedure to select stereo data suitable for structural mapping in different areas. The first step determines the average slope length of an area of interest and is represented by the two histograms on the left in Fig. 6.28. The histogram (Fig. 6.28 a) shows the lengths of 50 sample slopes measured from existing surface geological maps in the Delémont Basin of the Swiss Jura mountains. Figure 6.28 b shows a sample of 50 slopes measured from aerial photography in the Yates Field of West Texas. The histograms show an average lenght of 113 m from the slopes measured in the Jura mountains and 266 m for the slopes measured in West Texas.

The second step is illustrated on graphs depicted in Fig. 6.28 c and d. These graphs express the relationship among the following three variables:

1. The vertical axis represents the average length of slopes in the area as measured from geological maps, aerial photography, or other surface data.
2. The horizontal axis reflects the elevation accuracy of different stereo data (i. e., ± 2 m and ± 4 m, which was calculated for low- and high-altitude aerial photography, repectively, and ± 7 m for SPOT, as reported by Gugan and Dowman 1988 and Konency et al. 1987).
3. The diagonal lines (θ) represent the degrees of accuracy derived from the ratio of error in relating elevation measurement to length of slope as seen in Fig. 6.27.

The dashed lines shown in Fig. 6.28 c, d represent the second step of the procedure. These lines are drawn horizontally from the y-axis (i. e., average slope length as determined in step one) to the intersection point with the diagonal accuracy line (θ), and then they are projected vertically to determine the type of stereo data needed to achieve the desired accuracy. For example, data from the Jura indicate that SPOT stereo will give an accuracy of 5°, whereas low-altitude photography will be accurate to within 1° and high-altitude photography to within 2° to 3°. In this setting, the expected improvement in structural detail that would be provided by aerial photography probably does not justify the considerable added effort and cost. However, in the Pecos Arch region, the expected accuracies are to within 2° to 3° for SPOT and to within 1° or less for aerial photography. Here, the expected improvement provided by the use of aerial photography could be critical for detailed mapping of subtle structures and, thus, justifies the added effort and cost.

6.8 Incorporating Stereo Capabilities

The availability and usefulness of stereo satellite data for geological applications raises questions as to how one incorporates stereoscopic capability into a remote-sensing program. Simple pocket and mirror stereoscopes with parallax bars are best used for quick or preliminary studies of small areas but are cumbersome and time-consuming for systematic, quantitative mapping. Analytical stereoplotters are more appropriate for large-scale, high-precision studies but are expensive and require a trained technical staff. This technology is best suited for service companies that focus on geological and topographic mapping activities.

The alternative approaches offering stereo capability in the digital environment are now available due to recent advances in display technology and three-dimensional computer graphics. These systems are extremely easy to operate, relatively inexpensive to install, and useful for all types of stereo data. Improvements in digital terrain models are making them more attractive to geologists as exploration tools. Disadvantages of all computer-based methods include the need for training explorationists to operate the system and the need for all data to be digitized before use. However, it seems that these types of systems are most suitable for exploration companies and will most likely become more prevalent in the future.

6.9 Summary

The availability of stereo satellite data reinforces the value of stereoscopic measurements to remote sensing exporation geology. SPOT and other satellites are now available, providing inexpensive global stereo image coverage in digital format, and the data can be digitally enhanced and displayed in stereo. Form-line surface structure (FSS) maps generated from stereo imagery data are particularly useful for supporting exploration in low-amplitude structures or highly deformed regions. In the early stages of exploration, such maps can be used to establish the style, trend, size, and spatial distribution of structures, as well as to investigate the development and geometries of key structural elements. At more advanced prospecting stages, such maps are best used to support the mapping of subsurface structures in areas of poor quality seismic data.

Despite the attractiveness of SPOT stereo images which represent the new generation of stereo data, their accuracy of dip measurement can be limited by the satellite's resolution (which limits vertical accuracy). This caveat decreases SPOT's usefulness for detailed mapping in some areas. In these cases, the explorationist would be required to use high-resolution aerial photography or airborne radar. Thus, the explorationist must initially analyze the slope conditions and the structural details required for the study to decide if using satellite data is appropriate.

References and Further Reading

Baraldi A (1992) A knowledge-based system for unsupervised classification of high resolution multispectral satellite images. In: Fritz LW, Jucas JR (eds) Int Symp of Photogramm and remote sensing, 17th Congr, vol 29, part B2 Commission 2. International Society of Photogrammetry and Remote Sensing, Bethesda, MD, p 59

Berger Z, Aghassy J (1983) Near surface moisture and evolution of structurally controlled drainage in soft sediments. In: LeFleur RG (ed) Groundwater as a geomorphic agent. 13th Ann Binghampton Geomorphology Symp, Allen and Unwin, Winchester, MA, pp 59–77

Berger Z, Corona FV (1986) Landsat structural analysis of the Rhine Valley and the Jura Mountain area, western Europe. Int Symp on Remote sensing of the environment, 5th Thematic Conf, Colorado Springs, CO. Environmental Research Institute of Michigan, Ann Arbor, pp 35–48

Berger Z, Williams TH, Anderson DW (1991) Geological stereo mapping of geological structures with SPOT satellite data. AAPG Bull 76 (1): 101–120

Brockelbank DC, Tam AP (1991) Stereo elevation determination techniques for SPOT imagery. Photogramm Eng Remote Sens 57 (8): 1065–1073

Bryant M (1974) Digital image processing: optronics. Inter Publ 146, Chelmsford, Mass

Burrough PA (1986) Principles of geographic information systems for land resource assessment. Clarenden Press, Oxford

Christner DG (1940) Todd Ranch (Oil) discovery, Crockett Country, Texas. AAPG Bull 24 (6): 1126–1127

Diebold P, Laubscher HP, Schneider A, Tschopp R (1960) Geologic atlas of Switzerland, Sheet 1085, St Ursanne

Galley JE (1958) Oil and geology in the Permian Basin of Texas and New Mexico. In: Weeks LG (ed) Habitat of oil. AAPG Sympo Ser 582: 395–446

Gess G, Chrowicz J, Becue B, Curnelle R, Deroin JP, Huger J, Perrin G, Ronfola D (1986) Methodology for the use of SPOT imagery in petroleum exploration. In: SPOT 1 image utilization assessment, results. Centre National d'Etudes Spatiales, Toulouse, France, pp 811–819

Gugan DJ, Dowman IJ (1988) Topographic mapping from SPOT imagery. Photogramm Eng Remote Sens 54 (10): 1409–1414

Hennen RV, Metcalf RJ (1929) Yates oil pool, Pecos County, Texas. AAPG Bull 13 (12): 1509–1556

Hopkins HR, Navail H, Berger Z, Merembeck BF, Brovey RL, Schriver JS (1987) Structural analysis of the Jura Mountains-Rhine Graben intersection for petroleum exploration using SPOT stereoscopic data. In: SPOT 1 image utilization assessment, results: Centre National d'Etudes Spatiales, Toulouse, France, pp 803–810

Jachinski J, Zielinski J (1992) Digital stereoplotting using the PC-SVGA monitor. In: Fritz LW, Lucas JR (eds) Int Symp for Photogramm and Remote Sensing, 17th Congr, vol 29, part B2, Commission 2. International Society of Photogrammetry and Remote Sensing, Bethesda, MD, p 127

Klaver J, Walker AS (1992) Entry level digital photogrammetry: latest development of the DVP. In: Fritz LW, Lucas JR (eds) Int Symp for Photogramm and Remote Sensing, 17th Congr, vol 29, part B2, Commission 2. International Society of Photogrammetry and Remote Sensing, Bethesda, MD, p 31

Koenig G, Lohmann P, Engle H, Kruck E (1992) A digital stereophotogrammetry system in a GIS environment. In: Fritz LW, Lucas JR (eds) Int Symp for Photogramm and Remote Sensing, 17th Congr, vol 29, part B2, Commission 2. International Society of Photogrammetry and Remote Sensing, Bethesda, MD, p 199

Konency G, Lohmann P, Engle H, Kruck E (1987) Evaluation of SPOT imagery on analytical photogrammetric instruments. Photogramm Eng Remote Sensing 53 (9): 1223–1230

Lattman LH (1959) Geomorphology applied to oil exploration. Min Indus 28 (6): 1–4

Lattman LH, Matzke RH (1961) Geological significance of fracture traces. Photogramm Eng 27: 435–438

LaPrade GL (1972) Stereoscopy – a more general theory. Photogramm Eng 38 (12) 1177–1187

LaPrade GL (1972) Stereoscopy – will data or dogma prevail? Photogramm Eng 39 (12): 1271–1275

LaPrade GL (1980) Stereoscopy. In: Slama CC (ed) Manual of photogrammetry, 4th edn. American Society of Photogramm, Falls Curch, VA, pp 519–545

Laubscher HP (1977) Fold development in the Jura. Tectonophysics 37: 337–362

Laubscher HP, Bernoulli D (1980) Cross-section from the Rhine-Graben to the Po Plain. In: Laubscher HP, Bernoulli D (eds) Geology of Switzerland, guidebook, part B. Wepf, Basel, pp 183–209

Mark RP, Franke K (1992) New interpretation instruments from Carl Zeiss Jena. In: Fritz LW, Lucas JR (eds) Int Symp for Photogramm and Remote Sensing, 17th Congr, vol 29, part B2, Commission 2. International Society of Photogrammetry and Remote Sensing, Bethesda, MD, p 75

Miller SB, Thiede JE (1992) A line of high-performance digital photogrammetry workstations – the synergy of general dynamics, Helava Assoc and Leica. In: Fritz LW, Lucas JR (eds) Int Symp for Photogramm and Remote Sensing, 17th Congr, vol 29, part B2, Commission 2. International Society of Photogrammetry and Remote Sensing, Bethesda, MD, p 87

Moffit FH (1985) Photogrammetry, 3rd edn. Harper and Row, New York

Ostrowski JA, Benmouffok D, He DC, Horler DNH (1989) Geoscience applications of digital elevation models. Statistical Applications in the Earth Sciences, ed. F. P. Agterberg and G. F. Bonham-Carter, Geological Survey of Canada, Paper 89-9, p. 33–37

Peltzer G, Armijo R, Tapponnier P (1987) Rate of slip on the Altyn Tagh fault (North Tibet, China). In: SPOT 1 image utilization assessment, results. Centre National D'Etudes Spatiales, Toulouse, France, p 709–729

Petrie G (1992) Trends in analytical instrumentation. ITC J 1992–94 Spec Issue, p 364

Ray RG (1960) Aerial photographs in geologic interpretation and mapping. Geol Surv Prof Pap 373

Renfro HB, Feray DE, Dott RH Sr, Bennison AP (1973) Geological highway map of Texas. AAPG United States Geological Highway Map Ser, Map 7

Rodriguez V, Gigord P, deGaujac AC, Munier P, Begni G (1988) Evaluations of the stereoscopic accuracy of the SPOT satellite. Photogramm Eng Remote Sens 54 (2): 217–221

Sabins FF (1987) Remote sensing: principles and interpretation. W. H. Freeman, San Francisco

Sasowsky KC, Peterson GW, Evans BM (1992) Accuracy of SPOT digital elevation model and derivatives: utility for Alaska's north slope. Photogramm Eng Remote Sens 58 (6): 815–824

Siegel BS, Gillespie AR (1980) Remote sensing in geology. John Wiley and Sons, New York

Slama CC (ed) (1980) Manual of photogrammetry, 4th edn. American Society of Photogramm, Falls Church, VA

Spicer A (1980) Tectonic map of Switzerland. Commission Geologue Suisse, scale 1 : 500 000

Star JL, Estes JE (1992) A system for the integration of remote sensing and GIS. In: Fritz LW, Lucas JR (eds) Int Symp for Photogramm and Remote Sensing, 17th Congr, vol 29, part B2, Commission 2. International Society of Photogrammetry and Remote Sensing, Bethesda, MD, p 255

Taranik JK (1978) Characteristics of the Landsat multispectral data system. USGS Open-File Rep 78–187, p 76

Taranik JK (1987) First results of international investigation of the applications of SPOT 1 data to geologic problems, mineral and energy exploration. In: SPOT 1 image utilization assessment, results. Centre National D'Etudes Spatiales, Toulouse, France, pp 701–708

Thompson MH (ed) (1966) Manual of photogrammetry, 3rd edn. American Society of Photogramm, Falls Church, VA

Tilley DG (1992) SAR-optical image fusion with Landsat, Seasat, Almaz and ERS-1 satellite data. In: Fritz LW, Lucas JR (eds) Int Symp for Photogramm and Remote Sensing, 17th Congr, vol 29, part B2, Commission 2. International Society of Photogrammetry and Remote Sensing, Bethesda, MD, p 512

Toth CK (1992) A GIS workstation-based analytical plotter. In: Fritz LW, Lucas JR (eds) Int Symp for Photogramm and Remote Sensing, 17th Congr, vol 29, part B2, Commission 2. International Society of Photogrammetry and Remote Sensing, Bethesda, MD, p 240

Toutin T, Carbonneau Y, St-Laurent L (1992) An integrated method to rectify airborne radar imagery using DEM. Photogramm Eng Remote Sens 58 (4): 417–422

Trümpy R (1980) An outline of the geology of Switzerland. In: Laubscher HP, Bernoulli D (eds) The geology of Switzerland, guidebook, part 2. Wepf, Basel, p 104

Welch R (1989) Desktop mapping with personal computers. Photogramm Eng Remote Sens 55 (11): 1651–1662

Williams TH, Thompson LV (1988) Digital display of SPOT stereo images. In: Remote sensing for exploration geology. Proc 6th Thematic Conf, Houston, TX. Environmental Research Institute of Michigan, Ann Arbor, pp 345–351

Wolf PR (1974) Elements of photogrammetry. McGraw-Hill, New York

Chapter 7 Structural Analysis of Sedimentary Basins

7.1 Introduction

At this stage of the book, all the aspects of imagery interpretation of different structural features have been covered and methods for constraining the analysis with surface and subsurface controls have been illustrated. The next step is to introduce methods to integrate the analysis of the imagery data into a conventional exploration program. Presented here is a systematic approach which begins with the analysis of the entire sedimentary basin and progresses towards the recognition of regional-scale features and their associated hydrocarbon leads. This process often leads to the recognition of new concepts and leads in both frontier and mature areas. As emphasized throughout this book, the analysis is limited to the recognition of geological features that can be constrained with other surface and subsurface exploration tools (seismic, gravity, magnetic, etc.). An additional step in this approach may include the analysis of small-scale linear features that cannot be constrained with subsurface data. This step is not illustrated here for the sake of brevity and the lack of encouraging results experienced by the author in this endeavor.

7.2 Approach

The basic concept of the approach promoted here is illustrated in Fig. 7.1. First, from a satellite imagery point of view, a basin is divided into two categories:

- The basin's margins, which consist predominantly of exposed structures, can be mapped and analyzed in detail from surface data (satellite, aerial photography and geological maps). Basin margin structures are represented on the interpreted data by solid lines.
- Basin interior structures are either obscured by vegetation and soil cover or are completely buried under unconsolidated sediments. These structures are recognized on imagery by their geomorphic expressions and are further constrained by other available subsurface data. These structures are represented by dashed lines or arrows at each end of the element.

The analysis starts in the basin's margins and then extends into the central parts. It proceeds in seven steps which are illustrated in Fig. 7.2.

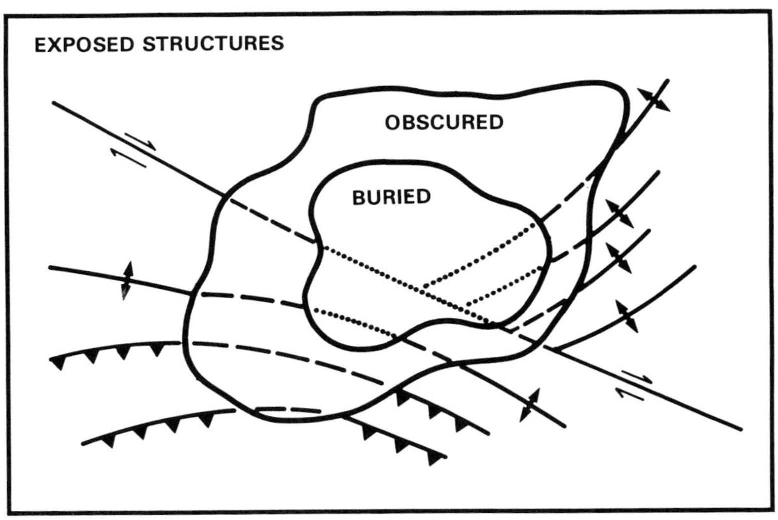

Fig. 7.1. Conceptual Model Showing spatial relationships among structures exposed in the basin's margins and buried and obscured structures in the basin's interior

Fig. 7.2. Procedures for integration of imagery data

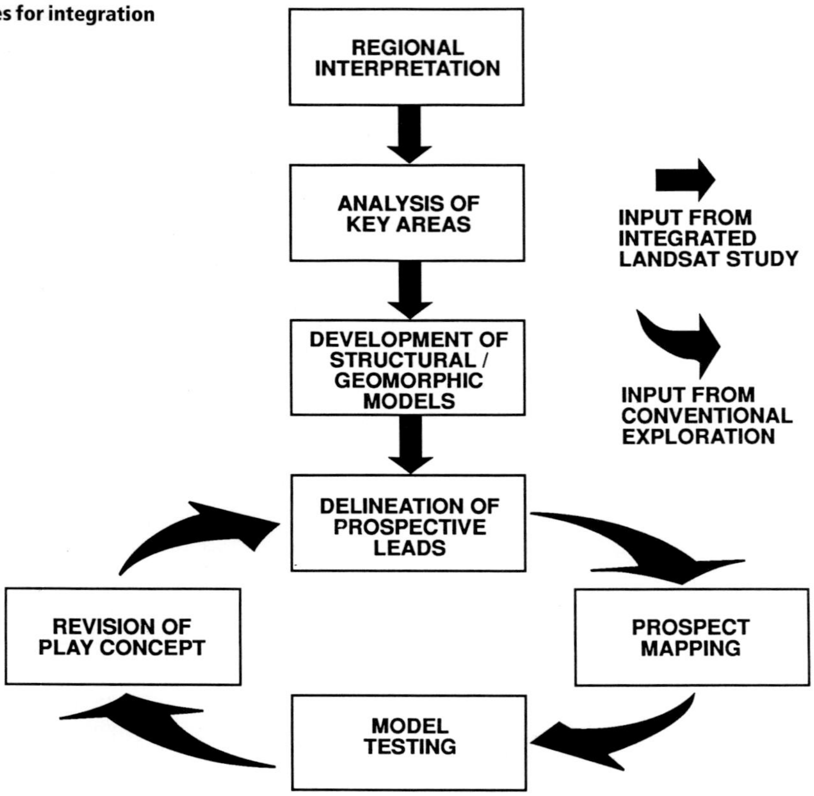

The first four steps are based on the integration of satellite imagery with other available data sets. The remaining three steps consist of conventional exploration programs including mapping prospects with new seismic data and testing and revising the new concepts and leads with a drilling program. Note, however, that after the first round of drilling has been completed and the structural models have been revised, the process of identifying leads with imagery data should also be reexamined.

At the outset of such studies, it is recommended to compile a regional data set which includes the following:

- selected structural and isopach maps
- a location map of seismic lines
- potential field data
- pool and show maps
- surface geological data, particularly surface structures and tectonic element maps.

These data can be prepared as a series of overlays which are at the same scale as the interpreted imagery or they may be digitized and superimposed with an image processing system. The latter choice, however, leads to increases in expense and effort which should be weighed against its usefulness. The analytical procedures involved in the integration of image data and the resulting products of such studies are briefly demonstrated in this chapter with two examples; a fracture play study in the Molasse Basin, Switzerland, and an investigation of buried Mesozoic basins in the Coastal Plain region of the eastern United States.

7.3 Analysis of Fractured Reservoir Trends in the Swiss Molasse Basin

7.3.1 Objectives

There are several areas within the sedimentary basins of western Germany and Switzerland where conventional exploration has had limited success so far. In these areas, the lack of commercial discoveries has been attributed to the overall tightness of the sedimentary section which limited the ability of hydrocarbons to migrate and accumulate in potential traps. Within these areas, however, a few commercial accumulations of hydrocarbons were found in places where reservoir quality was enhanced by natural fracture systems. These were attributed to the presence of active fault systems in the vicinity of the producing structures (Berger and Gundlach 1990).

The exploration department of BEB Erdgas und Erdöl GMBH, in cooperation with the remote sensing group of Exxon Production Research Co., has successfully utilized integrated satellite imagery studies as part of the exploration program for fracture reservoir plays in these areas. The approach used in these studies and their contributions to the exploration programs are demonstrated here with an example taken from a report on the Swiss Molasse Basin which was prepared by the author and T. Gundlach (1990).

7.3.2 Approach

Satellite imagery may be used to detect fracture-related reservoir plays in two different ways. The most common way is to interpret all the linear surface features that can be recognized on the imagery and to spatially relate their densities and trends to predict areas with enhanced reservoir properties by fracturing (Norman 1976; Peterson 1979). This approach provides the most comprehensive and unbiased interpretation of satellite data as well as a relatively inexpensive way to generate potential leads in large areas. It was very popular during the early days of remote sensing applications, particularly among small operators, but has been received with great skepticism by many geoscientists (Wise 1982; Sterns et al. 1988; see also discussion on linear features in Chap. 4). The most common criticism is that Landsat "lineament maps" (recall that linear features are often mislabeled as lineaments) are usually overcrowded with thousands of small-scale linear features virtually impossible to constrain with subsurface data. As a result, it is often not possible to resolve the tectonic origin, timing of reactivation and possible influence of these surface features on the fracture system at the reservoir level.

The alternative approach, which was used in this study, is to focus the interpretation of the imagery on the recognition of the fracture plays that develop along or at the intersection of major active or reactivated fault systems that constitute the primary structural framework of the basin. Recall that this will be considered, in most cases, as first-order magnitude lineaments. This approach limits the scope of the study to only one or a combination of two types of fracture plays – fault-related or force fold-dependent. It provides fewer prospective leads than the first approach, but is well received and utilized by explorationists for the following reasons:

- Structural elements identified by this approach can be overlooked during the routine mapping of subsurface structures in a basin. This is primarily because many of them have subtle expressions that may be hidden by other more dominant structural elements within the basin, particularly if these structures trend obliquely to the regional trend. (Remember that seismic lines are always oriented to capture the dominant structural features.)
- Once identified on the imagery, these structural elements can be analyzed further with other data sets such as gravity, magnetic and surface geology maps as well as seismic data.
- In many other sedimentary basins, hydrocarbon plays have been established along such deep-seated reactivated fault systems of first-order magnitude lineaments (see the discussion on the Scipio-Albion Trend in the Michigan Basin by Harding and Lowell 1979).

7.3.3 Geological Settings of the Molasse Basin

The Molasse is a large, arcuate foreland basin formed during Oligocene to Recent orogenic events of the adjacent Alps (Trümpy 1980). The basin can be divided into two sub-basins, the East Molasse (Germany and Austria) and the Swiss Molasse, which differ significantly in their structural style, reservoir characteristics and hydrocarbon potential (Fig. 7.3).

The East Molasse extends along the Alps from the Bohemian Massif in the west to Lake Constance in the east. To the north, the Molasse sediments are truncated against the Mesozoic sediments of the Swabian Alb, and to the south, it is overridden by the fold and thrust belt of the Alps. The structural characteristics of the East Molasse are generally related to crustal bending of the basin (flexural foreland basin) in response to loading of the Alps. This process resulted in the formation of high-angle normal fault systems that trend parallel to the structural axis of the basin and form small but effective traps (A-A' in Fig. 7.3; Betz and Wendt 1983). Continuing compressive deformation of the southern

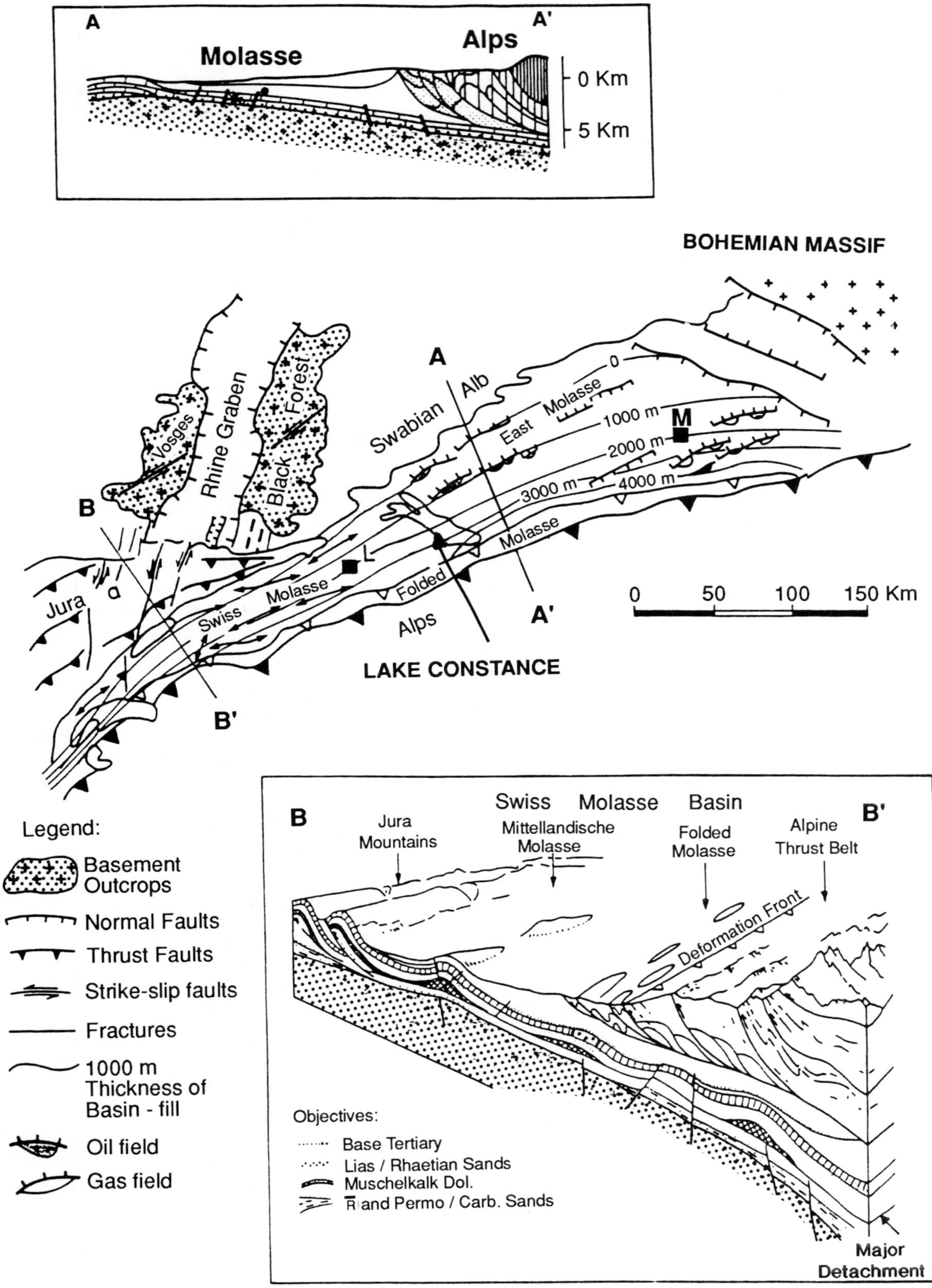

Fig. 7.4. Geological cross section of the east and west Molasse Basin
Showing the occurrences of hydrocarbon in various reservoir rocks and the distribution of organic-rich shales which are potential source rocks. (After Betz and Wendt 1983)

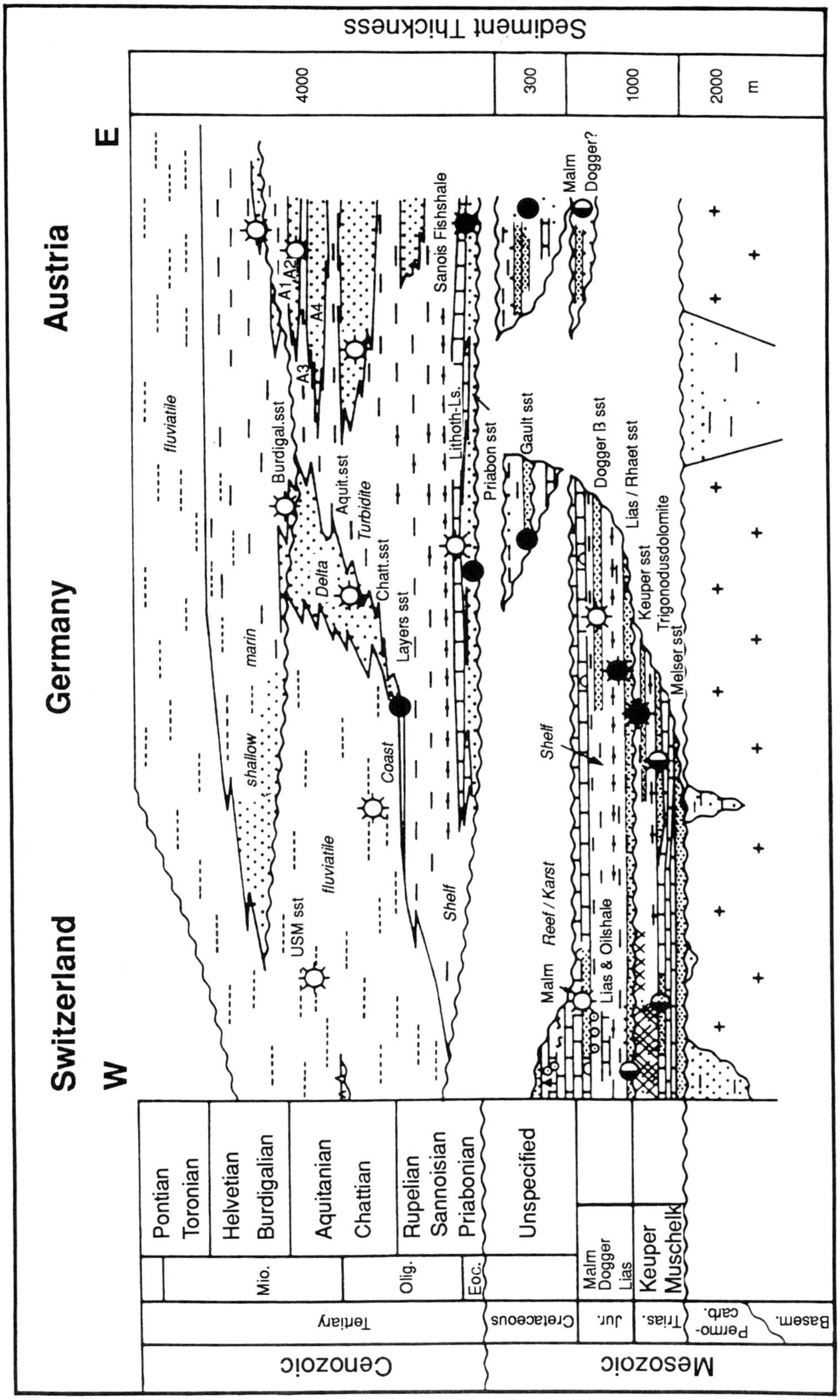

Fig. 7.3. Structural framework in the Molasse Basin

Showing the major tectonic elements and variations in structural styles between the East Molasse and the Swiss Molasse. (Compiled from various sources mentioned in the text)

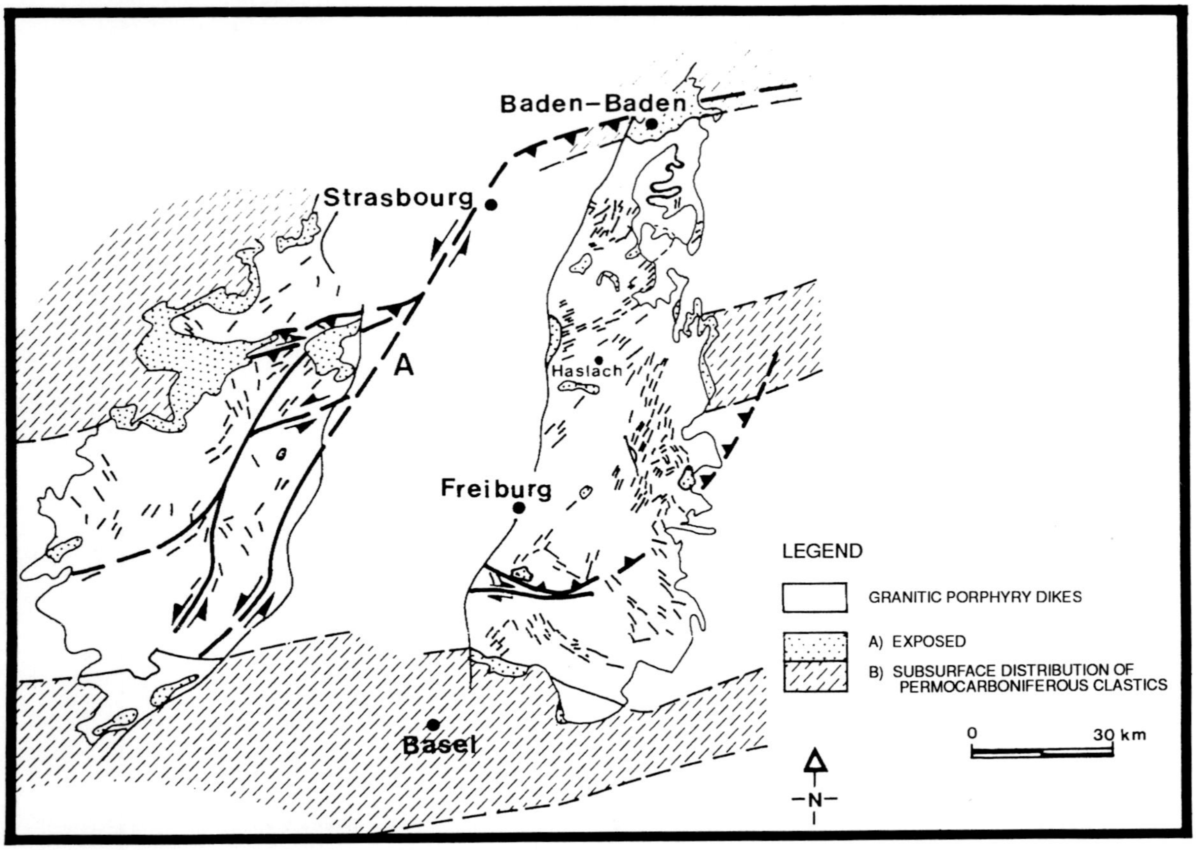

Fig. 7.5. Tectonic elements in the southern Upper Rhine Graben
Showing strike-slip elements of Carboniferous/Lower Permian age. (After Eisenbacher et al. 1989)

part of this basin resulted in folding of Molasse sediments in a narrow zone in front of the Alps, known as the Folded Molasse. This fold zone also exists in the West Molasse. The producing reservoirs include deltaic and shore sand bars of the Tertiary basin fill and pre-Molasse sand and carbonate units of the entire Mesozoic section. The potential source is considered to be the organic-rich Tertiary as well as Mesozoic shales and possibly some Carboniferous coal-bearing clastics (Fig. 7.4; Boigk 1981; Lemcke 1977).

Extending along the Alps from Lake Constance to the west, the Swiss Molasse Basin is bordered to its north and west by the arcuate fold and thrust belt of the Jura Mountains. The structural evolution of this sub-basin has been influenced by the presence of a Mesozoic evaporite sequence which, during continuous compressive deformation, supported the origin of a detachment surface triggering thrusting and folding of the Swiss Jura and the formation of isolated folds within the Molasse sediments (B-B' in Fig. 7.3). The location of the folds appears to be related to the presence of ramp-like structures associated with the block-faulted autochthonous floor.

Hydrocarbon trapping may either occur in these structures in folds above the salt or in uplifted blocks below the dêcollement surface. The stacking of these two potential targets increases the attractiveness of these structures for hydrocarbon exploration. However, facies changes from the east to the west Molasse within the same stratigraphic units resulted in a decline in reservoir quality towards the west (Betz and Wendt 1983).

The Tertiary Molasse sediments are of continental origin, lacking continuous porous sand bodies. The Mesozoic sands laterally change into more shaley units. Additional reduction of reservoir qualities might be expected in the West Molasse because of its higher compaction relative to the East Molasse, related to the overall shortening and burial of the entire section (Vollmayr 1989, pers. comm.). The Jurassic Malm and the Triassic Muschelkalk are considered to be the most potential reservoir

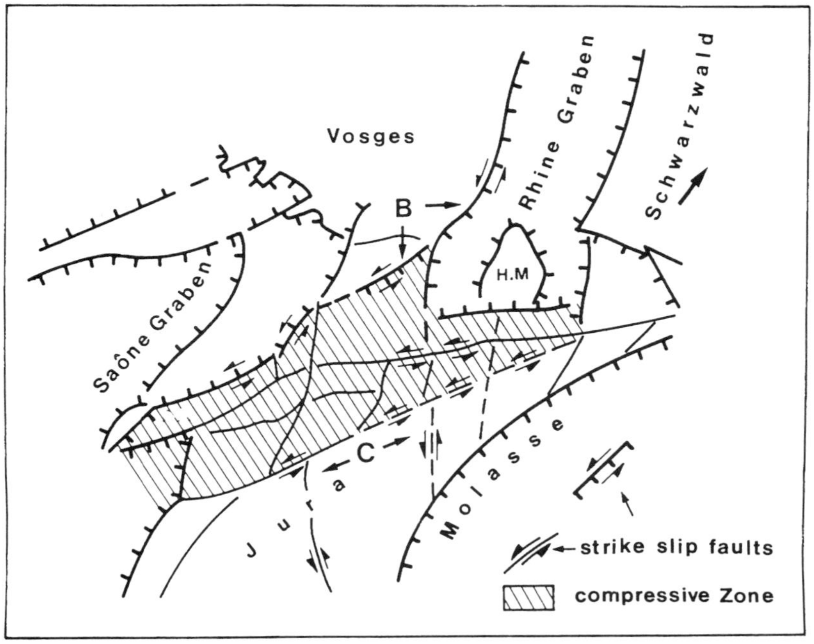

Fig. 7.6. Basin grain map of the southern Rhine Graben area
Showing locations of strike-slip and reactivated normal faults. (Prepared by T. Gundlach)

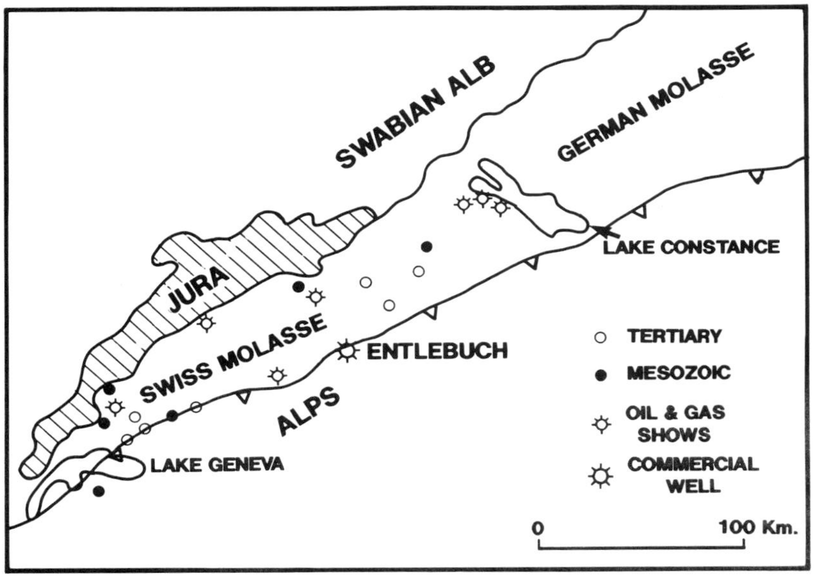

Fig. 7.7. Well locations in the Molasse Basin
Many wells had oil and gas shows, but only the Entlebuch well has produced commercial quantities of gas

rocks in this area (Lemcke 1977) because they are likely to be less affected by compaction processes. Source rocks are mainly of Mesozoic and possibly Tertiary and Carboniferous age.

Another important process that has affected the structural style and, perhaps, the hydrocarbon potential of this sub-basin is related to the superposition of a wrench fault system of at least Upper Carboniferous age (A in Fig. 7.5; Ziegler 1982; Krohe and Eisbacher 1988; Eisbacher et al. 1989). This fault system was reactivated during the main rifting event of the Rhine-Bresse System predominantly in a normal sense followed by shear movements since Pliocene time (Illies et al. 1981; Laubscher and Bernoulli 1980; B in Fig. 7.6).

The complex movements of the reactivated basement grain have been well documented from surface geology data in the Swiss Jura as well as in the Molasse (C in Fig. 7.6; Laubscher and Bernoulli 1980; Rigassi 1980; Trümpy 1980). However, the in-

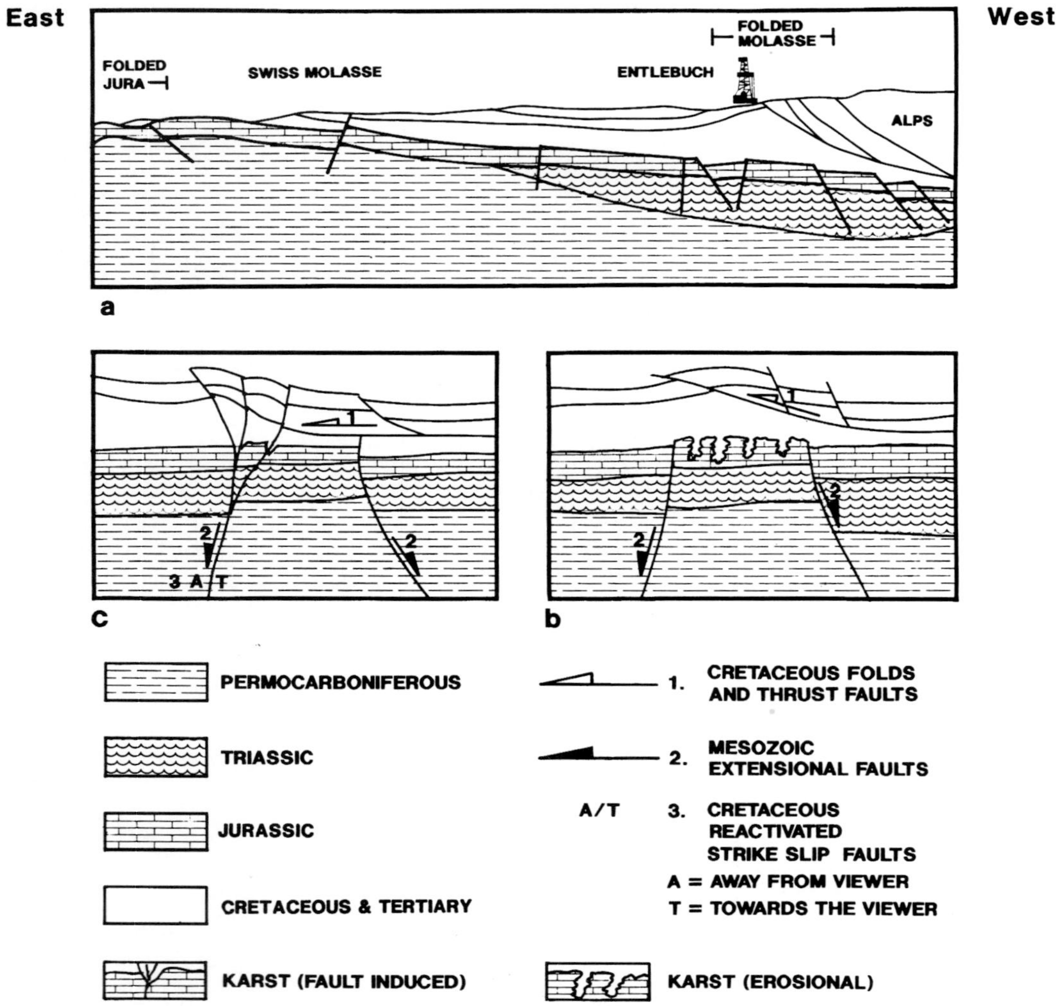

fluence of this fault system on the trapping mechanism of hydrocarbons has not been well established, especially the aspect of fractured reservoirs and available sealing lithologies.

7.3.4 Exploration in the Swiss Molasse

Since 1912, 35 wells have been drilled for different targets in the Swiss Molasse (Fig. 7.7). A large percentage of these wells had oil and gas shows, but only one well, Entlebuch 1, can be considered to have commercial potential of accumulated gas in Malm carbonates that show extensive karstification. This phenomenon occurred when Malm carbonates were subjected to subareal erosion. The karstification was possibly intensified along a pre-existing fracture system. The well was drilled through the thrust sheets of the folded Molasse into

Fig. 7.8. Interpreted seismic lines across the Entlebuch gas field
a shows regional structural interpretation and the producing well, which is located on an uplifted basement block beneath the subalpine molasse (after Vollmayr, pers. comm.). **b** and **c** show two different ways to interpret the typical seismic profile across the field. The first interpretation relates the producing closure to an uplifted block, whereas the second suggests the superimposition of a reactivated wrench fault

a large and complex compressional structural element in the Mesozoic rocks which was interpreted as a deep-seated thrust complex, related to the late stage of deformation of the Alps (Vollmayr and Wendt 1987; Fig. 7.8 a, b). These authors point out, however, that other types of interpretation of this structure have been postulated, linking its complexity to superimposition of a reactivated wrench fault system (Fig. 7.8 c).

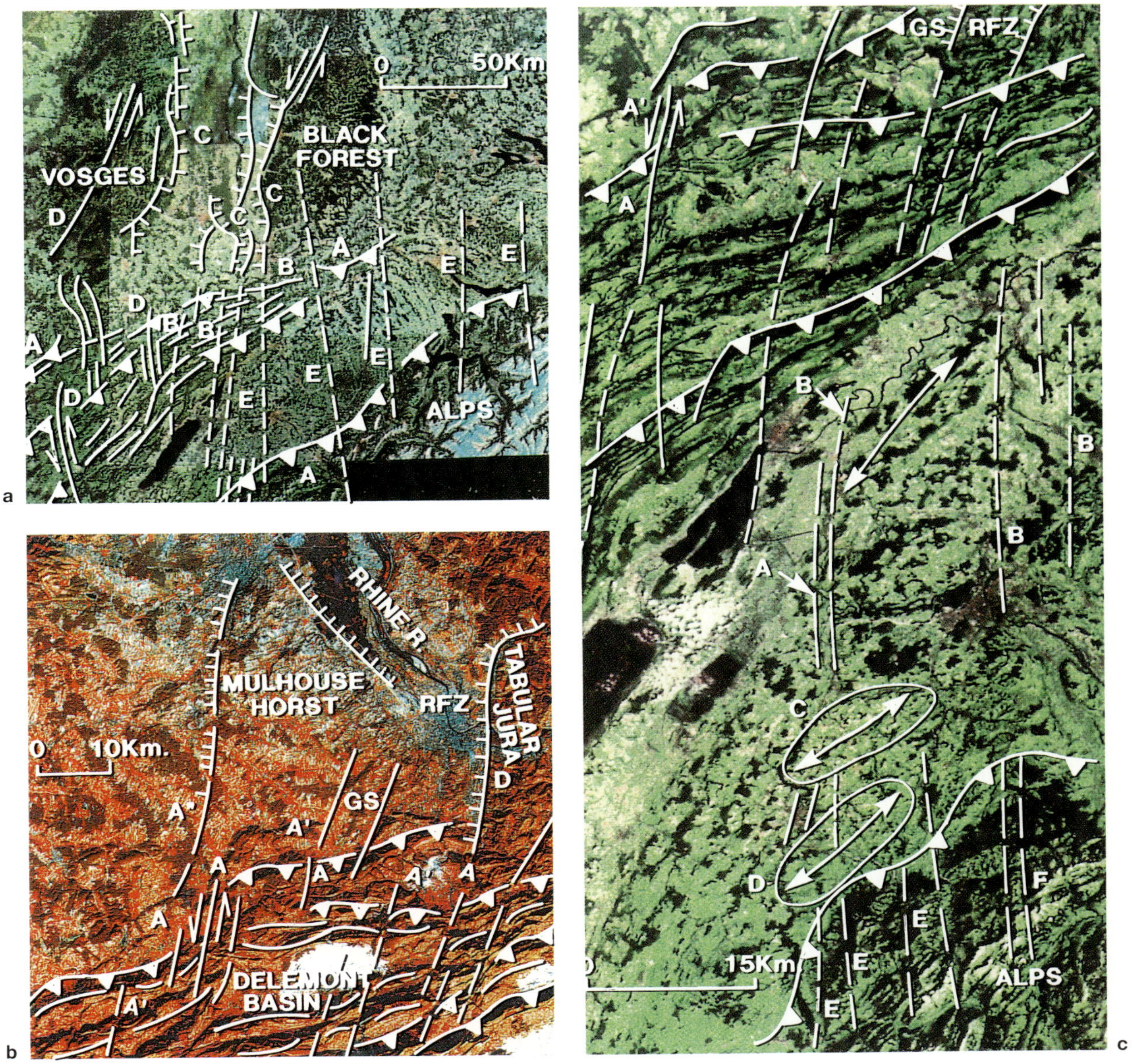

Fig. 7.9. **Interpreted imagery of the Molasse Basin**
Showing a regional structural interpretation (**a**), detailed analysis of the Jura Mountains (**b**) and detailed interpretation of the Rhine Valley Fracture Zone (RFZ) element (**c**). *Letters* are explained in the text

7.3.5 Regional Analysis of Satellite Imagery

The first step of regional interpretation is illustrated by the portion of a Landsat imagery mosaic covering the entire Swiss Molasse and adjacent areas (Fig. 7.9 a; see also the enlarged image of the same area in Fig. 3.22 and 6.21). The thrust faults of the Alps and the Swiss Jura exhibit irregular to sinuous FLTs which commonly trend perpendicular to the direction of maximum stress (A in Fig. 7.9 a). The folds in the Jura are well expressed as topographic ridges that trend parallel to the thrust faults (B in Fig. 7.9 a). High-angle dip-slip faults of the southern Rhine Graben consist of short segments of linear faults that trend at oblique angles and form a zigzag pattern. These faults, unless severely eroded, are well expressed at the surface where the upthrown side is consistently expressed as a topographic high (C in Fig. 7.9 a). Strike-slip faults in the Black Forest, the Vosges as well as in the Jura are expressed as relatively straight to curvilinear valleys where neither side is consistently higher than the other (D in Fig. 7.9 a). Alignments of north-northeast- to north-trending individual faults and fractures that cut and offset structures in the Jura and the Alps and, in some cases, seem to continue to manifest more subtle expressions in the Molasse Basin are interpreted to be the surface of large-scale, reactivated buried fault systems (E in Fig. 7.9 a).

7.3.6 Analysis of the Basin's Margins

High-resolution imagery of the junction between the Rhine Graben and the Swiss Jura were used to investigate the cross-cutting relationship among the different structures identified in the regional analysis (Fig. 7.9 b; see more detailed imagery and interpretation in Figs. 3.22 and 3.23 b). It is evident from the imagery as well as from surface geology data that the Jura Mountains are divided into several stacks of thrust sheets that are bounded by northeast-trending zones of faults and fractures of different scale with a "tear" fault character. These complex boundaries have been described from surface data in the case of the western termination of the Delémont Basin (Laubscher and Bernoulli 1982). From a Landsat point of view, this boundary is a sharp and linear to curvilinear topographic element which in some cases separates structures with a different amount of deformation on both sides of the element (A in Fig. 7.9 b), whereas in other segments it cross-cuts the individual folds within the Jura (A' in Fig. 7.9 b). This relationship suggests that this fault system was active during the entire deformation event of the Jura Mountains, predominantly as a strike-slip fault but with only minor vertical displacement. The detachment surface can be traced further to the north into the Rhine Graben fill where it forms the faulted western boundary of the Horst of Mulhouse (A" in Fig. 7.9 b) after having been offset to the west along east-west-trending transcurrent elements which were recognized as part of the Rhine-Bresse shear system (Illies and Greiner 1978).

Two other parallel to subparallel vertical detachment surfaces of the same magnitude can also be correlated to major Rhine Graben features. One trends into a narrow graben system that forms the eastern boundary of the Horst of Mulhouse – the graben of Sierentz-Wolschwiller (GS) – whereas the second one becomes part of the Rhine Valley Flexure Zone (RFZ; Fig. 7.9 b). Element GS is made up of two fault systems, the eastern one has strong surface expressions which can be traced through the Delémont Basin where the dipping limb of the syncline has been offset by it. The western part only exhibits subtle surface expressions with no observable offsets of the surrounding rock units. Element RFZ consists of a single through-going linear feature with an extremely conspicuous surface expression in the areas where it forms the faulted boundary of the Tabular Jura.

The interpretation of the Landsat imagery data indicates that the pre-Molasse basin floor, at least in the area of the north-trending detachment surfaces of the Jura, was affected by a continuous deformation restricted to distinct narrow zones which are triggered by the presence of an autochthonous normal fault system of regional scale.

7.4 Basin Interior

Elements GS and RFZ can be recognized clearly as surface topographic elements in the Molasse sediments (Fig. 7.9 c). They manifest linear topographic ridges which often control the location of the stream valleys. In some cases the valleys follow the topographic scarp, whereas in other cases they show an abrupt deflection when crossing these elements (A and B in Fig. 7.9 c). In places, the expression of the elements diminishes and subtle oblique anticlines are observed along and in the vicinity of its traces. The anticlines are outlined by unique drainage patterns which reveal their sizes (C an D in Fig. 7.9 c). Note that these elements can be traced from across the southern border of the Molasse Basin, and further into the Alpine nappes where they form quite similar features as in the exposed Jura Mountains (E in Fig. 7.9 c). These observations indicate the very recent strike-slip movement along some of these elements.

Additional north-trending young elements can be observed in other parts of the basin. One of these elements cuts and offsets the anticline in the folded Molasse where the Entlebuch 1 drill site was located (Fig. 7.10).

The Landsat imagery interpretation suggests that these elements consist of a zone of open fractures which is manifested by the intense breaching of the structures (Fig. 7.10). These cross-cutting fractures are well manifested on imagery, showing, in places, minor horizontal offsets of marker beds (Fig. 7.10).

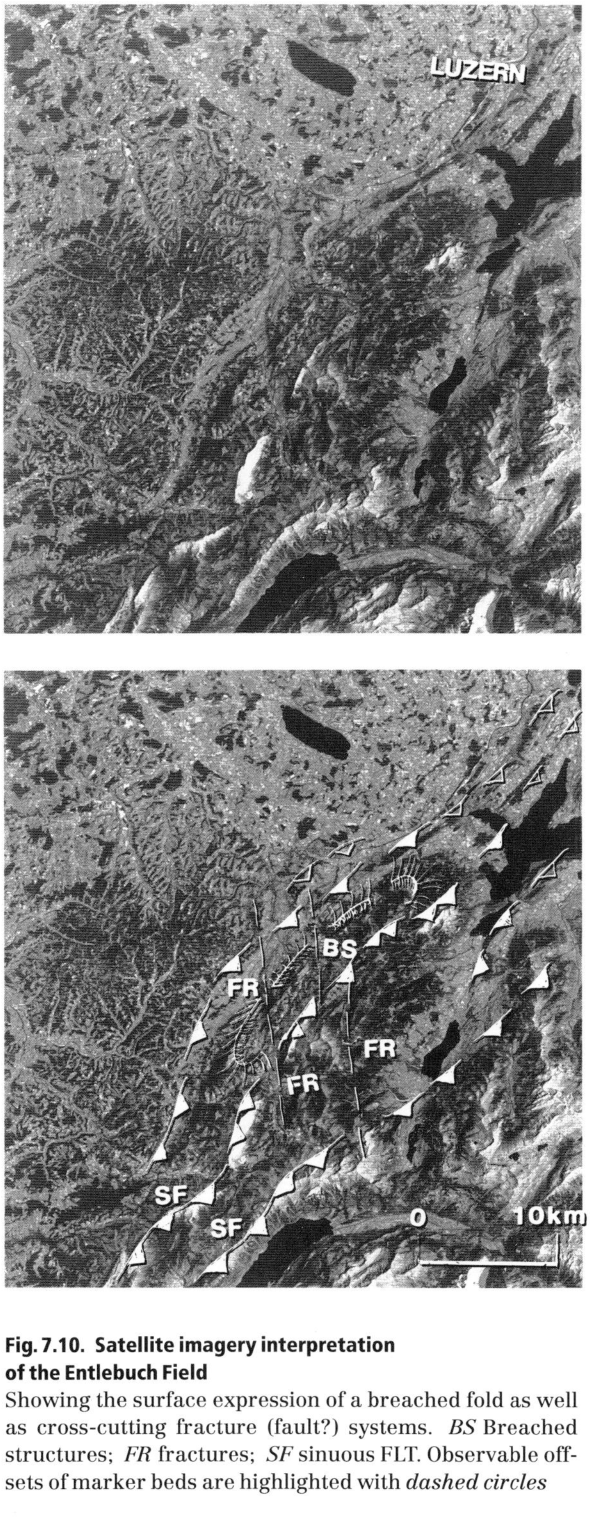

Fig. 7.10. Satellite imagery interpretation of the Entlebuch Field
Showing the surface expression of a breached fold as well as cross-cutting fracture (fault?) systems. *BS* Breached structures; *FR* fractures; *SF* sinuous FLT. Observable offsets of marker beds are highlighted with *dashed circles*

7.4.1 Subsurface Data

Element RFZ, described in the detailed Landsat analysis, is expressed on the residual gravity map as an approximately north-trending negative anomaly which appears to be superimposed on the northeast-trending grain of the basin (A in Fig. 7.11 a). Seismic data across this feature show the presence of a graben feature bounded by a series of north-trending fault systems that show the general characteristics of strike-slip faults (i. e., "flower" structures). The graben is narrow in the northern portion, showing predominately a compressional structural style and becoming wider and more diffuse to the south showing more extensional characteristics. The compressional characteristic of the graben in the north is attributed to its proximity to the Jura fold and thrust belt, whereas the extensional nature of the graben and its diffuse nature are attributed to the presence of subthrust normal horst blocks and related normal fault systems as shown in the cross section of the Entlebuch Field in Fig. 7.8.

7.4.2 Swiss Molasse Play Concept

The tectonic evolution of the Molasse basin and its potential hydrocarbon trapping mechanisms must be viewed in terms of the different stress conditions and related deformation styles which existed below and above the décollement surface in Triassic evaporites. This evolution might be described as a continuing and interactive process evolving the typical formation of a flexural foreland basin with the superimposition of a preexisting fault system. Due to stacking of the thrust sheets of the Alpine nappes, a flexure-like deformation of the basin floor occurred, while some areas in the proximity of the deformation front were uplifted.

The preexisting normal faults beneath the detachment surface were reactivated as reverse faults forming the local uplifts that built ramps for later thrusting. The process of thrusting, loading and inverting the basin floor into ramps occurs in succession. The last inverted structures are expected to be situated in the vicinity of the leading edge of the deformation front. This process of inverting normal faults that border basement blocks into reverse faults has been very well documented from a variety of foreland basins. The tectonic pattern of the Swiss Molasse basin floor was dominated by the presence of north-, northeast-southwest- and east-trending elements.

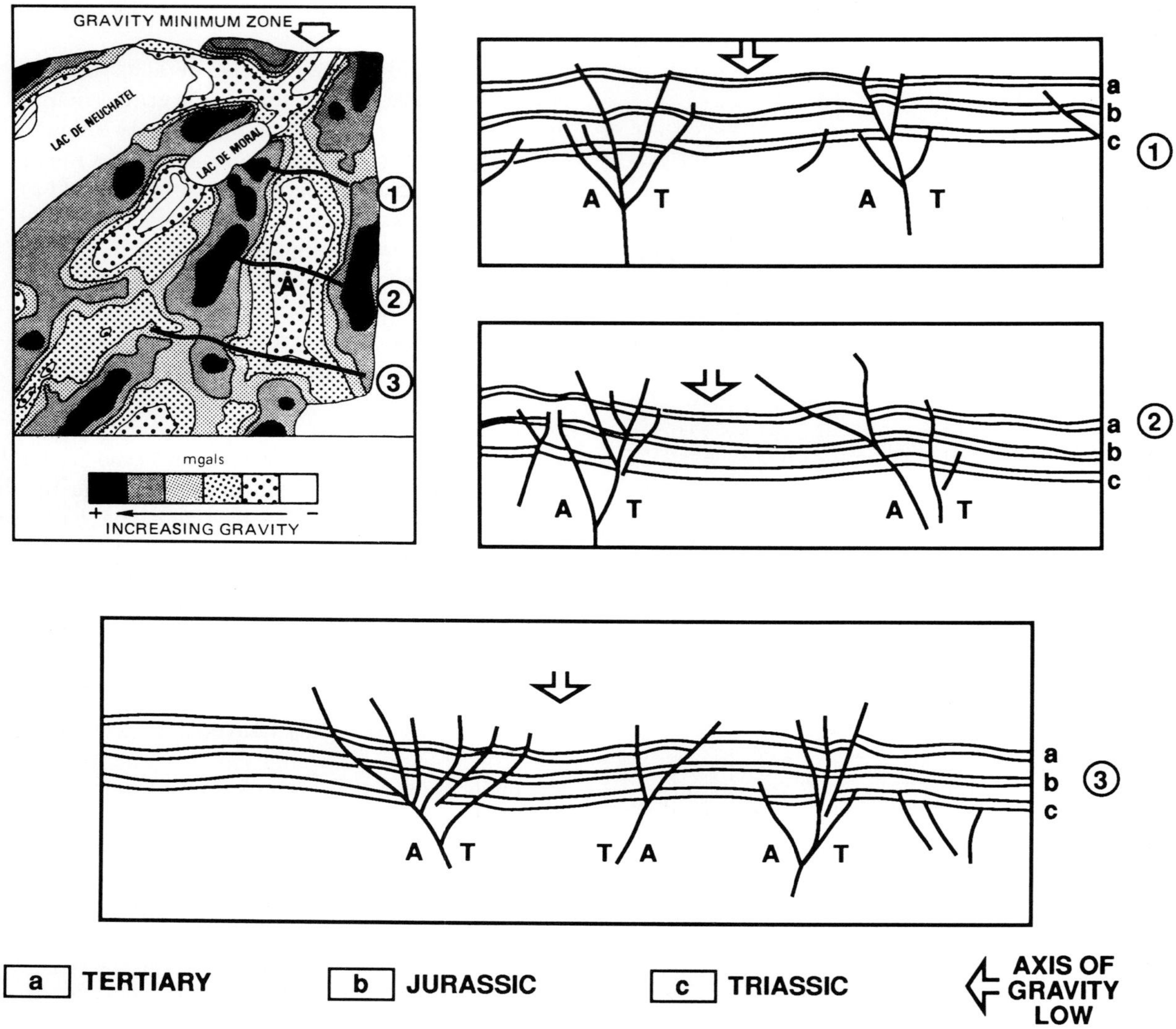

Fig. 7.11. Subsurface expression of the RFZ element
Gravity and generalized seismic interpretation of the RFZ element showing its size and possible structural style. (Data courtesy of BEB)

Within the area under investigation the north-trending elements appear to have exerted the most significant influence on the basin's prospective targets by young reactivation. Based on the existing literature, the north-trending elements must be related to major shear zones of at least upper Carboniferous age (Ziegler 1982). These elements have been reactivated twice, once as normal faults during early stages of rifting of the Rhine-Bresse System and for the second time during loading of the basin floor by continuous thrusting and the latest shear-dominated rifting stage of the Rhine Graben. The resulting fault pattern from this latest movement is illustrated in the tectonic map of Fig. 7.12.

A belt of extensional regime present in front of the deformation front represents the zone where bending, arching, and inverting of existing fault systems of the overridden autochthonous occurs. The structural characteristics of this belt are dominated by different principal stress orientations above and underneath the décollement surface (Blümling 1983; Müller and Hsü 1980). These kinds of tectonic processes led to the development of inverted blocks bounded by the major north-trending elements below the salt and to the formation of anticlinal features cross-cut by basement-induced shear movements above the décollement. Further north, however, the region is dominated by compression and the old fault systems have been closed

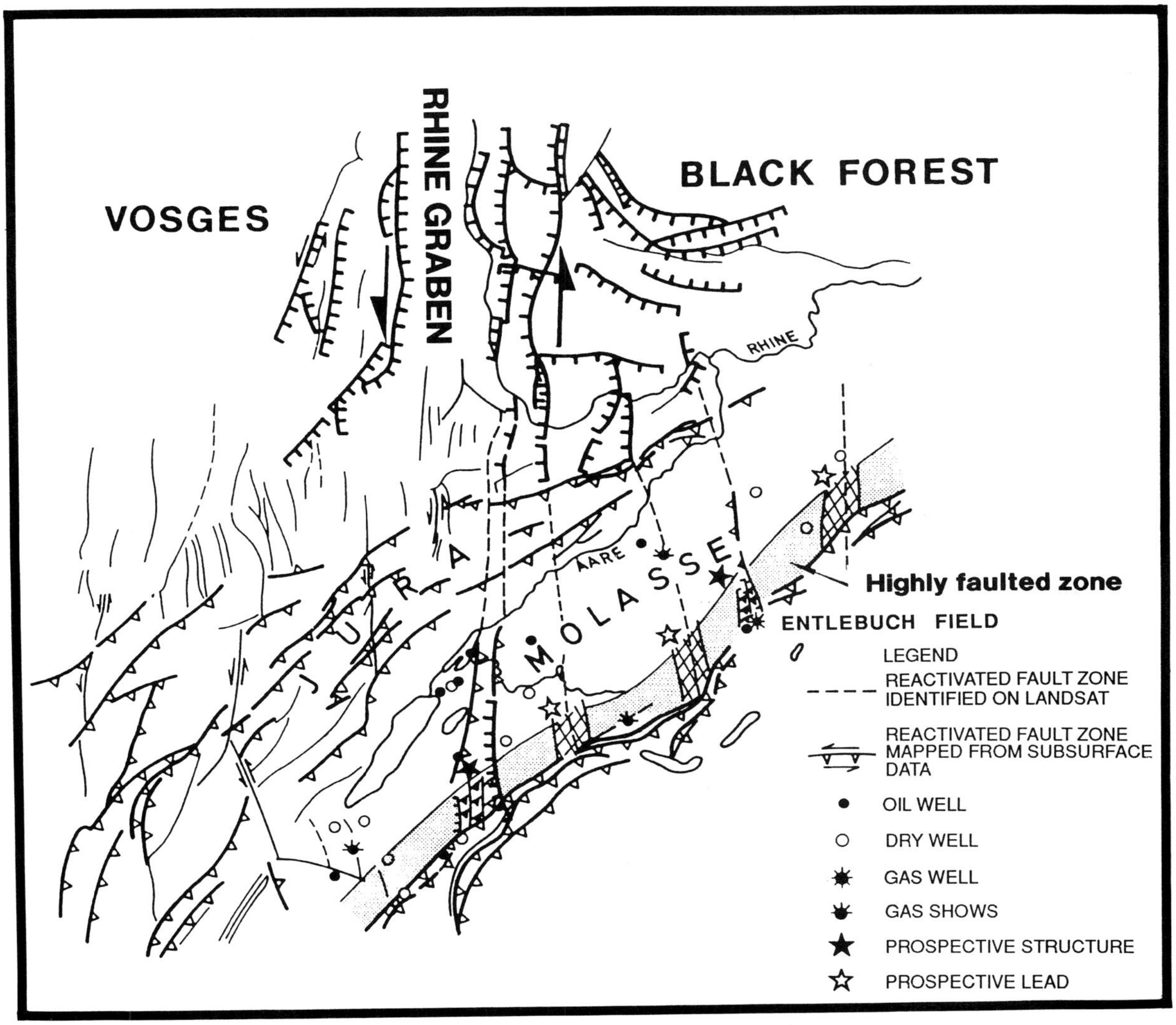

Fig. 7.12. Tectonic map of the Molasse Basin
Showing location of highly faulted zones and their relation to wells, prospects and leads

and the affected structures inverted. From a fracture play point of view, structural traps within the "extensional fault belt" and particularly those that are cross-cut by reactivated fault systems are the most prospective leads.

7.4.3 Prospective Leads from the Swiss Molasse

Based on the concept presented here and on the available data, it is possible to identify three different levels of prospective areas (Fig. 7.12) for further subsurface investigations:

1. Prospective structure – the potential areal extent of the Entlebuch Field must be further examined, particularly along the reactivated fault systems which were delineated by the imagery data. The large anticlinal feature detected at the surface strongly suggests the presence of a profound, ramp-like, positive structure below the dêcollement surface. Exploitation of this well is particularly warranted because of the extensive seismic coverage available in this area.

2. Prospective leads – this is the large surface anticline which is cut and offset by the normal fault system of element GS. This structural lead requires further confirmation by subsurface mapping of additional seismic data.
3. Prospective zone – these are the southernmost segments of structural elements which were identified by Landsat imagery interpretation and were not constrained with subsurface information. For further evaluation, regional seismic lines must be compiled and interpreted.

7.5 Detection of Buried Triassic Basins in the Coastal Plains of the Eastern United States

7.5.1 Objectives

The oil industry has had long-standing interest in the hydrocarbon potential of the coastal plain region of the eastern United States (Swain 1947; Spangler 1950). In the last decade or so, this interest was focused primarily on the hydrocarbon potential of Triassic graben fill sediments that were accumulated in a series of buried Mesozoic basins which were encountered in at least 12 different areas of the coastal plain regions (Fig. 7.13). An integrated satellite imagery study of this region was initiated by the exploration department of Exxon Co., USA, Eastern Division in 1987 (Berger et al. 1987). The study, which was conducted by the author and his colleagues (F. V. Corona, J. S. Schriver, R. M. Eppihimer and R. L. Brovey) had two main objectives: (1) to provide structural details in areas of interest where well and seismic data show the presence of Mesozoic basins and (2) to search for new sites where such basins are most likely to be found. The study incorporated a set of satellite images covering the entire region, available gravity and magnetic data as well as seismic and well information.

The integrated satellite imagery study followed, more or less, the same procedures that were illustrated earlier for the Molasse basins. Thus, only the key results of this study are illustrated here.

7.5.2 Geological Setting of the Mesozoic Basins

Upper Triassic and Lower Jurassic nonmarine sedimentary rocks in eastern North America accumulated in a series of discrete, elongated, half-grabens that are superimposed on the core of the Paleozoic Appalachian orogenic belt from South Carolina to offshore Nova Scotia (Sanders 1963; Rodgers 1970; Manspeizer 1981; see Figs. 7.14 and 7.15).

These basins formed as a result of extension during the Late Triassic to Early Jurassic, prior to the opening of the Atlantic Ocean during the Early to Middle Jurassic (Manspeizer et al. 1978; Swanson 1982). The orientation of border faults in these basins appear to be controlled primarily by Paleozoic "basement grain" (Lindholm 1978; Ratcliffe et al. 1986; Swanson 1986). Both normal and strike-slip offsets have occurred along these faults (Manspeizer 1981).

Exposed Mesozoic basins contain strata referred to as the Newark Supergroup (Olsen 1978; Froelich and Olsen 1985). The supergroup consists mainly of continental clastic sedimentary rocks ("redbeds"), lacustrine deposits, basaltic flows, sills, and dikes. Lacustrine black shales are commonly rich in organic matter and are thought to be a potential hydrocarbon source (Hatcher and Romankiw 1985; Olsen 1985; Pratt et al. 1985; Robinson 1985; Spiker 1985).

Field studies of individual Mesozoic basins reveal that these basins are typically bounded on one side by a zone of normal faults with relatively large displacements, referred to as the "border fault", and on the opposite side by a series of antithetic normal faults with minor displacements that step down into the basin (e. g., see Reinemund 1955; Van Houten 1969; Randazzo et al. 1970; Thayer 1970 a, b; Lee 1978, 1980; Turner-Peterson and Smoot 1985; Fig. 7.15). Local highs occur within the basins and are defined by small horst blocks or formed by differential offset along a series of downfaulted blocks.

Many of the border faults appear to follow preexisting Paleozoic structures such as thrust faults, reverse faults, or possible strike-slip faults (e. g., Ratcliffe 1971; Ratcliffe and Burton 1985; Ratcliffe et

Fig. 7.13. Map of the Appalachian and eastern coast region, USA ▶
Showing the location of exposed Mesozoic basins of the lowland area and wells that penetrated Triassic strata of buried basins in the coastal plain region. The location of the Potomac River and the Taylorsville Basin, which will be discussed later, have been added to the original figure. (Adopted by Marine and Siple 1984 from McKee 1959 and Cohee 1962. Published with permission of the Geological Society of America)

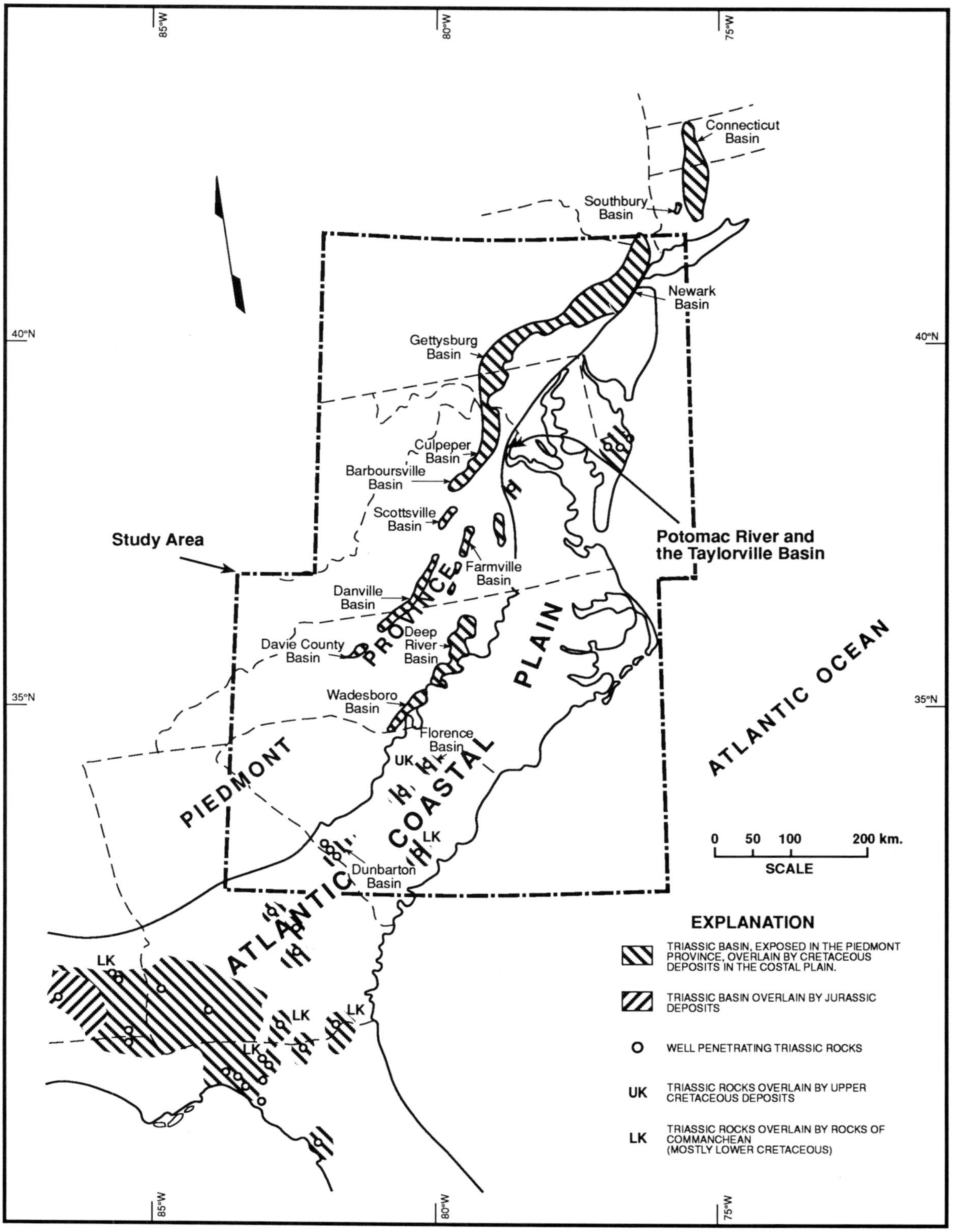
85°W
80°W
75°W
40°N
35°N
Connecticut Basin
Southbury Basin
Newark Basin
Gettysburg Basin
Culpeper Basin
Barboursville Basin
Scottsville Basin
Farmville Basin
Danville Basin
Deep River Basin
Davie County Basin
Wadesboro Basin
Florence Basin
Dunbarton Basin
Study Area
Potomac River and the Taylorville Basin
PIEDMONT PROVINCE
ATLANTIC COASTAL PLAIN
ATLANTIC OCEAN
UK
LK
0 50 100 200 km.
SCALE
EXPLANATION
TRIASSIC BASIN, EXPOSED IN THE PIEDMONT PROVINCE, OVERLAIN BY CRETACEOUS DEPOSITS IN THE COSTAL PLAIN.
TRIASSIC BASIN OVERLAIN BY JURASSIC DEPOSITS
WELL PENETRATING TRIASSIC ROCKS
UK TRIASSIC ROCKS OVERLAIN BY UPPER CRETACEOUS DEPOSITS
LK TRIASSIC ROCKS OVERLAIN BY ROCKS OF COMMANCHEAN (MOSTLY LOWER CRETACEOUS)

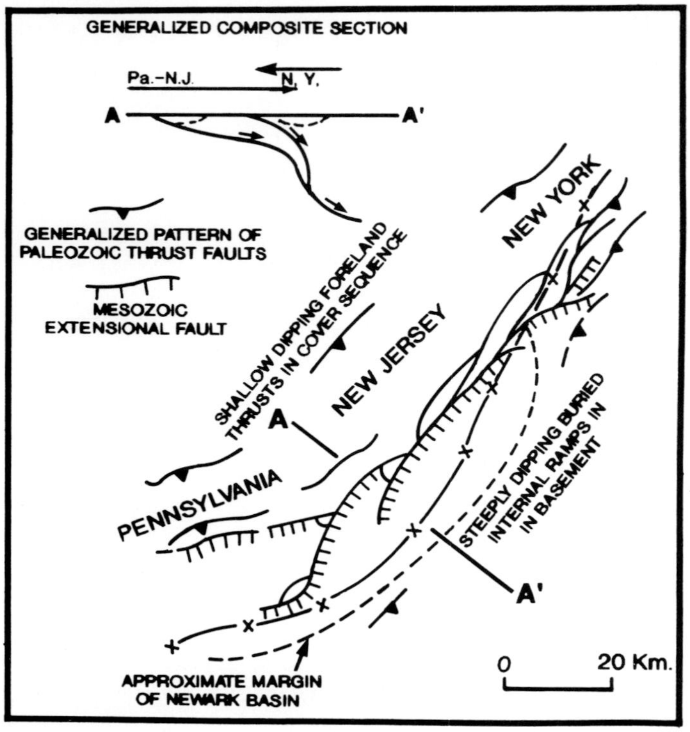

Fig. 7.14. Geological setting of the Newark Basin
Position of the Newark Basin in relation to reactivated Paleozoic thrust faults. In the diagrammatic section A-A', the New York and Pennsylvania-New Jersey parts of the basins are shown in relation to their position on a hypothetical thrust-ramp system. (Ratcliffe and Burton 1985. Published with the permission of the Geological Society of America)

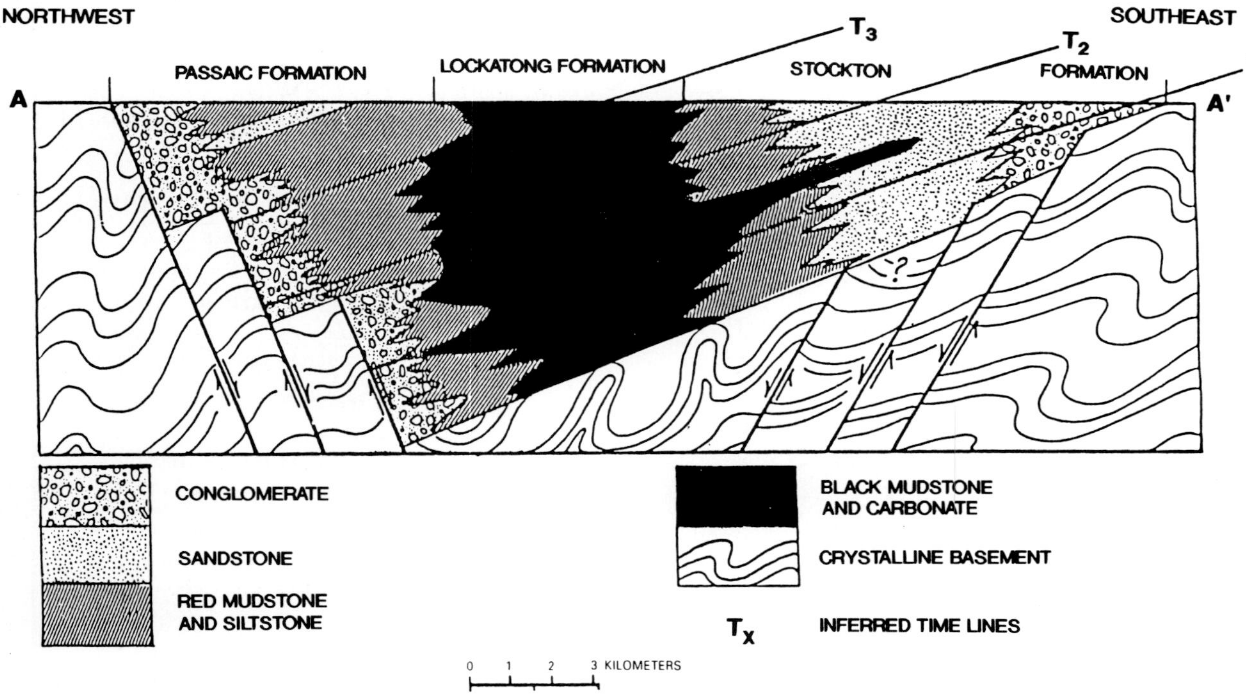

Fig. 7.15. Schematic cross section of buried Mesozoic basins
Showing structural and facies relationships within the Newark Basin with T1, T2 and T3 as inferred time lines. (Turner-Peterson and Smoot 1985. Published with the permission of the USGS)

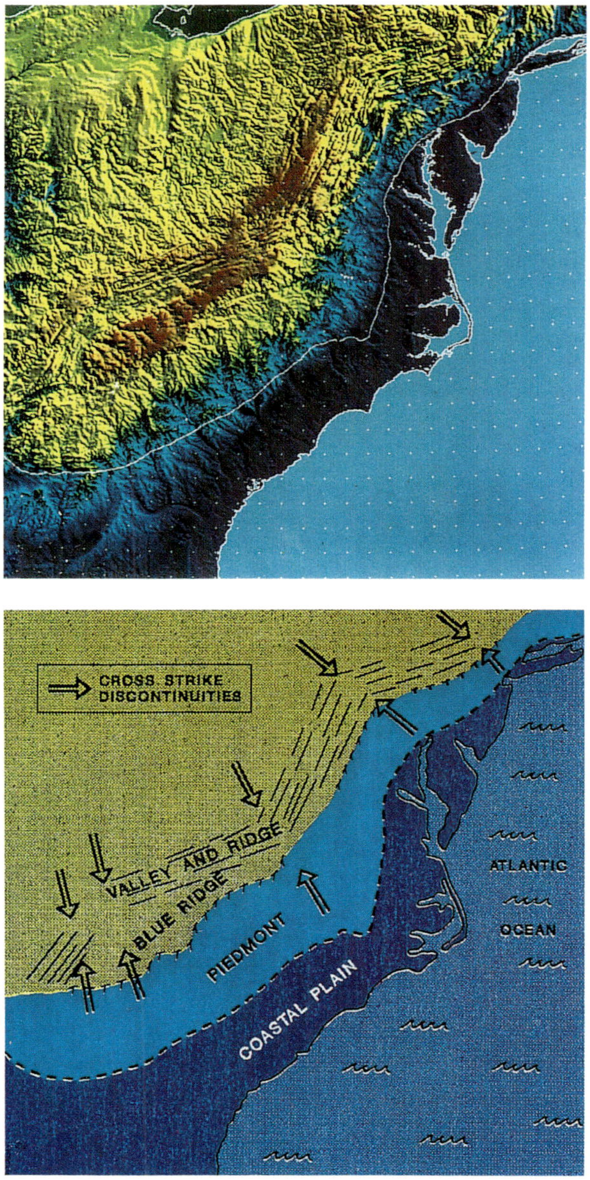

Fig. 7.16. Digital relief map of the eastern United States
a Produced by an image processing system using USGS 30-m grid topographic data, nicely illustrates the presence of major cross-cutting discontinuities in this region (prepared by R. L. Brovey). **b** Interpretative sketch showing major cross-strike discontinuities

al. 1986; Swanson 1986). COCORP seismic reflection studies support the correlation of Mesozoic basin border faults with reactivated Paleozoic thrust faults (e. g., Cook et al. 1981; Petersen et al. 1984).

Contractional structures, particularly transverse folds and strike-slip faults within and along the boundaries of the Mesozoic basins, have been described in the literature (e. g., Sanders 1963; Manspeizer 1981). Origin of these structures is not clear. Some authors postulate that these structures are associated with postrifting, strike-slip shearing during the Jurassic to Early Cretaceous along northeast-trending fault zones which may coincide with some of the Mesozoic border faults (e. g., Sanders 1963; Manspeizer 1981). Regardless of their origin, the contractional structures found in these basins may be important in forming potential hydrocarbon traps.

Large structural features trending at high angles to the Appalachian tectonic grain have been described by many geologists as cross-strike discontinuities (Fig. 7.16). According to Pohn (1985), such features are old, involving Precambrian basement, and were active as lateral ramps during Paleozoic contractional deformation. If extrapolated eastward, many of these features are coincident in strike and nearly coincident in spacing with ocean transform faults in the North Atlantic. These features are also coincident with aeromagnetic anomalies off the North American coast (Van Houten 1969; Grow and Sheridan 1981).

Many of the Appalachian cross-strike discontinuities have a proper orientation to be reactivated as "transform faults" between rift basins during the Late Triassic-Early Jurassic extensional event. In places, cross-strike discontinuities appear to segment or bound some of the Mesozoic basins indicating possible reactivation of these old features during the formation of the basins.

Much of the eastern portion of North America is covered by Cretaceous and Tertiary Coastal Plain sediments. Coastal Plain sediments also cover many Mesozoic basins (e. g., Marine and Siple 1974; Ballard and Uchupi 1975; Grow and Sheridan 1981; Daniels et al. 1983; Nadon and Middleton 1984; Newell 1984). Some of the buried basins have been identified with seismic data (e. g., Ballard and Uchupi 1975), and others with potential field data (e. g., Daniels et al. 1983). Tertiary compressional deformation in the eastern Coastal Plain region makes definitive identification of Mesozoic structures very difficult. Tertiary structures consist mainly of northeasterly trending reverse faults (Fig. 7.17). Some of

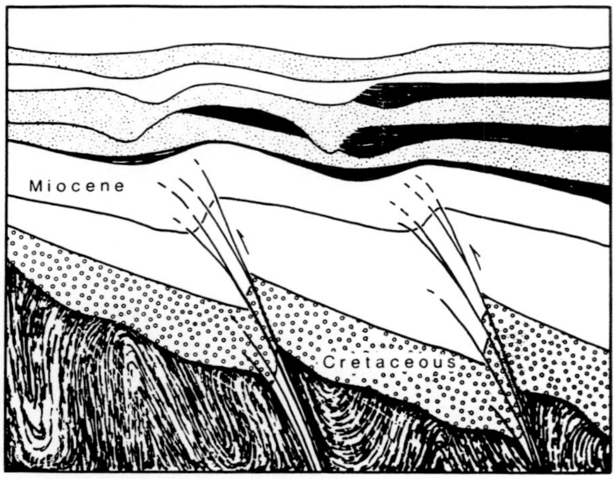

Fig. 7.17. Reactivated structures in the Coastal Plain region Schematic cross section showing the style of deformation of subsurface strata interpreted from outcrop, drill holes and geophysical data in the Virginia coastal plain (Newell 1984). Northwest-trending, high-angle, basement-involved reverse faults offset Cretaceous strata and die out within the Miocene formation near the surface. Subsiding troughs and rising arches overlie the fault zone. Figure not to scale

the Mesozoic normal faults may have been inverted during this deformation (Mixon and Newell 1977; Newell 1984).

7.5.3 Satellite Imagery Interpretation

Satellite images were used, in conjunction with available surface geologic information, to map the surface expressions of exposed, obscured and buried structures in the study area (Fig. 7.18).

The imagery covered, from west to east, five different tectonic provinces:

- the Paleozoic fold and thrust belt of the Valley and Ridge Province.
- the Precambrian-Eocambrian metamorphic complex of the Blue Ridge Province.
- the exposed rift basins in the Triassic lowland region.
- the Precambrian crystalline rock complex of the Piedmont Plateau.
- the Cretaceous and younger sediments of the Coastal Plain region.

Several different types of structures were recognized and analyzed on imagery and are highlighted on the imagery mosaic in Fig. 7.18. The exposed fold and thrust faults of the Valley and Ridge Province manifest strong expressions on imagery. The entire sedimentary section of this region has been deformed into a series of tight folds that were differentially eroded to form a unique pattern of narrow topographic ridges and intervening valleys. The thrust folds manifest the typical sinuous FLTs and several other diagnostic features such as truncated folds and splaying and branching of secondary thrust faults. The exposed Triassic basins in the lowland regions were subjected to intense erosional processes which severely modified their surface expressions. In most cases, only the major bounding faults of the Triassic basin can be detected on imagery. These structures manifest subtle linear topographic features that reflect the surface expressions of the eroded FLTs of the master faults.

Two different types of first-order lineaments are detected on the imagery. The first type consists of large-scale, northwest-trending cross-strike discontinuities that cut, offset and breach the exposed structures in the Piedmont Region and continue to manifest subtle topographic expressions in the Coastal Plain area. The second type of first-order lineaments consist of large-scale north-northeast-trending linear features which form the extension of the border faults of the Triassic grabens or trend parallel or subparallel to these faults. These lineaments exert significant control on the major stream valleys of the Lowland and the Coastal Plain areas which exhibit several structurally controlled patterns along them.

Two different types of obscured and buried structures are recognized on imagery in the Coastal Plain region and are illustrated in Fig. 7.19. The first type consists of subtle, elongated topographic ridges (drape folds?) that trend parallel to the north-northeast lineaments and are often outlined by marginal, circular stream valleys and local drainage divides. The second type consists of restricted, topographically low areas with poorly drained conditions (grabens?) that are located adjacent to the topographic ridges and manifest dark anomalies on the imagery.

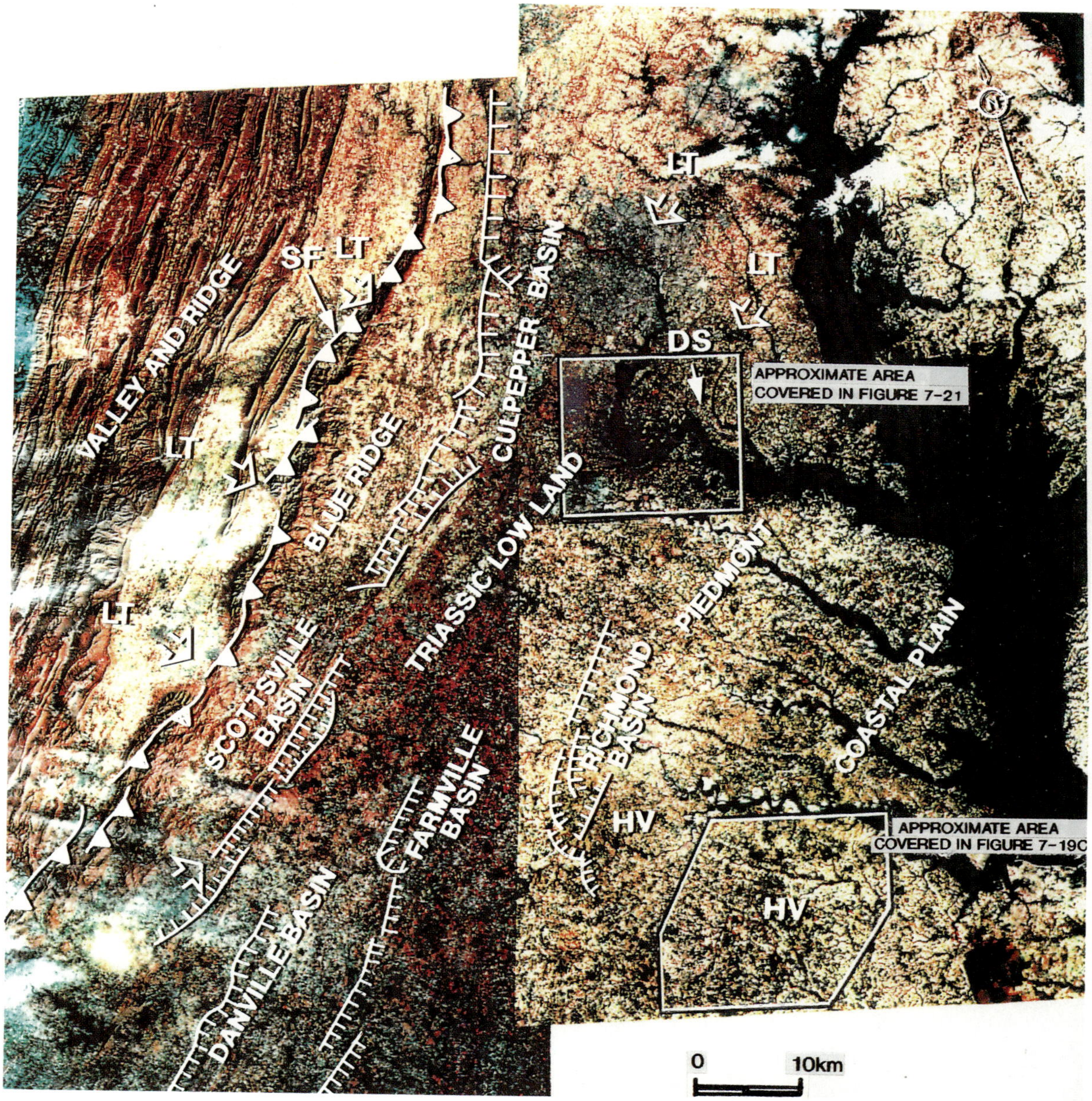

Fig. 7.18. Regional structural interpretation
An interpreted Landsat image mosaic of the northern half of the Coastal Plain region showing the expression of major tectonic and structural elements. *HV* Heavy vegetative cover; *DS* deflected streams; *SF* sinuous FLT; *LT* lineament

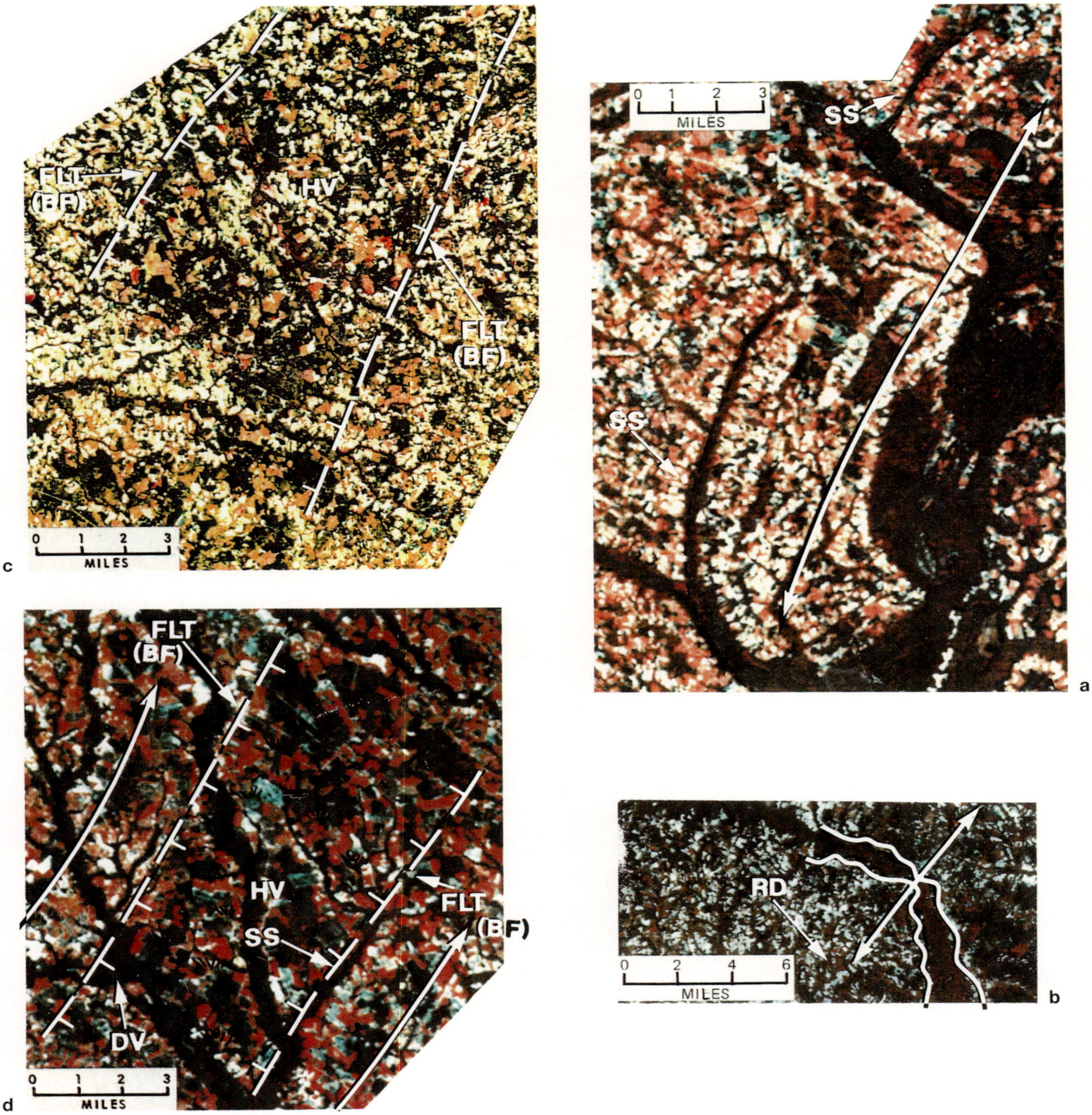

Fig. 7.19. Surface expression of buried structures in the Coastal Plain region
Showing the possible geomorphic expression of buried reactivated drape folds (**a** and **b**) and graben features (**c** and **d**). *FLT* Fault-line trace; *HV* heavy vegetative cover related to excessive ground moisture; *SS* subsequent stream; *DV* deflected valley; *RD* radial drainage

7.5.4 Regional Gravity and Magnetic Data Interpretation

Regional gravity and magnetic data which were available for this study included the following:

- Several regional geological/geophysical profiles that were constructed from the regional data and

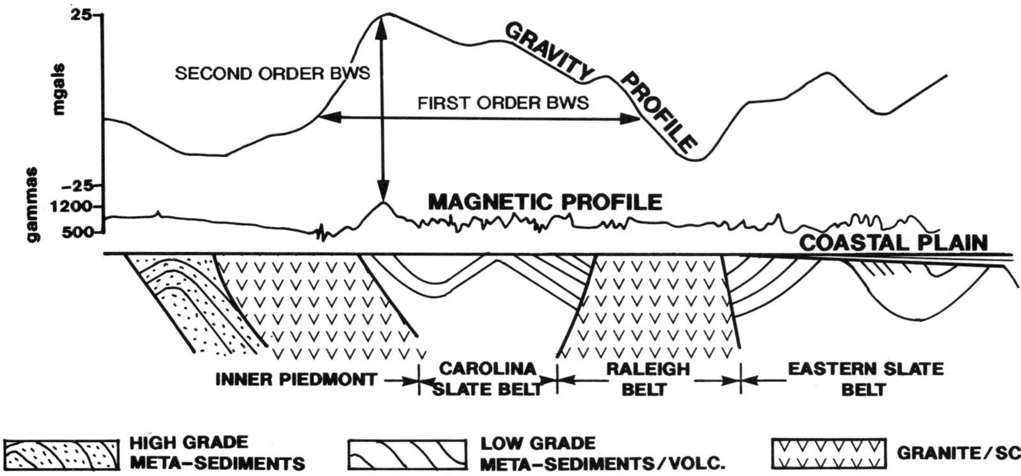

Fig. 7.20. Gravity and magnetic profiles across the eastern United States
Showing the expression of major basement units. Examples of first- and second-order magnitude anomalies are highlighted

other available magnetic surveys and surface geological map.

- A 4 km total intensity aeromagnetic grid acquired from the Electric Power Research Institute (EPRI) and displayed and enhanced by Exxon Production Research Company image processing computer.
- A digital tape containing all Defense Mapping Agency (DMA) gravity data of the eastern USA. These data consist of evaluated segments (data checked by DMA personnel for errors and leveled and adjusted to adjacent surveys) as well as some unevaluated segments. Data spacing varies from less than a mile over most of South Carolina and Virginia to 2.5 to 3 miles in North Carolina. The data were gridded at intervals of 15 000 feet and mapped at two milligals in color.

The cross section in Fig. 7.20 illustrates the gross characteristics of potential field data in this region. First-order magnitude basement structures are clearly manifested by unique characteristics of the magnetic signature and positive or negative anomalies of the gravity data. Second-order magnitude features are also noticeable. It is evident from this data set that the strong relationships between basement structures and gravity anomalies may obscure the expression of Triassic basins on the regional gravity data, and their detection may require higher-resolution data sets.

A typical example of the regional gravity and magnetic data sets available for the study are shown in Fig. 7.21 b, c, which focus on the basement characteristics of the Piedmont and Coastal Plain area in the vicinity of the Taylorsville Basin. This region is dominated by high-amplitude, northeast-trending anomalies that reflect the structural fabric of the magnetic basement. The edges of the northeast-trending anomalies are spatially coincident with first-order magnitude northeast lineaments interpreted on imagery, whereas cross-strike discontinuities are manifested on both data sets by local offsets and "breaching" of the regional trends as well as by the presence of smaller cross-trending anomalies. The interpretation of this data set will be further illustrated in the next section.

7.6 Detailed Analysis of Known Buried Basins

A detailed analysis of all available surface and subsurface data was conducted in several of the known exposed and buried basins. The objective of this effort was twofold: (1) to better understand the structural/topographic expressions of the basins and (2) to establish a methodology for mapping these and other suspect basins through the integrations of all available data sets.

Examples of the procedures used in this step and the resulting products are shown with data from the Taylorsville Basin in Fig. 7.21 and Table 7.1. The upper left figure (a) is a block diagram that illustrates the relationships between the subsurface expressions of the basin as interpreted on seismic and

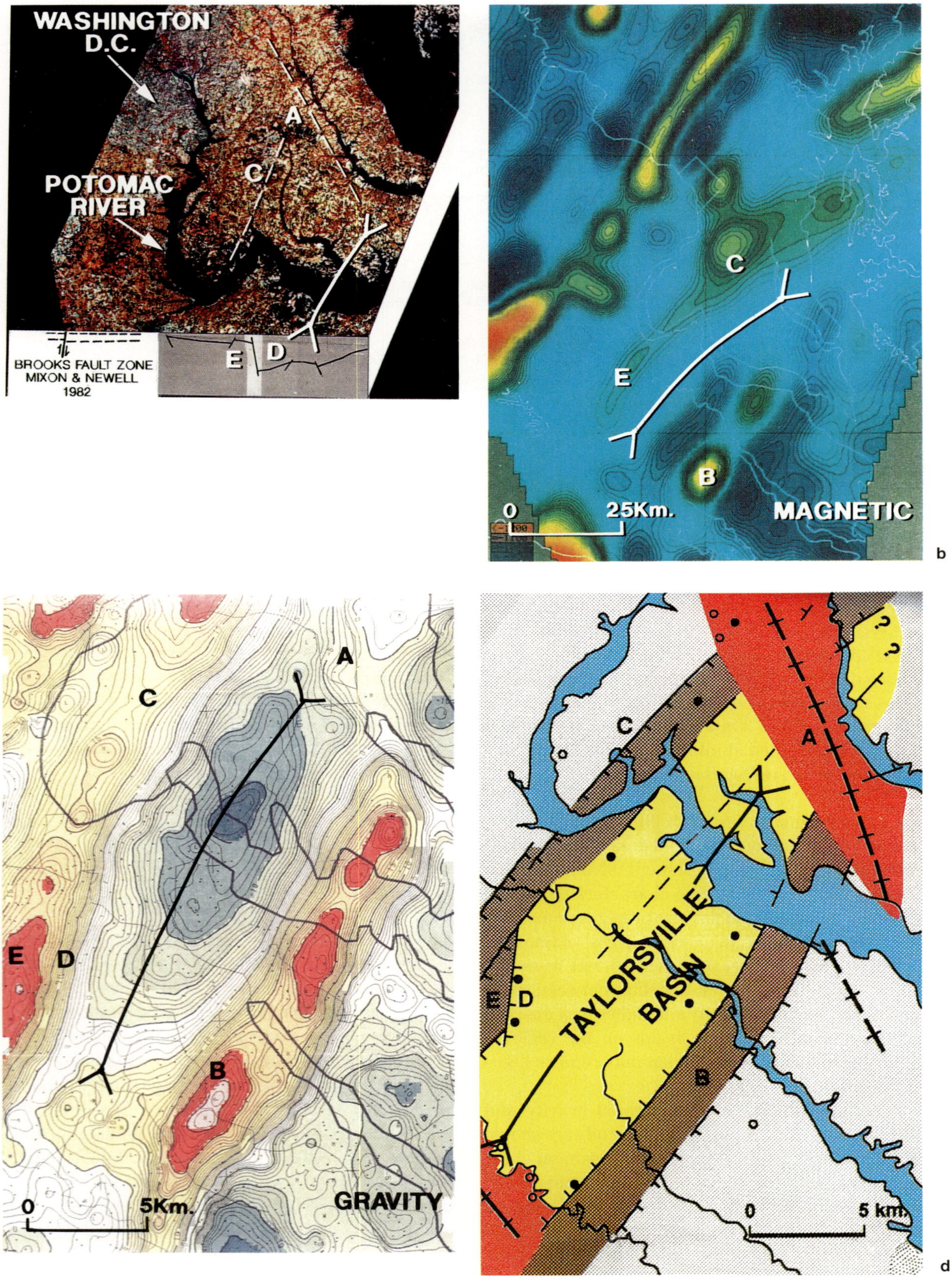
a
WASHINGTON D.C.
POTOMAC RIVER
A
C
E
D
BROOKS FAULT ZONE
MIXON & NEWELL
1982
b
C
E
B
0
25Km.
MAGNETIC
c
A
C
E
D
B
0
5Km.
GRAVITY
d
C
A
E
D
TAYLORSVILLE BASIN
B
0
5 km

Table 7.1. Integrated structure interpretation of the Taylorsville Basin
Showing an example of the legend used in this study

	LEGEND
	Possible subsurface fault, hachures on probable downthrown side
	Possible outline of a Mesozoic basin
? ? ? ?	Possible outline of a Mesozoic basin with evidence of a Mesozoic basin, areal extent is uncertain
	Possible outline of a structurally low area
	Possible outline of a structurally high area
	Possible cross strike discontinuity
●	Well location, with Mesozoic penetration
○	Well location, no Mesozoic penetration, bottomed in basement

◀ **Fig. 7.21. Surface and subsurface expression of the Taylorsville Basin**
a A block diagram showing the relationship between surface and subsurface structures; **b** and **c** gravity and magnetic data with interpreted basement structures; **d** an integrated structural map of the basin. See Table 7.1 for explanation. (Seismic data courtesy of Exxon Exploration Co.; magnetic data courtesy of the Electric Power Research Institute; gravity data courtesy of the US Defense Mapping Agency)

surface structures interpreted on imagery. Key features to observe here are (1) the elongated topographic ridge that outlines the western boundary fault of the graben which can be interpreted as a fault-edge flexure; (2) a narrow internal horst block that exerts significant control on the orientation and width of the Potomac River; and (3) a linear northeast-trending tributary that reflects the surface expressions of the cross-strike discontinuity forming the northeastern boundary of this basin. Figure 7.21 d shows how all the available data sets, including well information, were integrated to produce a map of the buried graben and its possible constraining basement features. Note that corresponding features on all data sets are labeled by the same letters.

7.6.1 Geological/Geophysical Model of Buried Triassic Basins

A conceptual model that illustrates the most important relationships between basement, subsurface and surface structures in the study area was then constructed (Fig. 7.22). This model can be summarized as follows. The Paleozoic basement in the Coastal Plain region contains features that are similar to those observed in the nearby, exposed Piedmont region. That is, the basement of the Coastal Plain is divided into several belts consisting of blocks with contrasting basement lithologies and structural fabrics that trend parallel or subparallel to the structural grain of the Appalachian Orogenic Belt and are bounded by major fault systems. The region is cut and locally offset by west- to northwest-trending, regional-scale, cross-strike discontinuities. The buried basins are localized along the major basement faults that can be offset or truncated by cross-strike discontinuities. The faulted boundaries of the basement structures are often characterized by the presence of drape folds that are manifested as subtle topographic ridges, whereas the buried basins are usually expressed as topographically low areas with poor drainage conditions. Structurally controlled stream valleys as well as tonal features related to variations in moisture conditions are the prime surface indicators for detection of buried structures in this region.

7.6.2 Structural Lead Map

The results of this study were compiled into a structural lead map that shows the location of all the known basins as well as areas where topographically low features seem to be situated in the same structural/basement setting as the known basins (Fig. 7.23). For each of the areas shown with letters A–J, an integrated map was prepared in the same manner as the one shown for the Taylorsville Basin in Fig. 7.22 d. The structural lead map can be used by explorationists as a guide for high grading areas where buried Triassic basins are likely to be found. At the same time, explorationists should be aware of possible constraints inherent to the data used; these are discussed in the next section.

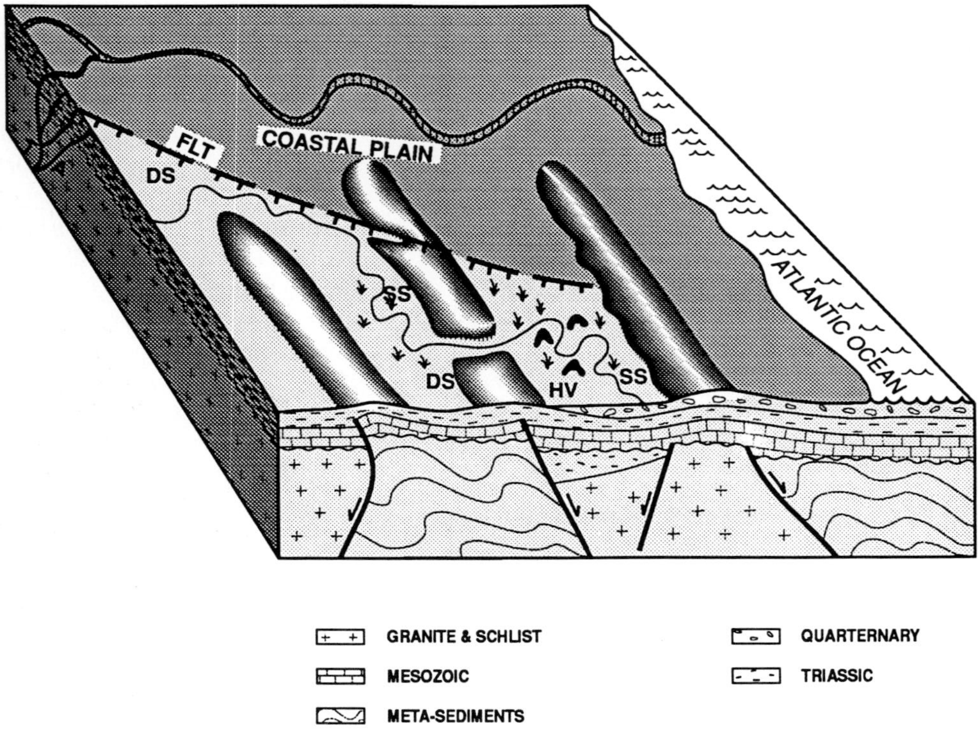

Fig. 7.22. Conceptual structural/geomorphic model
Showing the possible stack relationships among basement structures, buried Triassic basins and their surface/geomorphic expression. *FLT* Fault-line trace; *HV* heavy vegetative cover; *DS* deflected stream; *SS* subsequent stream

Fig. 7.23. Structural lead map for the Coastal Plain region ▶
Showing the location of known Triassic basins (shown by *numbers*) and proposed buried basins (shown by *letters*). Letters are in order from north to south and not by prospect ranking (see Table 7.1)

7.6.3 Conclusions and Recommendations

The results of this study strongly indicate that an integrated analysis of Landsat imagery, gravity, and magnetic data can be used to detect and analyze buried structures in the Coastal Plain region of the eastern United States. Differential compaction, erosional processes related to movement of groundwater and surface water in the vicinity of buried structures, and most importantly, structural reactivation are believed to be the reasons for the strong correlation that exists between basement structures interpreted from gravity and magnetic data and surface features that are observed with Landsat imagery.

Various types of basement features produce characteristic expressions on Landsat, gravity, and magnetic data in this region. These features include major lithologic and faulted boundaries of various basement blocks, local intrusive bodies, and dikes and fracture zones. With only Landsat, gravity, and magnetic data available, it is possible to detect the location of reactivated fault systems, but it is not possible to distinguish Mesozoic-age faults from Tertiary-age fault zones in the region, or to determine definitively the presence of Mesozoic basins. With higher-resolution data sets and application of more sophisticated interpretation techniques, it may be possible to increase the probability of delineating Mesozoic faults and associated basins in the Coastal Plain region.

This study is based on the assumption that the orientation and location of Mesozoic basins in the Coastal Plain region is controlled by basement structures in the same manner as the Mesozoic basins in the Piedmont region. Basement structures in this region are likely to be reactivated during Triassic-Jurassic time because extension was approximately normal to the Paleozoic grain. One must realize, however, that rift basins may deviate from basement structural controls and trend partly or completely oblique to the basement grain (e. g. the Rhine Valley developed oblique to the Paleozoic

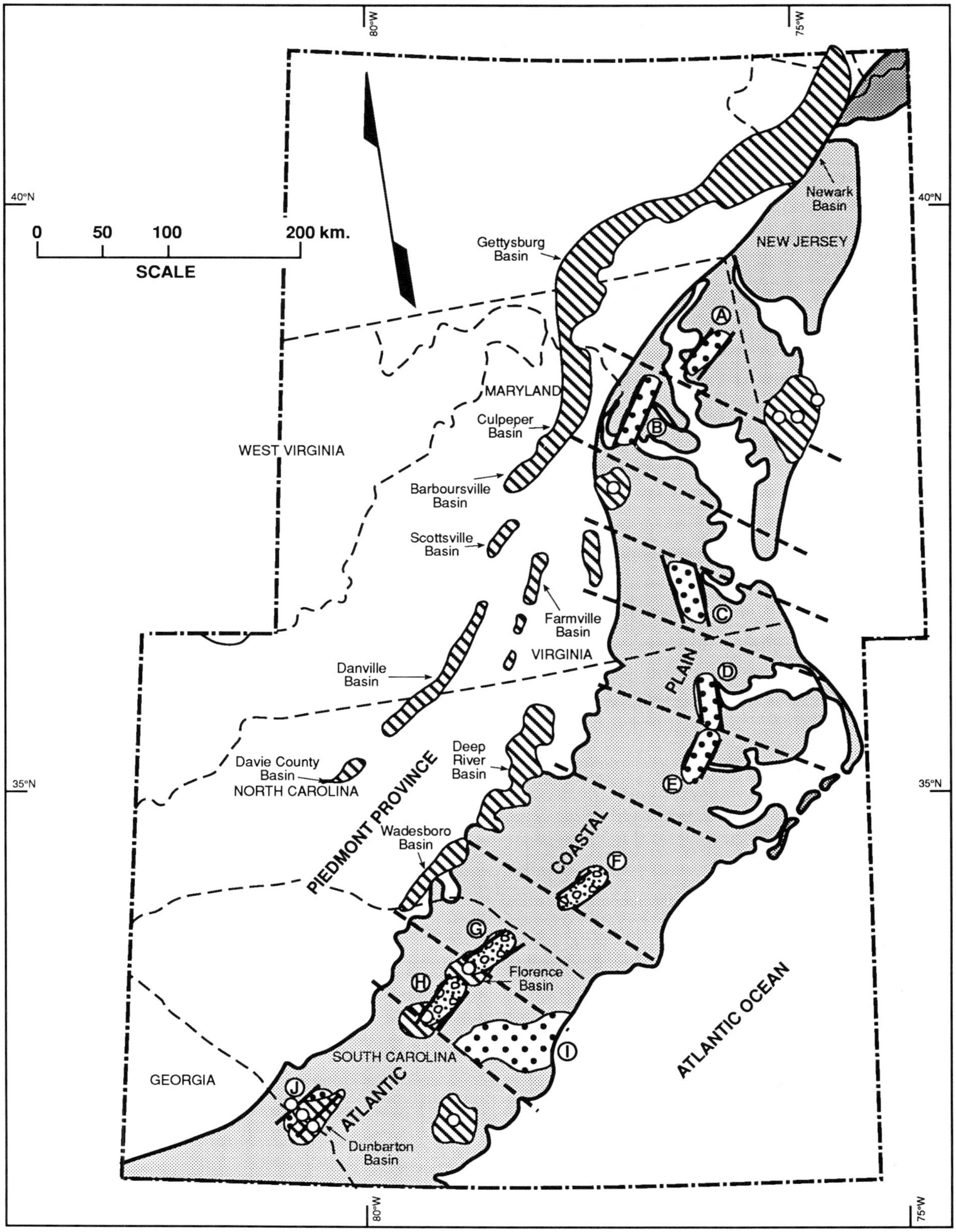

80°W
75°W
40°N
35°N
0 50 100 200 km.
SCALE
Newark Basin
NEW JERSEY
Gettysburg Basin
MARYLAND
Culpeper Basin
WEST VIRGINIA
Barboursville Basin
Scottsville Basin
Farmville Basin
VIRGINIA
Danville Basin
Deep River Basin
Davie County Basin
NORTH CAROLINA
PIEDMONT PROVINCE
Wadesboro Basin
COASTAL
PLAIN
Florence Basin
SOUTH CAROLINA
GEORGIA
ATLANTIC
Dunbarton Basin
ATLANTIC OCEAN
A
B
C
D
E
F
G
H
I
J

grain, as illustrated in Chap. 3). If such basins exist in the Coastal Plain region, their expressions may be detected on Landsat imagery or gravity data, but their significance will be substantially reduced because they will not correlate with regional trends of the mangetic data.

Detailed analysis of selected areas indicates that the interpretation of cross-strike discontinuities as single, linear features represents an extreme simplification of wide, disrupted, and structurally complex features. One must recognize that these features are more abundant in the area than initially realized and presented in this study. The geological models presented herein should be regarded as generalized and tentative and should be further examined with seismic and well data and through the application of modern concepts related to the tectonic evolution of North America as well as to processes of crustal rifting.

7.7 Summary

Integrated analysis of satellite imagery data begins with regional interpretation of exposed structures at the basin's margins and progresses towards the analysis of obscured and buried structures in the basin's interior. It is focused primarily on the recognition of structural elements that are large enough to be constrained by other available exploration tools. The analysis consists of seven steps which include regional interpretation, analysis of key areas, development of a structural/geomorphic model, delineation of prospective leads, prospect mapping, model testing and revision of the structural/geomorphic model. The first four steps represent an input from the integrated imagery study, whereas the remaining three consist of conventional exploration activities. Two examples were used to illustrate this process. A study of the Molasse Basin led to the recognition of prospects and leads which develop along reactivated fault systems. A study of the buried Triassic basins in the eastern United States provided a methodology for improved subsurface mapping of known basins as well as delineation of potential new ones.

References and Further Reading

Ballard RD, Uchupi E (1975) Triassic rift structure in Gulf of Maine. AAPG Bull 59 (7): 1041–1072

Berger Z (1984) Structural analysis of low relief basins using Landsat data. Proc Int Symp on Remote sensing of the environment, 3rd Thematic Conf, Colorado Springs, CO. Environmental Research Institute of Michigan, Ann Arbor, pp 251–271

Berger Z, Gundlach T (1990) The contribution of satellite imagery interpretation to exploration of fracture reservoirs in West Germany. BEB, Internal Rep

Berger Z, Corona FV, Schriver JS, Eppihimer RM (1987) Detection and analysis of buried structures in the coastal plain region, eastern United States. Exxon Production Research Company, Internal Rep

Betz D, Wendt A (1983) Neuere Ergebnisse der Aufschluß- und Gewinnungstätigkeit auf Erdöl und Erdgas in Süddeutschland. Bull Ver Schweiz Petroleum-Geol Ing 49 (117) Oktober 1983: 9–36

Blümling P (1983) Bohrlochauskesselungen und ihre Beziehung zum regionalen Spannungsfeld. Sonderforschungsbereich 108, Spannung und Spannungsumwandlung in der Lithosphäre. Universität Karlsruhe (TH), Berichtsband für die Jahre 1981–1983, S 313–323

Cohee GV, Committee Chairman (1962) Tectonic map of the United States. USGS and AAPG, 1 : 2,500,000 scale

Cook FA, Brown LD, Kaufman S, Oliver JE, Petersen TA (1981) COCORP seismic profiling of the Appalachian orogen beneath the coastal plain of Georgia. Geol Soc Am Bull, part I, 92: 738–748

Daniels DL, Sietz I, Popenoe P (1983) Distribution of subsurface lower Mesozoic rocks in the southeastern United States, as interpreted from regional aeromagnetic and gravity maps. USGS Prof Pap 1313-K

Eisbacher GH, Lüschen E, Wickert F (1989) Crustal scale thrusting in the Hercynian Schwarzwald and Vosges, central Europe. Tectonics 8 (1): 1–21

Froelich AJ, Olsen PE (1985) Newark supergroup, a revision of the Newark Group. In: Robinson GR Jr, Froelich AJ (eds) Eastern North America. Proc 2nd USGS Worksh on the Early Mesozoic basins of the eastern United States. USGS Circ 946, pp 1–3

Grow JA, Sheridan RE (1981) Deep structure and evolution of the continental margin off the eastern United States. Oceanol Acta 1981. Proc 26th In Geol Congr, Geology of continental margins Symp, Paris, France, pp 11–19

Gürler B, Hauber L, Scwander M (1987) Die Geologie der Umgebung von Basel mit Hinweisen über die Nutzungsmöglichkeiten der Erdwärme. Beiträge zur Geologischen Karte der Schweiz. Landeshydrologie und Geologie und schweizerische Geologische Kommission Lieferung 160

Harding TP, Lowell JD (1979) Structural styles, their plate-tectonic habitats, and hydrocarbon traps in petroleum provinces. AAPG Bull 63: 1016–1058

Hatcher PG, Romankiw LA (1985) Nuclear magnetic resonance studies of organic-matter-rich sedimentary rocks of some early Mesozoic basins of the eastern United States. In: Robinson GR Jr, Froelich AJ (eds) Proc 2nd USGS Worksh on the Early Mesozoic basins of the eastern United States. USGS Circ 946, pp 65–70

Illies JH, Greiner G (1978) Rhinegraben and the Alpine system. Geol Soc Am Bull 89: 770–782

Illies JH, Baumann H, Hoffers B (1981) Stress pattern and strain release in the Alpine foreland. Tectonophysics 7: 157–172

Krohe A, Eisbacher GH (1988) Oblique crustal detachment in the Variscan Schwarzwald, southwestern Germany. Geol Rdsch 77 (1): 25–43

Laubscher HP, Bernoulli D (1980) Cross-section from the Rhine-Graben to the Po Plain. In: Laubscher HP, Bernoulli D (eds) Geology of Switzerland, guidebook, part B. Wepf, Basel, pp 183–209

Laubscher HP, Bernoulli D (1982) History and deformation of the Alps. In: Hsü KJ (ed) Mountain building processes. Academic Press, London, pp 169–180

Lee CH (1978) Preliminary study of the uranium potential of the Triassic Sanford Basin and Colon cross structure, North Carolina. US Dept Energy, Grand Junction, CO, Rep GJBX – 8(78)

Lee KY (1980) Triassic-Jurassic geology of the southern part of the Culpeper Basin and Barboursville Basin, Virginia. USGS Open-File Rep 80–468, 7 pp

Lemcke K (1977) Erdölgeologisch wichtige Vorgänge in der Geschichte des süddeutschen Alpenvorlandes. Erdöl–Erdgas Z 93 (Sonderausgabe): 50–56

Lindholm RC (1978) Triassic-Jurassic faulting in eastern North America – a model based on pre-Triassic structures. Geology 6: 365–368

Manspeizer W (1981) Early Mesozoic basins of the central Atlantic passive margins. In: Bally AW et al. (eds) Geology of passive continental margins; history, structure and sedimentologic record. AAPG Continuing Education Course Note Ser, no 19, p 60

Manspeizer W, Puffer JH, Cousminer HL (1978) Separation of Morocco and eastern North America: a Triassic-Liassic stratigraphic record. Geol Soc Am Bull 89: 901–920

Marine IW, Siple GE (1974) Buried Triassic basin in the central Savannah River area, South Carolina and Georgia. Geol Soc Am Bull 85: 311–320

McKee ED (1959) Paleotectonic map of the Triassic System. USGS Misc Geol Inv Map 1–300

Mixon RB, Newell WL (1977) Stafford fault system: structures documenting Cretaceous and Tertiary deformation along the fall line in northeastern Virginia. Geology 5: 437–440

Müller WH, Hsü KJ (1980) Stress distribution in overthrusting slabs and mechanics of Jura deformation. In: Schneidegger AE (ed) Rock mechanics, supplement 9. Tectonic stresses in the Alpine-Mediterranean region. Springer Berlin Heidelberg New York, pp 219–232

Nadon GC, Middleton GV (1984) Tectonic control of Triassic sedimentation in southern New Brunswick: local and regional implications. Geology 12: 619–622

Newell WL (1984) Architecture of the Rappahannock estuary – neotectonics in Virginia. In: Morisawa M, Hack JT (eds) Tectonic geomorphology. Proc 15th Ann Binghamton Geomorphology Symp. Allen and Unwin, Boston, pp 321–342

Norman JW (1976) Photogeological fracture trace analysis as a subsurface exploration technique. Inst Mining Metall, Trans Sect B, 85: B52–B61

Olsen PE (1978) On the use of the term Newark for Triassic and Early Jurassic rocks of eastern North America. Newslett Stratigr 7 (2): 90–95

Olsen PE (1985) Distribution of organic-matter-rich lacustrine rocks in the early Mesozoic Newark supergroup. In: Robinson GR Jr, Froelich AJ (eds) Proc 2nd USGS Worksh on the Early Mesozoic basins of the eastern United States. USGS Geol Surv Circ 946, pp 61–64

Peterson RM (1979) Oil and gas exploration by pattern recognition. In: Shahrokhi F, Paludan T (eds) Remote sensing of earth resources. The University of Tennessee Space Institute, Tullahoma, Tennessee, vol 8, pp 28–60

Petersen TA, Brown LD, Cook FA, Kaufman S, Oliver JE (1984) Structure of the Riddleville Basin from COCORP seismic data and implications for reactiviation tectonics. J Geol 92: 261–271

Pohn HA (1985) Proposed model for intersection between basement structures and folds in cover rocks of central and southern Appalachians. AAPG Abstr 69 (9): 1445

Pratt LM, Vuletich AK, Daws TA (1985) Geochemical and isotopic characterization of organic matter in rocks of the Newark supergroup. In: Robinson GR Jr, Froelich AJ (eds) Proc 2nd USGS Worksh on the Early Mesozoic basins of the eastern United States. USGS Circ 946, pp 74–78

Randazzo AF, Swe W, Wheeler WH (1970) A study of tectonic influence on Triassic sedimentation – the Wadesboro Basin, central Piedmont. J of Sed Petrol 40 (3): 998–1006

Ratcliffe NM (1971) The Ramapo fault system in New York and adjacent northern New Jersey: a case of tectonic heredity. Geol Soc Am Bull 82: 125–142

Ratcliffe NM, Burton WC (1985) Fault reactivation models for origin of the Newark Basin and studies related to eastern US seismicity. In: Robinson GR Jr, Froelich AJ (eds) Proc 2nd USGS Worksh on the Early Mesozoic basins of the eastern United States. USGS Circ 946, pp 36–45

Ratcliffe NM, Burton WC, D'Angelo RM, Costain JK (1986) Low-angle extensional faulting, reactivated mylonites, and seismic reflection geometry of the Newark basin margin in eastern Pennsylvania. Geology 14: 766–770

Reinemund JA (1955) Geology of the Deep River coal field, North Carolina. USGS Prof, Pap 246, p 159

Rigassi D (1980) Map of the Molasse Basin. In: Laubscher HP, Bernoulli D (eds) The geology of Switzerland – a guidebook, part B. Wepf, Basel

Robinson GR Jr (1985) Organic geochemistry in sedimentary basins and ore deposits – the many roles of organic matter. In: Robinson GR Jr, Froelich AJ (eds) Proc 2nd USGS Worksh on the Early Mesozoic basins of the eastern United States. USGS Circ 946, pp 27–29

Rodgers J (1970) The tectonics of the Appalachians. Wiley-Interscience, New York

Sanders JE (1963) Late Triassic tectonic history of northeastern United States. Am J Sci 261: 501–524
Spangler WB (1950) Subsurface geology of the Atlantic Coastal Plain of North Carolina. AAPG Bull 16: 460–480
Spiker EC (1985) Stable-isotope characterization of organic matter in the early Mesozoic basins of the eastern United States. In: Robinson GR Jr, Froelich AJ (eds) Proc 2nd USGS Worksh on Early Mesozoic basins of the eastern United States. USGS Circ 946, pp 70–73
Sterns DW, Berger Z, Hopkins HR, Nelson RA (1988) The contribution of remote sensing data to exploration of fractured reservoirs. 6th Thematic Conf on Remote sensing for exploration Geol Houston, Tx. Environmental Research Institute of Michigan, Ann Arbor, pp 357–371
Swain FM (1947) Two recent wells in the Coastal Plain of North Carolina. AAPG Bull 31: 2054–2060
Swanson MT (1982) Preliminary model for an early transform history in central Atlantic rifting. Geology 10: 317–320
Swanson MT (1986) Pre-existing fault control for Mesozoic basin formation in eastern North America. Geology 14: 419–422
Thayer PA (1970 a) Geology of Davie County Triassic basin, North Carolina. Southeastern Geol 11 (3): 187–198
Thayer PA (1970 b) Stratigraphy and geology of Dan River Triassic Basin, North Carolina. Southeastern Geol 12 (1): 1–31
Trümpy R (1980) An outline of the geology of Switzerland. In: Laubscher HP, Bernoulli D (eds) The geology of Switzerland, guidebook, part 2. Wepf, Basel
Turner-Peterson CE, Smoot JP (1985) New thoughts on facies relationships in the Triassic Stockton and Lockatong formations, Pennsylvania and New Jersey. In: Robinson GR Jr, Froelich AJ (eds) Proc 2nd USGS Worksh on the Early Mesozoic basins of the eastern United States. USGS Circ 946, pp 10–17
Van Houten FB (1969) Late Triassic Newark Group, north central New Jersey and adjacent Pennsylvania and New York. In: Subitzky S (ed) Geology of selected areas in New Jersey and eastern Pennsylvania and guidebook of excursion. Geol Soc Am Ann Meet, Rutgers University Press, New Brunswick, pp 314–347
Vollmayr T, Wendt A (1987) Die Erdgasbohrung Entlebuch 1, ein Tiefenaufschluß am Alpennordrand. Bull Ver Schweiz Petroleum-Geol Ing 53 (125) Oktober 1987: 67–79
Wise DH (1982) Linesmanship and the practice of linear geo-art. Geol Soc Am Bull 93: 886–888
Ziegler PA (1982) Geological atlas of western and central Europe. Elsevier, Amsterdam

Chapter 8 Other Applications

8.1 Introduction

The main objective of this book is to review techniques for the analysis of geological structures with satellite imagery. Such techniques are the main application of satellite imagery in hydrocarbon exploration geology today. Any comprehensive treatment of remote sensing would not be complete, however, without mentioning several other related uses of this technology. Covered here are

- detection of surface alterations which directly indicate the presence of hydrocarbons in the subsurface;
- emerging concepts of outcrop mapping which are being used to establish the stratigraphic framework of sedimentary basins and to detect levels of maturation of source rocks;
- the extensive use of satellite image data for various types of logistical and environmental applications associated with hydrocarbon exploration activities.

8.2 Principle of Direct Detection of Hydrocarbons

The objective of this section is twofold: first, to briefly describe the concept of surface alteration related to hydrocarbon leakages and their unique spectral signatures, and second, to review results of current remote sensing studies in this field. Most of the material covered in this section is based on reports written by T. E. Townsend (deceased) who devoted much of his career to the evaluation of this unique phenomenon.

Hydrocarbon reservoirs may be marked at the surface by anomalous distributions of clay and ferric oxide minerals. The distribution of these minerals is often the result of fluids moving through fault or fracture systems. Models have been formulated to explain the nature of the alteration, its relationship to the fault and fracture systems, and its spatial association with the hydrocarbon reservoirs. Thus, it was assumed that delineation of the mineral alteration phenomena can be used as an exploration tool to locate hydrocarbon reservoirs.

Ferric oxide and clay minerals have distinctive reflectance characteristics and, when present, dominate the reflectance of rocks and soil in the visible and near-infrared (see Figs. 1.11 and 1.12). These characteristics allow ferric oxides and clay minerals to be recognized and mapped by remote sensing techniques from air- and space-based platforms. The extent to which the reflectance properties of these alteration minerals can be used for exploration applications is highly dependent on (1) the exposure of the alteration minerals at the surface and the intensity of their development; (2) overburden, soil or vegetation cover; and (3) cultural or agricultural interference.

8.2.1 Prototype Example – The Velma Field, Oklahoma

8.2.1.1 Background Information

Hydrocarbons leaking from reservoirs at depth have altered minerals in the surface rocks of the Velma, Cement, Eola, and Chickasha oil fields in the Anadarko Basin of south-central Oklahoma. At these fields, a zone of mineralogic alteration with well-defined vertical boundaries, called a chimney, extends from the surface downward for more than 1000 feet to the base of the Permian section. Study of these chimneys allows the construction of exploration models based on alteration phenomena related to hydrocarbon reservoirs. For example, they define which type of alteration one should attempt to detect. Also, they illustrate factors responsible for the surficial distribution of the alteration and its relationship to the productive zones. The occurrence of these alteration minerals at the Velma Field is described here as an example because of the detailed

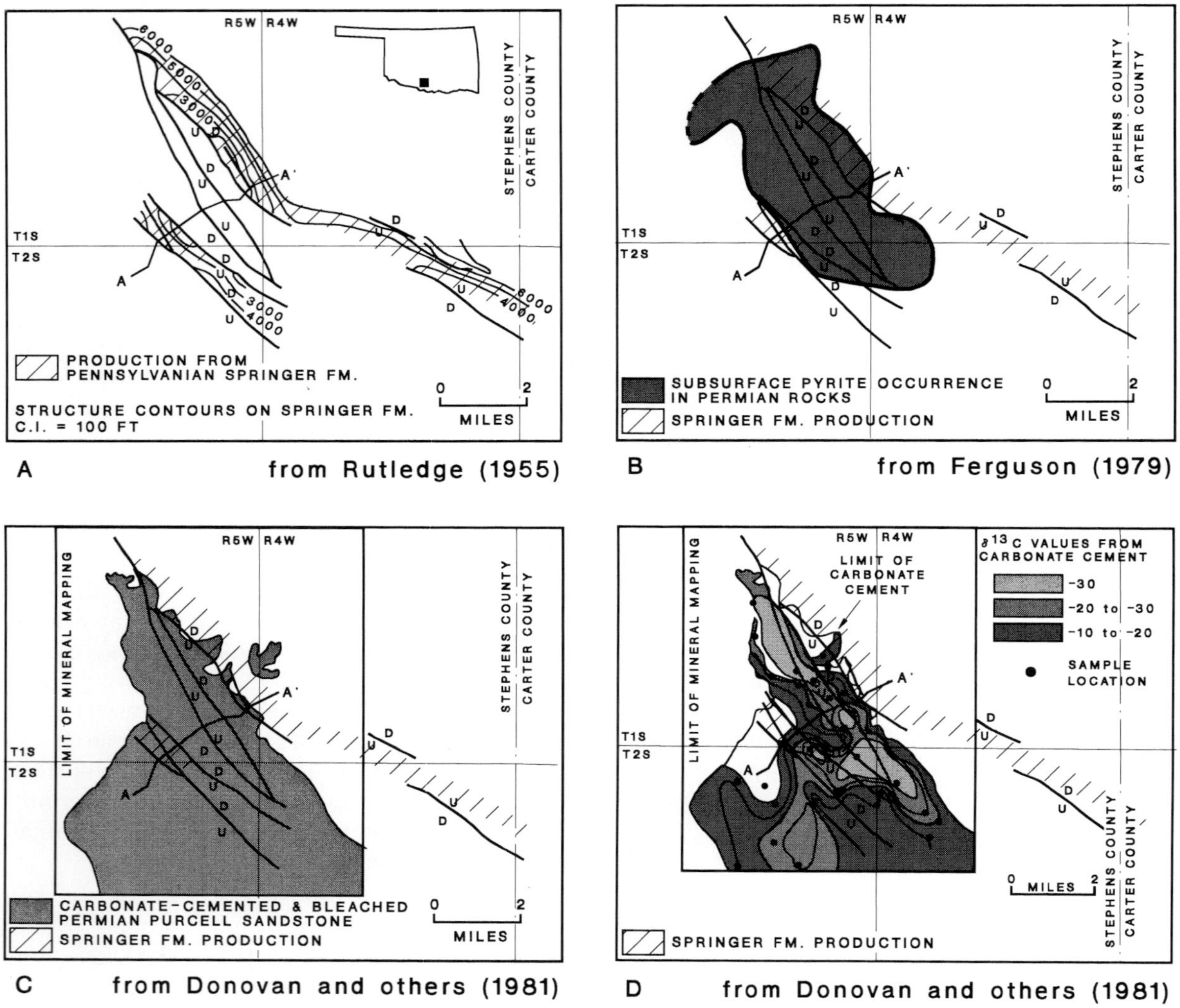

surface mapping available there. The structure and stratigraphy of all of these fields are very similar.

The geological setting of Velma is complex (see the plan view map, Fig. 8.1 A, and cross section A-A', Fig. 8.2 a). Major structures are asymmetric anticlinal folds affecting Pennsylvanian and older rocks. These are broken up by high-angle faults that form a series of horst and graben features. Gently folded Permian redbeds rest unconformably on these older rocks and are not faulted. Surface mapping, initiated because of oil seeps in the area, revealed an anticline at Velma leading to the initial discovery of oil in 1917. Earliest production was from the Permian rocks, but deeper drilling in the 1940s developed much more productive Ordovician through Pennsylvanian reservoirs, of which the Pennsylvanian Springer Formation is the most prolific. The pro-

Fig. 8.1. Distribution of mineralogic alteration at the Velma oil field

A compilation of data from the Velma oil field, Stephans County showing the productive area of the Pennsylvanian Springer Formation, the most prolific horizon at Velma (**A**), subsurface distribution of pyrite (**B**), distribution of bleaching and carbonate cement in surface exposure of the Permian Purcell sandstone (**C**), and variation of $\delta^{13}C$ values within the carbonate cement (**D**). The spatial distribution of the pyrite and the lowest $\delta^{13}C$ values are clearly related to the pre-Permian faults and not to production. (Compiled by Townsend 1984)

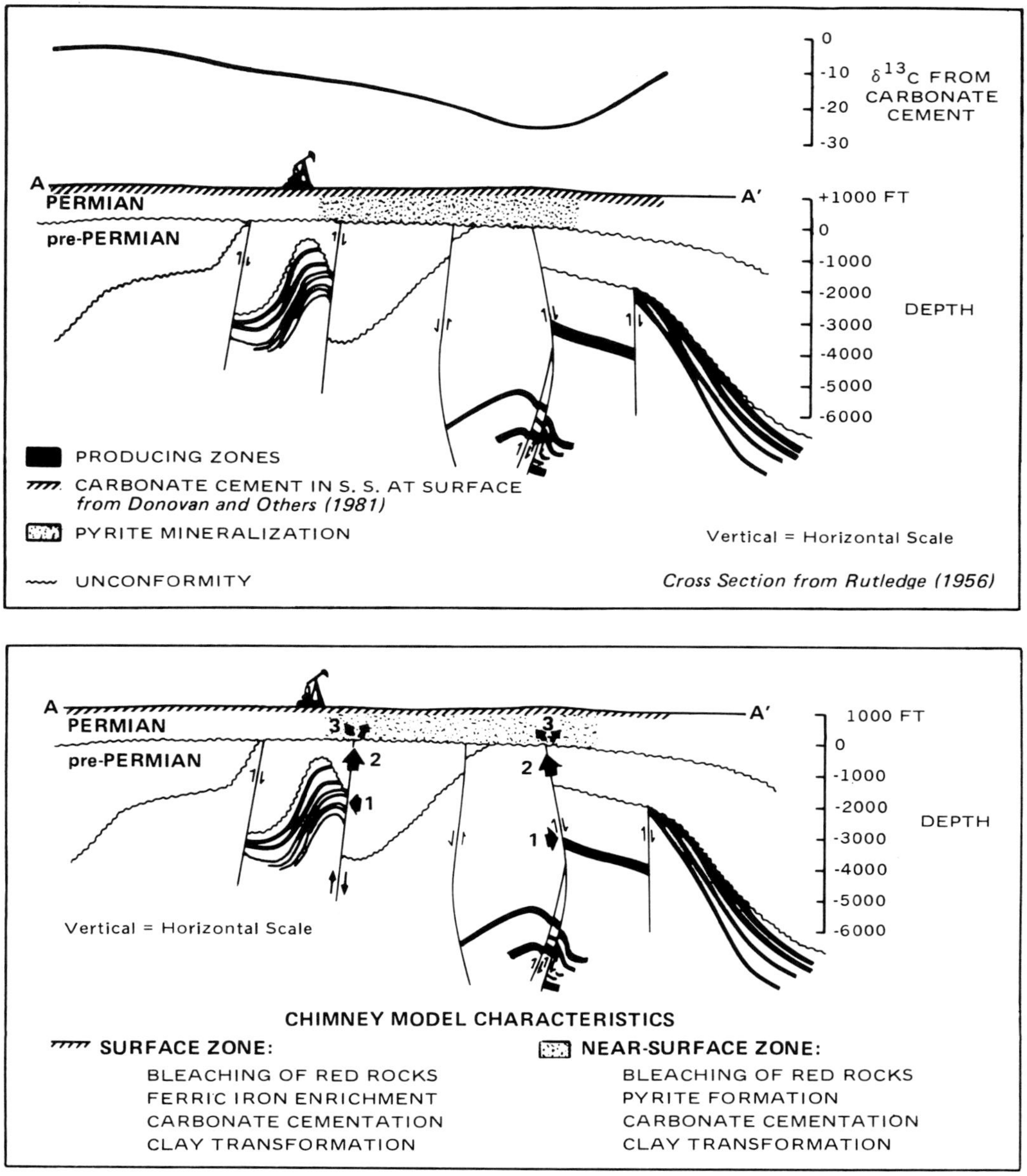

Fig. 8.2. Basic structural and chimney models at Velma
A generalized cross section of the Velma Field showing relationship of producing zones and pre-Permian faults to alteration of Permian rocks (**a**), and basic elements of the chimney model illustrated by the Velma cross section as a base (**b**). (Compiled by Townsend 1984)

ductive area of the Springer corresponds to subunconformity and tight anticlinal fold traps.

Pyrite cement occurs in a zone extending vertically from the base of the Permian section to the surface (see Fig. 8.2 a) and is laterally elongated along the structural trend overlying the pre-Permian fault (see Fig. 8.1 B). The average pyrite content of sandstones within the zone ranges from 2 to 4 % but pyrite is rare or absent outside the zone (Ferguson 1979). In addition, detailed petrologic studies of cores and well cuttings from the Cement Field (Lilburn and Al-Shaieb 1983) have shown that illite clay has been transformed to smectite and kaolinite clays. This is also probably true of the altered zone at Velma, although the presence of clay minerals there has not been established.

8.2.1.2 Surface Manifestations

At the surface of Velma, normally red clastic rocks are bleached to hues of grey, white or yellow and the usually uncemented Permian Purcell sandstone is impregnated with a carbonate cement (Donovan et al. 1981). The lateral distribution of this alteration in relation to the pre-Permian faults is shown in Fig. 8.1 C. A contour map of the isotopic composition of the carbon, reported as a $\delta^{13}C$ value, in this carbonate cement (Fig. 8.1 D) shows that the lowest $\delta^{13}C$ values fall within the – 29.4 to – 30.4 range determined for oil from the Pennsylvanian reservoirs (Young 1984, pers. comm.).

Alteration of the Permian sandstones by hydrocarbons is evidenced by (1) the bleaching of the redbeds; (2) hydrocarbon residues still present in the rocks; and (3) isotopic values indicating that the carbon in the carbonate was derived from a crude oil source. The mineralogic effects described above are predicted by geochemical models in which hydrocarbons are responsible for reducing conditions (Ferguson 1979). In addition, the bleaching of rocks by hydrocarbons in the Colorado Plateau has been described by Campbell and Ritzma (1979). The mineralogic alteration present at the Velma, Eola and Chickasha fields is summarized in Table 8.1.

The Oklahoma chimneys are unique because mineralogic alteration at the surface has been directly linked to vertical migration of hydrocarbons from reservoirs at depth. Vertical migration and the resulting chimneys, however, may be a common occurrence. For example, preferential leakage of gas along fault planes has been documented by Vinkovetsky (1982, pers. comm.) in the Gulf Coast region.

Also, Jones and Drozd (1983) have shown that the composition of gas in soil-gas anomalies in the Utah-Wyoming Overthrust Belt is strongly correlated with the composition of reservoir gas. Furthermore, they found that a strong relationship exists between fault and fracture systems and the larger gas seeps. Richers et al. (1982) also report abnormally high amounts of soil gas associated with surface lineaments at the Patrick Draw oil fields, in the Green River Basin of Wyoming.

These observations form the basis for a model of the vertical migration of hydrocarbons and the resulting mineralogic alteration (Ferguson 1979). The cross section of Velma (Fig. 8.2 b) is used to illustrate the basic elements of this chimney model. Gases such as methane and light-end hydrocarbons are dissolved in water (1 in Fig. 8.2 b). These fluids migrate upward by effusive movement along high-angle faults, or through “microfractures” by hydrodynamic and chemical potential drive (2 in Fig. 8.2 b). Oxidizing conditions within the chimney determine whether the ferric iron minerals will be reduced to mobilize the iron or precipated from iron carried in solution by reducing fluids. The first case leads to bleaching of the host rocks, while the second leads to a local surficial enrichment of ferric iron. In either case an anomalous surface alteration is the result.

8.2.1.3 Imagery

Several techniques to display the surficial mineral distribution with imagery have been developed and used to enhance detectability of the alteration phenomenon. Several of these techniques, which were implemented by Townsend (1983, 1984), are shown with images in Fig. 8.3. Townsend concluded that the alteration phenomena cannot be identified

Table 8.1. Mineralogic effects observed at the surface due to hydrocarbon leakage

Observation	Example(s)[b]
Pervasive bleaching of red clastic rocks by reduction and removal of iron[a]	Cement, Eola, Velma, Chickasha [1], [2], [3]
Sporadic enrichment of ferric iron minerals[a]	Velma [3]
Precipitation of ferroan carbonate cements[a]	Cement, Eola, Velma, Chickasha [1], [2], [3]
^{13}C in carbonate cements which is indicative of a hydrocarbon source	Cement, Eola, Velma, Chickasha [1], [2], [3]
Transformation of illite to smectite and kaolinite clays[a]	Cement [4]

[a] These phenomena have distinctive reflectance characteristics in the visible and near infrared.
[b] [1] Donovan (1974); [2] Ferguson (1979); [3] Donovan et al. (1981; [4] Lilburn and Al-Shaieb (1983).

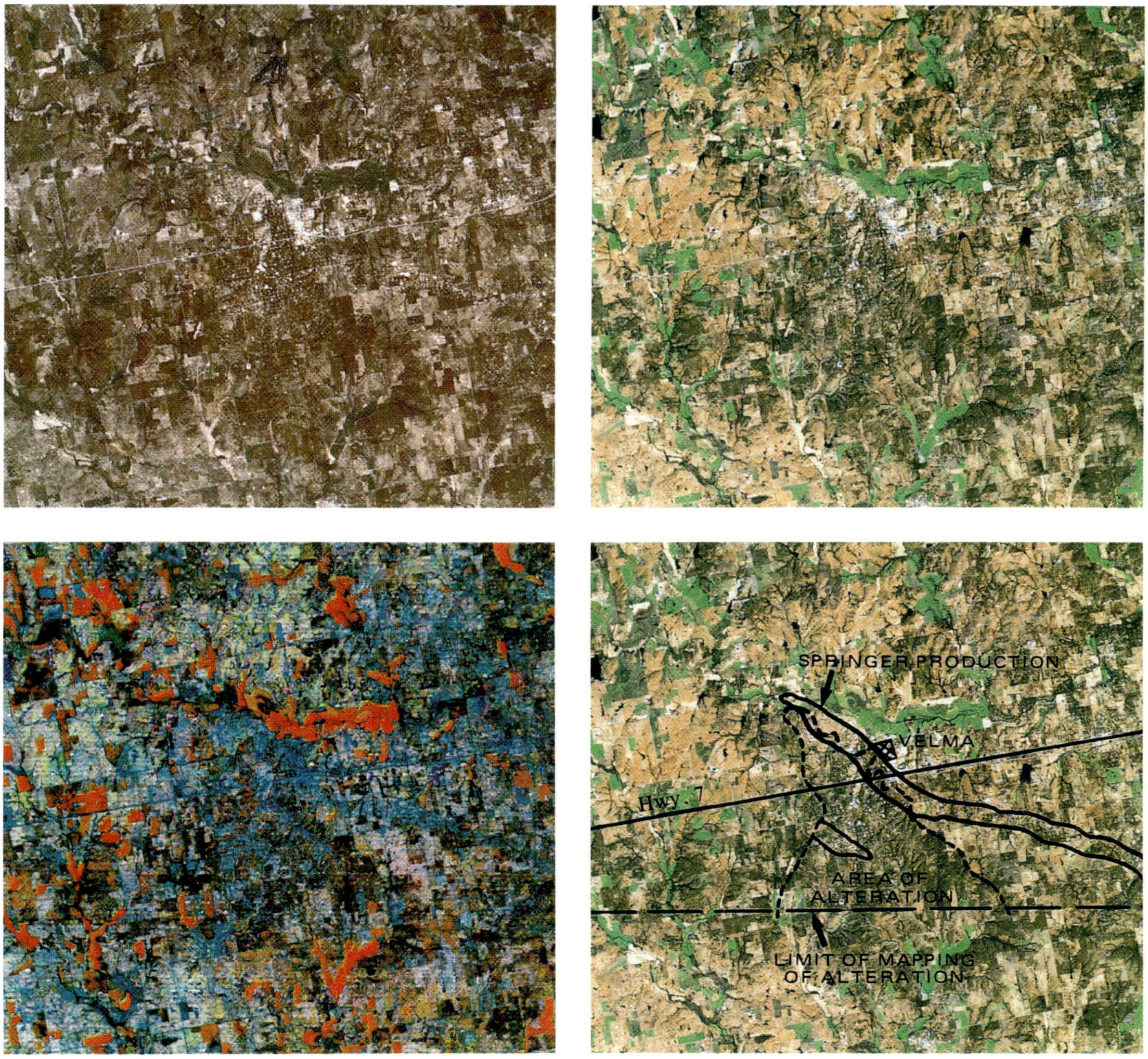

Fig. 8.3. Surface expression of the Velma Field
Landsat 4 TM images of the Velma Field, Oklahoma, showing **a** a natural-color composite (blue, green and red assigned to bands 1, 2 and 3, respectively); **b** an infrared-color composite (blue, green, and red assigned to bands 1, 4, and 5, respectively); **c** a color-ratio composite formed by assigning blue, green and red to the 3/4, 5/1 and 5/7 bandwidth ratios, respectively; and **d** the second image repeated with an outline of the producing field and extent of surficial mineralogic alteration. (Map data from Donovan et al. 1981. Compiled by Townsend 1984)

uniquely on imagery regardless of the type of enhancements employed. He noted, however, that the presence of structurally controlled stream valleys and tonal anomalies reveal the presence of the buried producing structure in this area. Similar conclusions were reached by Townsend in examining imagery data over other fields which are known to exhibit surface alteration phenomena, particularly the Cement Field in Oklahoma. Donovan et al. (1979) also failed to recognize the altered rocks at Cement using Landsat MSS imagery data. They stated that alteration is not uniform over the field commonly occurring in patches with exposed areas of less than 50 m^2. The remaining altered rocks are difficult to detect at the surface because they are covered largely by unconsolidated sediments and vegetation.

Sabins (1987) summarizes this effort stating that no unique spectral signatures of color and mineralogical alteration patterns have been reported for the Cement and Velma Fields. Failure to detect these surface manifestations are believed to come from two classes of limitations. First, limitations of the remote sensing systems such as spatial, spectral or radiometric resolution and, second, scene-related limitations such as vegetation and soil cover, culture and agriculture development. The scene-related limitations at Cement and Velma place severe constraints on the remote sensing required to detect the subtle reflectance anomalies that exist there. Apparently, both the radiometric and spatial resolutions of TM are insufficient to identify this feature.

8.2.2 Other Related Studies

Several other types of direct detection phenomena have been studied with remote sensing data with no significant success to date. In the Anadarko Basin in the Texas Panhandle and western Oklahoma, it was reported by Everett and Petzel (1973) that some oil and gas fields are manifested on imagery data as unique features which were referred to as "hazy anomalies". On some imagery data, the surface expressions of these fields appeared to be smudged or erased. Comprehensive studies of this phenomena were carried out by several researchers but they could neither find an explanation for these irregularities nor identify similar features in other fields around the globe (e. g. Short 1975).

Other comprehensive studies that attempted to detect surface alteration phenomena were carried out by a joint NASA/GEOSAT project using MSS and simulated TM data (Lang 1985). This project focused on three oil and gas fields, the Coyanosa Field in West Texas, the Lost River in West Virgina and the Patrick Draw in Wyoming. The result of the investigation in the Coyanosa Field area is similar to that reported for the Velma and Cement Fields. No direct evidence of hydrocarbon seepage was recognized, but the buried structures of the field apparently produced detectable geomorphic expressions.

At the Lost River Field, which is located in the forested terrain of the Appalachian Mountains, a supervised classification of TM imagery data indicated concentrations of maple trees (as compared to oak and hickory trees) over parts of the field area. At Patrick Draw, which is mostly covered by sagebrush, TM images show the presence of anomalous vegetation growth which, in ground investigations, was found to reflect sagebrush with stunted growth and small leaves. In both cases, the phenomena of "stressed" vegetation over the fields were attributed to concentrations of gas in the soil related to chimney phenomena. This relationship among hydrocarbon traps, gas in the soil, and stressed vegetation, although quite promising, has not yet proven consistent enough to provide a reliable exploration tool.

8.2.2.1 Conclusions

One might wonder why so much attention has been given here to direct detection studies when the results have been largely negative so far. There are several reasons that such efforts merit discussion. First and foremost, it is felt that the theory behind the chimney phenomenon and related alteration processes is valid and well documented. The negative results obtained so far are related not to the lack of surface alteration phenomena but limitations in the detection and image enhancement equipment. In fact, recent studies at JPL, for example, have demonstrated that considerable improvements in spectral mapping of outcropping units can be achieved by integrating satellite imagery data with other airborne sensors (see discussion below). Given this, it is recommended here that remote sensing efforts in this field should focus on development of new equipment and enhancement technology rather than promoting the use of available data for direct detection. Because the structural mapping aspects of remote sensing have reached a mature stage, this aspect should provide the main incentive for further improvements in remote sensing technology.

8.3 Outcrop Studies

The use of remote sensing data for detailed mapping of exposed bedrock strata is a long-standing concept that began with early applications of aerial photography (e. g. Krumbein and Sloss 1951; Ray 1960). Recent improvements in the spatial and particularly spectral resolution of remote sensing image data and the strong emphasis on integrated basin studies for hydrocarbon exploration renewed interest in the application of this technology. Sgavet-

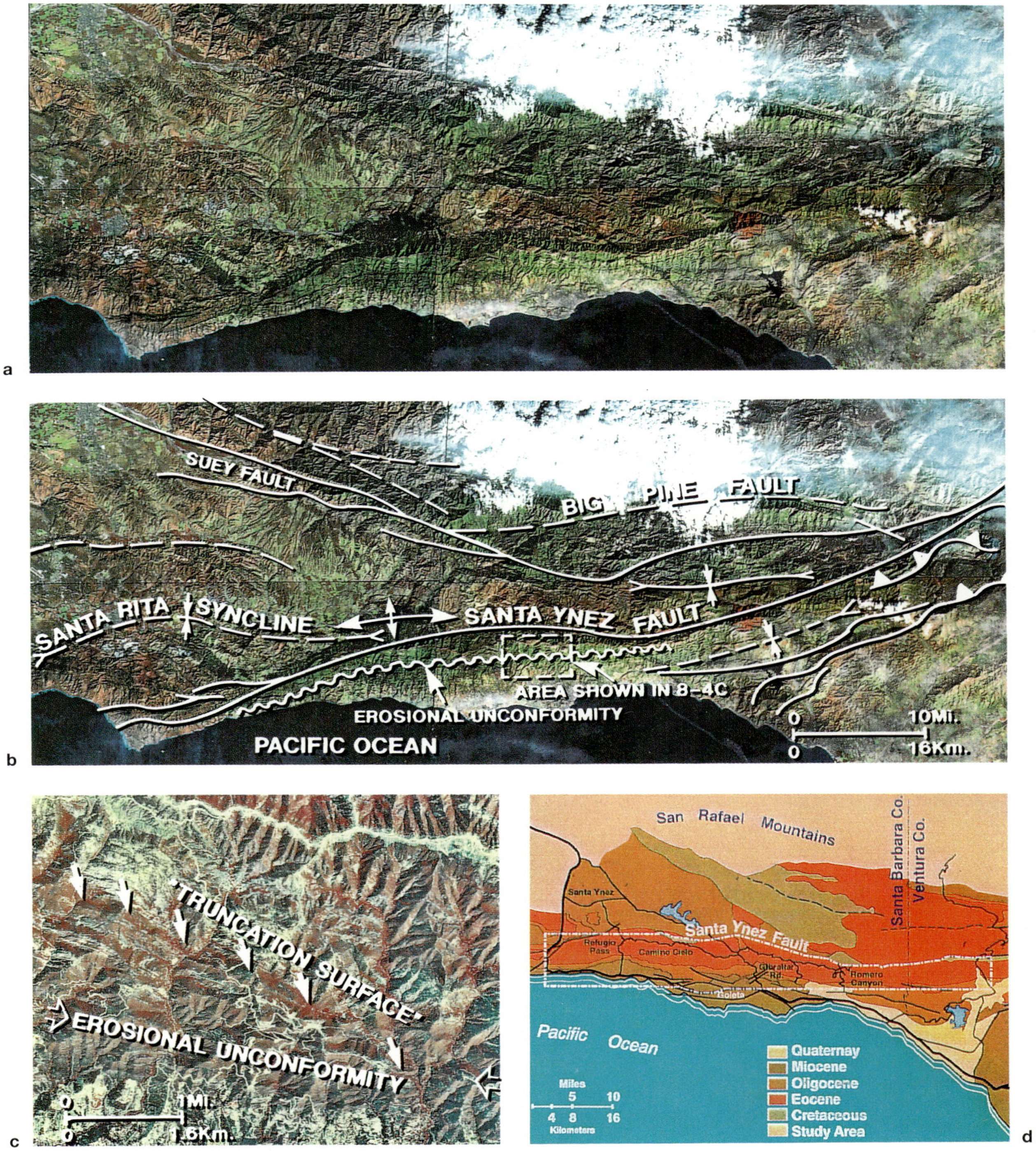

Fig. 8.4. Photostratigraphic interpretation of the Transverse Range

a and **b** show interpreted and uninterpreted satellite imagery of the western portion of the Transverse Range, respectively; **c** a close-up look of an erosional surface with high-altitude infrared photography; **d** a generalized geological map of the study area. (Compiled from Campion et al. 1988, 1992)

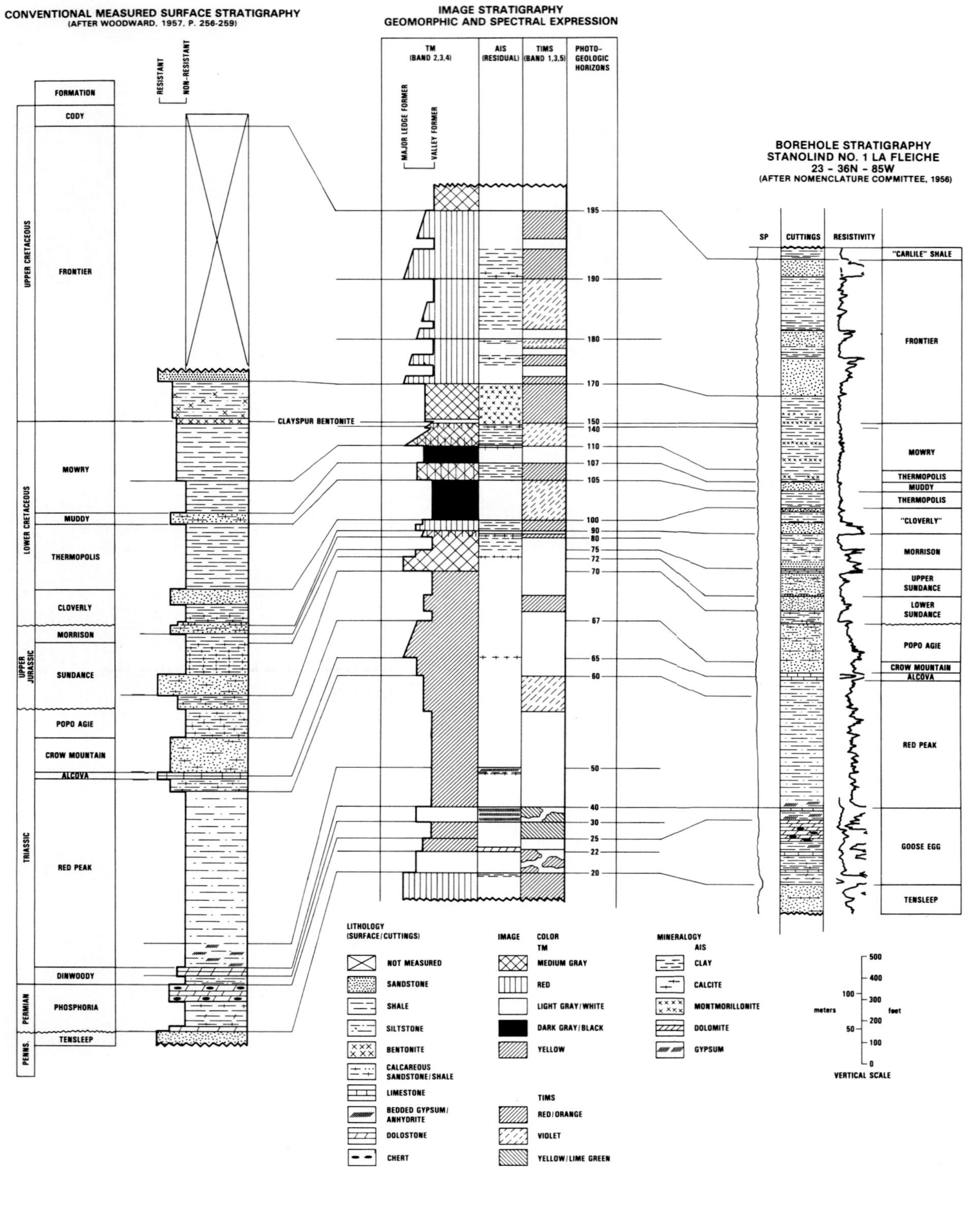

CONVENTIONAL MEASURED SURFACE STRATIGRAPHY
(AFTER WOODWARD, 1957, P. 256-259)
IMAGE STRATIGRAPHY
GEOMORPHIC AND SPECTRAL EXPRESSION
BOREHOLE STRATIGRAPHY
STANOLIND NO. 1 LA FLEICHE
23 - 36N - 85W
(AFTER NOMENCLATURE COMMITTEE, 1956)
FORMATION
RESISTANT
NON-RESISTANT
CODY
FRONTIER
MOWRY
MUDDY
THERMOPOLIS
CLOVERLY
MORRISON
SUNDANCE
POPO AGIE
CROW MOUNTAIN
ALCOVA
RED PEAK
DINWOODY
PHOSPHORIA
TENSLEEP
UPPER CRETACEOUS
LOWER CRETACEOUS
UPPER JURASSIC
TRIASSIC
PERMIAN
PENNS.
CLAYSPUR BENTONITE
TM (BAND 2,3,4)
AIS (RESIDUAL)
TIMS (BAND 1,3,5)
PHOTO-GEOLOGIC HORIZONS
MAJOR LEDGE FORMER
VALLEY FORMER
195
190
180
170
150
140
110
107
105
100
90
80
75
72
70
67
65
60
50
40
30
25
22
20
SP
CUTTINGS
RESISTIVITY
"CARLILE" SHALE
FRONTIER
MOWRY
THERMOPOLIS
MUDDY
THERMOPOLIS
"CLOVERLY"
MORRISON
UPPER SUNDANCE
LOWER SUNDANCE
POPO AGIE
CROW MOUNTAIN
ALCOVA
RED PEAK
GOOSE EGG
TENSLEEP
LITHOLOGY (SURFACE/CUTTINGS)
NOT MEASURED
SANDSTONE
SHALE
SILTSTONE
BENTONITE
CALCAREOUS SANDSTONE/SHALE
LIMESTONE
BEDDED GYPSUM/ ANHYDRITE
DOLOSTONE
CHERT
IMAGE
COLOR
TM
MEDIUM GRAY
RED
LIGHT GRAY/WHITE
DARK GRAY/BLACK
YELLOW
TIMS
RED/ORANGE
VIOLET
YELLOW/LIME GREEN
MINERALOGY
AIS
CLAY
CALCITE
MONTMORILLONITE
DOLOMITE
GYPSUM
500
400
300
200
100
0
100
50
meters
feet
VERTICAL SCALE

◀ Fig. 8.5. Correlation of stratigraphic columns in the Wind River Basin
The stratigraphic column in the middle was prepared through integrated analysis of remote sensing data. Lithostratigraphic horizons were then correlated with a measured surface section and borehole information. (After Lang et al. 1987)

ti (1992), using examples from the south-central Pyrennes, demonstrated several techniques that can be used to establish the sequence stratigraphic framework of the exposed Eocene units in this region. Sgavetti proposed the term "photostratigraphy" to describe this analytical approach which relies, more or less, on similar concepts and terminologies used for sequence stratigraphic studies of subsurface seismic and well data (e. g. Vail 1987) as well as outcrop mapping (e. g. Van Wagoner et al. 1992).

A typical example of photostratigraphic studies of outcropping units and their potential contribution to basin studies is illustrated in Fig. 8.4. The upper two images show a structural stratigraphic interpretation of the exposed Eocene rock units of the Transverse Range in California. Highlighted on the image are the FLTs of the San Andreas wrench fault system and a major mid-Eocene erosional unconformity which is interpreted parallel to these FLTs. The surface expression of the unconformity is nicely displayed with a high-altitude infrared color photograph shown at the bottom left. The photo depicts, in greater detail, an angular and erosional discontinuity surface with typical erosional truncations and terminations by onlap (see also Sgavetti 1992, Fig. 5). The analysis of the imagery data was conducted by Campion et al. (1988) as part of a regional tectono-stratigraphic investigation of the Transverse Range and the adjacent areas. The recognition of these regional-scale boundaries provided new interpretation regarding the sequence stratigraphic framework of the region and its relationship to rate of uplift, subsidence, and local tectonic activities (Campion et al. 1992).

Because photostratigraphy and sequence stratigraphy use similar concepts and terms, they are easy to integrate for the analysis of regional stratigraphy and basin evolution. Many aspects of this approach were recently documented by a group of geoscientists from the Jet Propulsion Laboratory (JPL) who participated in the NASA-funded Multispectral Analysis of Sedimentary Basins Project (Lang 1985). For example, a remote sensing data set from the Wind River Basin, which included data from Landsat TM, airborne visible and infrared imaging spectrometer (AVIRIS) and thermal infrared multispectral scanner (TIMS), was used together with digital topographic data to map lithologic contacts, measure dips and strikes and develop a stratigraphic column that is correlated with conventional surface and subsurface sections (Lang et al. 1987; Fig. 8.5). In other related studies from the Wind River Basin, Stucky and Krishtalka (1991) and Lang et al. (1991) have demonstrated the effective use of remote sensing imagery investigation for the mapping and recognition of biostratigraphic markers in areas of poorly exposed bedrocks. The effective use of the multispectral capabilities of the remote sensing data for investigation of organic-rich outcrops has also been demonstrated recently by Rowan et al. (1992) who mapped thermal maturity in the Chainman Shale, Nevada with Landsat TM images.

These studies, and many others, clearly demonstrate the potential use of remote sensing data for early investigations of basins, stratigraphy, source rock, and other important aspects related to the hydrocarbon potential of both frontier and mature areas.

8.4 Logistical and Environmental Applications

Although not directly related to geological investigations and prediction of hydrocarbon accumulations, logistical and environmental applications must be mentioned as other significant roles for satellite imagery interpretation. These applications may be divided into two major categories, marine and onshore.

Satellite images of near-shore areas provide significant clues for detection of natural oil seeps as well as monitoring of spills associated with hydrocarbon activities. Exxon's Valdez disaster, for example, has been extensively monitored and studied using a wide range of satellite imagery and other remote sensing data. In onshore activities, satellite imagery data have been routinely used to update existing topographic maps and support the planning of seismic programs. Other onshore applications include monitoring the impact of drilling activities on the environment, and assessing reclamation activities as well as detecting hazardous conditions such as active fault systems, areas of extensive collapse features related to karst or mud slides and frequently flooded regions.

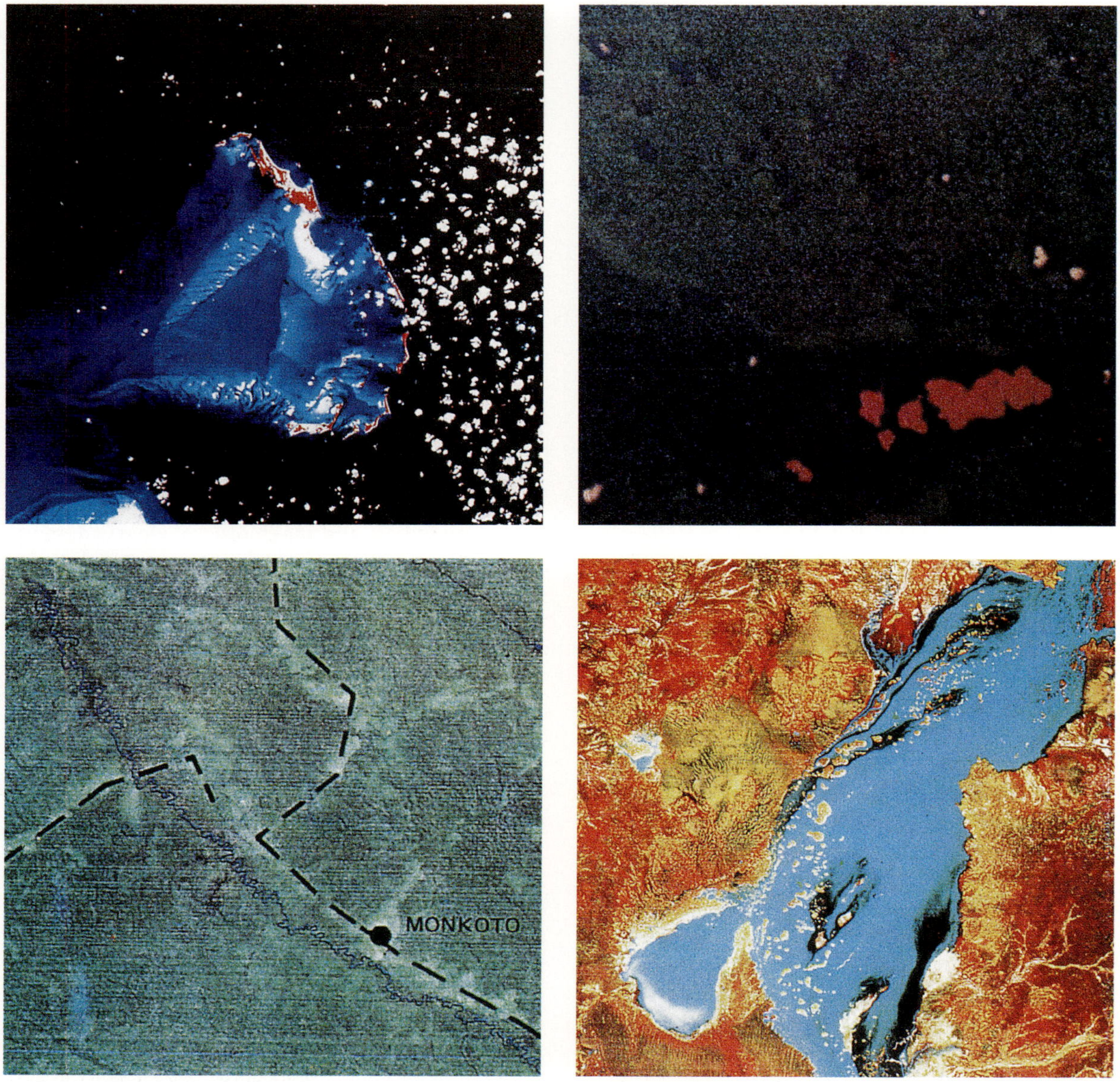

a b c d

Several typical examples of logistical and environmental applications of remote sensing data are illustrated in Fig. 8.6. The upper left image is from the northern Bahama Islands. It was enhanced to provide maximum water penetration over the Grand Bahamas Bank. The highly reflective carbonate bottom and clear, quiet water allow direct detection of bottom features in water depths up to 30 m.

The upper right image shows a subscene from the southwestern Gulf of Mexico centered on a major well blowout. The well and the oil spill that it generated were monitored periodically by a Landsat sat-

Fig. 8.6. Logistical application of satellite imagery data
a Distribution of shallow carbonate banks (blue and purple) in the Grand Bahamas Bank; **b** detection of a well blowout (red) in the southwestern Gulf of Mexico; **c** identification of old seismic cuts in the forests of Zaire; **d** delineation of maximum flooding potential for wadies of Australia. (Prepared by R. L. Brovey 1983)

ellite. The spill is visible as a dark blue and pink band across the lower portion of the image. The slick is mappable because the heavy paraffin-based oil has a yellowish brown color that produces a higher reflective value in band 5, compared with the adjacent, open marine water.

The bottom left image shows the most common usage of satellite imagery for logistical onshore applications. Here, Landsat imagery data from Zaire was used to locate the trail of forest-cutting left by previous seismic surveys.

The bottom right image shows another seismic-related investigation of satellite data. The imagery, which is from an arid area in Australia, was taken during a wet season where wadies are at maximum flooding stage. These data were proven critical in designing and scheduling the seismic programs in this area.

8.5 Summary

Three major types of satellite imagery interpretation which are not directly related to structural mapping of geological structures are nonetheless of importance to hydrocarbon exploration. The first is related to the belief that the hydrocarbon chimney phenomenon can be observed on image data and, hence, used for direct detection of hydrocarbons. There is not yet any evidence, however, to substantiate this claim. Current studies indicate that such efforts have been hampered by the resolution limitation of the remote sensing equipment and possibly the lack of adequate enhancement techniques. Future research and development in this field are highly warranted.

The second utilization of image data focuses on the analysis of exposed outcrop units for the purpose of establishing their sequence stratigraphic nature and source rock potential. Such studies have shown a great deal of promise and are being pursued further.

Logistical and environmental studies of marine and onshore areas are the third use for imagery data. These applications have been found useful for detection of natural oil seeps and man-made oil spills as well for investigation of surface conditions related to seismic programs, geological hazards and environmental monitoring and reclamation activities. With the emergence of environmental awareness, the importance of such studies is increasing with their demand for repeated satellite data overshadowing even exploration projects.

References and Further Reading

Brovey RL (1983) Logistical applications of satellite imagery. Exxon Production Research Company, Internal Rep

Buckingham WF, Sommer SE (1983) Mineralogical characterization of rock surfaces formed by hydrothermal alteration and weathering – application to remote sensing. Econ Geol. 78: 664–674

Campbell JA, Ritzma HR (1979) Aeromagnetic detection of diagenetic magnetite over oil fields – discussion. AAPG Bull 63: 1538–1539

Campion KM, Morgan SR, Berger Z (1988) Recognition of a Middle Eocene unconformity in transverse ranges, utilizing satellite imagery and high-altitude photography. AAPG Bull 72 (3): 376 (Abstr)

Campion KM, Morgan SR, Lohmar JM (1992) Paleogene sequence stratigraphy of the transverse ranges (southern California) and comparison to the Gulf of Mexico and Hampshire Basins. 26th Annu Meet of the Geol Soc Am, Houston, Tx, vol 24, No 1, p 5 (Abstr)

Donovan TJ (1974) Petroleum microseepage at Cement, Oklahoma – evidence and mechanism. AAPG Bull 58: 429–446

Donovan TJ, Friedman I, Gleason JD (1974) Recognition of petroleum-bearing traps by unusual isotopic compositions of carbonate-cemented surface rocks. Geology 2: 351–354

Donovan TJ, Termain PA, Henry ME (1979) Late diagenetic indicators of buried oil and gas, II. Direct detection experiment at Cement and Garza oil fields, Oklahoma and Texas, using enhanced Landsat 1 and 2 images. USGS Open File Rep 79–243

Donovan TJ, Roberts AA, Dalziel MC (1981) Epigenetic zoning in surface and near-surface rocks resulting from seepage-induced redox gradients, Velma oil field, Oklahoma – a synopsis. Shale Shaker 32 (3): 1–7

Everett JR, Petzel G (1973) An evaluation of the suitability of ERTS data for the purposes of petroleum exploration. 3rd Earth resources technology satellite Symp. NASA SP-356, pp 50–61

Ferguson JD (1979) The subsurface alteration and mineralization of Permian red beds overlying several oil fields in southern Oklahoma. Shale Shaker 29 (8): 172–178, continued in no 9: 200–208

Jones VT, Drozd RJ (1983) Predictions of oil or gas potential by near-surface geochemistry. AAPG Bull 67: 932–952

Krumbein WC, Sloss LL (1951) Stratigraphy and sedimentation. Freeman, New York

Lang HR (ed) (1985) Report of the workshop on geologic applications of remote sensing to the study of sedimentary basins, Lakewood, Colorado, January 10–11. JPL Publ Pasadena, CA

Lang HR, Adams SL, Conel JE, McGuffie BA, Paylor ED, Walker RE (1987) Multispectral remote sinsing as stratigraphic and structural tool, Wind River Basin and Big Horn Basin areas, Wyoming. AAPG Bull 71 (4): 389–402

Lang HR, Frerichs WE, McGugan A, Paylor ED (1991) Biostratigraphic significance of a new unit, mapped remotely with multispectral thermal infrared data, Late Cretaceous Cody Shale, southern Bighorn Basin, Wyoming. Mountain Geol 28 (2/3): 67–73

Lilburn RA, Al-Shaieb Z (1983) Geochemistry and isotopic composition of hydrocarbon-induced diagenetic aureole (HID), Cement Field, Oklahoma. Shale Shaker 34 (4): 40–56 and continued in no 5: 57–67

Ray RG (1960) Aerial photographs in geologic interpretation and mapping. Geol Surv Prof Pap 373

Richers DM, Reed RJ, Horstman KC, Michels GD, Baker RN, Lundell L, Marrs RW (1982) Landsat and soil-gas geochemical study of Patrick-Draw Oil Field, Sweetwater County, Wyoming. AAPG Bull 66: 903–922

Rowan LC, Wetlaufer PH (1975) Iron-absorption band analysis for the discrimination of iron-rich zones. USGS, Type III Final Rep

Rowan LC, Goetz AFH, Ashley RP (1977) Discrimination of hydrothermally altered and unaltered rocks in visible and near-infrared multispectral images. Geophysics 42: 522–535

Rowan LC, Pawlewicz MJ, Jones OD (1992) Mapping thermal maturity in the Chainman Shale, near Eureka, Nevada, with Landsat thematic mapper images. AAPG Bull 76 (7): 1008–1023

Rutledge RB (1955) The Velma Oil Field, Stephens County, Oklahoma. In: Moore CA (ed) Proc 4th Subsurface geological Symp, Univ Oklahoma, Norman, Oklahoma, March 1–2, pp 49–66

Rutledge RB (1956) The Velma Oil Field, Stephens County, Oklahoma. In: Petroleum geology of southern Oklahoma, a symposium. AAPG, Tulsa, OK, Ardmore Geol Sect 1: 260

Sabins FF (1987) Remote sensing: principles and interpretation. W. H. Freeman, San Francisco

Sgavetti M (1992) Criteria for stratigraphic correlation using aerial photographs: examples from the southcentral Pyrennes. AAPG Bull 76 (5): 708–730

Short NM (1975) Exploration for fossil and modern fuels from orbital altitudes: remote-sensing energy related studies. Hemisphere Publishing Corp, Washington DC, pp 189–232

Stucky RK, Krishtalka L (1991) The application of geologic remote sensing to vertebrate biostratigraphy: general results from the Wind River Basin, Wyoming. Mountain Geol 28 (2/3): 75–82

Townsend TE (1983) Discrimination of iron alteration minerals in remote sensing data. PhD Thesis, Stanford University

Townsend TE (1984) The significance of iron oxides and clays in petroleum and minerals exploration. Exxon Production Research Company, Internal Rep

Vail PR (1987) Seismic stratigraphy interpretation using sequence stratigraphy. Part 1. Seismic stratigraphy interpretation procedure. In: Bailey AW (ed) Atlas of seismic stratigraphy. AAPG Stud Geol 27: 1–10

Van Wagoner JC, Mitchum RM, Campion KM, Rahmanian VD (1992) Siliclastic sequence stratigraphy in well logs, cores and outcrops. AAPG Methods Explor Ser 7

Closing Remarks

Remote sensing brings new and exciting tools to surface mapping which are constantly undergoing substantial improvements. Towards the end of the century, it is expected that most of the satellite images used in this book will be deemed obsolete and replaced by higher resolution images that are more easily obtained. Some of the new imaging systems that are currently being designed include a new generation of SPOT and Landsat satellites as well as a wide range of space radar data which include the Japanese JERS-1 and the Canadian Radarsat. Russian imaging capabilities, previously unavailable, have now entered the common market with the introduction of the ALMAZ satellite.

At an even greater pace, computer technologies are being developed to process, enhance, merge and display data. Current trends to decrease cost and increase user friendliness are likely to continue in the future, eliminating the two greatest barriers to the technology in the past.

On the interpretation side, the remote sensing community has undergone a cycle typical to the adoption of any new technology. This cycle began with the introduction of satellite imaging which brought forth grandiose statements of the ability to easily detect hydrocarbons anywhere on the planet. This notion triggered a flood of young geologists pursuing careers in the straightforward analysis of satellite data. In 1980, when I joined the Exxon Production Research Co., the company was in a process of assembling a large group of geoscientists totally dedicated to the evaluation of this technology. Many other companies followed suit in their quest to obtain the competitive edge in exploration.

The second stage began when the initial euphoria dissipated as experience showed the limitation of this technology for direct detection of hydrocarbons. In addition, the use of satellite imagery for structure mapping had proven to be quite limited without support from other data sets. Shortly, individuals from the team, realizing these limitations, either left the company or requested a transfer to different departments in order to pursue other avenues of exploration. It is interesting to note that all of these individuals, despite the varied paths they took in geoscience, maintained an interest in remote sensing technology and continued to search for its implementations. For example, while working exploration programs in the Western Canada Basin, primarily with seismic and well data, I kept finding myself analyzing imagery data and noting its possible contributions to exploration.

At this point, it may be said that a third stage of the cycle began wherein satellite imagery analysis has found its role in exploration. This role places it as a reconnaissance tool for identification of structures in both frontier and mature areas. In frontier regions, satellite imagery provides an excellent tool for identification of structural style and trend and can be used to plan seismic data collection as well as for early evaluation of the hydrocarbon potential of these areas (size, spacing and possible geometries of hydrocarbon traps). In mature areas, satellite imagery data can be used to identify and map structure leads in places where the subsurface imaging of structures is either hampered by noise or complicated by the nature of the structure observed. As a rule of thumb, one should consider spending no more than 10 % of the data collection and interpretation budget for the combined use of all reconnaissance tools including gravity, magnetic and satellite imagery data. Also, during this stage, I came to the realization that a book that emphasized this integrated approach would be beneficial.

So where does the future of remote sensing lie? Foremost, we must continue to improve satellite imaging technology that is directed towards surface mapping of different lithological units as well as alteration phenomena. This is, by far, the most important area of research and development and can open the door to a wider range of applications, including the direct detection of hydrocarbons. The structural interpretation aspect of imagery data has reached a stage of maturity, and I do not foresee that improvements in technology will lead to significant advances in this endeavor. The crucial element in the proper application of this technology will remain in the abilities of interpreters to integrate other subsurface information and creatively implement the advantages of remote sensing technology. This aspect is mostly dependent on the training programs of future interpreters.

Training must begin at the university level and current trends in this respect are rather alarming. Remote sensing programs seem to focus more and more on imagery technology, spectrometry and computer manipulation techniques before the basics of structural analysis, geomorphic processes and stratigraphic correlations are covered. This trend must be stopped by providing remote sensing interpreters with a thorough understanding of the processes that lead to the development of those surface features which are recognized on image data and the ability to relate them to subsurface geology. We also must introduce the exploration community, via books, articles and short courses, to the idea of integrated remote sensing interpretation techniques and their appropriate role in exploration.

Part 2

Additional Examples of Remote Sensing Interpretation and Integration

Introduction

While in Part 1 of this book, techniques for satellite imagery interpretation and integration were introduced, this part provides the reader with more detailed examples of interpreted imagery and related surface and subsurface constraints. In selecting these examples, an attempt was made to cover several different types of tectonic regions and climatic settings where structures exhibit different levels of bedrock exposure. The examples are presented in a specific order beginning with the interpretation of well-exposed structural regions and progressing to areas of buried and obscured structures. Furthermore, examples in this part of the book may be used by the reader as exercises to practice interpretation skills. The areas selected are the following:

- Salt Flat Graben, West Texas – a well-exposed tectonic feature that exhibits an extensional fault block structural style and the presence of reactivated basement faults.
- Death Valley Region, Nevada and California – a well-exposed tectonic feature that exhibits several different structural styles, including wrench fault assemblages, complex fold and thrust belts and high- and low-angle normal faults.
- Fort St. John Graben, Western Canada – a buried graben feature characterized by the presence of subtle drape structures that can be mapped and analyzed by their geomorphic expression as well as stereo imagery data.
- Canadian Foreland Fold and Thrust Belt, British Columbia – typical structurally complex fold and thrust belt that is best analyzed with radar imagery.
- The Paris Basin, France – a large, low relief intracratonic sag characterized by the presence of obscured and buried structures related to reactivated basement warps and large-scale fault systems.
- East Texas Region – obscured and buried structures in a predominately extensional setting of a passive margin including an example of a huge basement warp structure (the Sabine Uplift).

Students and explorationists who wish to expand their interpretation skills and test some of the concepts that are illustrated in this part of the book should follow the training instructions which are presented for each chapter. In these chapters, the figures are presented in a specific order which is designed to lead the reader through the most common process of interpretation and integration of satellite imagery data. It begins with regional background information and continues with site-specific investigations.

As a first step in a training session, it is recommended to carefully review all the material presented and use clear overlays for comments and additional interpretations. Many of the maps are presented at the same scale as the imagery so that the same overlay can be used to compile information from different data sets. Rivers, major highways, shorelines, and other prominent features should be traced from the imagery and be used as registration marks. The procedural steps are listed on the first page of each chapter. Also included is a list of questions that are designed to guide the reader to the key issues which are covered in each chapter.

Chapter 9 The Salt Flat Graben, West Texas

9.1 Background

The Salt Flat Graben is a north-trending, extensional feature located in the western part of West Texas and the southern tip of New Mexico (Fig. 9.1). The graben was formed during Late Cenozois Basin and Range extensional events. The faulting is believed to be mainly Paleozoic and younger, with evidence of Holocene offsets on some faults (Goetz 1980). The development of the graben and some of its structural features were locally influenced by the presence of preexisting, west-northwest-trending high-angle basement-involved reverse faults and monoclinal flexures of the Diablo Platform (see Fig. 9.1).

9.2 Objectives

The graben, located in an arid area, is characterized by well-exposed structures that manifest clear expressions in outcrops and on remote sensing data. These structures have also been well mapped at the surface mostly by King (1949, 1965). The main aspects to observe here are: (1) the typical structural and topographic expressions of extensional fault-block style (see Figs. 9.2, 9.5), (2) the influence of preexisting reactivated structures and superimposed styles on the development of individual traps within the graben (see Figs. 9.3, 9.4); and (3) the unique surface expression of different FLTs (see Figs. 9.5–9.8).

Most of the data shown here were taken from a field guidebook which was prepared by Phelps and Harding (1987).

9.3 Training Instructions

9.3.1 Interpretation Procedures

1. Place a clear overlay on top of Fig. 9.5 and map in detail the faulted margins of the Salt Flat Graben.
2. Use the tectonic map in Fig. 9.2 and the outcrop information from Figs. 9.6–9.8 to recognize (and mark on the overlay) different fault-line traces.
3. Use the total intensity magnetic map in Fig. 9.3 and the Precambrian surface structure map in Fig. 9.4 to identify the trends of preexisting basement structures.
4. Highlight on the interpretative overlay basement trends that seems to control the structural setting of the graben.
5. Identify (with differently colored pencils) other trends of faults in the area.
6. Compare the results of your mapping with the data shown on the tectonic map in Fig. 9.2.

9.3.2 Questions

Assuming that the Salt Flat Graben was buried under sediments and was identified as an effective hydrocarbon play:

1. Which side of the graben should be considered more prospective?
2. What type of hydrocarbon traps may exist in this area?
3. Where would you plan your initial seismic program?
4. Where would you anticipate drilling the first exploration well?
5. How would you apply the observations made here to the exploration of other graben features including those which are deeply buried.
6. List several prolific hydrocarbon plays that exhibit the same structural style as the Salt Flat Graben.

References and Further Reading

Goetz LK (1980) Quaternary faulting in Salt Basin Graben, West Texas. In: Dickerson PW, Hoffer JM (eds) Trans-Pecos region southeastern New Mexico Geological Society 31st field conference guidebook, pp 83–92

Harding TP (1984) Graben hydrocarbon occurrence and structural style. AAPG Bull 68: 333–362

Harding TP, Lowell JD (1979) Structural styles, their plate-tectonic habitats and hydrocarbon traps in petroleum provinces. AAPG Bull 63: 1016–1058

King PB (1949) Regional geologic map of parts of Culberson and Hudspeth counties. USGS Oil and Gas Investigations, Preliminary Map 90 (reprinted 1980)

King PB (1965) Geology of Sierra Diablo Region, Texas. USGS Professional Pap 480

Phelps PW, Harding TP (1987) Extensional fault blocks guidebook of Salt Graben, West Texas. Exxon Production Research Company, Internal Rep

Shepard TM, Walper JL (1983) Tectonic evolution of Trans-Pecos, Texas. In: Meador-Robert SJ (ed) Geology of Sierra Diablo and southern Hueco Mountains, West Texas. SEPM Permian Basin Section guidebook, pp 131–140

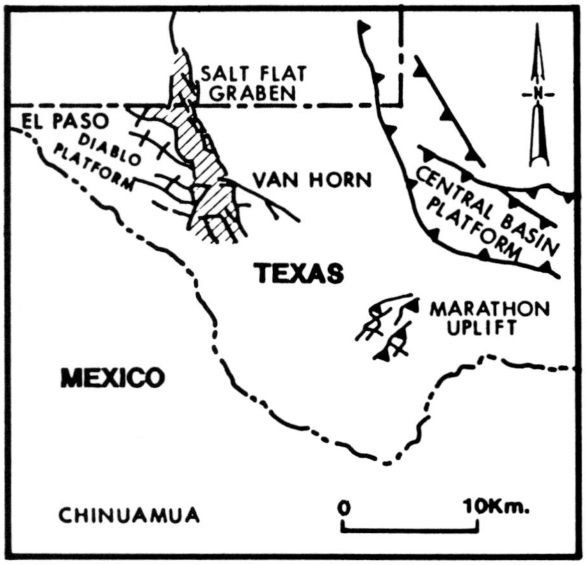

Fig. 9.1. Late Paleozoic structures in West Texas
A generalized tectonic map showing the location of major Paleozoic tectonic elements and the superimposed Salt Flat Graben. (After Phelps and Harding 1987)

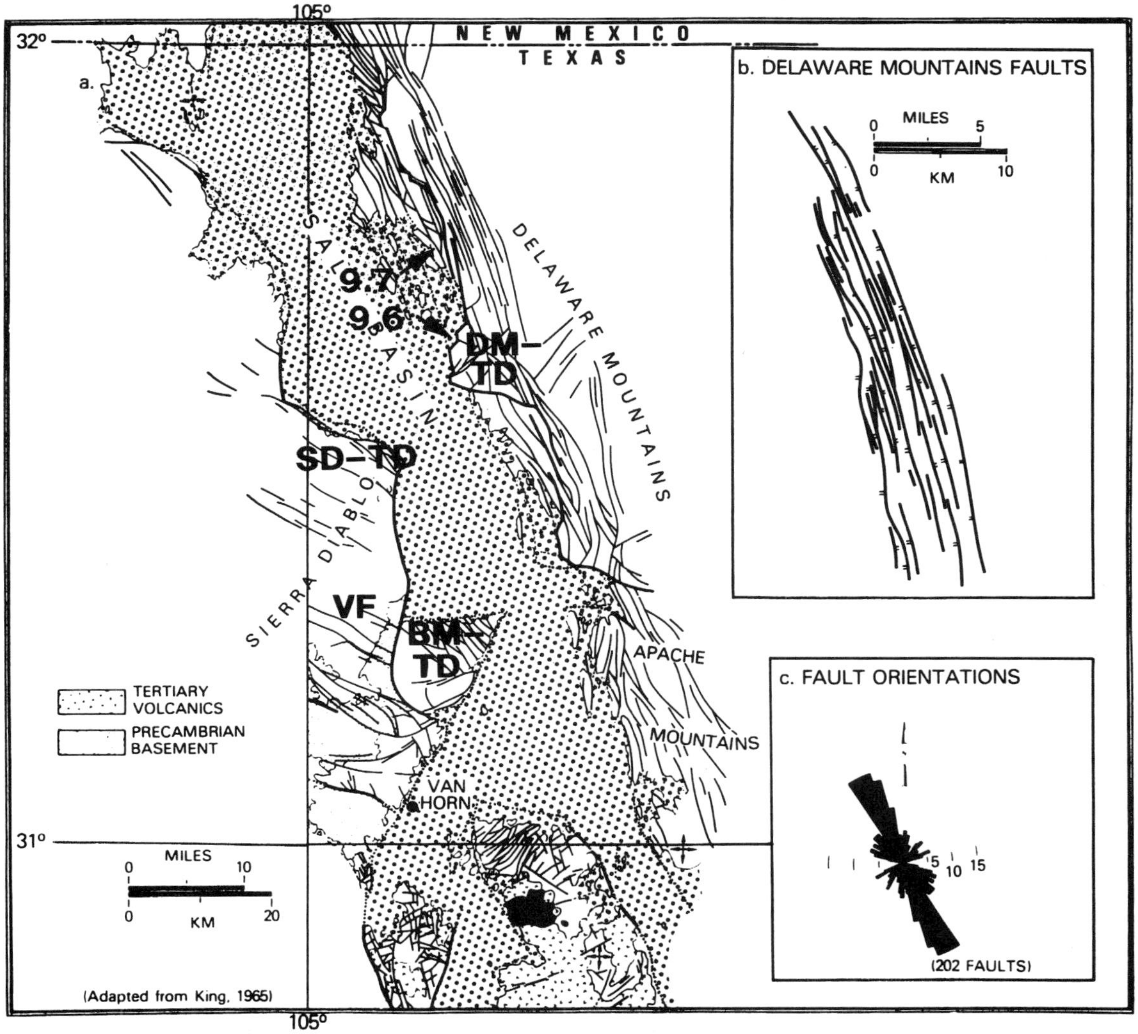

Fig. 9.2. Major structural elements of the Salt Flat Graben
a Map of extensional fault blocks; **b** enlargement of longitudinal relay fault patterns at the east flank of the graben; **c** plot of number of fault orientations. The map shows location of figures and key structural elements. *DM-TD* Delaware Mountains trapdoor; *BM-TD* Baylor Mountain trapdoor; *VF* Victorio flexure; *SD-TD* Sierra Diablo trapdoor. (Phelps and Harding 1987)

Fig. 9.3. Total intensity magnetic map of the Salt Flat Graben ▶
Showing the relationship between the major faults of the graben (generalized by *dashed lines*) and basement structures. Preexisting basement (Late Paleozoic) structures such as the Victorio flexure exert significant control on the orientation and segmentation of the eastern margin of the graben leading to the development of the Sierra Diablo and Baylor Mountains trapdoor structure. (After Phelps and Harding 1987)

Fig. 9.4. Precambrian surface of the Diablo Plateau ▶▶
Structure contour map on top of Precambrian surface showing the development of trapdoor structures that were formed at the intersection of two sets of faults, west-northwest-trending (Late Paleozoic) reactivated structures and north-trending (Late Cenozoic) graben-age faults. (After King 1965; published with permission of USGS)

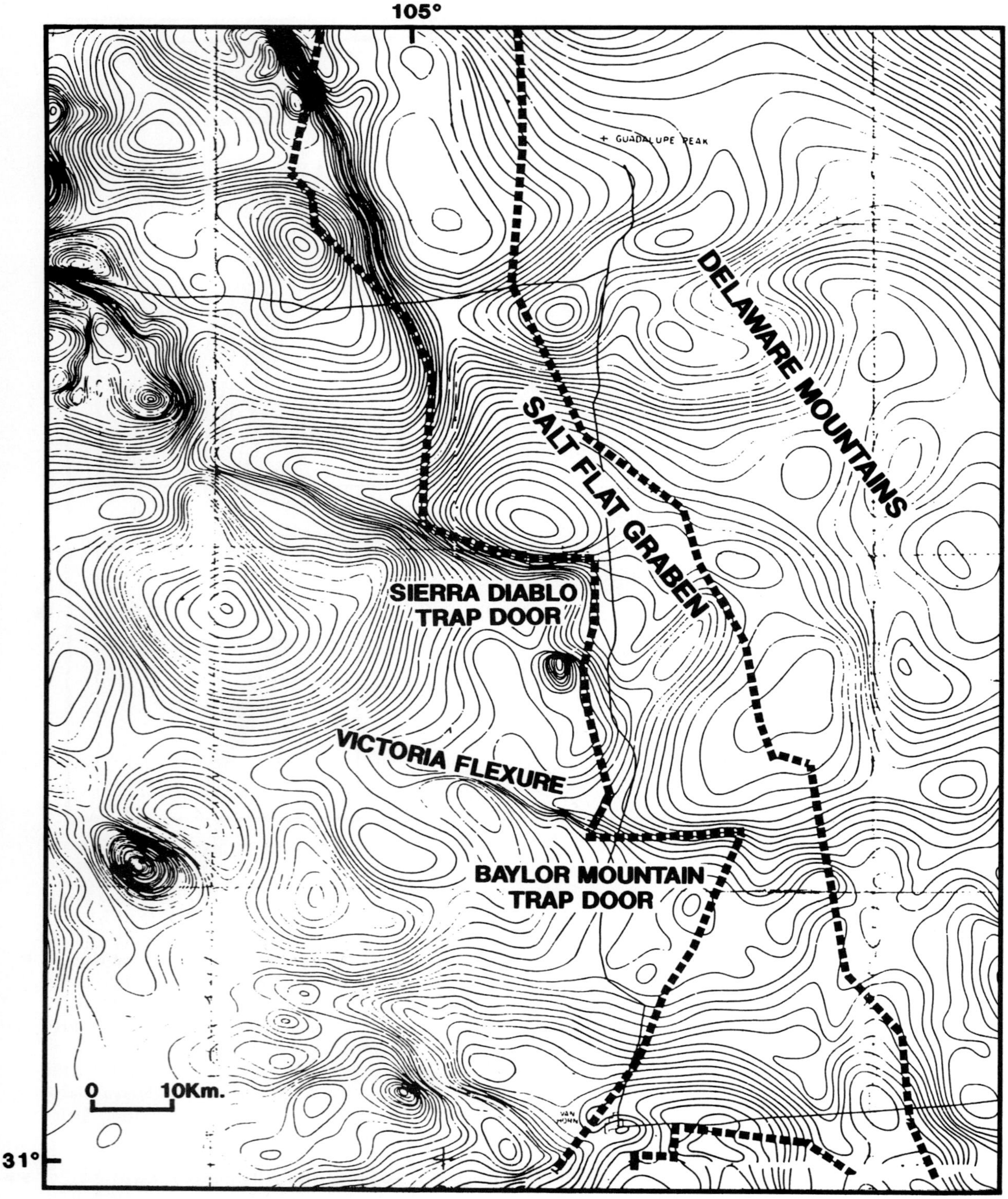
105°
+ GUADALUPE PEAK
DELAWARE MOUNTAINS
SALT FLAT GRABEN
SIERRA DIABLO
TRAP DOOR
VICTORIA FLEXURE
BAYLOR MOUNTAIN
TRAP DOOR
0
10Km.
31°

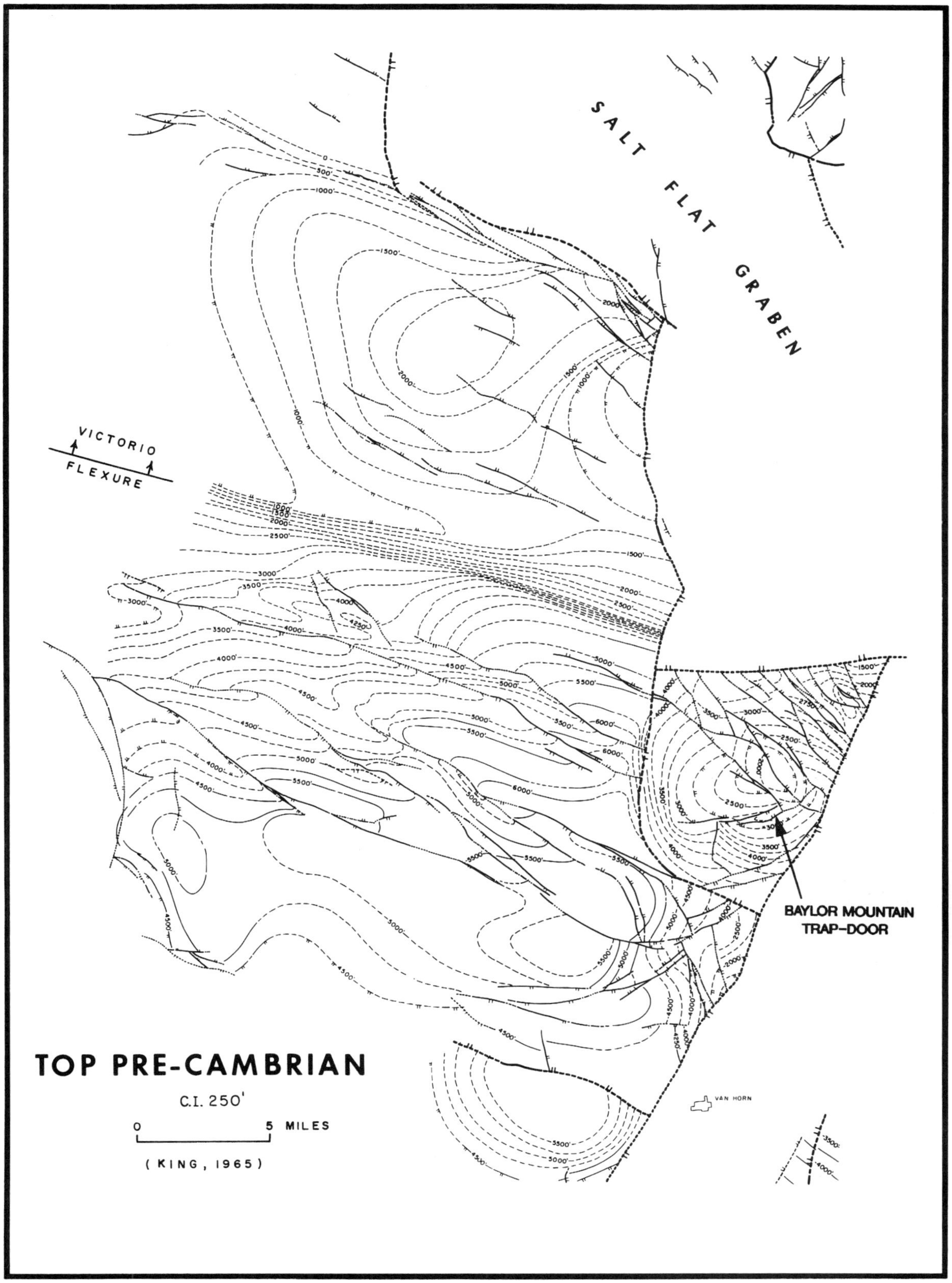
SALT FLAT GRABEN
VICTORIO FLEXURE
BAYLOR MOUNTAIN TRAP-DOOR
VAN HORN
TOP PRE-CAMBRIAN
C.I. 250'
0 5 MILES
(KING, 1965)

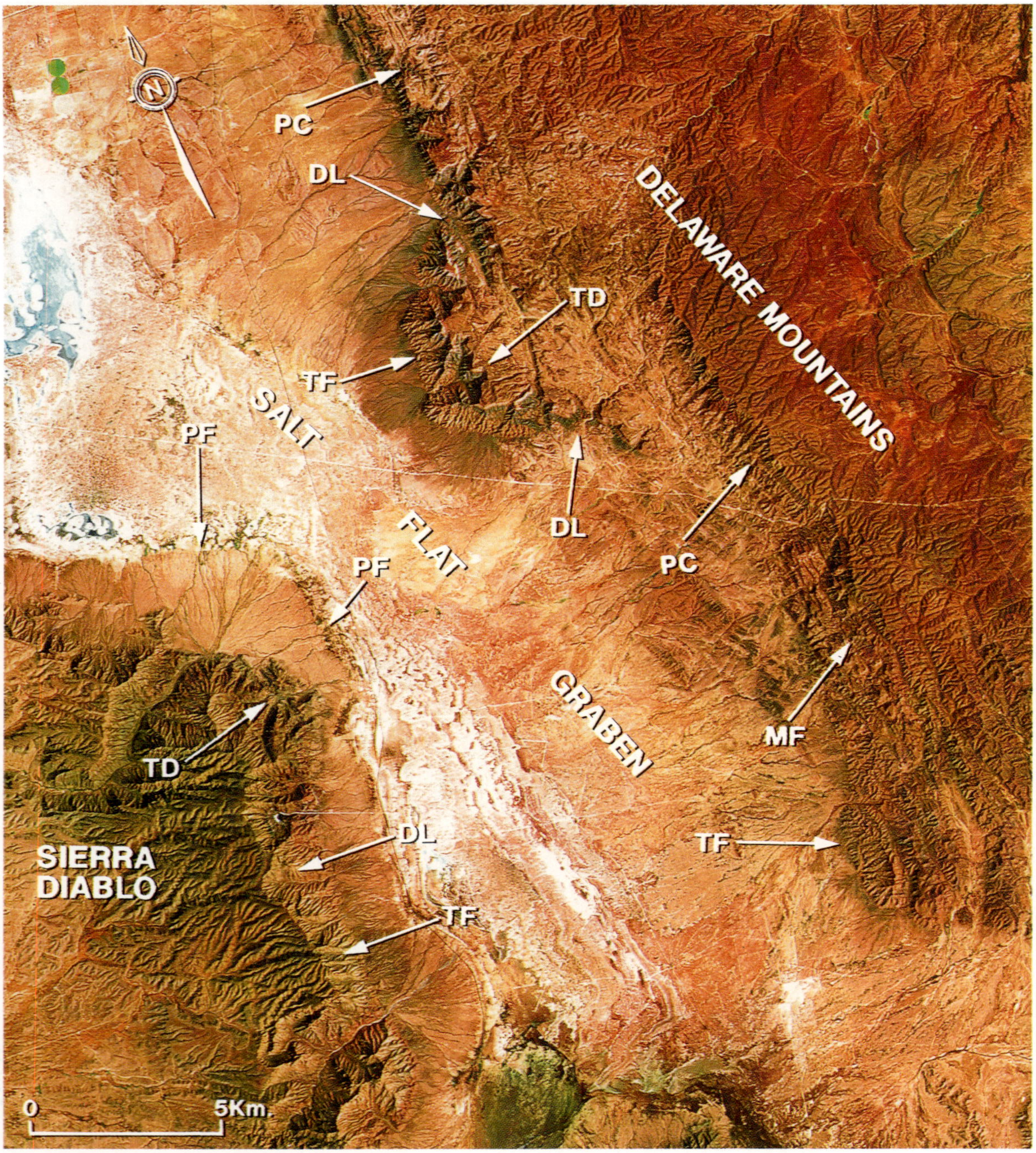

Fig. 9.5. TM imagery of the central part of the Salt Flat Graben Typical surface expressions of extensional fault blocks and related FLTs are shown. *DL* Dogleg feature; *TD* trapdoor; *MF* multidirectional positive FLTs; *PF* pseudo-FLTs; *TF* triangular facets; *PC* parallel composite FLT. Note the profound expression of the pseudo-FLTs which are formed at the eastern side of the graben by the alignment of the alluvial fans and related vegetative patterns. Locations of key figures are also indicated

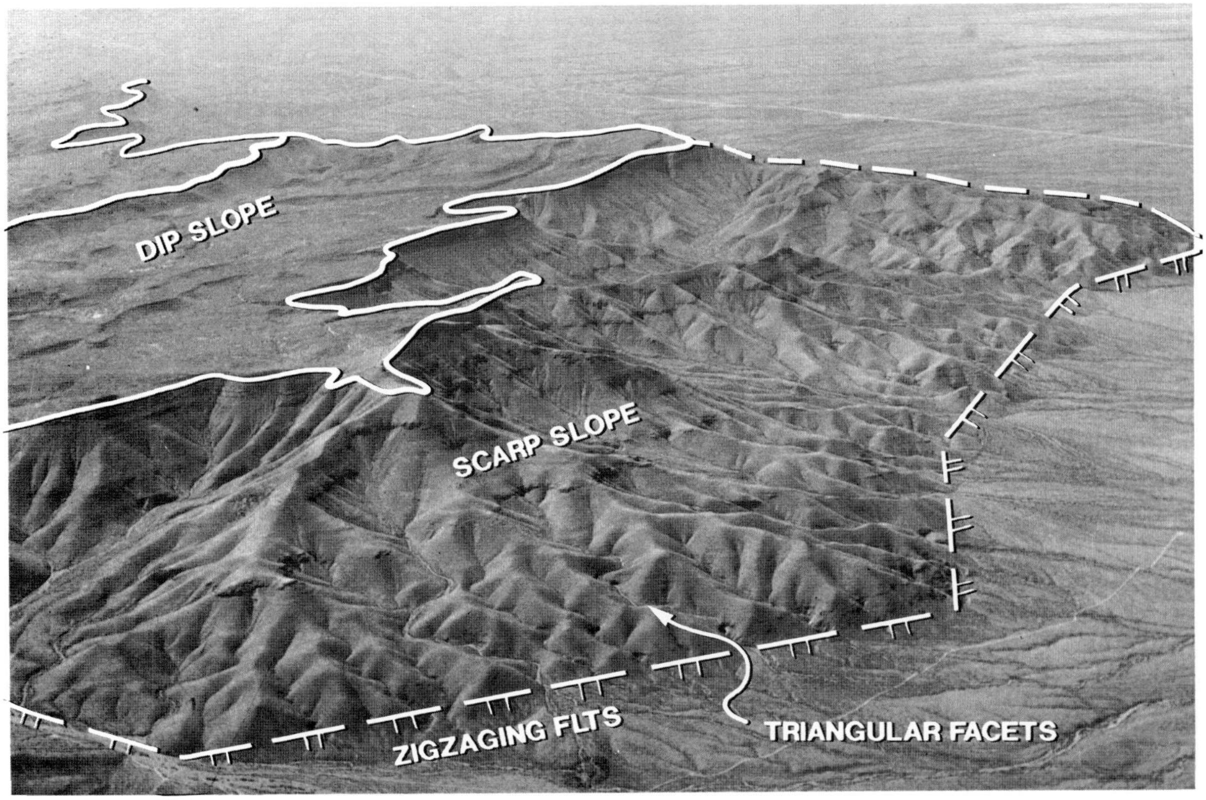

Fig. 9.6. Delaware Mountains trapdoor structure
Oblique aerial photograph of the Delaware Mountains shows the typical expression of a trapdoor structure. Note the development of small triangular facets along the FLTs which indicates a recent movement along these faults. The zigzagging nature of the trapdoor boundaries is also quite apparent. The photograph also demonstrates the profound differences between the surface expression of scarp slopes versus dip (isoclinal) slopes. (Photographed by T. P. Harding)

Fig. 9.7. Surface expression of multiple scarps
Photograph of complex, normal faults and associated flexures along the western margin of the Salt Flat Graben. On imagery data, these features manifest profound multiple FLTs. (Photographed by D. W. Phelps)

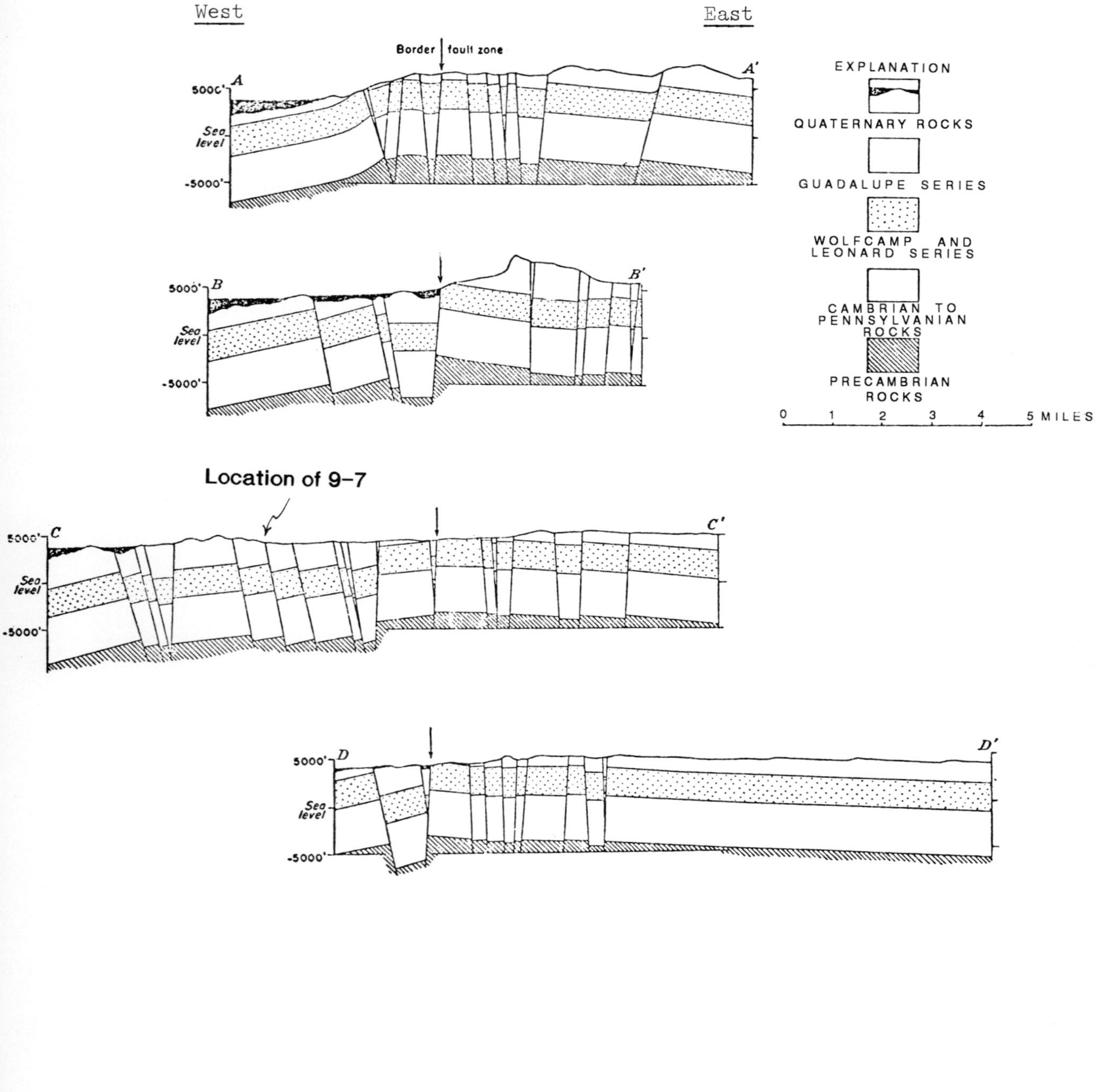

Fig. 9.8. Structural expression of graben edge
Four schematic cross sections representing the change in style from north to south along the northeastern border zone of the Salt Flat Graben. On imagery, such changes will be reflected by alternating positive and multiple FLTs. (King 1949; published with permission from USGS)

Chapter 10 Death Valley Region, Nevada and California

10.1 Background

Death Valley represents a complex tectonic region that developed at the intersection of three major deformational belts (Figs. 10.1, 10.2): (1) the Late Cenozoic Basin and Range province that is predominately extensional (Fig. 10.3); (2) a diffuse zone of Late Cenozoic, northwest-striking, right-slip wrench fault systems and west to southwest-striking left-slip faults that were active between the North American and the Pacific lithospheric plates (Fig. 10.4); and (3) a belt of Mesozoic folds and thrusts, localized near the hinge zone of the Late Proterozoic and Paleozoic miogeoclinal wedge (Stewart 1978; Fig. 10.5). Figure 10.1 illustrates the present structural setting of Death Valley which is characterized by the presence of both major wrench and extensional fault systems. The map also shows the location of three unique structural features known as turtlebacks which have been quite controversial in their origin (Wright et al. 1974).

10.2 Objectives

Structures in this region are well exposed and manifest clear expression on imagery data, aerial photography and outcrops. The area has also been well mapped in the field and studied by several investigators who provided detailed syntheses of the tectonic evolution of this area (e.g. Troxel 1974; Stewart 1983). The main aspects illustrated here are (1) the use of satellite imagery as an effective tool for the detection of structural style (see Fig. 10.6); (2) closer examination of the surface expression of different FLTs related to wrench faults, high-angle normal faults and thrust faults (Figs. 10.7–10.11); and (3) unique expression of turtleback structures and their possible structural origins (Figs. 10.12–10.14).

A significant portion of the examples shown here are taken from an EPR field guidebook by Christie-Blick et al. (1987).

10.3 Training Instructions

10.3.1 Interpretation Procedures

1. Place a clear overlay on top of the TM imagery of Fig. 10.6 and map all structural elements that can be recognized from these data.
2. Use the generalized structure map in Fig. 10.2 as well as the list of symbols used in the interpretation of satellite data which is shown in Appendix A to indicate the style of deformation of each structure.
3. Use the information on individual structures which is shown in Figs. 10.7–10.14 to further improve the interpreted overlay. At this stage it should include information related to different types of FLTs, axes of individual folds, and locations of recent geomorphic features which are offset by faulting, etc.
4. Use the tectonic map in Figs. 10.3–10.5 to recognize the style and timing of deformation of different fault systems. Mark on the overlay each of the fault systems in differently colored pencils.
5. Compare the results of your mapping with the structure map in Fig. 10.2.

10.3.2 Questions

Assuming that Death Valley was buried under sediments and was identified as a prospective target for exploration:

1. What type of structural and stratigraphic traps may be found in this area?
2. Rank these traps in terms of their potential size, frequency of occurrence and predictability.
3. Which side of the graben is likely to have more stratigraphic traps?
4. List several prolific hydrocarbon plays that exhibit, more or less, similar structural and stratigraphic settings.

References and Further Reading

Christie-Blick N, Phelps DW, Harding TP (1987) Superimposed and hybrid structural styles guidebook to the Death Valley Region. Exxon Production Research Company, Internal Rep

Fleck RJ (1970) Age and tectonic significance of volcanic rocks, Death Valley area, California. Geol Soc Am Bull 81: 2807–2816

Stewart JH (1978) Basin-range structure in western North America: a review. In: Smith RB, Eaton GP (eds) Cenozoic tectonics and regional geophysics of the western Cordillera. Geol Soc Am Mem 152, pp 1–31

Stewart JH (1983) Extensional tectonics in the Death Valley area, California: transport of the Panamint Range structural block 80 km north westward. Geology 11: 153–157

Troxel BW (1974) Geologic guide to the Death Valley region, California and Nevada. In: Troxel BW, Wright LA (co-leaders) Guidebook: Death Valley region, California and Nevada: Field trip 1, Geol Soc Am Cordilleran Sect Meet, pp 3–16

Troxel BW, Wright LA (eds) (1976) Geologic features, Death Valley, California. California Div Mines Geol Spec Rep 106

Wright LA, Otton JK, Troxel BW (1974) Turtleback surfaces of Death Valley viewed as phenomena of extensional tectonics. Geology 2: 53–54

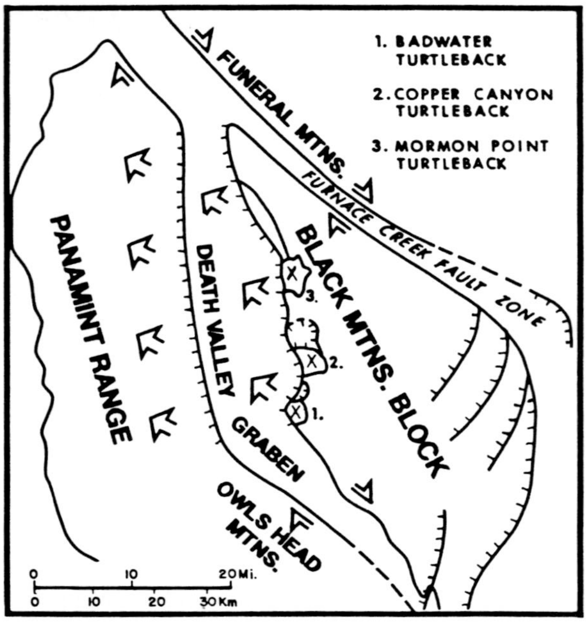

Fig. 10.1. Structure map of the Death Valley region
Showing the locations of major structural elements and turtlebacks. *Hachured lines* mark the positions of major normal faults. *Full arrows* show inferred direction of crustal extension. *Half-arrows* show relative displacement of strike-slip fault zones. Locations of figures and key features which are illustrated in this section also are indicated. (Wright et al. 1974; published with permission from Geology)

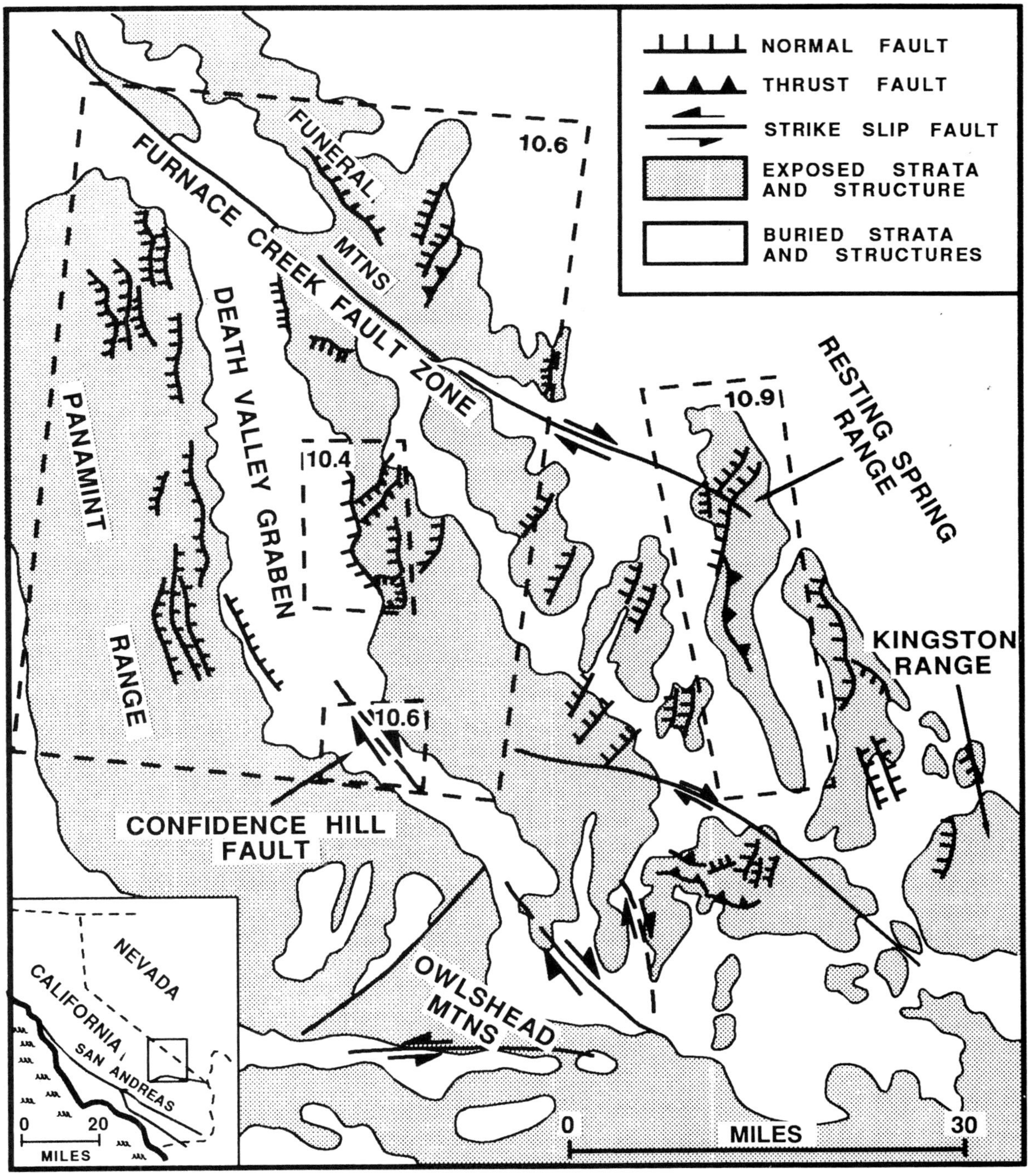

Fig. 10.2. Generalized structure map of Death Valley
Showing locations of major structural elements and surface features which are illustrated in this section. (Generalized from Troxel and Wright 1976; published with the permission of the authors)

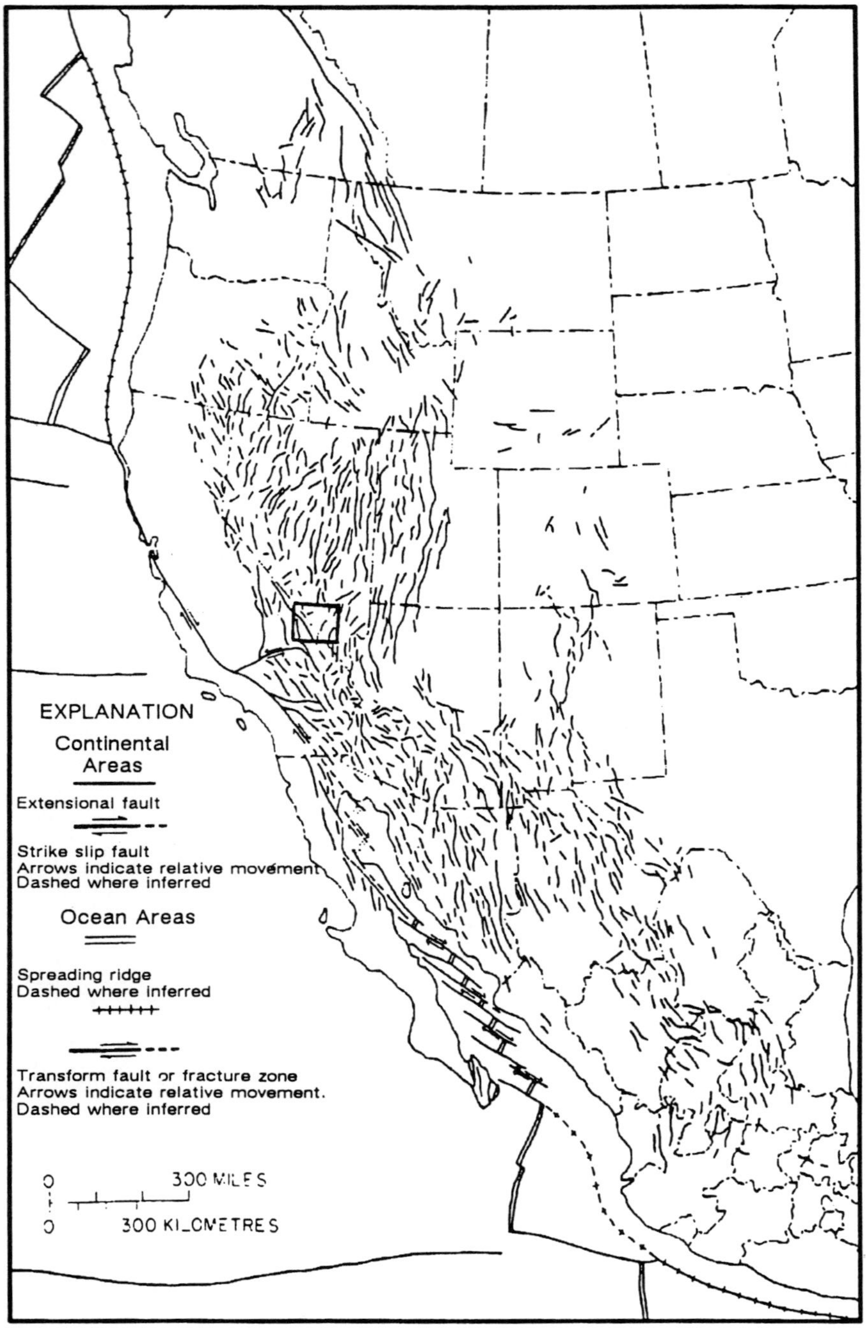

Fig. 10.3. Late Cenozoic tectonic elements in western North America

The map shows the location of Death Valley (in *box*) in relation to distribution of Late Cenozoic extensional faults and several regional wrench faults. (Stewart 1978; published with permission from the Geological Society of America)

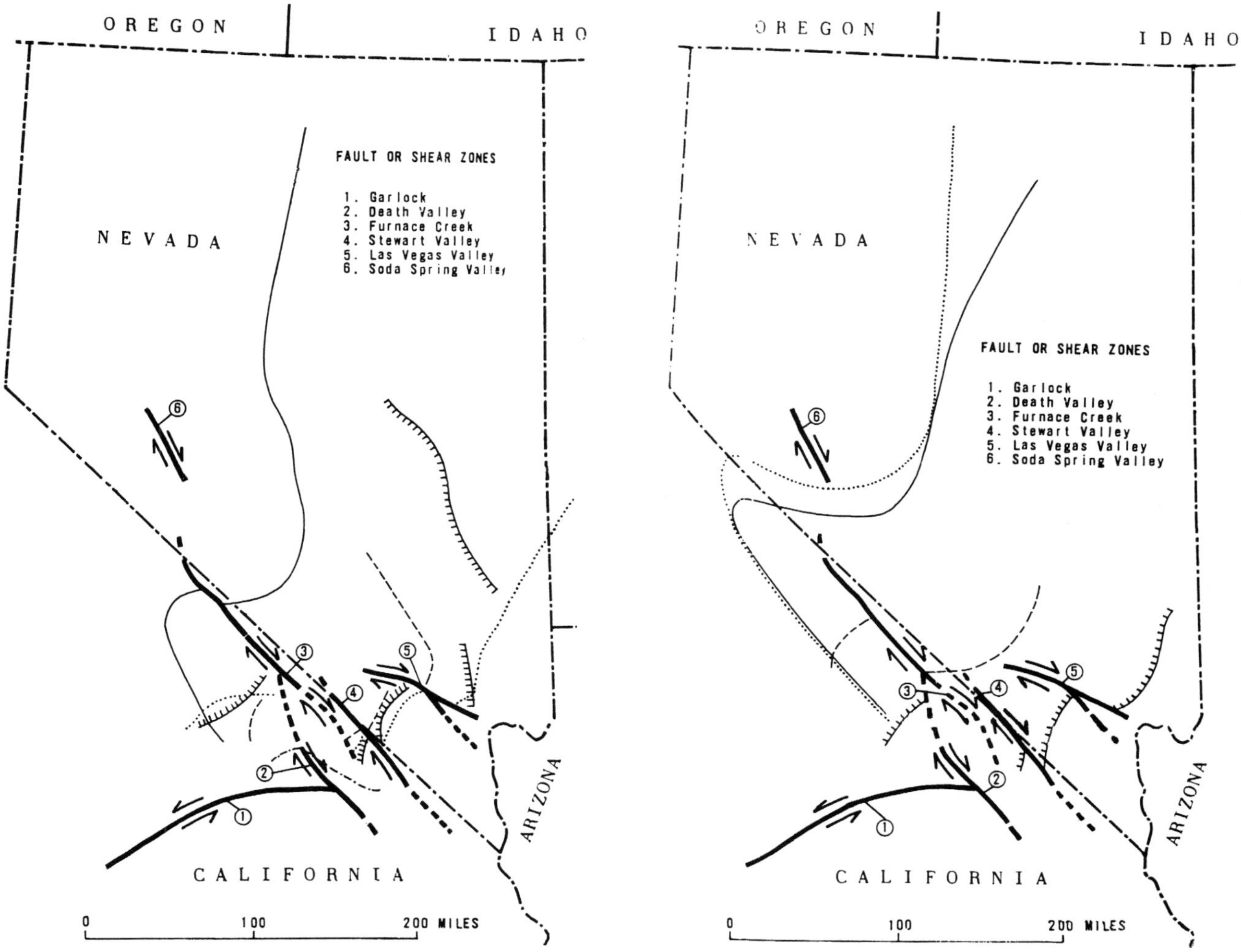

EXPLANATION (a)

.... 1500 - FOOT ISOPACH OF DEVONIAN SYSTEM (POOLS AND OTHERS, 1967, FIG. 6)

ııııııı EASTERN LIMIT OF LOWER DEVONIAN "CHERTY ARGILLACEOUS UNIT" OF OSMOND (1962) (POOLE AND OTHERS, 1967. FIG. 7)

—— WESTERN EDGE OF ORDOVICIAN CARBONATE FACIES (ALBERS. 1967. PL. 1

- - - - 3000 - FOOT ISOPACH OF PRECAMBRIAN STIRLING QUARTZITE (STEWART, 1967. FIG. 2)

-·-·- SOUTHWESTERN AND SOUTHERN LIMIT OF "ALGAL" DOLOMITE MEMBER OF PRECAMBRIAN NOONDAY DOLOMITE (WRIGHT AND TROXEL. 1967. FIG.7)

EXPLANATION (b)

.... EASTERN LIMIT OF UPPER MIDDLE AND UPPER TRIASSIC SEDIMENTARY ROCKS. (ALBERS, 1967. PL. 2)

—— EASTERN LIMIT OF PERMIAN, TRIASSIC AND JURASSIC VOLCANIC ROCKS (ALBERS. 1967. PL. 3)

ııııııı SOUTHEASTERN LIMIT OF UPPER MISSISSIPPIAN (LATE CHESTER) SHALE UNIT (POOLE, THIS REPORT)

- - - - WESTERN LIMIT OF LOWER MISSISSIPPIAN LIMESTONE UNIT (POOLE, THIS REPORT)

Fig. 10.4. Late Cenozoic wrench fault systems in the Death Valley region

The maps show the presence of five major wrench fault systems that dominated the structural deformation of Death Valley during Late Cenozoic time. **a** Precambrian to Devonian isopach and facies trends; **b** Mississippian to Mesozoic isopach and facies trends. (Stewart 1978; published with the permission of the Geologic Society of America)

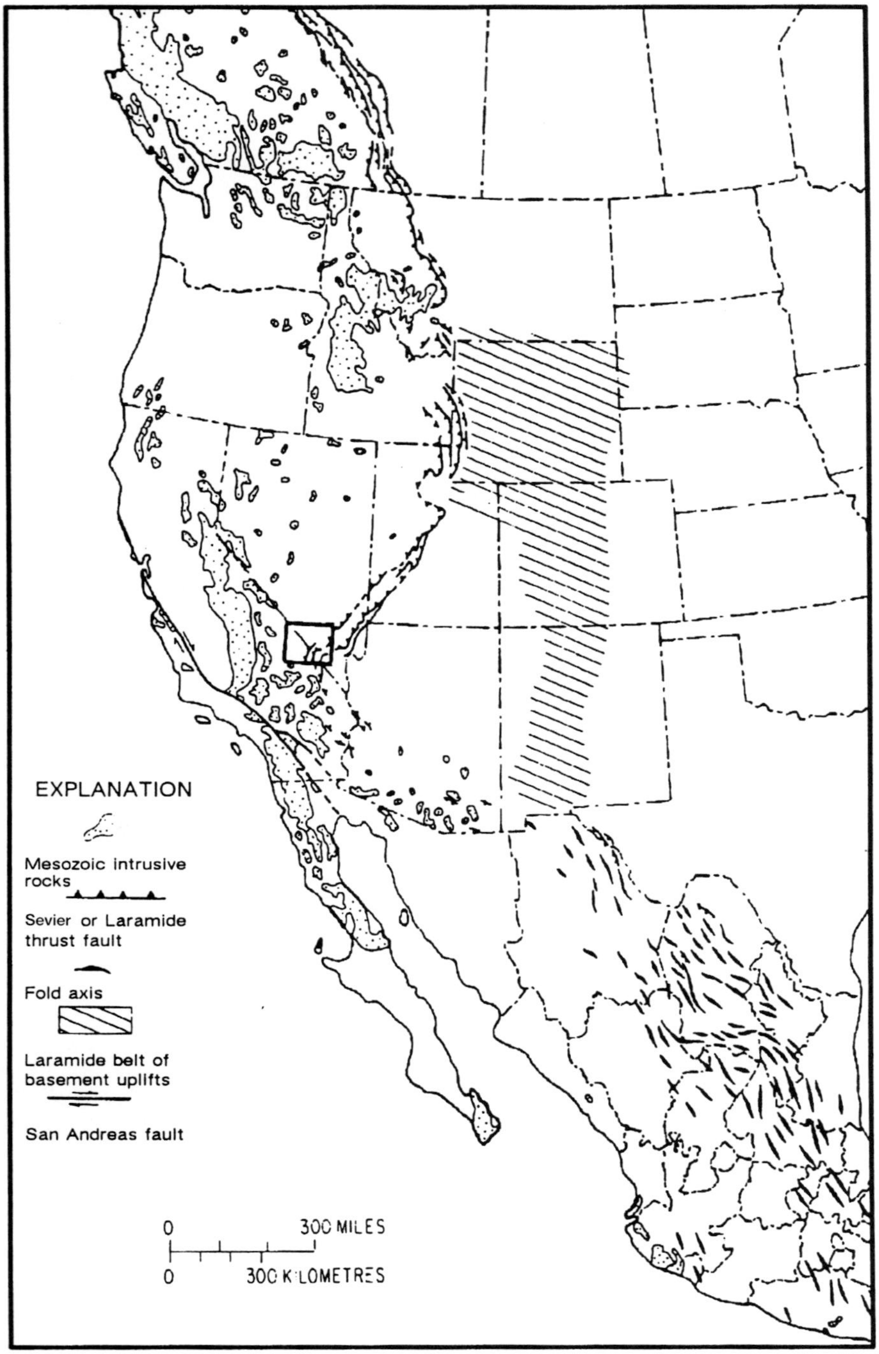

Fig. 10.5. Mesozoic and Early Cenozoic tectonic elements
The map shows the location of Death Valley (in *box*) in relation to the Late Mesozoic and Early Cenozoic deformation belts which are dominated by the presence of thrust fault systems. (Stewart 1978; published with the permission of the Geologic Society of America)

Fig. 10.6. Landsat TM imagery of the Death Valley region
Major faults and related folds are highlighted and interpreted in terms of unique expression of their FLTs. *TD* trapdoor; *DL* dogleg; *TF* triangular facets; *CSP* compressional splays with oblique anticlines; *LN* linear negative scarps; *PF* pseudo-FLTs; *SF* sinuous FLTs; *TB* turtleback structures

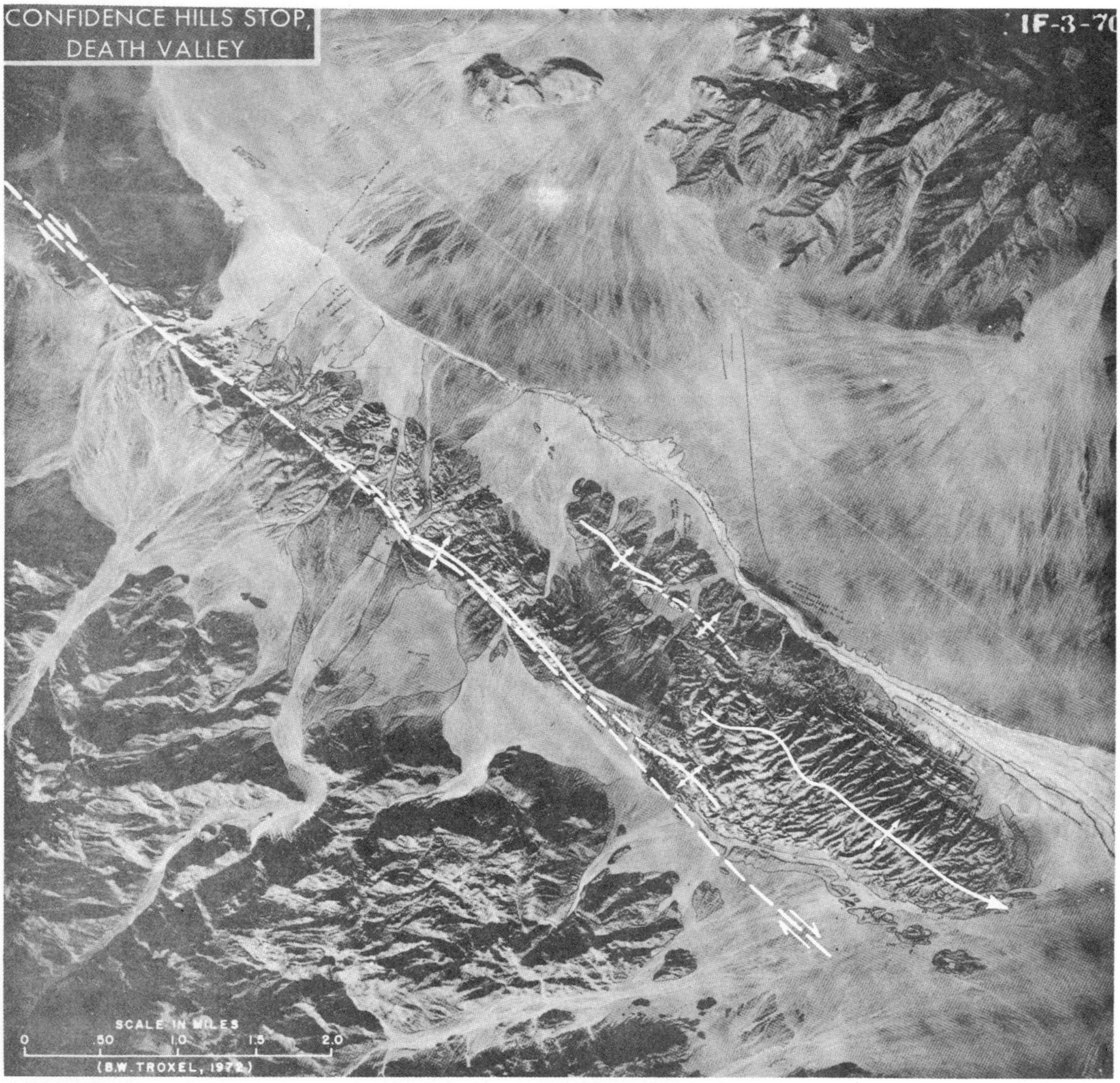

Fig. 10.7. The Confidence Hills wrench fault system
Low-altitude, aerial photography showing the negative FLT of the Confidence Hill wrench fault system and related compressional folds. Note that the anticlinal features which developed in recent unconsolidated sediments are severely eroded, exhibiting a "badland" type topography which may indicate the presence of piping and sapping processes. (Troxel 1974; published with the permission of the Geologic Society of America)

Fig. 10.8. Pliocene and Pleistocene folds ▶
An example of a young fold developed along active wrench fault systems in Death Valley. Note that the oblique fold is highly dissected but still manifests strong expression on the imagery. (Photographed by D. W. Phelps)

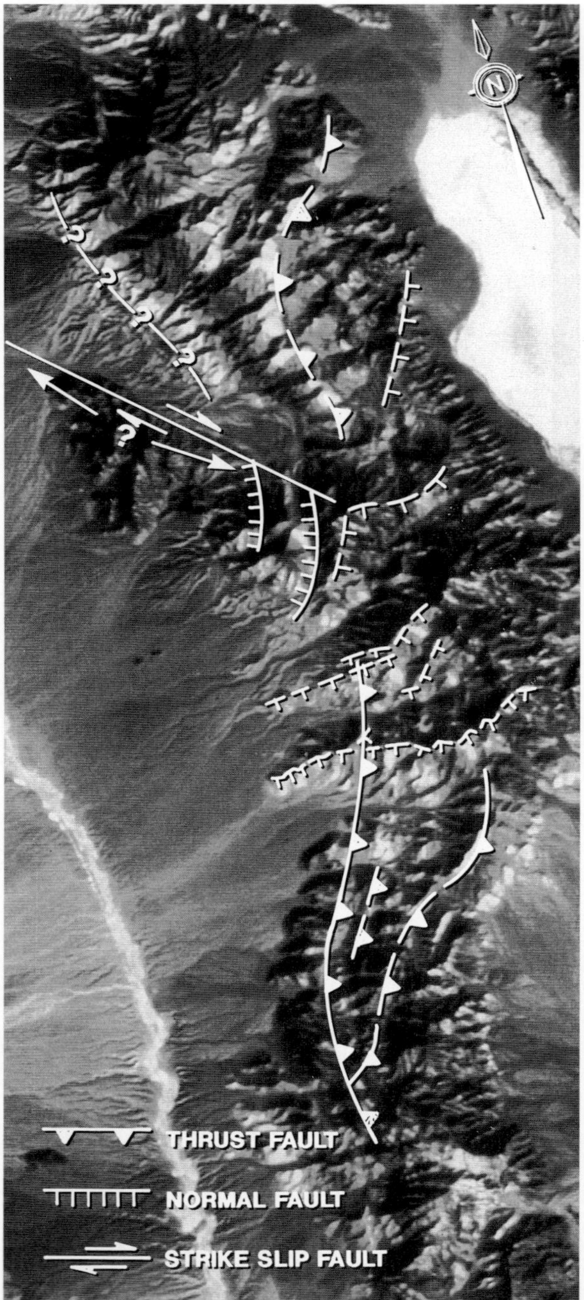
?
?
?
?
?
THRUST FAULT
NORMAL FAULT
STRIKE SLIP FAULT

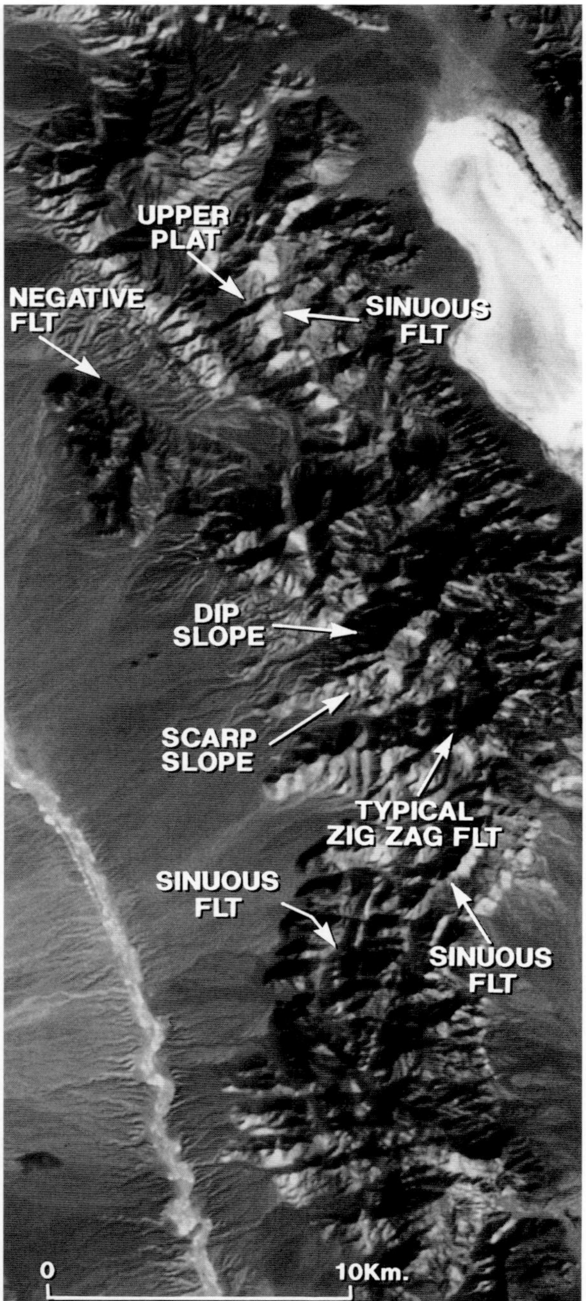
UPPER PLAT
NEGATIVE FLT
SINUOUS FLT
DIP SLOPE
SCARP SLOPE
TYPICAL ZIG ZAG FLT
SINUOUS FLT
SINUOUS FLT
0
10Km.

◀ **Fig. 10.9. Surface expression of different faults**
High-resolution TM imagery of the southeastern portion of the study area illustrating the surface characteristics of thrust, normal and strike-slip faults and related sinuous FLTs. Note the strong differences in the surface expressions of the upper and lower thrust plates as well as the presence of profound triangular facets along the normal faults. These surface characteristics are illustrated further in Figs. 10.10 and 10.11

Fig. 10.10. Surface expression of thrust faults.
Example of thrust-related contacts from the study area. The FLT is characterized by a sinuous appearance where upper and lower plates manifest significant contrast in topographic dissection which are manifest on imagery as a significant contrast in texture and color. (Photographed by D. W. Phelps)

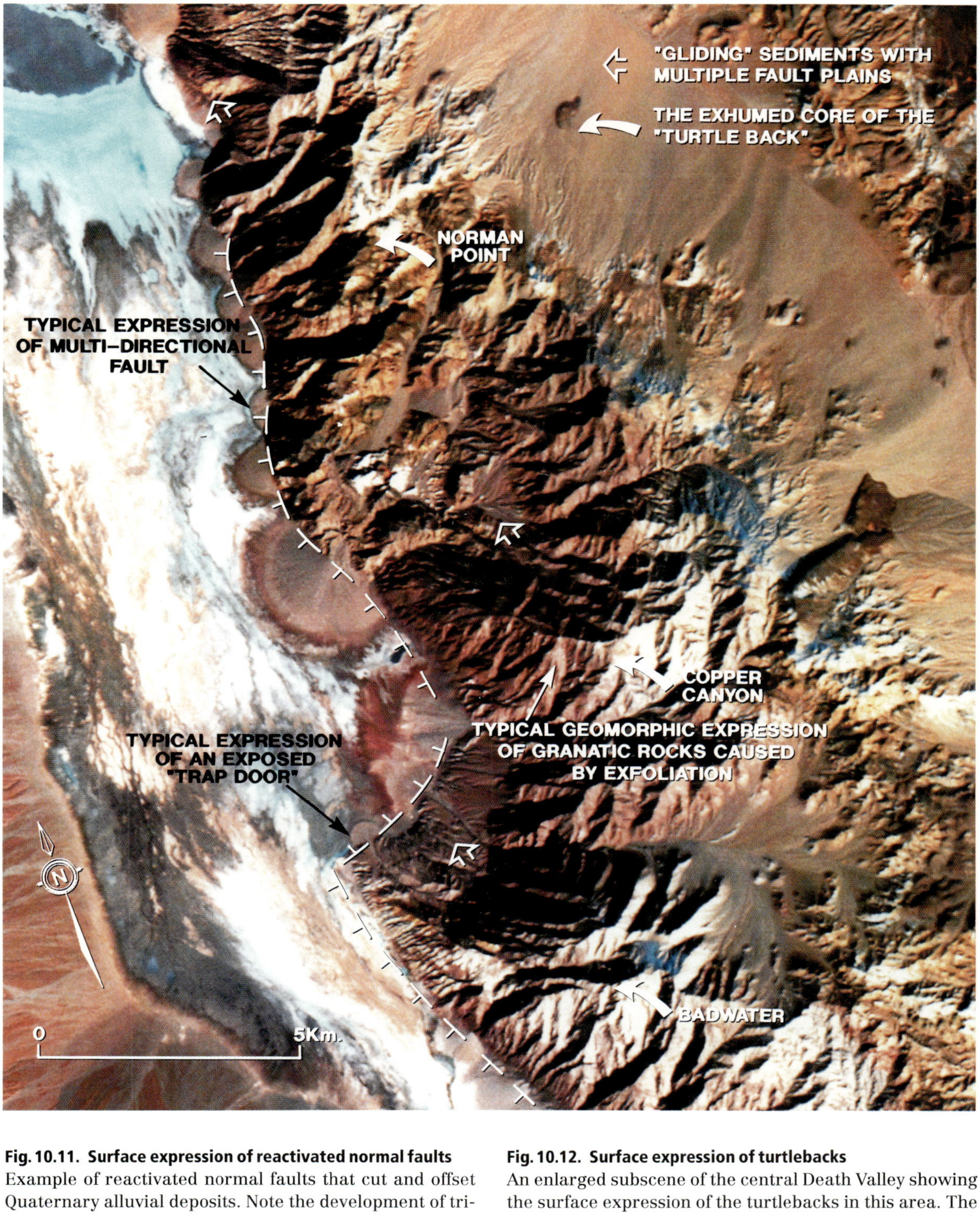

◀ **Fig. 10.11. Surface expression of reactivated normal faults**
Example of reactivated normal faults that cut and offset Quaternary alluvial deposits. Note the development of triangular facets along the fault-line trace which enhances their expressions on imagery. (Photographed by D.W.-Phelps)

Fig. 10.12. Surface expression of turtlebacks
An enlarged subscene of the central Death Valley showing the surface expression of the turtlebacks in this area. The imagery suggest that the turtleback structures represent a Precambrian surface which was exhumed by a series of normal fault systems which are manifested as multiple scarps. This interpretation supports the geological model proposed by Wright et al. (1974) and is shown in Fig. 10.13

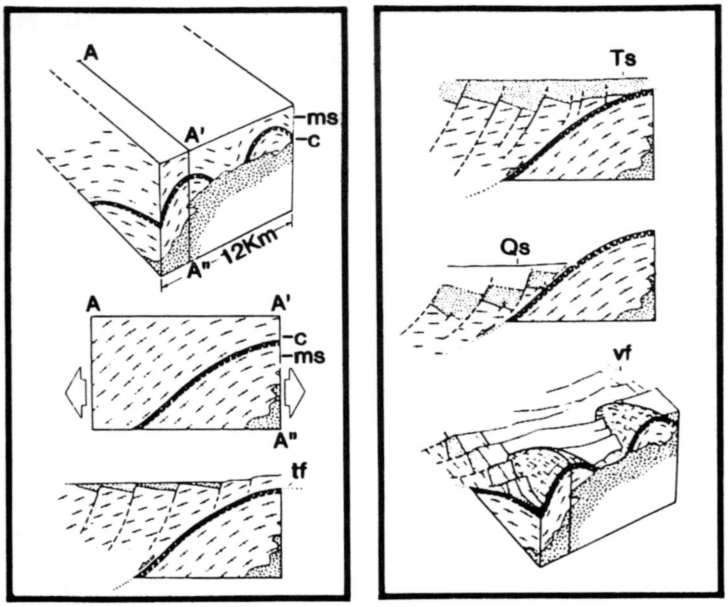

Fig. 10.13. Structural evolution of turtlebacks
An idealized block diagram and cross section illustrating the development of turtlebacks. A series of low-angle normal faults exhumes an old Precambrian surface during a pull-apart process. This mechanism, which was proposed by Wright et al. (1974), appears to be supported by imagery interpretation. Several other explanations to this structure have been proposed including compressional folding, differential erosion and up-arching relating to the intrusion of shallow plutons

Fig. 10.14. Copper Canyon turtleback
An arcuated erosional surface of Precambrian rocks produce a unique feature that looks like a giant turtleback. The *three arrows* show the location of FLTs that reflect the surface expression of low-angle listric normal faults which have exhumed this structure and are shown in Fig. 10.12 as multiple FLTs. (Photographed by D. W. Phelps)

Chapter 11 The Fort St. John Graben, Western Canada

11.1 Background

The Fort St. John Graben (FSJG) is a regional-scale, west-trending, fault system located at the center of the Peace River Embayment of the Western Canada Sedimentary Basin (Fig. 11.1). The FSJG forms the central arm of a larger fault system known as the Dawson Creek Graben Complex which was filled with sediments in several stages during Carboniferous to Permian times (Figs. 11.2–11.4), Barclay et al. 1990). Although the Dawson Creek Graben Complex was formed during a long period of extension, its bounding fault systems often display alternate zones of both compressional and extensional structural styles. The unique structural characteristics of the graben complex have been attributed to Carboniferous to Permian strike-slip faulting as well as to late reactivation and inversion processes associated with the development of the nearby Rocky Mountain Fold and Thrust Belt (R. D. Oggy, pers. comm.).

11.2 Objectives

The FSJG is completely buried under Cretaceous rock units and its bounding fault systems have not been recognized by surface mapping. Furthermore, the Cretaceous outcropping units are severly obscured by thick vegetative cover as well as glacial and fluvioglacial sediments. In spite of these impediments, the main structural features of the graben manifest detectable expressions on remote sensing data (Figs. 11.5–11.13). The main aspects to observe here are (1) detection of buried and obscured structures on imagery; (2) effective use of stereo data and FSS mapping in areas of poor seismic data; and (3) integration analysis of surface and subsurface data for the recognition of different structural styles.

11.3 Training Instructions

11.3.1 Interpretation Procedures

1. Place a clear overlay on the TM imagery of Fig. 11.8 and trace geomorphic features that appear to be related to the presence of buried and obscured structures. These features should include structurally controlled stream valleys, linear topographic scarps, circular drainage patterns, and topographic features.
2. Use the structure contour map of Fig. 11.6, the generalized fault map of Fig. 11.7 and the list of symbols for buried and obscured structures in Appendix A to improve your structural/geomorphic interpretations.
3. Make a distinction between surface features that can be constrained with surface and subsurface data from those which can only be recognized on the imagery.
4. Perform the same procedures to interpret the TM imagery of Fig. 11.11.

11.3.2 Questions

Assuming that an exploration program was initiated in the graben area and that the only available data were the satellite images of Figs. 11.8 and 11.11 and the two corresponding cross sections of Fig. 11.7 and 11.12:

1. Which of the known pools could have been detected with this data set?
2. What portion or segments of the graben can be mapped with this data set?
3. Is it possible to determine the structural style of the faulted margins of the graben?
4. Based on this information alone, how would you plan your initial seismic program?
5. Where would you use stereo data to improve the surface mapping?

References and Further Reading

Barclay JE, Krause FF, Campbell RL, Utting J (1990) Dynamic casting and growth faulting: Dawson Creek Graben Complex, Carboniferous-Permian Peace River Embayment, Western Canada. Bull Can Pet Geol 38 A: 115–145

Harding TP (1984) Graben hydrocarbon occurrences and structural style. AAPG Bull 68: 333–362

Henderson CM (1989) Absaroka Sequence – the lower Absaroka Sequence: Upper Carboniferous and Permian. In: Ricketts BD (ed) Western Canada sedimentary basin – a case history. Can Soc Pet Geol, Spec Publ 30, Calgary, Alberta, pp 203–217

O'Connell SC, Dix GR, Barclay JE (1990) The origin, history and regional structural development of the Peace River Arch, Western Canada. Bull Can Pet Geol 38 A: 4–24

Podruski JA, Barclay JE, Hamblin AP, Lee PJ, Osadetz KG, Procter RM, Taylor GC (1988) Conventional oil resources of Western Canada (light and medium), part 1, resource endowment. Geol Surv Can Pap 87–26, pp 1–125

Richards BC (1989) Upper Kaskaskia Sequence: Uppermost Devonian and Lower Carboniferous. In: Ricketts BD (ed) Western Canada sedimentary basin – a case history. Can Soc Pet Geol, Spec Publ 30, Calgary, Alberta, pp 165–20

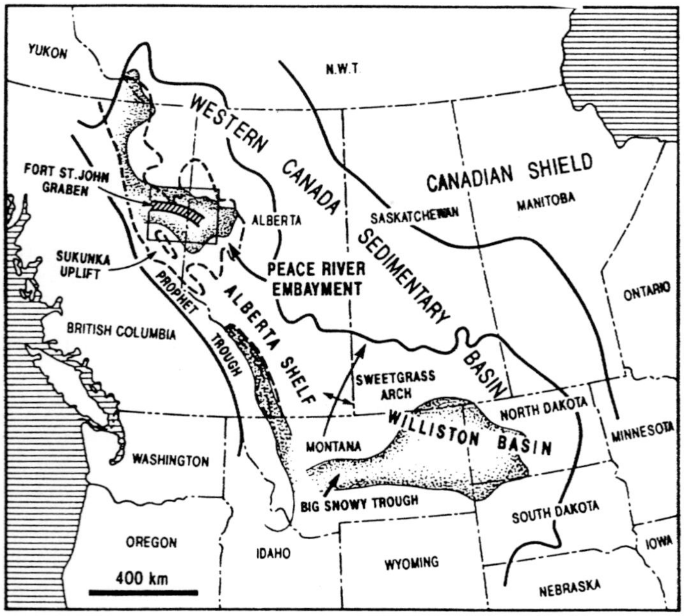

Fig. 11.1. Carboniferous-Permian geological and tectonic elements in the Western Canada Basin
Showing the location of the Fort St. John Graben and the Peace River Embayment. (After Barclay et al. 1990; published with the permission of the Canadian Society of Petroleum Geologists)

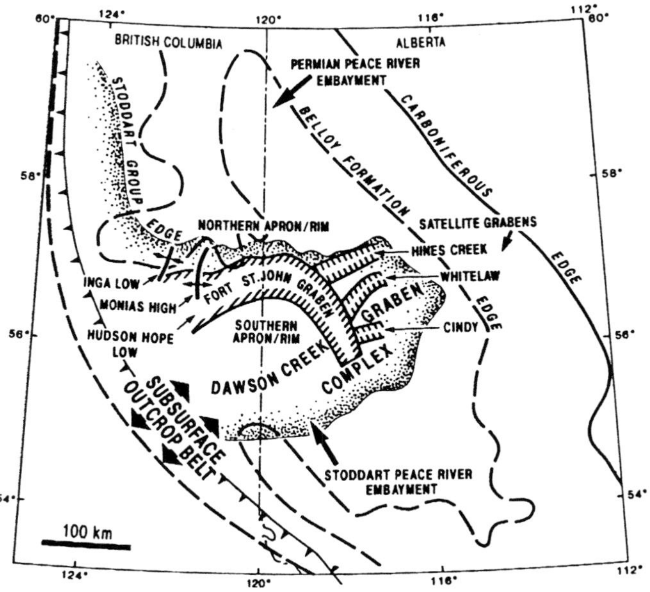

Fig. 11.2. Tectonic map of the Peace River embayment
Showing the location of the Fort St. John Graben and other related satellite graben features that form in the Dawson Creek Graben Complex. (After Barclay et al. 1990; published with the permission of the Canadian Society of Petroleum Geologists)

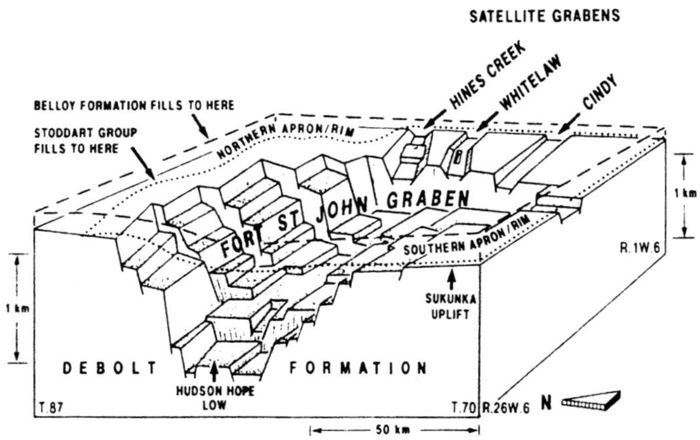

Fig. 11.3. Sketch of Carboniferous-Permian Dawson Creek Graben Complex
The sketch is based on cross sections and on isopach and structure maps. (After Barclay et al. 1990; published with the permission of the Canadian Society of Petroleum Geologists)

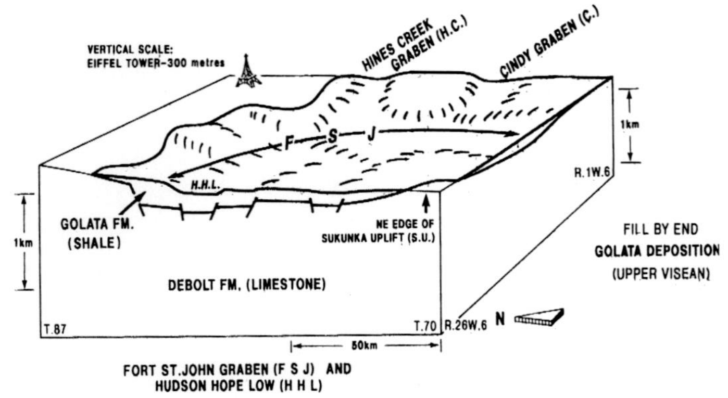

A Top Golata Formation time.

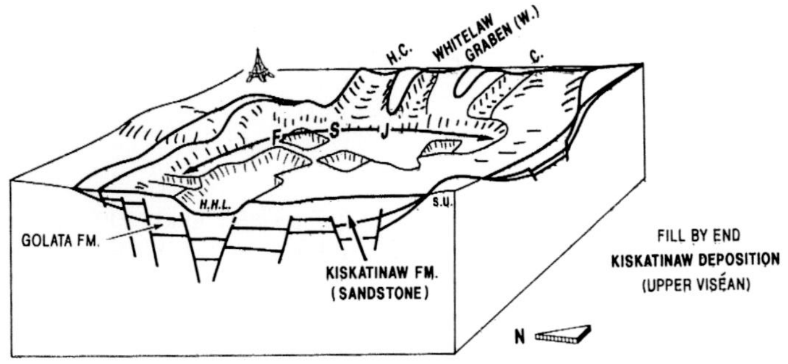

B Top Kiskatinaw Formation time.

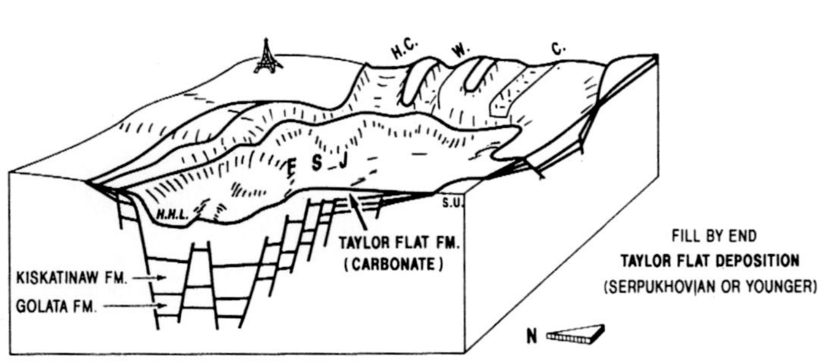

C Top Taylor Flat Formation time.

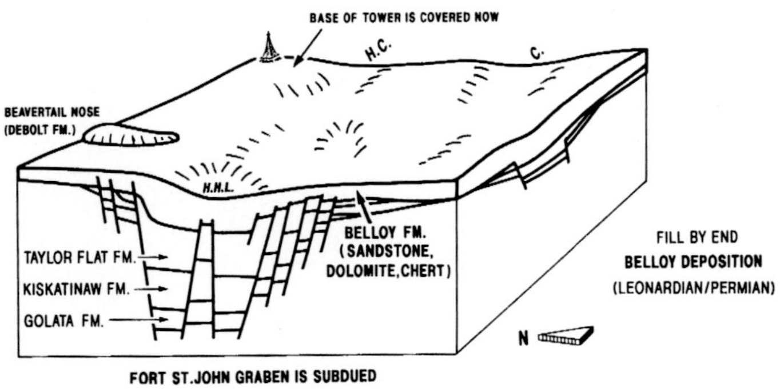

Fig. 11.4. Evolution of the Dawson Creek Graben Complex
Shown are several stages in the development of the graben complex during Carboniferous to Permian time. Eiffel Tower shown for comparative purposes; valid for vertical dimension only. (After Barclay et al. 1990; published with the permission of the Canadian Society of Petroleum Geologists)

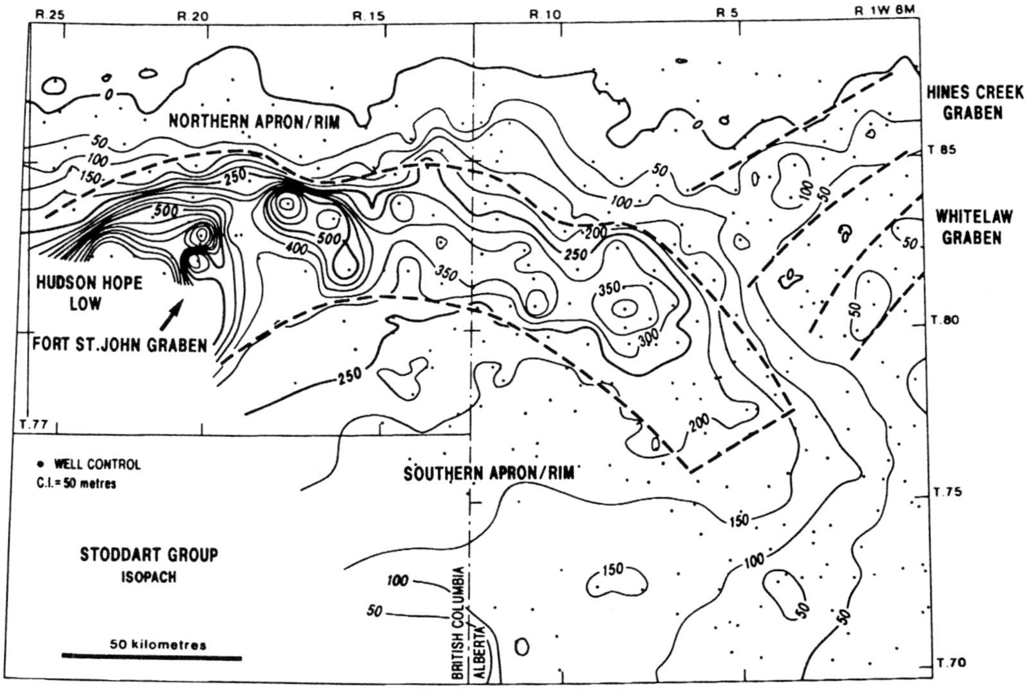

Fig. 11.5. Stoddart group isopach map
Showing the main Carboniferous sediments that filled the Ft. St. John Graben and the location of the major bounding faults. The area noted as the Hudson Hope Low which constitutes the less-explored portion of the graben will be further analyzed in subsequent figures. (After Barclay et al. 1990; published with the permission of the Canadian Society of Petroleum Geologists)

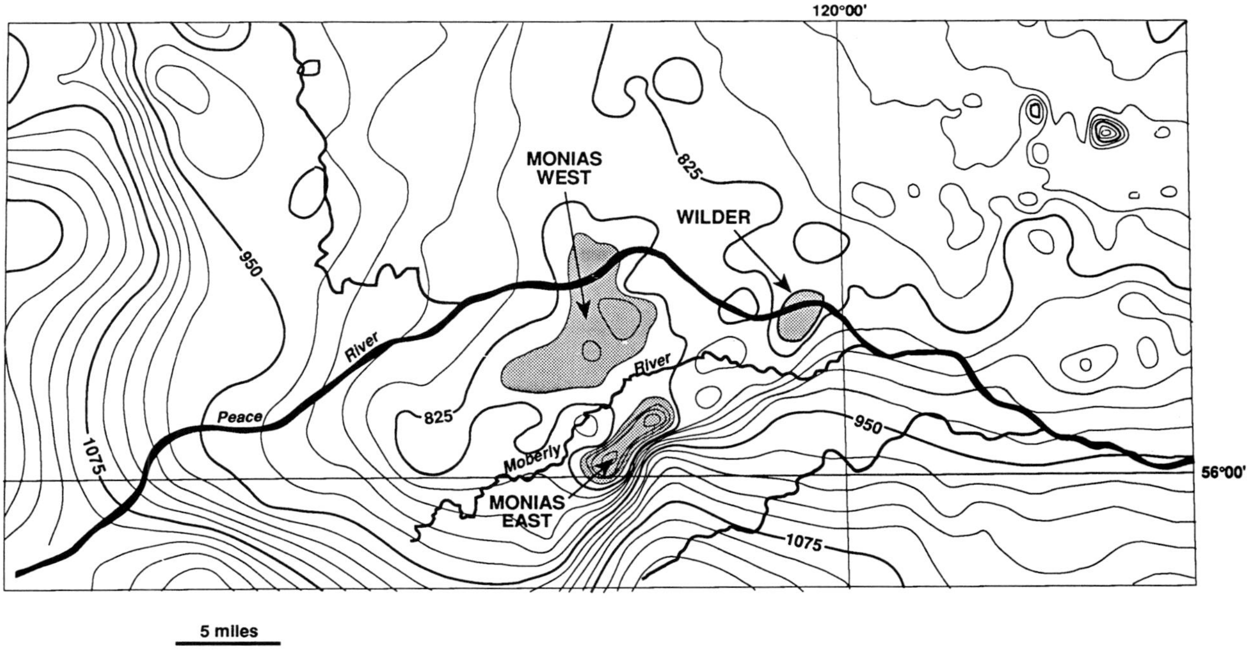

Fig. 11.6. Triassic structure contour map
Showing main structural features of the Hudson Hope Low area. Main producing structural fields include the Monias East, Monias West and Wilder. These fields produce from Triassic reservoir sediments trapped in various types of graben-related folds which are formed over deeper faults

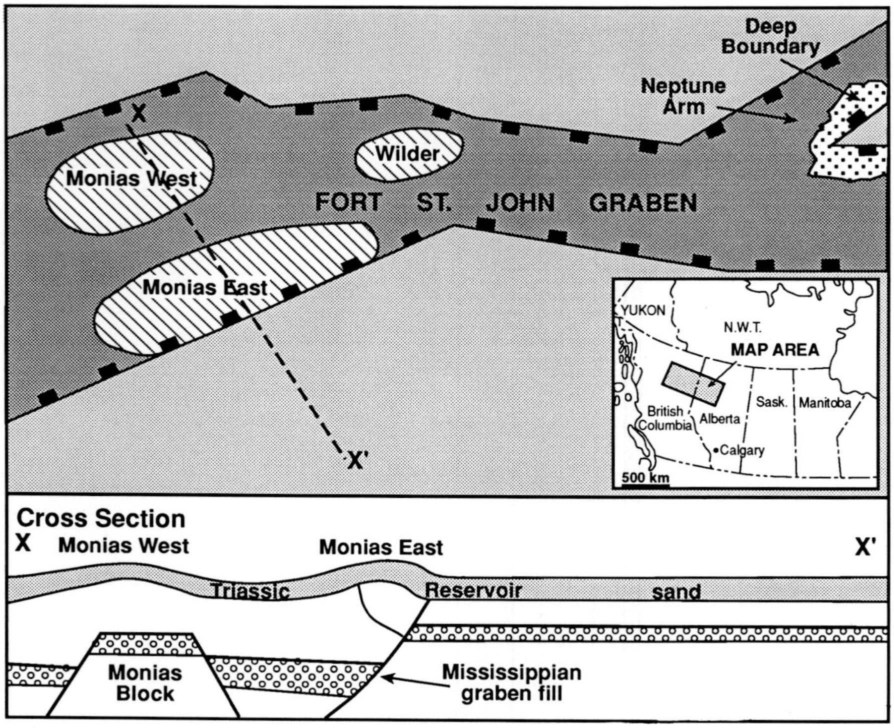

Fig. 11.7. Fault map of the Fort St. John Graben
The sketch shows the location of major bounding fault systems of the graben in British Columbia as interpreted through integrated analysis of satellite imagery and subsurface data. The cross section illustrates the structural style of fields that produce from purely structural traps. Note the presence of the Deep Boundary Field which is a stratigraphic trap formed along the faulted margin of the Neptune Arm

Fig. 11.8. Landsat image of the Hudson Hope Low area ▶
Shown is the southwest-trending arm of the graben. Note that all three of the producing fields mentioned above are highlighted by structurally controlled stream valleys. Furthermore, the northern margin of the graben is manifested by a series of topographic scarps that reflect the fault-line traces of the deeper faults

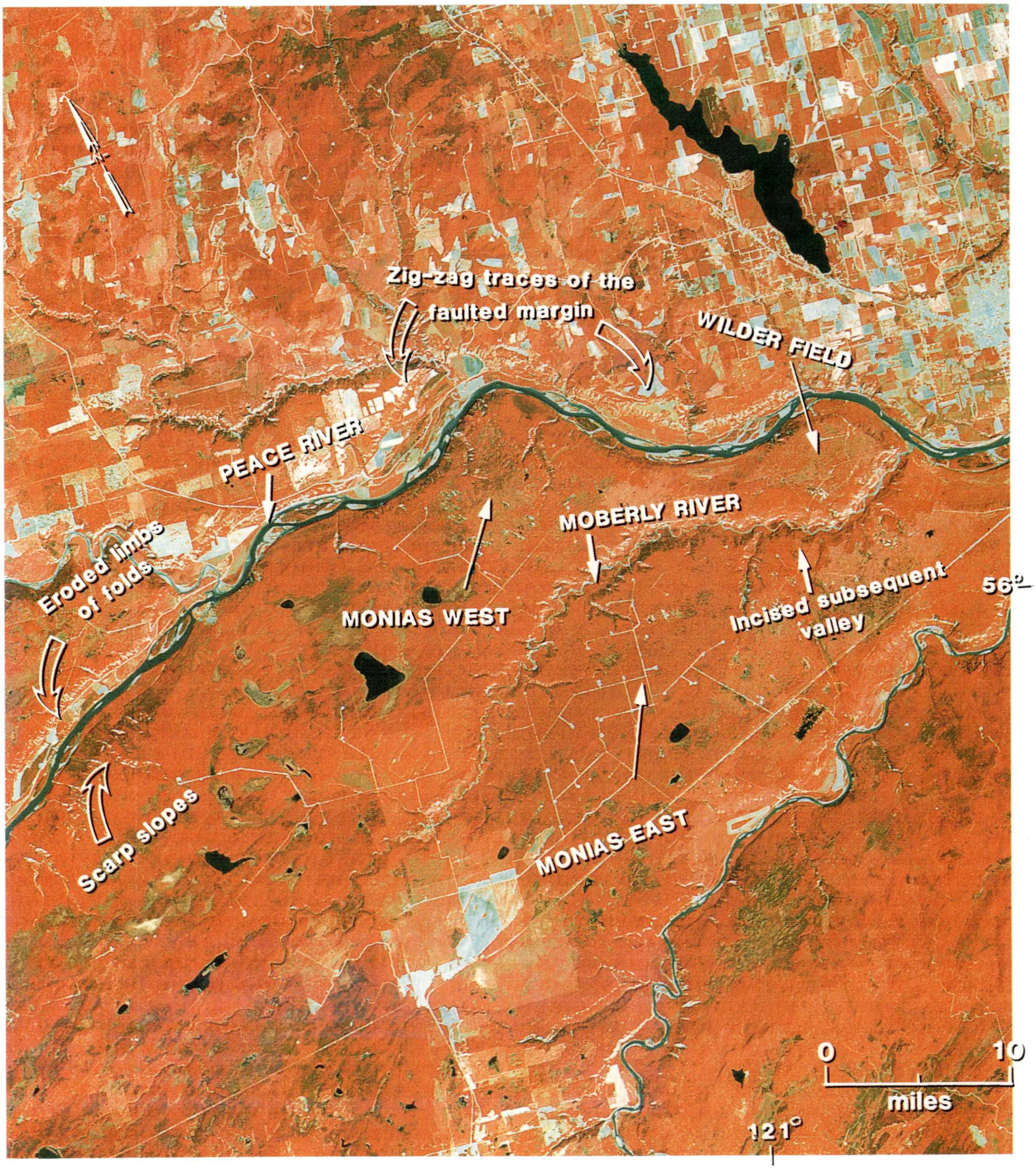
Zig-zag traces of the faulted margin
WILDER FIELD
PEACE RIVER
MOBERLY RIVER
Eroded limbs of folds
MONIAS WEST
Incised subsequent valley
56°
Scarp slopes
MONIAS EAST
0
10
miles
121°

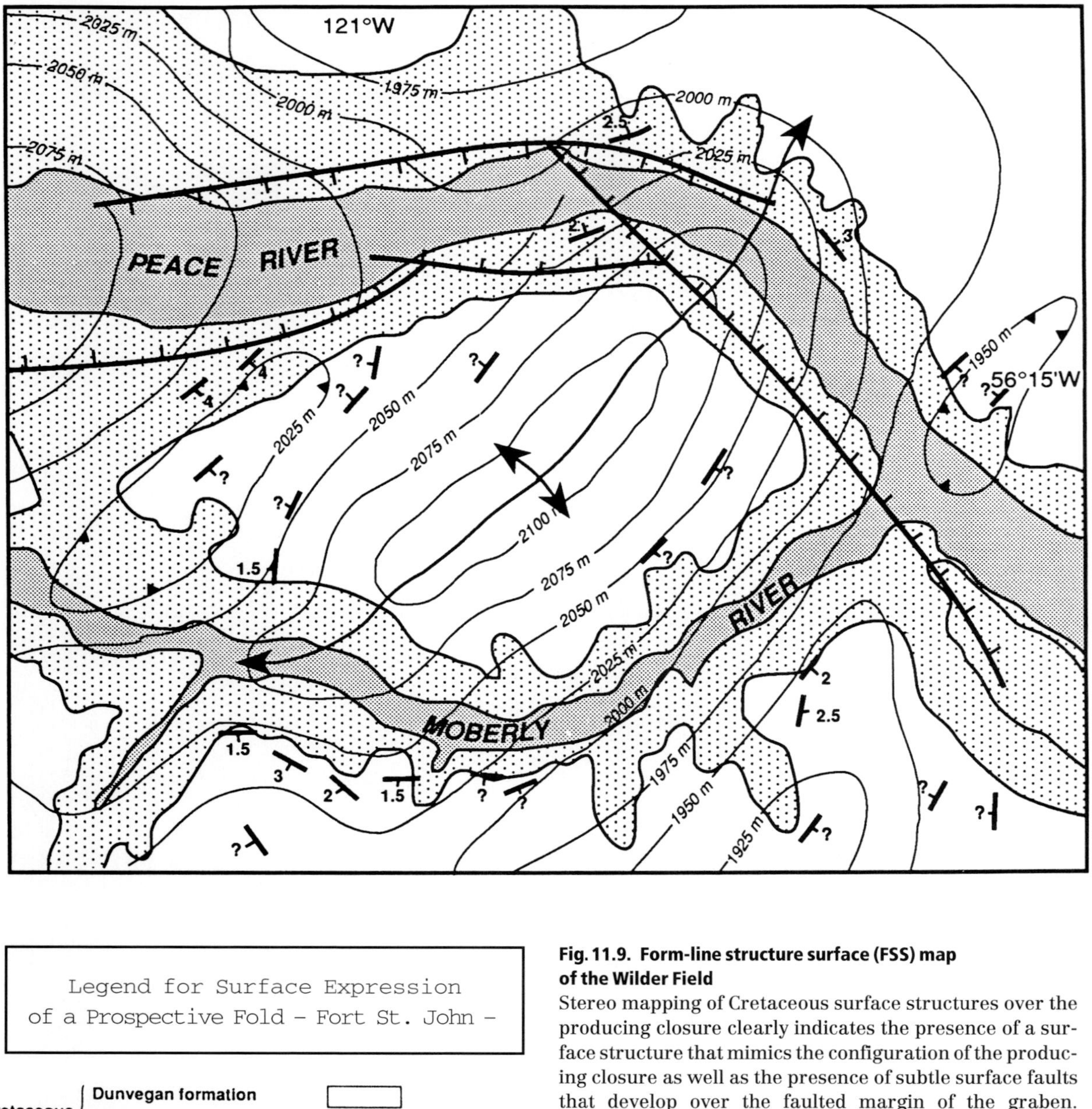

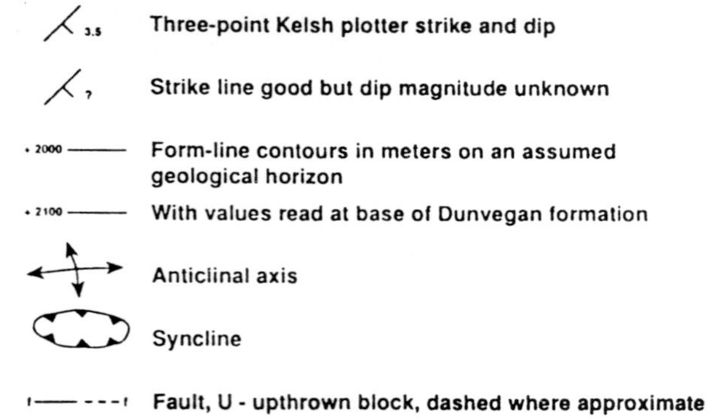

Fig. 11.9. Form-line structure surface (FSS) map of the Wilder Field
Stereo mapping of Cretaceous surface structures over the producing closure clearly indicates the presence of a surface structure that mimics the configuration of the producing closure as well as the presence of subtle surface faults that develop over the faulted margin of the graben. (Mapped by D. Anderson)

◀ **Fig. 11.10. Form-line structure surface map of the Wilder Field**
Shown is the legend for Fig. 11.11 as well as other form-line structure surface maps used in this study

Fig. 11.11. TM imagery of the Neptune Arm ▶
The extent of the graben and the Neptune Arm is clearly manifested on imagery by the cultivated areas indicating the presence of rich soil which fills the topographic depression of the graben. Bounding fault systems are manifested by structurally controlled stream valleys as well as tonal anomalies. *DD* Drainage divide; *DV* deflected valley. The correspondence between surface features and the graben is further demonstrated in Fig. 11.12

Trends of buried faults that form the graben's margin
NORTHERN RIM
DD
NEPTUNE ARM
DV
SS
FLORAL HIGH
Peace River
0
10
miles

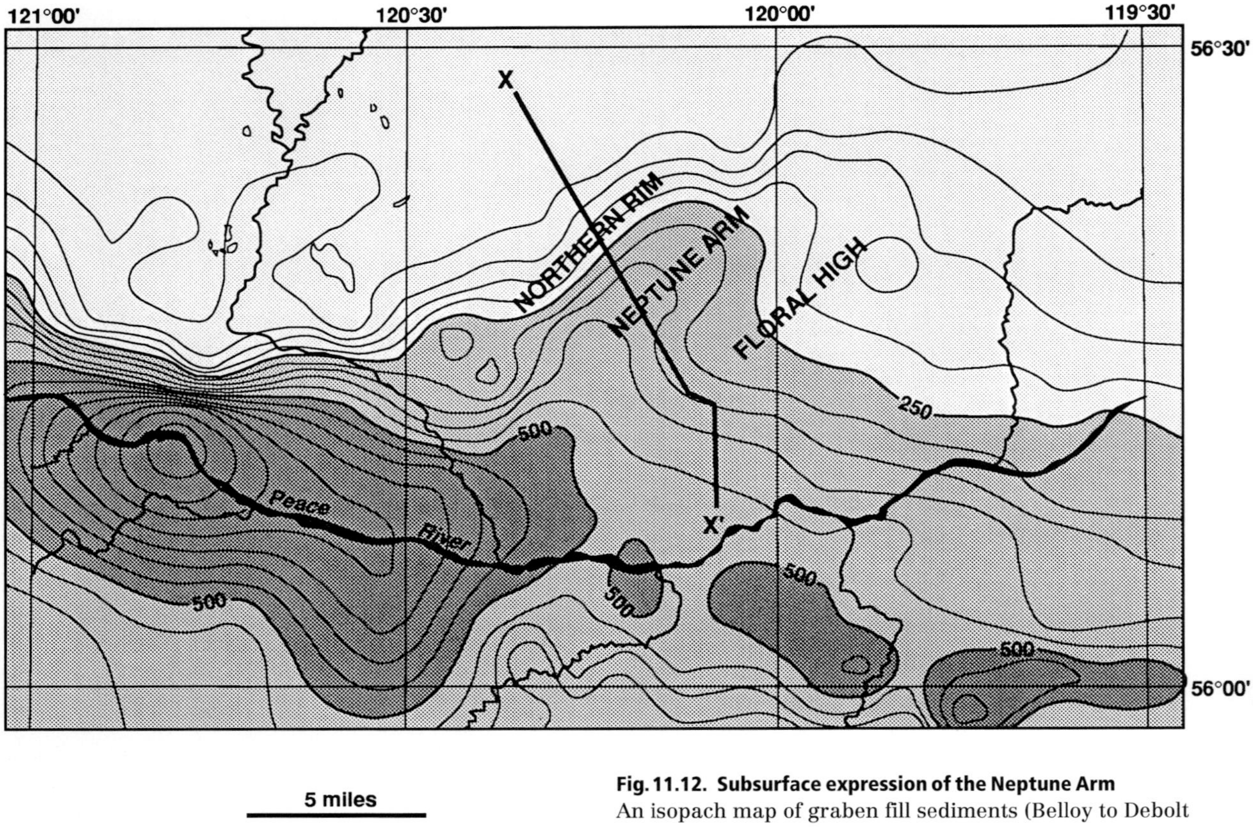

Fig. 11.12. Subsurface expression of the Neptune Arm
An isopach map of graben fill sediments (Belloy to Debolt formations) showing the extent of the Neptune Arm and its bounding margins

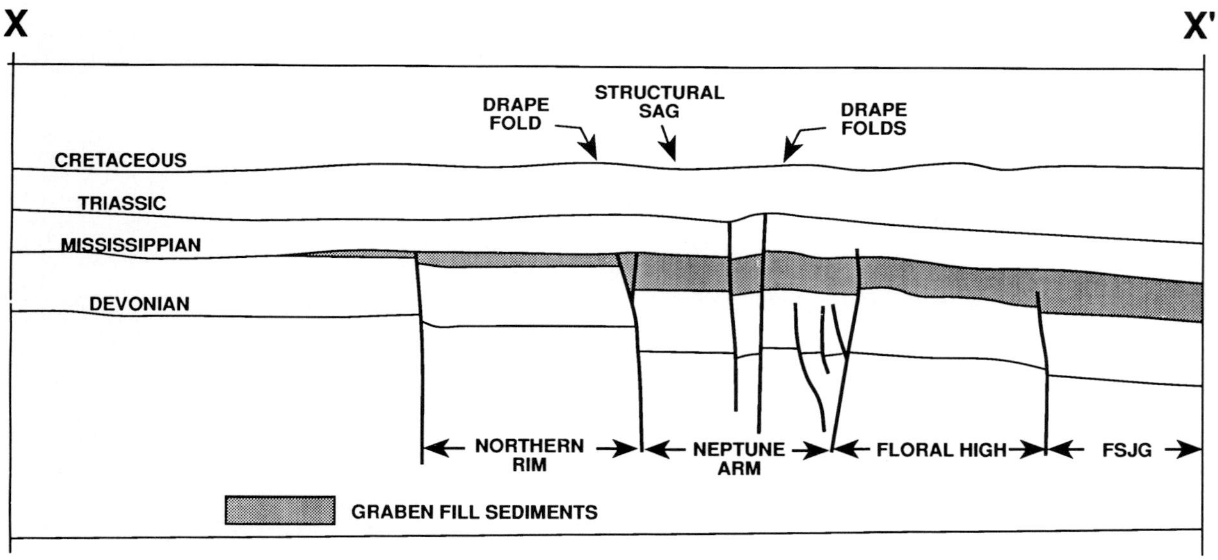

Fig. 11.13. Seismic expression of the Neptune Arm
Shown are the major fault systems that form the Neptune Arm and the expression of the northern and southern rims. Many of the fault systems that manifest surface expressions can be detected on the satellite image shown in Fig. 11.11

Chapter 12 The Canadian Foreland Fold and Thrust Belt, Northern British Columbia

12.1 Background

The Foreland Fold and Thrust Belt of northern British Columbia represents a small segment of the much larger Cordilleran Belt that extends 10 000 km from the Yukon-Alaska boundary in the north and south into Mexico (King 1959). The fundamental stratigraphic and structural elements of this long belt change little along its strike and the Canadian portion of the Cordillera, which is shown here, provides many of the geological attributes that can be observed with remote sensing data.

The area shown consists of the Halfway River map sheet which was studied to a great level of detail by both Irish (1970) and, more recently, Thompson (1989). Their maps and cross section have provided the basis for evaluating the data shown here (Fig. 12.1).

12.2 Objectives

Three subprovinces are represented in the map area and are shown in Fig. 12.2. The Northern Mountains subprovince is characterized by the presence of folded thrust sheets of Ordovician to Middle Devonian carbonate strata. The structural characteristics of this subprovince will not be discussed here. Chevron or box-style folds, with limb dips of 60° or more, characterize the structural style of the Foothills subprovince. Anticlines occur individually and as anticlinorial complexes. Separating them are broad, simple, flat-bottomed synclines. Thrust faults generally display various types of bifurcation, relay and en echelon map patterns, whereas "tear" faults or other cross-trending structures are generally absent (Figs. 12.3–12.9). The detachment surface is considered to be Devonian Carbonate rocks of the Besa River Shale (Thompson 1989).

Terminating the Foothills subprovince is a profound topographic escarpment of Cretaceous rock units that marks the western boundaries of the Western Canada Basin. Further to the south, this area is known as the Alberta Syncline, but, here, where it is characterized by the presence of low-amplitude, elongated folds, it is referred to as the Fore-Foothills Trend (Figs. 12.10– 12.12). In fact, many of these low-amplitude folds form effective hydrocarbon traps in the Lower Cretaceous section and many of these fields were discovered in early days by surface mapping. (The area is also known locally as the Jeddney-Bubble Trend, after two of the producing fields.)

The Halfway River map sheet consists of topography ranging from smooth ridges in the east to rugged peaks in the west. Because of the heavy vegetative cover, structural features in this area do not manifest clear expressions on satellite imagery, but are excellent candidates for "treetop" geological mapping with radar images (see Chap. 1). The main aspects to observe here are (1) the effective use of airborne radar as a tool for mapping structures in folds and thrust belts; (2) typical mapping patterns and surface expressions of thrust-related structures; (3) evaluation of ERS-1 radar as compared to airborne radar and TM imagery data.

12.3 Training Instructions

12.3.1 Interpretation Procedures

1. On a clear overlay interpret the surface expressions of geological structures as they are manifested on the airborne radar imagery of Fig. 12.3.
2. Use the tectonic map of Fig. 12.2, the interpreted radar data of Fig. 12.4, the map pattern of thrust faults which is shown in Fig. 12.7 and the cross sections of Fig. 12.6 to improve the interpretation of the radar data.
3. Make a distinction between structural features that were previously recognized by field mapping from those that can only be recognized on the radar data.

4. Repeat these procedures to interpret all the airborne and spaceborne radar images shown in this chapter.

12.3.2 Questions

Assuming that geological field mapping in this area has never been conducted:

1. What type of structures could have been recognized by using radar data as the sole source of information in this area?
2. Make a list of the most common patterns of thrust faults in the area and compare it with the fault pattern shown in Fig. 12.7.
3. Are there any profound “cross trends” or “tear faults”?
4. Is there any indication that the basement is involved in structural reactivation?
5. What is the maximum structural/topographic relief that can be imaged by the radar data without severe interference by the “shadowing effect”?
6. To what level does the resolution limitation of the spaceborne radar or the layover effect hamper the use of this data set for geological mapping in this area as well as for logistical purposes?
7. Based on the radar information alone, how would you design your initial seismic program?

References and Further Reading

Fitzgerald EL (1968) Structure of British Columbia Foothills, Canada. AAPG Bull 52: 641–664

Irish EJW (1970) Halfway River map area, British Columbia. Geol Soc Can Pap 69–11

King PB (1959) The evolution of North America. Princeton University Press, New Jersey

Lingrey SH (1983) Exploration in fold and thrust belts. I. Geometry of faults. Exxon Production Research Company, Internal Rep

McMechan ME, Thompson RI (1989) Structural style and history of the Rocky Mountain fold and thrust belt. In: Ricketts BD (ed) Western Canada sedimentary basin – a case history. Can Soc Pet Geol, pp 47–71

Stott DF (1967) Fernie and Minnes strata north of Peace River, foothills of northeastern British Columbia. Geol Surv Can Pap 67–19, parts A, B

Thompson RI (1979) A structural interpretation across part of the northern Rocky Mountains, British Columbia, Canada. Can J Earth Sci 16: 1228–1241

Thompson RI (1989) Stratigraphy, tectonic evolution and structural analysis of the Halfway River map area (94 B), northern Rocky Mountains, British Columbia. Geol Surv Can, Mem 425

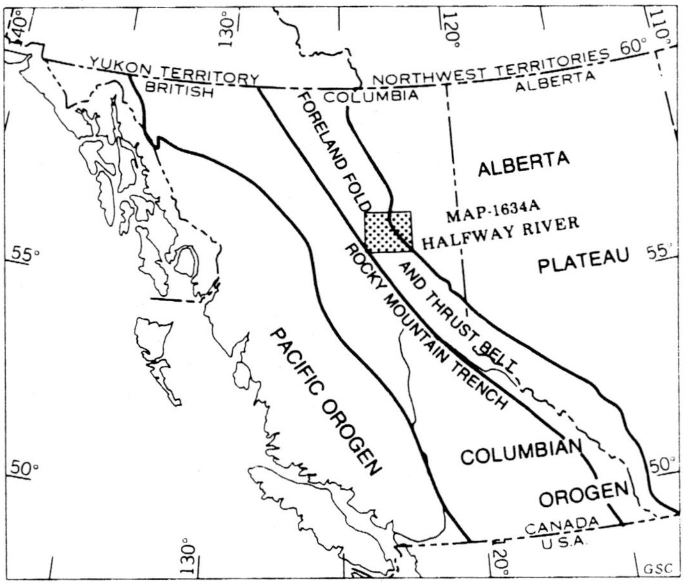

Fig. 12.1. Location map of the study area
Shown is the location of the Halfway River map area and its position relative to the major geological subdivision of the Canadian Cordillera. (After Thompson 1989; published with the permission of the Geological Survey of Canada)

Fig. 12.2. Tectonic map of the northern Canadian Rocky Mountains
Shown are major subprovinces and related structures. The Halfway River map area is outlined. Chowade River section is shown as C-C’ in Fig. 12.5. (After Thompson 1989; published with permission from the Geological Survey of Canada)

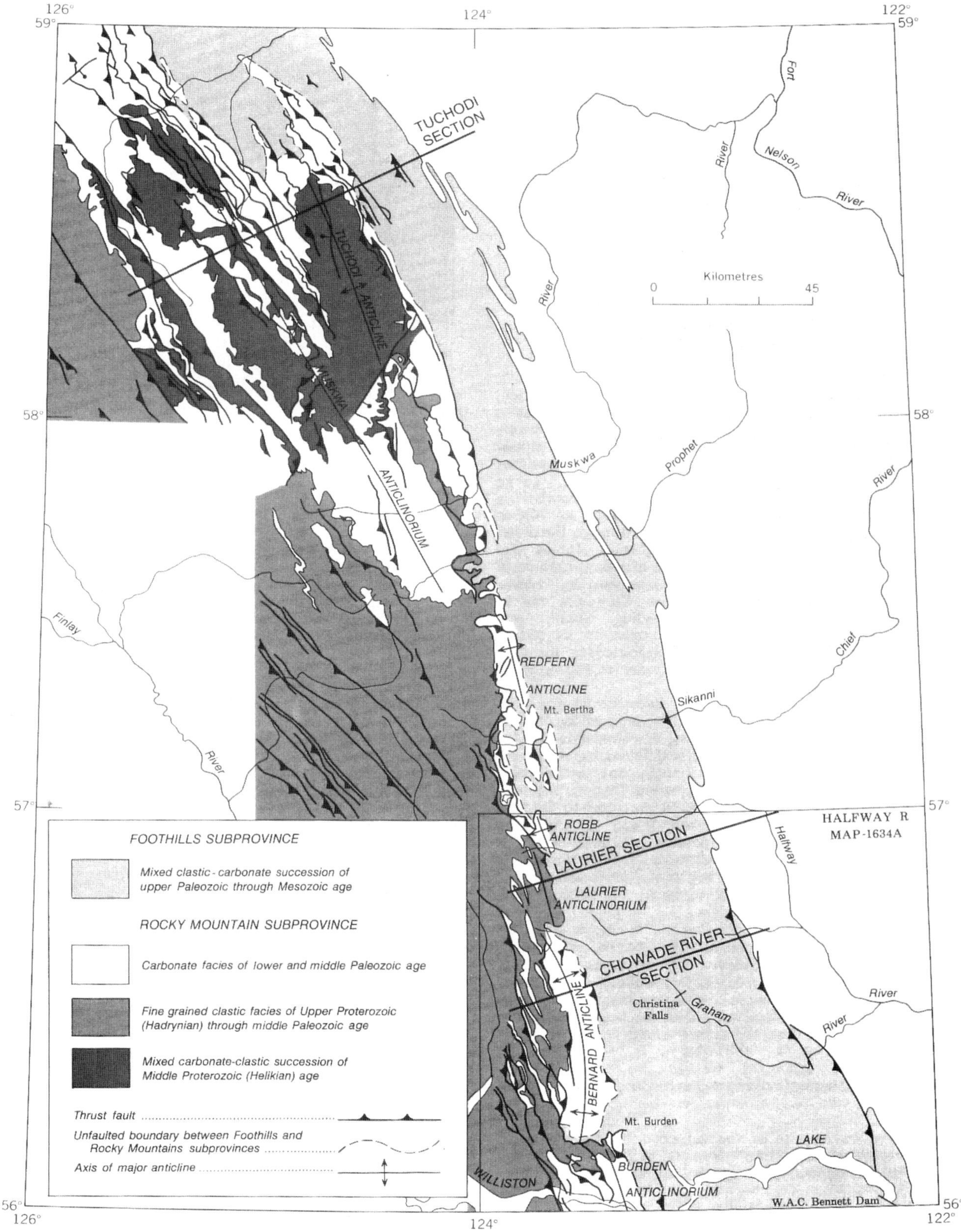
126°
124°
122°
59°
58°
57°
56°
TUCHODI SECTION
TUCHODI ANTICLINE
MUSKWA
ANTICLINORIUM
Fort
River
Nelson
River
River
Kilometres
0
45
Muskwa
Prophet
River
Finlay
River
REDFERN
ANTICLINE
Mt. Bertha
Sikanni
Chief
ROBB
ANTICLINE
LAURIER SECTION
HALFWAY R
MAP-1634A
Halfway
LAURIER
ANTICLINORIUM
CHOWADE RIVER SECTION
BERNARD ANTICLINE
Christina Falls
Graham
River
River
Mt. Burden
LAKE
WILLISTON
BURDEN
ANTICLINORIUM
W.A.C. Bennett Dam
FOOTHILLS SUBPROVINCE
Mixed clastic-carbonate succession of upper Paleozoic through Mesozoic age
ROCKY MOUNTAIN SUBPROVINCE
Carbonate facies of lower and middle Paleozoic age
Fine grained clastic facies of Upper Proterozoic (Hadrynian) through middle Paleozoic age
Mixed carbonate-clastic succession of Middle Proterozoic (Helikian) age
Thrust fault
Unfaulted boundary between Foothills and Rocky Mountains subprovinces
Axis of major anticline

Fig. 12.3. Airborne radar image of the Halfway River area
Shown is the structural and topographic expression of major subprovinces as well as key structural elements that are mentioned in the text. Cross sections A-A', B-B' and C-C' are shown in Fig. 12.6. (Courtesy of Intera, Inc)

Fig. 12.4. Airborne radar image of the Halfway River area
Shown are major exposed structure features in the area and their unique surface expression and structural patterns. Geology was modified from Irish (1970) and Thompson (1989)

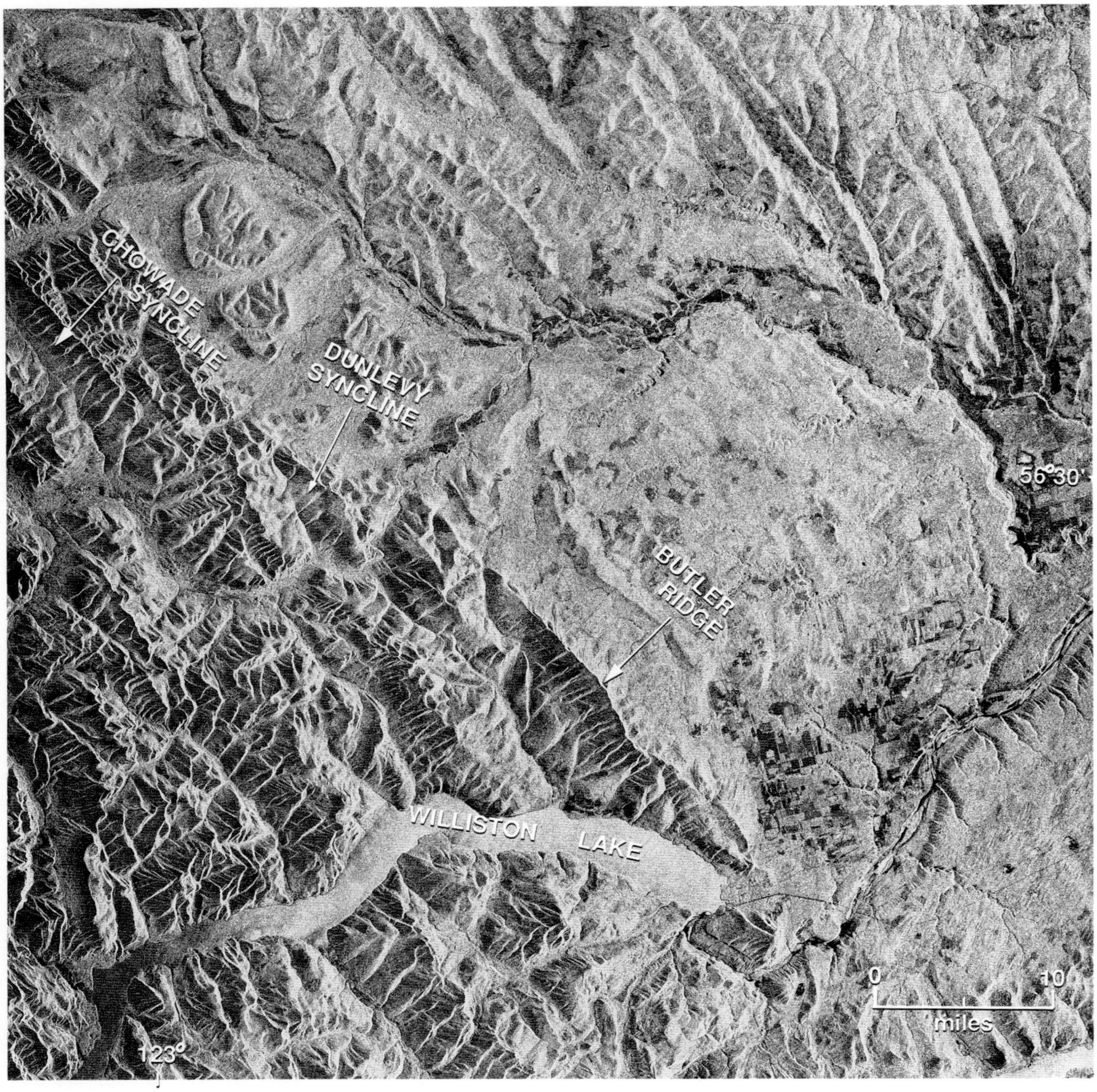
CHOWADE SYNCLINE
DUNLEVY SYNCLINE
BUTLER RIDGE
WILLISTON LAKE
56°30'
123°
0
10
miles

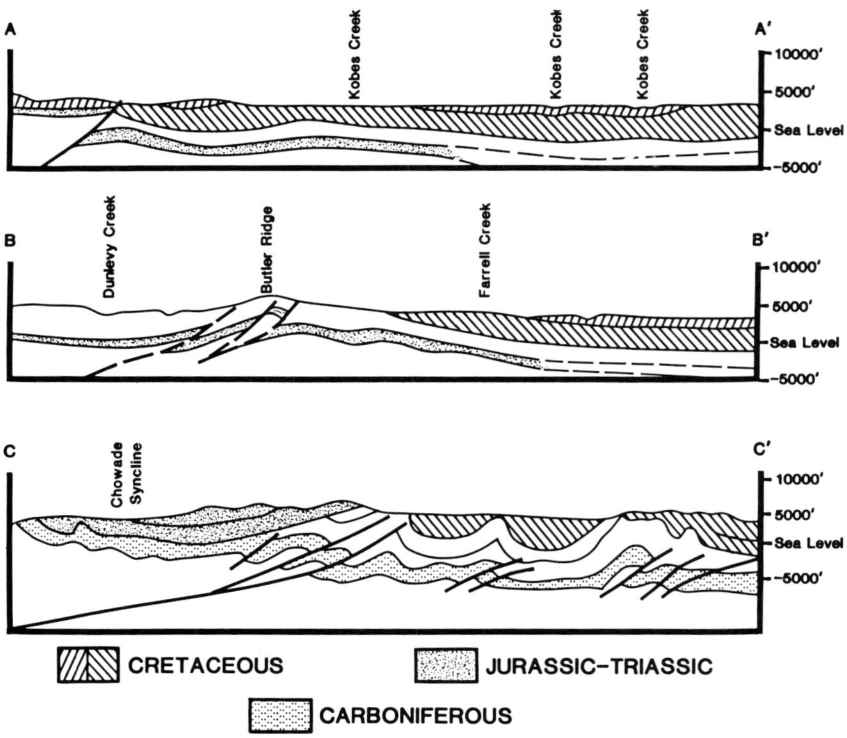

◀ **Fig. 12.5. ERS-1 radar of the Halfway River area**
These recently available radar data provide adequate regional views of the area with substantially reduced cost, as compared to airborne data. Note, however, that the exaggerated flatirons, caused by the layover effect, may hinder detailed interpretation in the mountainous area. It is possible to improve the interpretation of the mountainous area by obtaining images with flight paths further to the west. (Data courtesy of RSI)

Fig. 12.6. Cross sections from the Halfway River area
Shown are three typical cross sections including (1) the low-amplitude anticlines of the fore-foothills **(A-A')**; (2) the first thrust faults that form the eastern boundary of the Foothills subprovince **(B-B')**; and (3) the more complex patterns of the inner parts of this province **(C-C')**. **(A-A'** and **B-B'** are after Irish 1970; **C-C'** is after Thompson 1989)

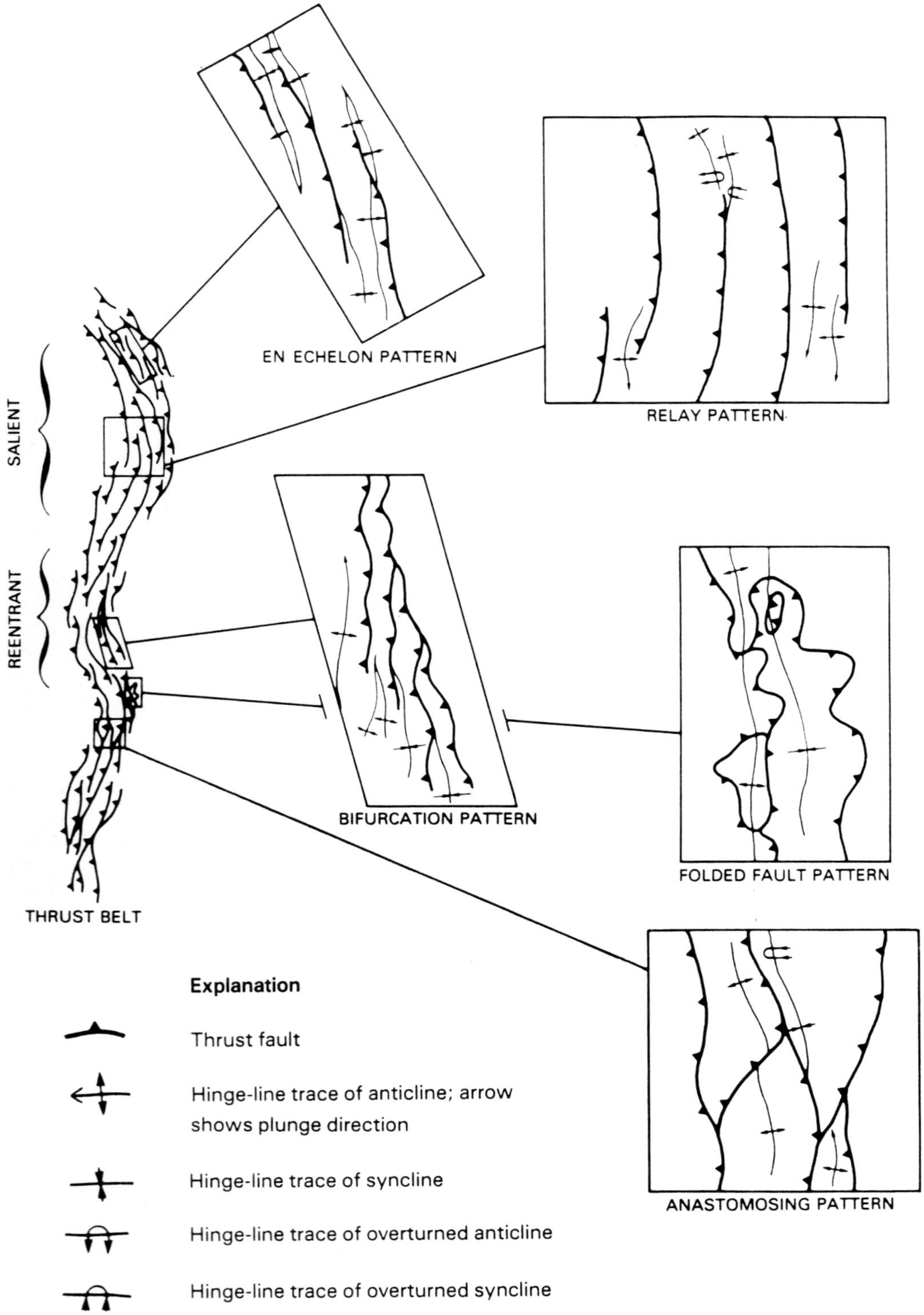

Fig. 12.7. Map patterns of thrust faults
Shown are different types of structural patterns found in the Canadian Cordillera fold and thrust belt. (Compiled by S. Lingrey 1983)

Fig. 12.8. Photographs of the Canadian Rockies
The upper photograph illustrates the first major physiographic expression of the Canadian Rocky Foothills subprovince in British Columbia. The remaining two photographs show the typical expression of the thrust faults. In the case shown, Jurassic Cretaceous Minnes and Fernie group thrust onto Lower Cretaceous Gething along the carbon thrust immediately south of the Peace Ridge of Williston Lake. Note that the area is characterized by heavy vegetative cover and lack of bedrock exposure. In spite of these conditions, radar systems still create clear images of structures using the principle of "treetop" geology. (Photos courtesy of A. LaRiviere)

Fig. 12.9. Airborne radar of the Butler Ridge area
Shown is the surface expression of thrust folds, narrow anticlinal ridges that are truncated against the thrust, and intervening, broad synclines. Note that the shadowing effect in the vicinity of steeper slopes causes a loss of information. This problem can be overcome by examining other images from this survey (i. e. different flight paths) that have more advantageous look angles. (Data courtesy of Intera)

Fig. 12.10. Airborne radar of the Jeddney-Bubble area
Shown are a series of elongated anticlinal ridges with intervening broad synclines. Most of the structural elements in this area have been breached, showing an early stage of topographic inversion. Note the clear definition of geological structure and topology and the lack of shadowing effect which is attributed to the relatively low topographic relief making radar imagery an ideal remote sensing tool in this area. (Data courtesy of Intera)

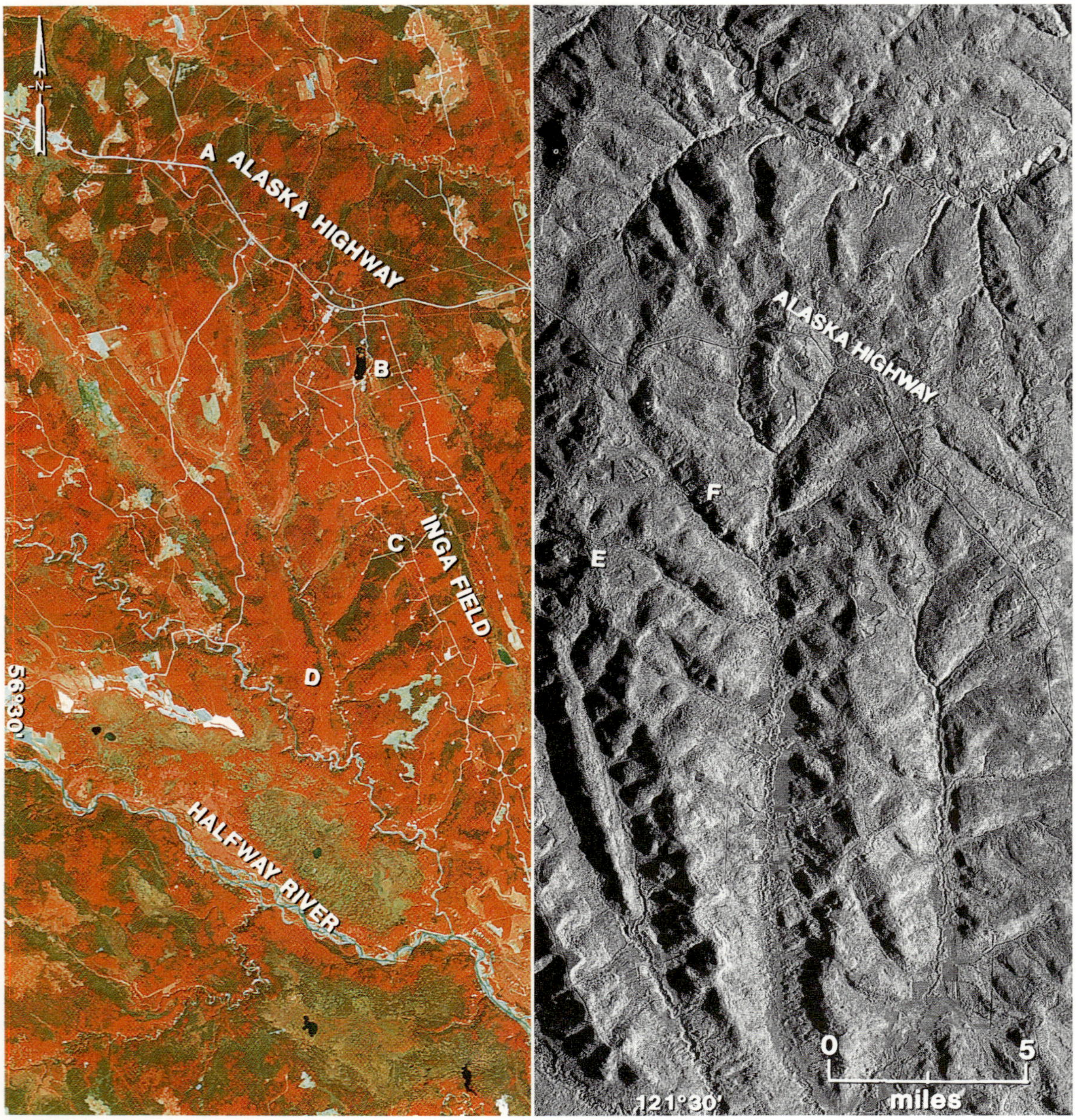

Fig. 12.11. Comparison of satellite imagery and radar
Shown is the Jeddney-Bubble trend area south of the Halfway River map sheet. The image is centered around the Inga Field which is located along an elongated anticlinal ridge, just south of the Alaska Highway. Note that the satellite data show very little information regarding the topographic expression of structures in this area, whereas the airborne radar accentuates these features. An attempt was made to mosaic the two images. Due to projection difference, though, there is a slight gap in the continuity of the imagery particularly on the northern half, along the Alaska Highway. Letters *A* through *F* indicate key areas which are shown in Fig. 12.12

Fig. 12.12. ERS-1 Radar image of the Inga Field area
Structures in this area have well-manifested surface expressions similar to those obtained with airborne radar. However, spatial resolution is very poor. Lithostratigraphic units and important cultural features (e. g. roads, seismic cut lines and drilling pads) cannot be easily identified. (Data courtesy of RSI)

Chapter 13 The Paris Basin, France

13.1 Background

The Paris Basin is a large intracratonic sag that was formed and filled with sediments during Triassic and Jurassic crustal extensional events that are related, in most parts, to the initial separation of the African and North American continents and the creation of the North Atlantic Ocean (Ziegler 1982; Pages 1987; Figs. 13.1 and 13.2). The basin exhibits structural styles that are common to many other intracratonic sags, meaning that the dominant structures are low-amplitude folds, domes, monoclinal flexures and other basement warp features (Harding and Lowell 1979; Figs. 13.3–13.14).

The latest phases of tectonic reactivation that influence the present geology and surface expression of structures in this basin include (1) a Late Eocene through Oligocene extensional event related to the development of the Rhine Graben (Illies 1970); (2) Miocene contractional events related to the deformation of the Alpine foreland (Megnien and Megnien 1980); and (3) a Mio-Pliocene-age uplift that gave the basin its distinct shape (Rutten 1969; Anderson 1978).

13.2 Objectives

The Paris Basin represents a low-relief region with subtle structures which are obscured by a thick cover of vegetation, soil and extensive man-made features. The main aspects to observe here are (1) recognition of structural styles and trends in the exposed margin of sedimentary basins and their influence on the basin's interior; (2) detection of basement warp structures through integrated analysis of satellite imagery, gravity and magnetic data; and (3) recognition of structural leads in such basins.

13.3 Training Instructions

13.3.1 Interpretation Procedures

1. Outline on a clear overlay, major tectonic units that are exposed around the Paris Basin as they are manifested on the imagery mosaic of the basin in Fig. 13.3. Trace the outcropping pattern of different lithostratigraphic units that are exposed in the basin as well as major linear features that can be extended from the exposed margins into the basin area. Mark on the overlay the location of structurally controlled features such as linear topographic scarps, breached topographic highs, structurally controlled stream valleys, etc.
2. Interpret geological and surface features of Figs. 13.5, 13.6, and 13.7.
3. Use the geological information from Figs. 13.4, 13.8, 13.10, 13.12, and 13.13 to constrain and improve the interpretation of the imagery.
4. Return to the regional mosaic of Fig. 13.3 and incorporate the interpretation of your detailed images onto the regional overlay.
5. Compare the results of your interpretation to the block diagram of Fig. 13.14.

13.3.2 Questions

1. What are the most dominant structural/topographic features in the Paris Basin area?
2. How many of these features are controlled by structural elements that can be traced in the exposed margins of the basin and/or in the adjacent tectonic units?
3. What is the structural style of the basin?
4. What are the most dominant structural trends in this basin and how are they related to the structural trends of the exposed margins?
5. What type of hydrocarbon traps may be found in such a basin?

References and Further Reading

Anderson JGC (1978) The structure of western Europe. Pergamon Pres, Oxford

Bureau de Recherches Geologiques et Minieres (1980 a) Carte geologique de la France: scale 1:1,500,000

Bureau de Recherches Geologiques et Minieres (1980 b) Carte tectonique de la France: scale 1:1,000,000

Bureau de Recherches Geologiques et Minieres (1981) Carte seismotectonique de la France: scale 1:1,000,000

Bureau de Recherches Geologiques et Minieres, various dates, various 1:320,000 scale geologic maps of France

Harding TP, Lowell JD (1979) Structural styles, their plate-tectonic habitats and hydrocarbon traps in petroleum provinces. AAPG Bull 63: 1016–1058

Illies JH (1970) Graben tectonics as related to crust-mantle interaction. In: Graben problems. E. Schweizerbart'sche Verlagsbuchhandlung, Stuttgart, pp 4–27

Megnien C, Megnien F (coordinators) (1980) Synthese geologique du bassin de Paris, vol 1, stratigraphie et paleogeographic. BRGM Mem 101, 466 pp

Pages L (1987) Exploration of the Paris Basin. In: Brooks J and Glennie K (eds) Petroleum geology of northwest Europe: Graham and Trotman, London, pp 87–93

Rutten MG (1969) The geology of western Europe. Elsevier, Amsterdam

Ziegler PA (1982) Geological atlas of western and central Europe. Elsevier, Amsterdam

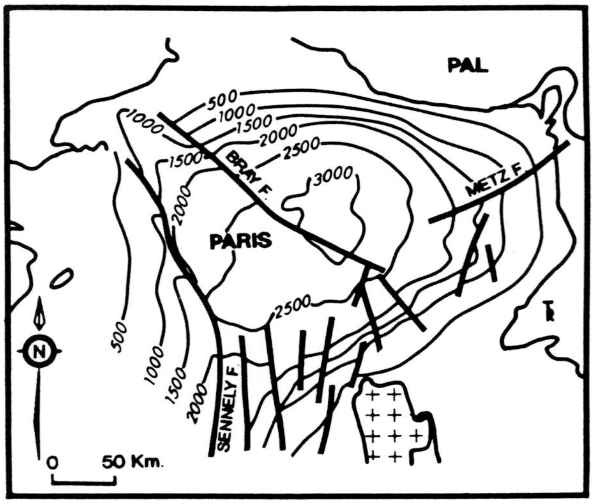

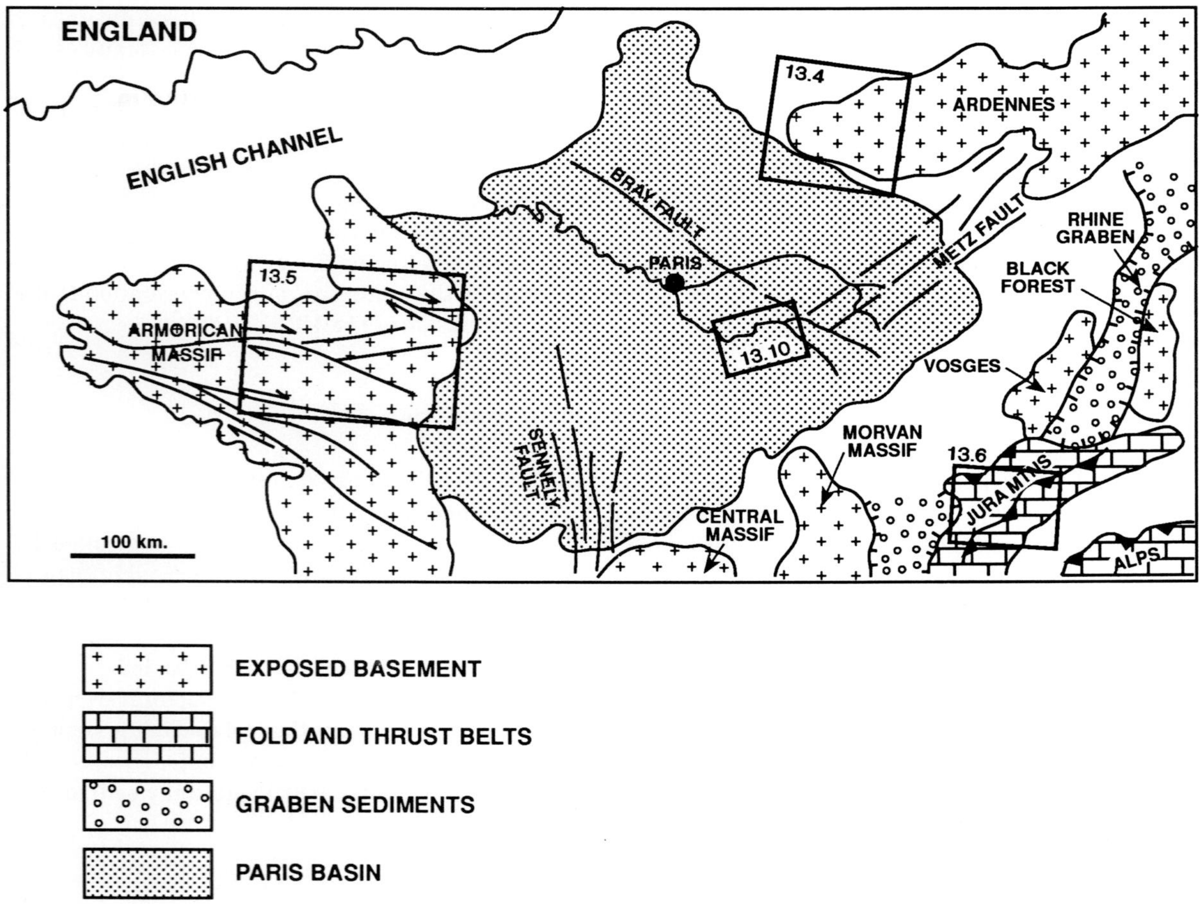

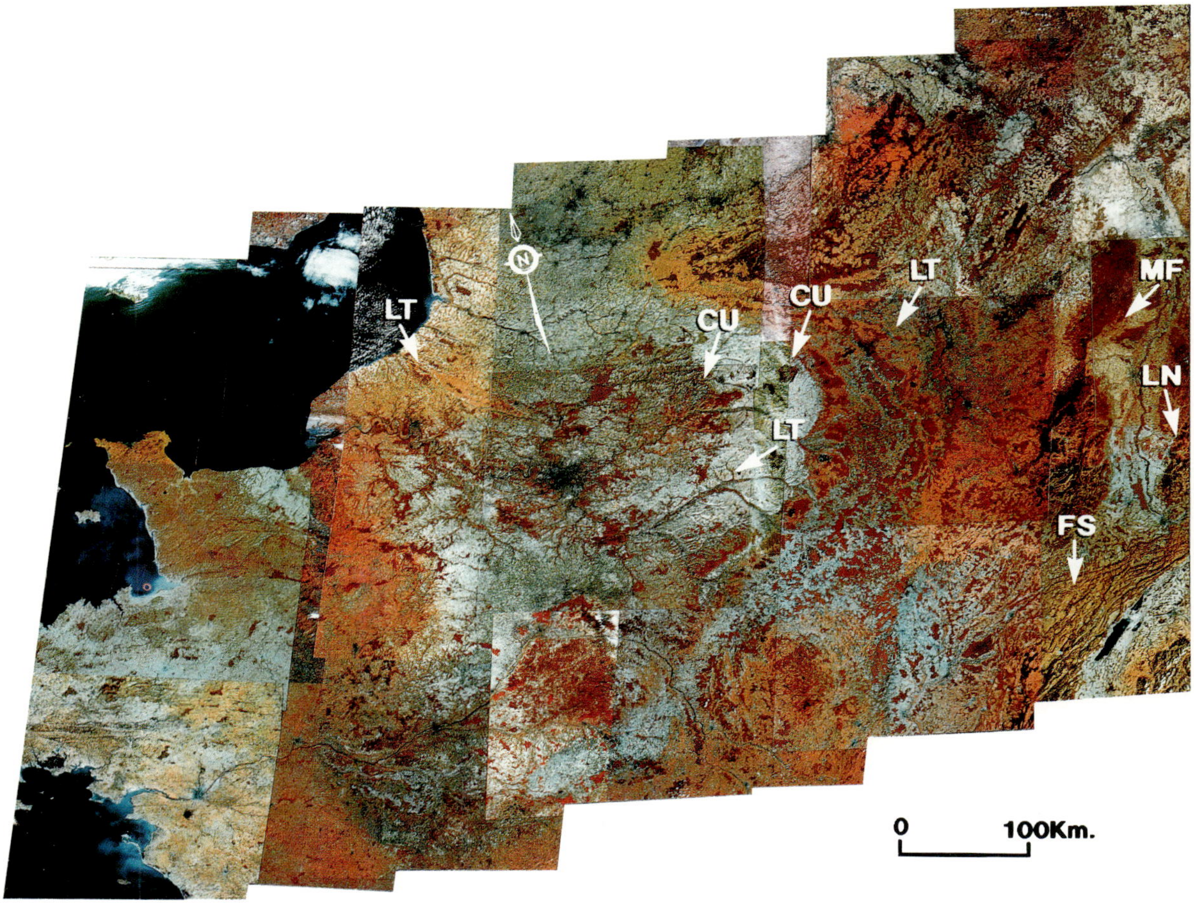

◀ **Fig. 13.1. The Paris Basin**
A generalized structure map of base of Triassic illustrating the general shape of the basin and the presence of major cross-cutting fault systems. (After Pages 1987)

◀ **Fig. 13.2. Generalized tectonic map of the Paris Basin**
The locations of major structural features that are illustrated in this section are shown. (Generalized from the Bureau de Recherches Geologiques et Minieres 1980 b)

Fig. 13.3. Satellite imagery mosaic of the Paris Basin region
The mosaic, which contains 30 images, showes the general shape of the basin, the tectonic units which are exposed around its margins and key surface features. *CU* cuesta; *LT* lineament; *LN* linear negative FLT; *SF* sinuous FLT; *MF* multidirectional FLT

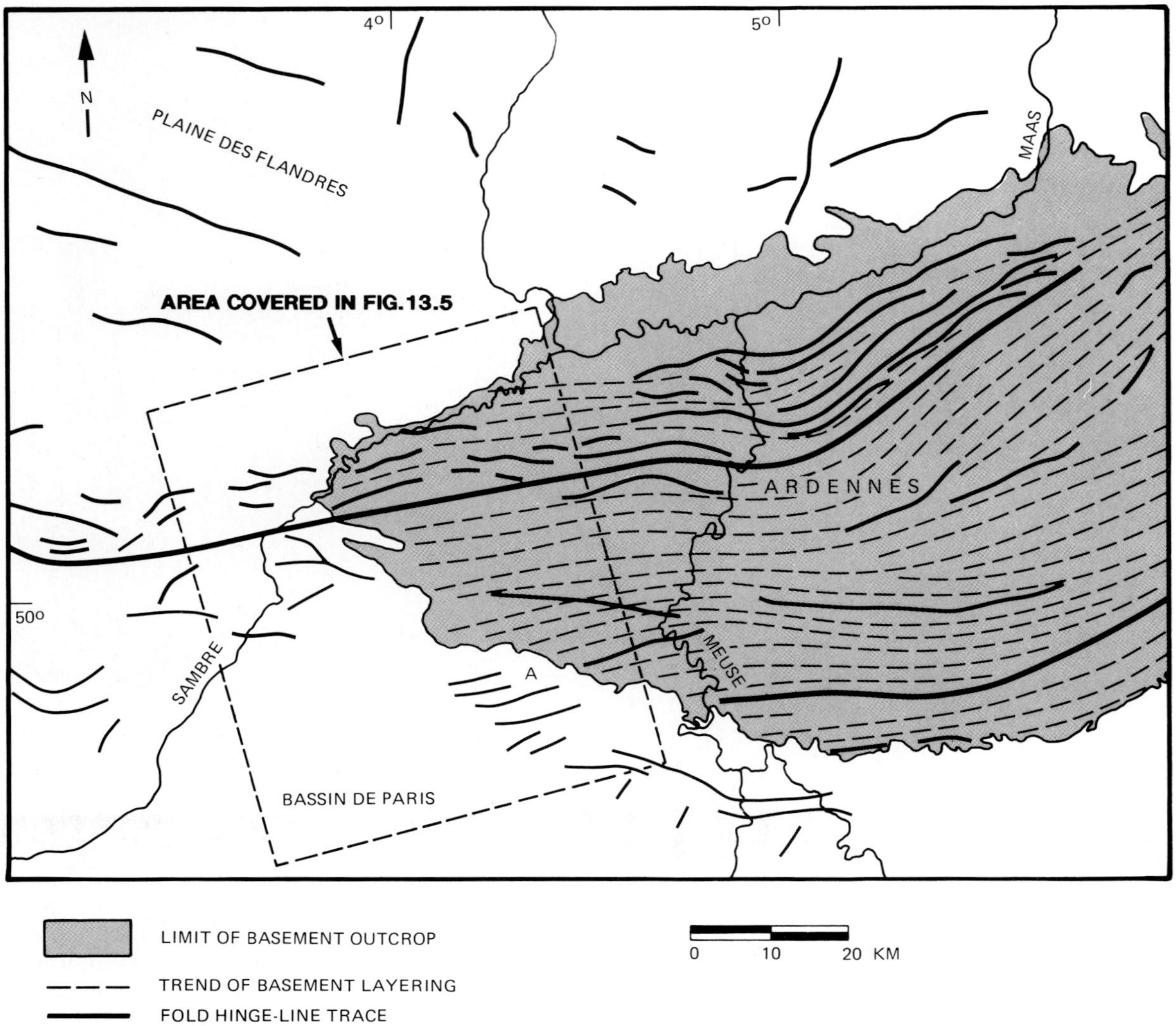

Fig. 13.4. Basement grain map of the Ardennes Mountains
The exposed basement grain is dominated by the presence of west-trending folds and basement layerings which can be traced with imagery into the basin area (see Fig. 13.5). (Generalized from Bureau de Recherches Geologiques et Minieres 1980 b)

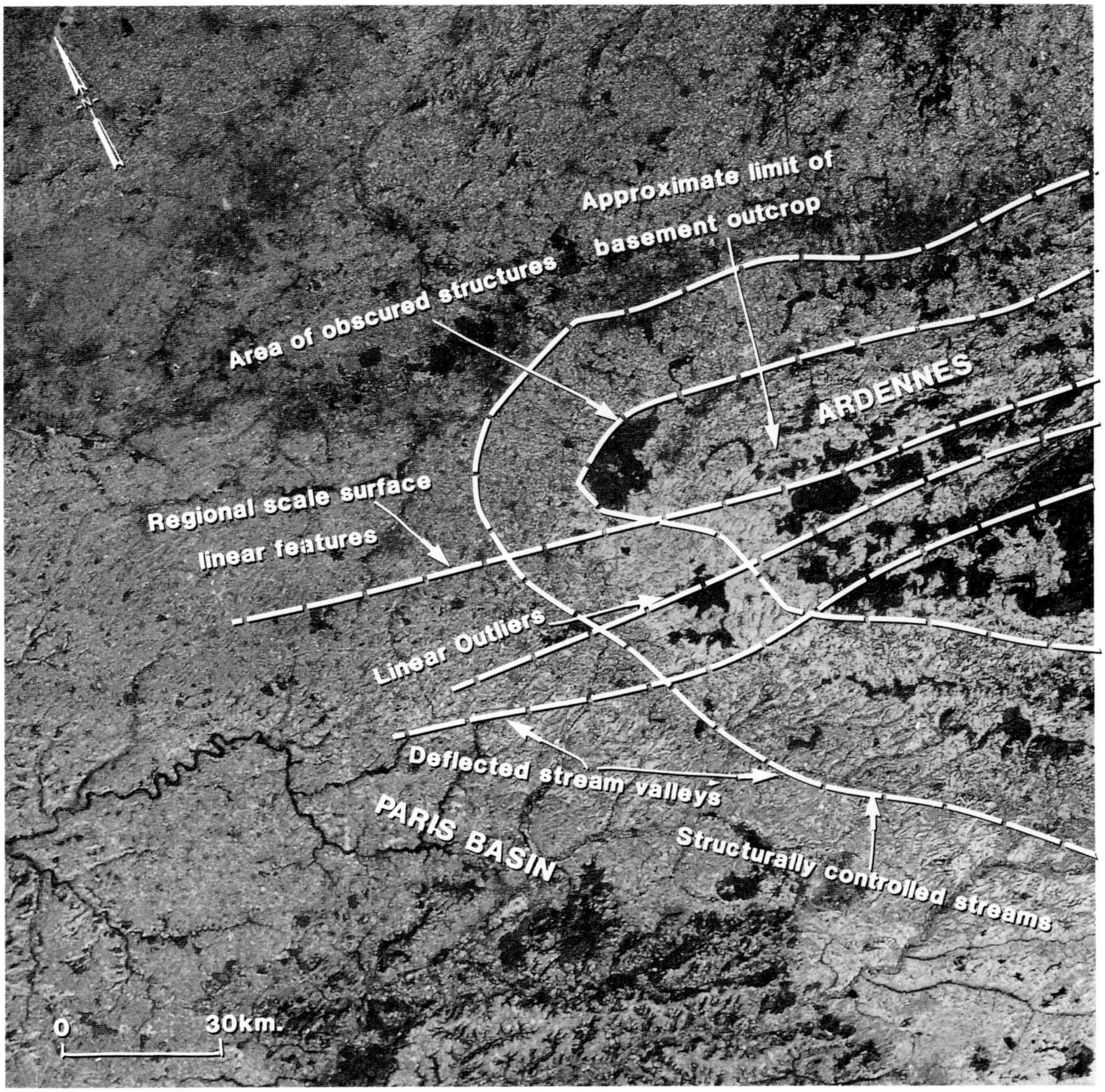

Fig. 13.5. Structural styles and trends of the Ardennes Mountains

Landsat imagery subscene and structural interpretation showing the surface expression of the exposed Ardennes Massif which is characterized by the presence of highly deformed, west-trending folds and strong basement layering trending in a similar direction. (Most reflect eroded layers of folded strata.) Many of the structures observed in the Ardennes continue to manifest surface expressions in the basin's interior which are manifested on imagery data as profound linear features that are spatially coincident with the exposed basement structures

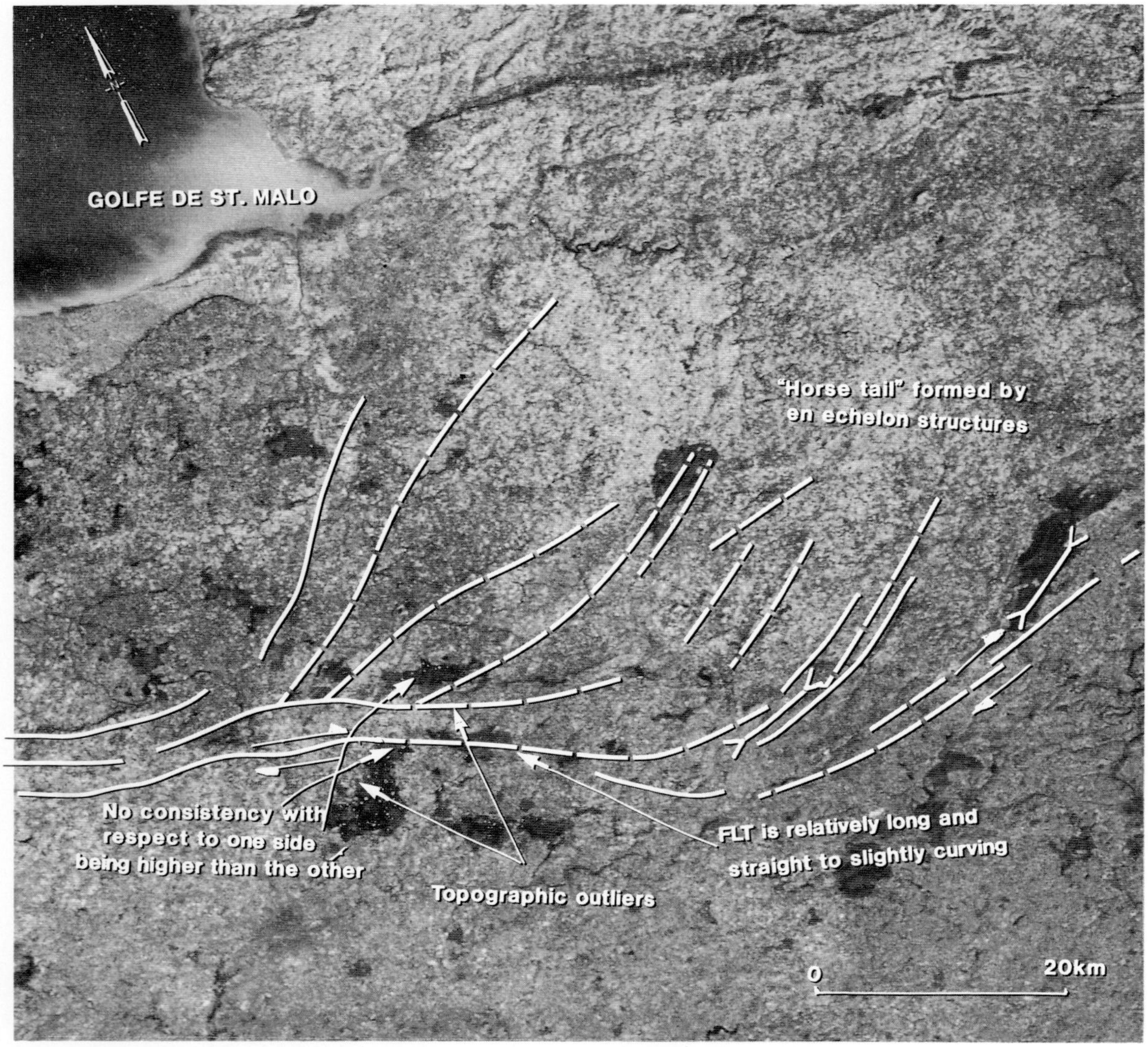

Fig. 13.6. Structural styles of the Armorican Massif
Landsat imagery subscene of the Amorican Massif in western France showing the expression of right-lateral strike-slip faults and associated oblique folds. These faults are characterized by long, slightly curvilinear FLTs. The FLTs occupy valleys and display lateral offset. However, no consistent vertical offset is usually observed along the FLTs. En echelon folds are often visible on either side of the fault and they truncate at oblique angles to the main FLT. The termination of these folds forms a classical compressional splay feature often referred to as a "horsetail". It is interesting to note that structures in this area manifest detectable expressions on imagery in spite of the intensive erosion and obliteration of the massif

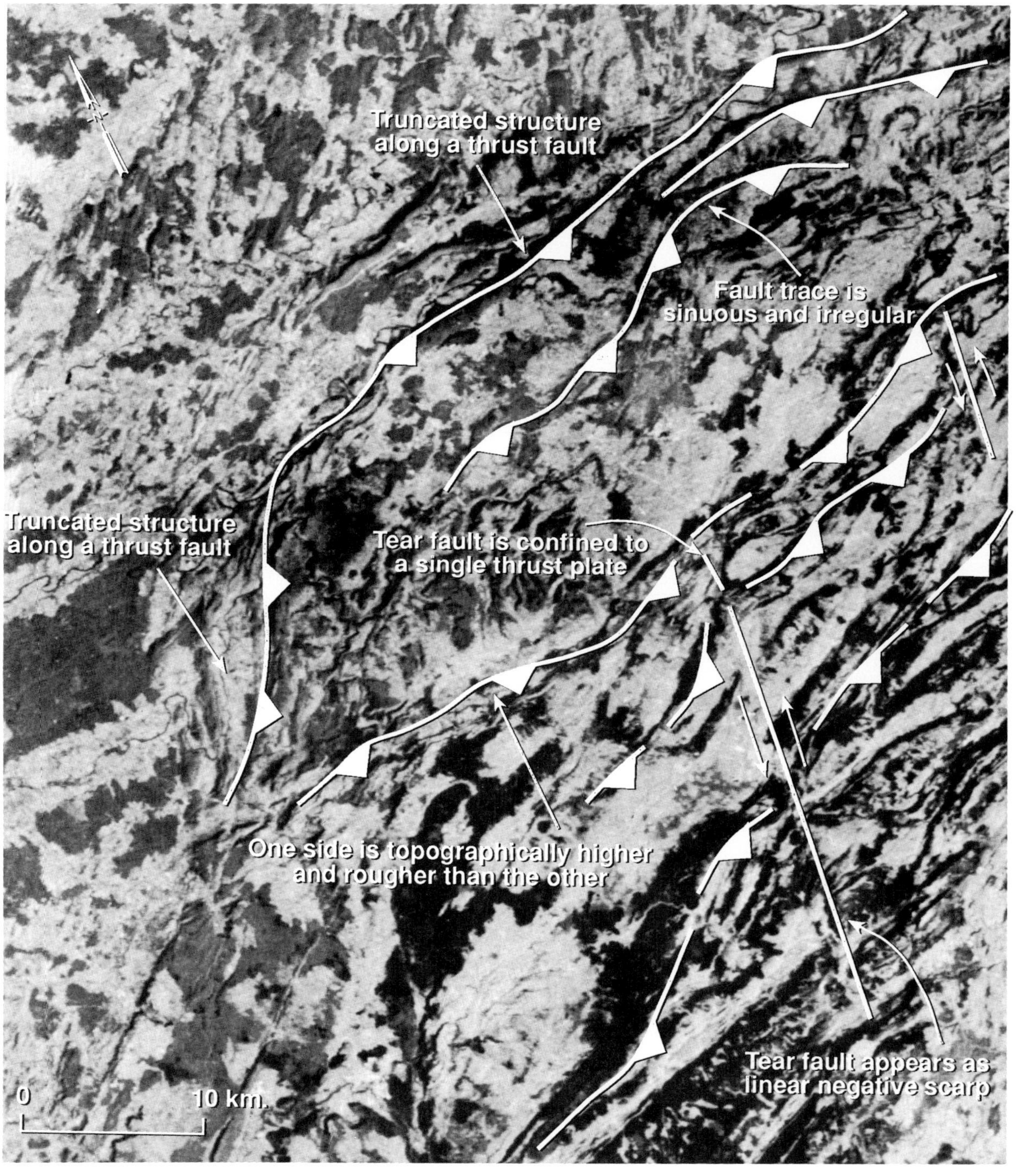

Fig. 13.7. Structural style of the Jura Mountains
Landsat imagery subscenes of the Jura Mountains region showing typical characteristics of sinuous thrust faults, oblique truncated anticlines and major "tear" faults. This portion of the Jura is believed to be completely detached from the basement showing very little correspondence to the structure trends in the basin. The deformation of the Jura, however, represents a contractional tectonic episode which reactivated many structures in this basin. As illustrated earlier in the book, many of the "tear" faults in the Jura reflect the reactivation of basement structures below the detachment surface

Fig. 13.8. Folds of the Jura Mountains
A photograph of a typical asymmetrical anticline that developed in association with thrust folds in the Jura region. The steep limb always points in the direction of the moving plates. Note the presence of a resistant carbonate layer (the Jurassic Malm) which protected the topographic expression of this fold, allowing many folds in the Jura to express a unique positive relief stage. The outstanding structural/topographic expression of folds in the Jura, which are so well manifested on imagery, is attributed to this phenomenon as well as to the relatively young age of this fold belt

Fig. 13.9. Surface expression of the Bray Fault

Landsat imagery subscene and structural interpretation of the Bray Fault and its associated structures. The FLT of this fault is quite complex, exhibiting typical characteristics of a reactivated wrench fault. The FLT is long and straight with oblique compressional splay features. The northwestern segment, however, forms the boundary of the Bray Anticline and exhibits strong expression of a positive FLT which characterizes high-angle normal faults. Seismic lines 1 through 5 demonstrate the gradual changes in the fault profile along strike which are manifested on the imagery by the changes in the characteristics of the FLTs

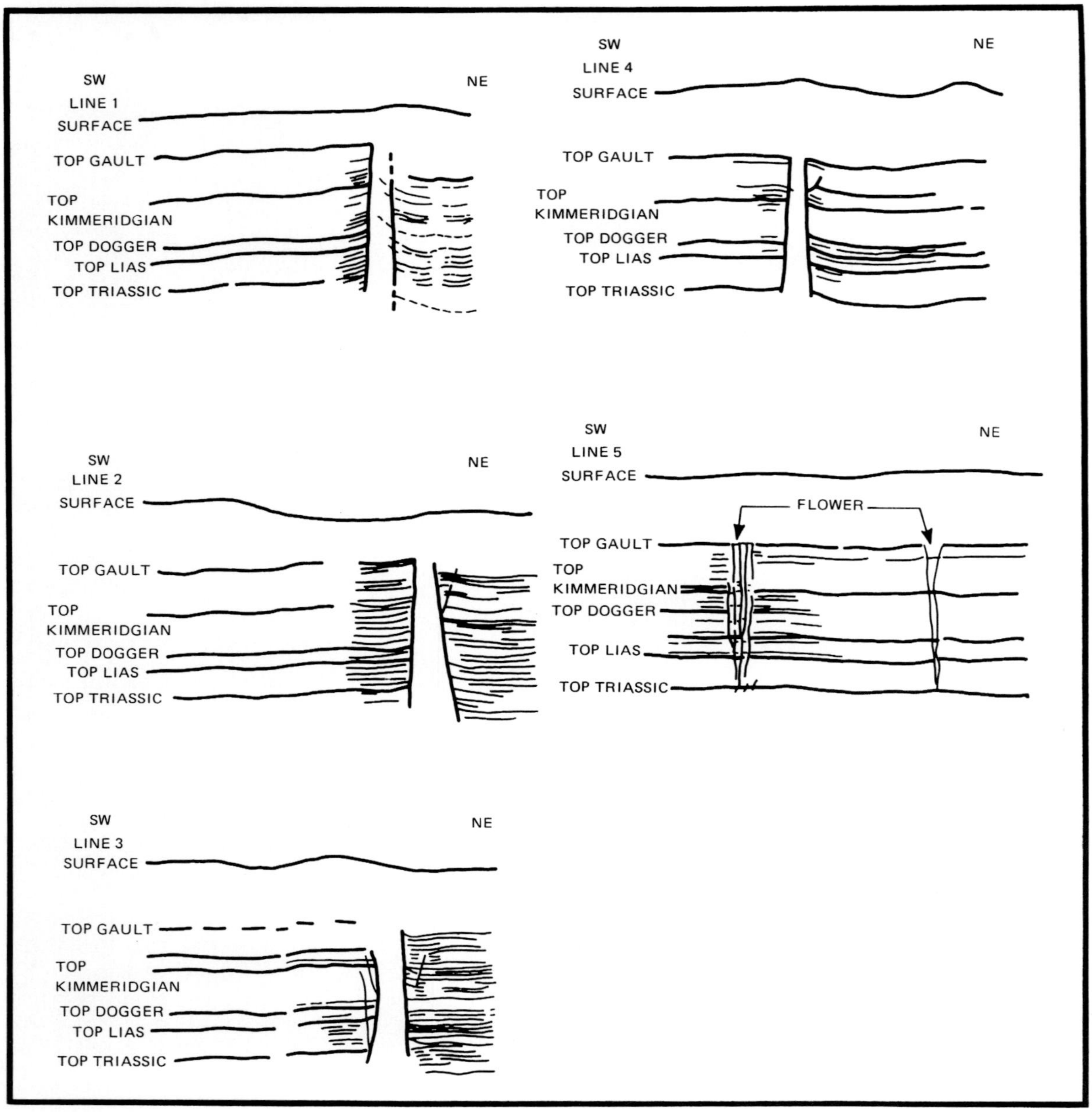
SW
NE
LINE 1
SURFACE
TOP GAULT
TOP
KIMMERIDGIAN
TOP DOGGER
TOP LIAS
TOP TRIASSIC
SW
NE
LINE 4
SURFACE
TOP GAULT
TOP
KIMMERIDGIAN
TOP DOGGER
TOP LIAS
TOP TRIASSIC
SW
NE
LINE 2
SURFACE
TOP GAULT
TOP
KIMMERIDGIAN
TOP DOGGER
TOP LIAS
TOP TRIASSIC
SW
NE
LINE 5
SURFACE
FLOWER
TOP GAULT
TOP
KIMMERIDGIAN
TOP DOGGER
TOP LIAS
TOP TRIASSIC
SW
NE
LINE 3
SURFACE
TOP GAULT
TOP
KIMMERIDGIAN
TOP DOGGER
TOP LIAS
TOP TRIASSIC

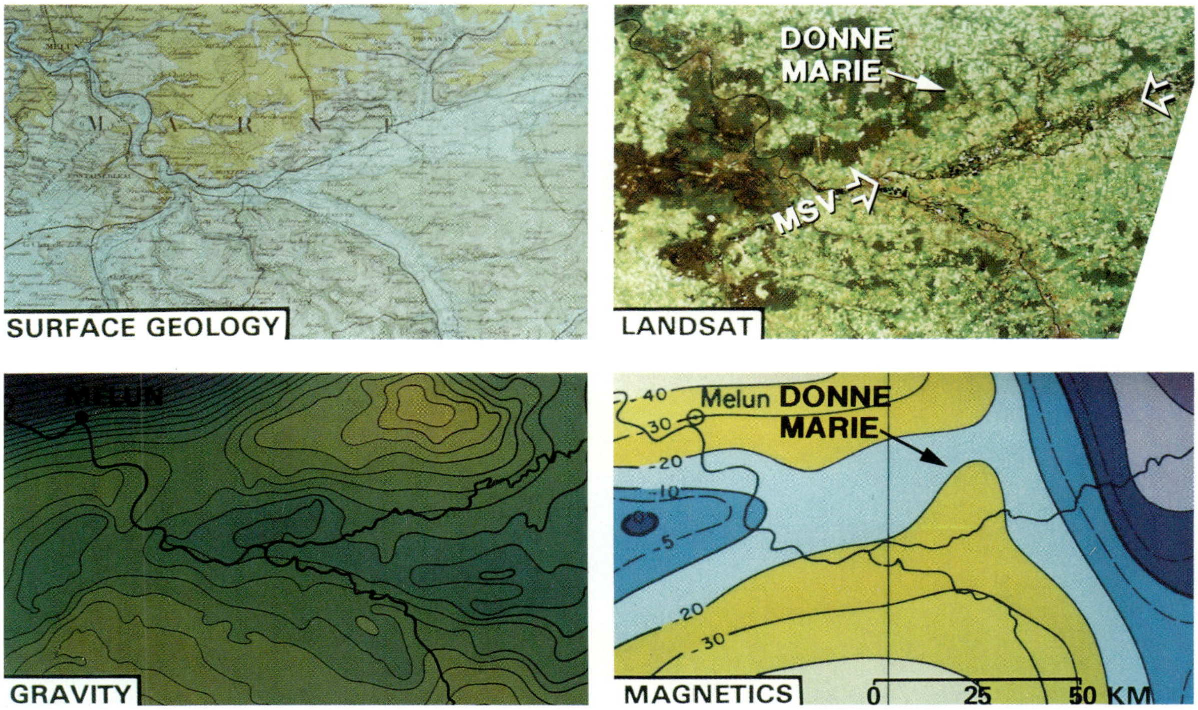

◀ **Fig. 13.10. Subsurface expression of the Bray Fault**
A series of interpreted seismic lines demonstrating the structural characteristics of the Bray fault and its possible wrench components. Note that the southwestern segments are characterized by the presence of high-angle dip-slip faults which gradually change to flower structures with little or no vertical displacement. (Data courtesy of Esso Rep, interpreted by C. C. Wielchowsky)

Fig. 13.11. Surface and subsurface expression of the Donne Marie Field area
Examples of surface and potential field data which were used to detect the extension of Metz-Seine Valley lineament and related individual folds. The Donne Marie Field is located along this major structural trend showing surface characteristics of a breached fold (illustrated in Fig. 13.12). The surface expression of this lineament, as well as the Donne Marie Field, has been demonstrated also in Fig. 4.22 through 4.25. Note the presence of an oblique, local magnetic nose which corresponds to the location of the Donne Marie Field. This magnetic anomaly may reflect the presence of a secondary basement fault that enhanced the local development of the producing structure; see also Fig. 13.13

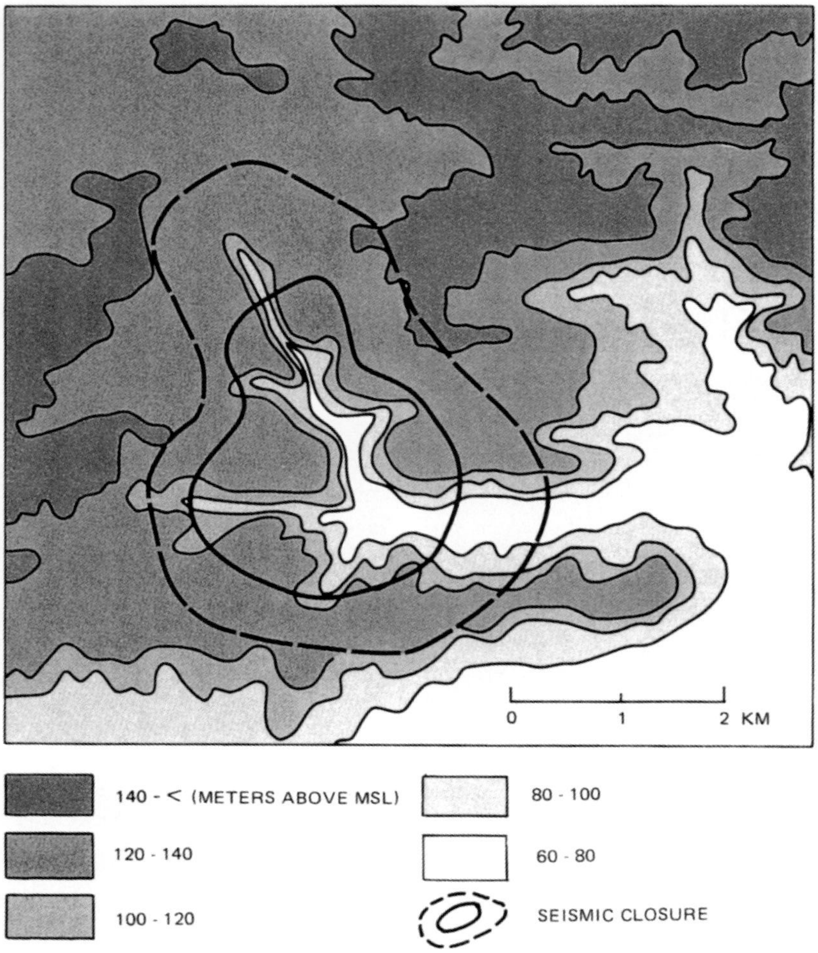

Fig. 13.12. Topographic expression of the Donne Marie Field
A topographic map of the Donne Marie Field illustrating the breaching of its crest and its correspondence with the producing closure

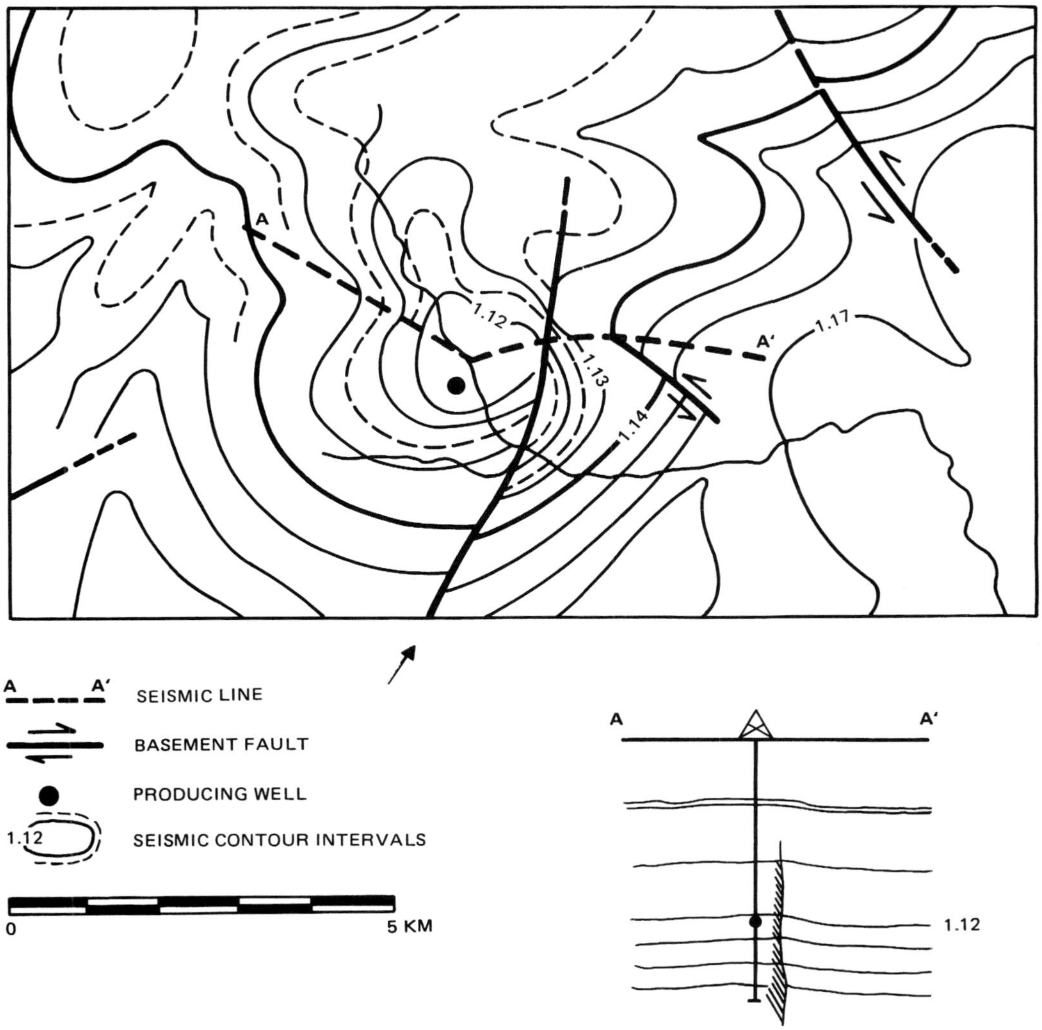

Fig. 13.13. Subsurface expression of the Donne Marie Field Structure contour map (top Triassic) and interpreted seismic line (A-A') of the Donne Marie Field. The possible position of the Metz-Seine Valley lineament suggests that the field is located along the edges of an uplifted basement fault block. The individual trap, however, must have been enhanced by the presence of secondary cross-trending faults which correspond to the local magnetic nose shown earlier in Fig. 13.11. (Data courtesy of Esso Rep)

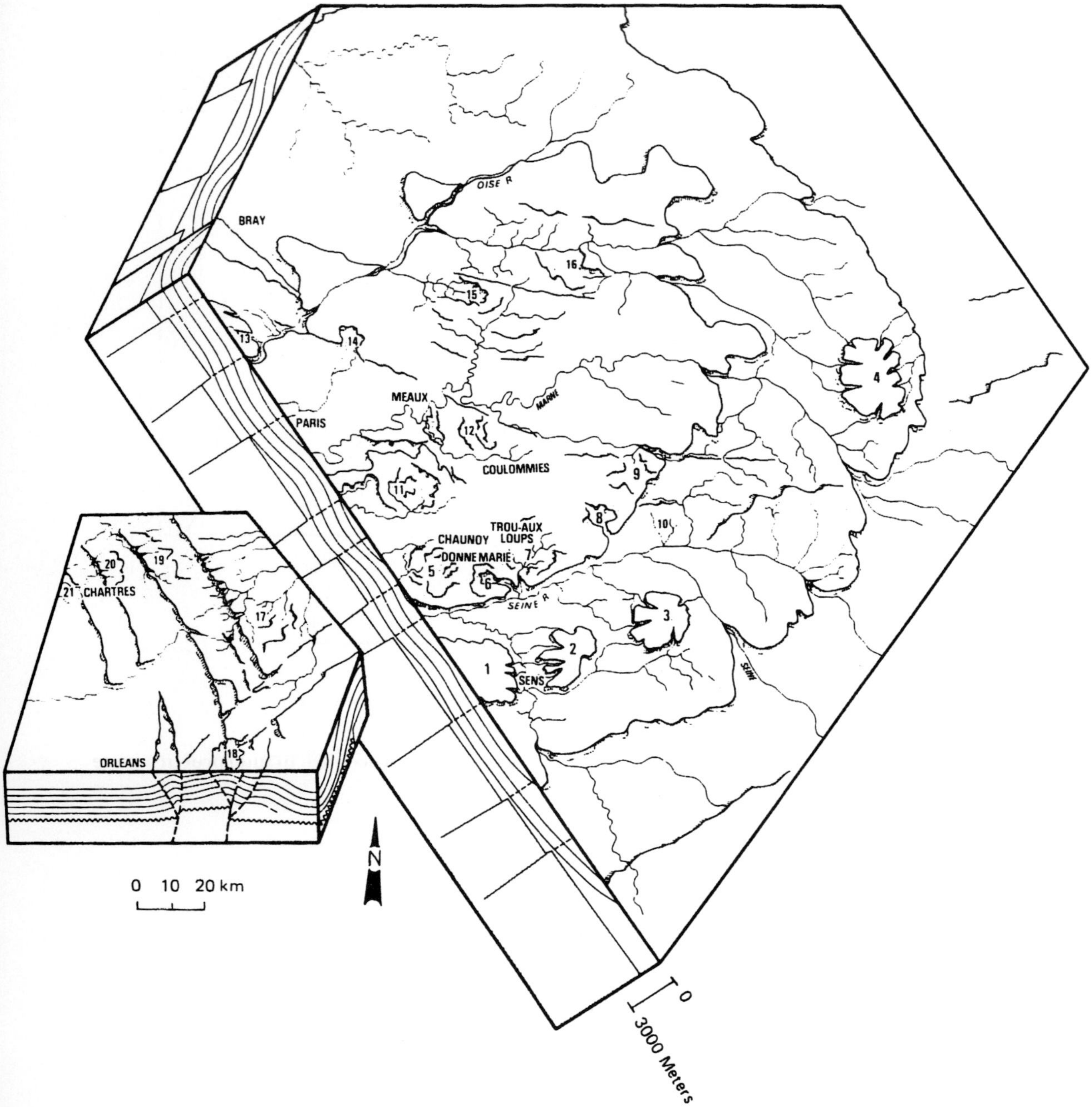

Fig. 13.14. Structure lead map for the Paris Basin
A summary block diagram showing the surface expression of several local structures which are identified in the Paris Basin through the integrated analysis of satellite imagery, gravity, magnetic and seismic data. The anomalies on the main block diagram mostly reflect the development of low-amplitude basement warp structures over uplifted basement fault blocks. The smaller block diagram shows surface expression of structural elements along the Sennely wrench fault (see Fig. 13.1). The number given to each lead does not reflect its ranking in terms of exploration significance

Chapter 14 The East Texas Region

14.1 Background

The East Texas Region is a passive continental margin characterized by extensional tectonics with various structures that develop within it (Figs. 14.1 and 14.2). These include large-scale basement warp structures such as the Sabine Uplift and the Angelina Coldwell flexure, extensional fault blocks and listric normal fault systems such as the Mount Enterprise and Mexia-Talco fault systems and salt-related structures of the northeast Texas basin (Figs. 14.3–14.14; Nichols 1964).

14.2 Objectives

Structures in this basin are either completely obscured by vegetative cover or buried under unconsolidated sediments. Outcropping units are extremely difficult to identify in the field, yet many of these structures manifest detectable expressions on satellite imagery. The main aspects to observe here are (1) detection of large-scale basement warp structures (the Sabine Uplift) through integrated analysis of satellite imagery and potential field data with specific applications of the domal-topographic models; (2) the use of structurally controlled drainage and tonal patterns for detection of buried and obscured faults in this area; (3) reconstruction of salt dome topography and related drainage slope and drainage assemblages; and (4) identification of structural trends and leads.

14.3 Training Instructions

14.3.1 Interpretation Procedures

1. On a clear overlay, trace the stream channels and valleys that are depicted on the satellite imagery of Fig. 14.4. Identify and map major linear features that cross-cut the stream valley causing the development of structurally controlled drainage features such as deflected stream valleys, abrupt changes in floodplain configuration, etc.
2. Place the interpreted overlay on top of Figs. 14.5 and 14.6 and trace the location of exposed and buried fault systems.
3. Repeat the same procedures for the interpretation of the satellite imagery of Fig. 14.8 and the surface and subsurface data of Figs. 14.9 and 14.10.
4. Repeat the same procedure to interpret drainage and slope elements of Fig. 14.12 and the surface data of Fig. 14.13.
5. Use data of Figs. 14.2 and 14.3 to determine the origin of the structural features which were identified in the previous steps. Indicate the origin of these features on the interpreted overlays (e. g., Mount Enterprise fault, salt-related domal feature, etc.
6. Use the data from Fig. 14.2 to make a composite overlay that integrates all your observations.

14.3.2 Questions

1. What are the most dominant surface features that can be used in this area to detect the presence of buried structures?
2. What is the level of correspondence between surface features identified on satellite imagery and surface and subsurface structures that were identified by other mapping techniques?
3. What are the main reasons for potential discrepancies between these features?
4. How can the analysis of the remote-sensing data influence exploration in this region?

References and Further Reading

Berger Z (1982) The use of Landsat data for detection of buried and obscured geologic structures in the East Texas Basin, U. S. A. Int Symp on Remote sensing of the environment, 2nd Thematic Conf, Dalas, TX. Environmental Reserved Institute at Michigan, Ann Arbor, pp. 577–589

Berger Z, Aghassy J (1980) Geomorphic manifestations of salt dome stability. In: Craig RG, Craft JL (eds) Applied geomorphology. 11th Annu Binghampton Geomorphology Symp, 1980. Allen and Unwin, Winchester, MA, pp 72–84

Bunn JR (1949) Lone Star Field. In: Occurrence of oil and gas in northeast Texas. Texas Bureau of Economic Geology, Univ Texas, Austin, pp 195–200

Flawn PF (1968) Palestine sheet (1 : 250 000) Geologic atlas of Texas. Bureau of Economic Geology, Univ Texas, Anstin

Halbouty MT, Halbouty JJ (1982) Relationship between East Texas Field and Sabine Uplift in Texas. AAPG Bull 66 (8): 1042–1054

Nichols PH (1964) The remaining frontiers for exploration in northeast Texas. Trans Gulf Coast Assoc Geol Soc, 14th Conv, pp 7–21

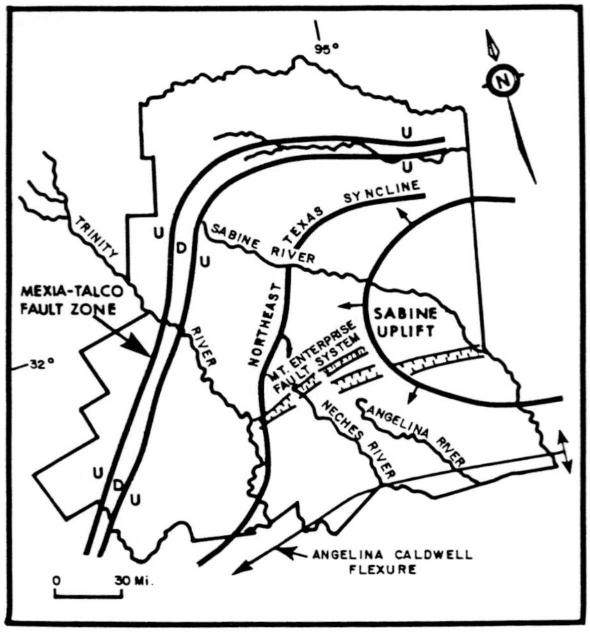

Fig. 14.1 Generalized tectonic map of the East Texas region (Compiled from Nichols 1964; Flawn 1968)

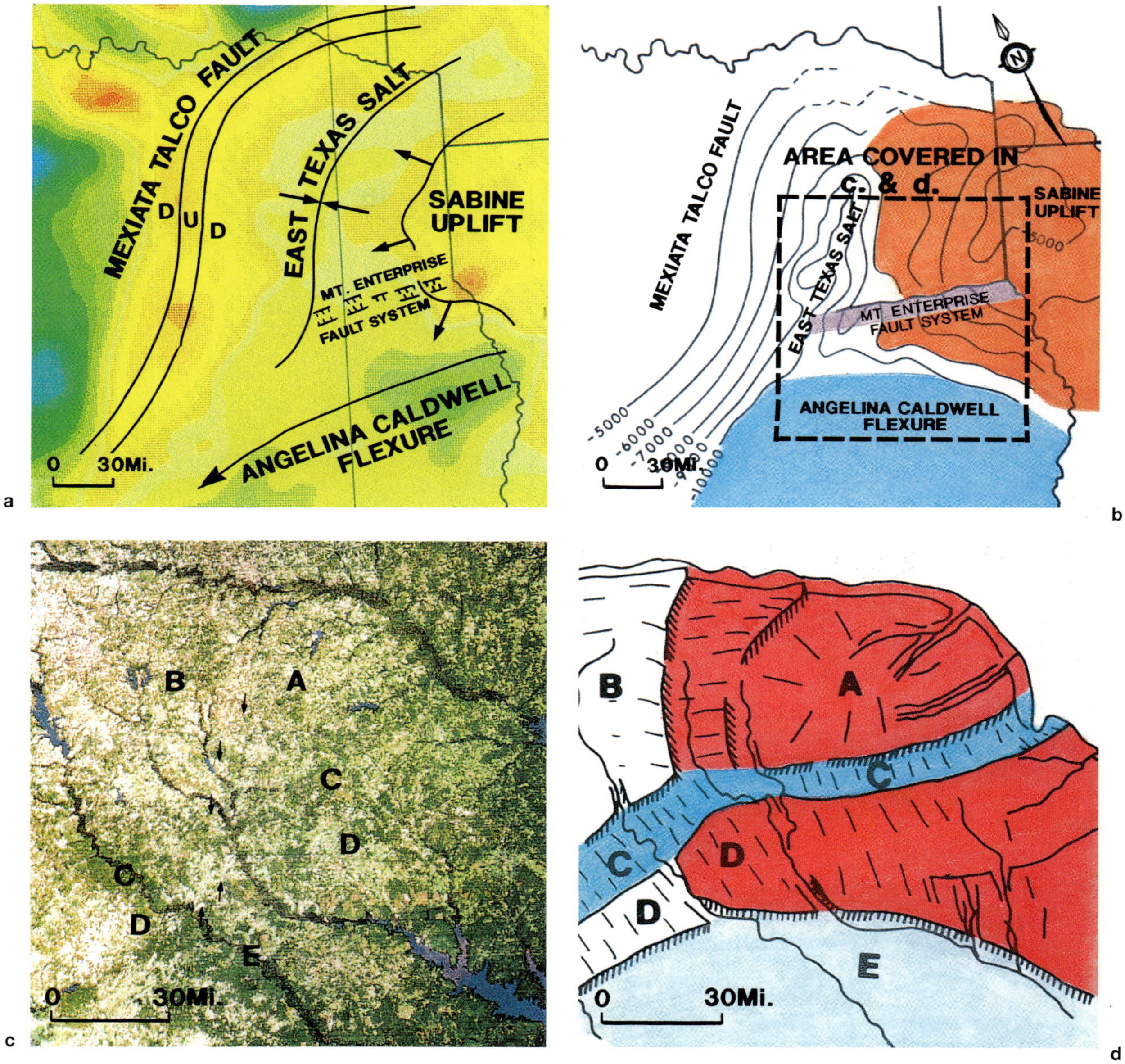

Fig. 14.2. Major tectonic features in the East Texas region
A set of surface and subsurface data which was used to detect the expression of tectonic features in the study area. **a** A reduced tectonic map of the area; **b** Bouguer gravity; **c** satellite imagery; and **d** an integrated surface structure map. *A* The crest of the Sabine Uplift; *B* the tilted limb of the East Texas Salt Basin, *C* the Mount Enterprise fault system; *D* the crest of the Angelina Caldwell flexure; and *E* the outermost topographic scarp of the Gulf of Mexico Coastal Plain region. Note that the Sabine Uplift manifests a typical expression of a domal feature at its early stage of breaching. Also shown on the imagery data, by *small arrows*, are several north-northeast-trending linear features which have not been mapped as structural elements but appear to manifest strong surface expressions. Their possible subsurface origins will be investigated later

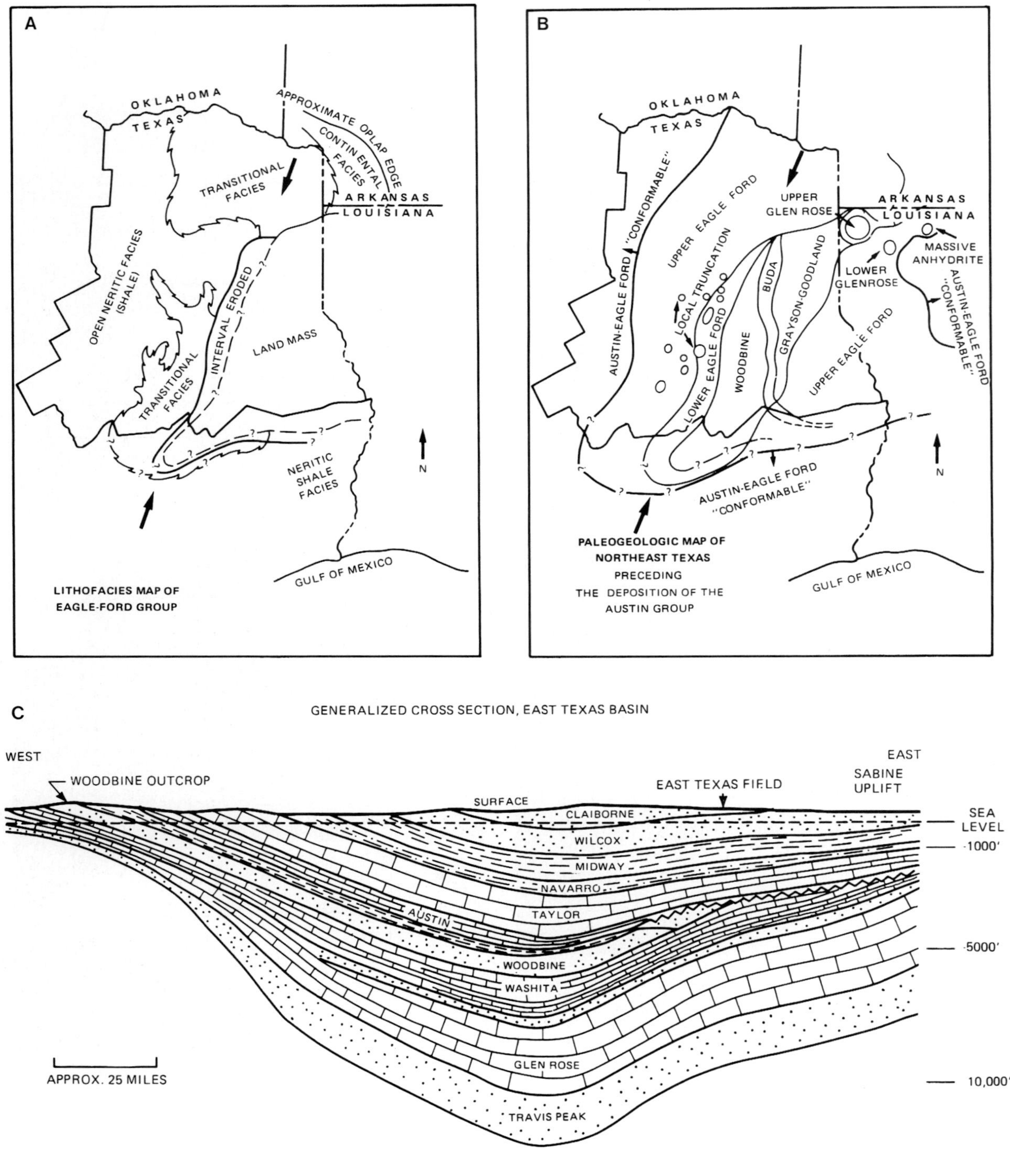

Fig. 14.3. North-northeast-trending geological features in the East Texas region

A Lithofacies map, **B** paleogeologic map; and **C** cross section of northeast Texas show the presence of north-northeast-trending unconformities and related truncated surfaces (*arrows*) developed in response to the rise of the Sabine Uplift during mid-Cretaceous time (Nichols 1964; Halbouty and Halbouty 1982). The north-northeast-trending linear features observed on the imagery in Fig. 14.2 appear to be related to these geological trends. These will be investigated further in the following figures

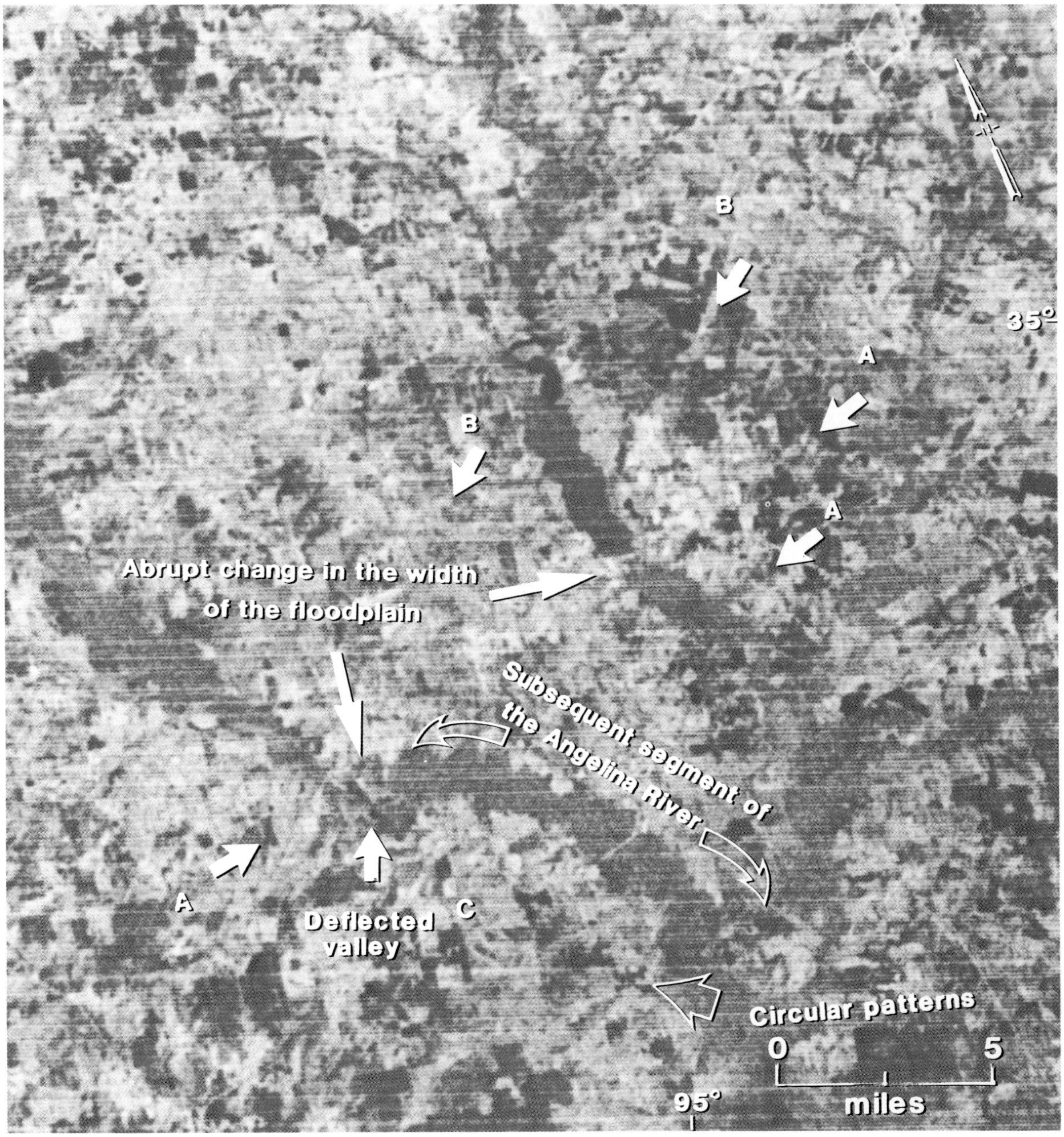

Fig. 14.4. Landsat image interpretation of the Lone Star Field area
A Surface expression of buried and obscured FLTs of the Mt. Enterprise fault zone; **B** a set of linear features believed to reflect the surface manifestation of regional unconformities and related truncated surfaces; **C** a domal structure that is believed to have developed by salt-related processes. Note the abrupt changes in the width of the floodplain and valley of the Angelina River and Striker Creek in places where they are cross-cut by the FLTs of the Mt. Enterprise fault system. The salt-related domal features are depicted by the deflection of the Angelina River which was forced to circum-navigate its northern limbs to form classical marginal streams

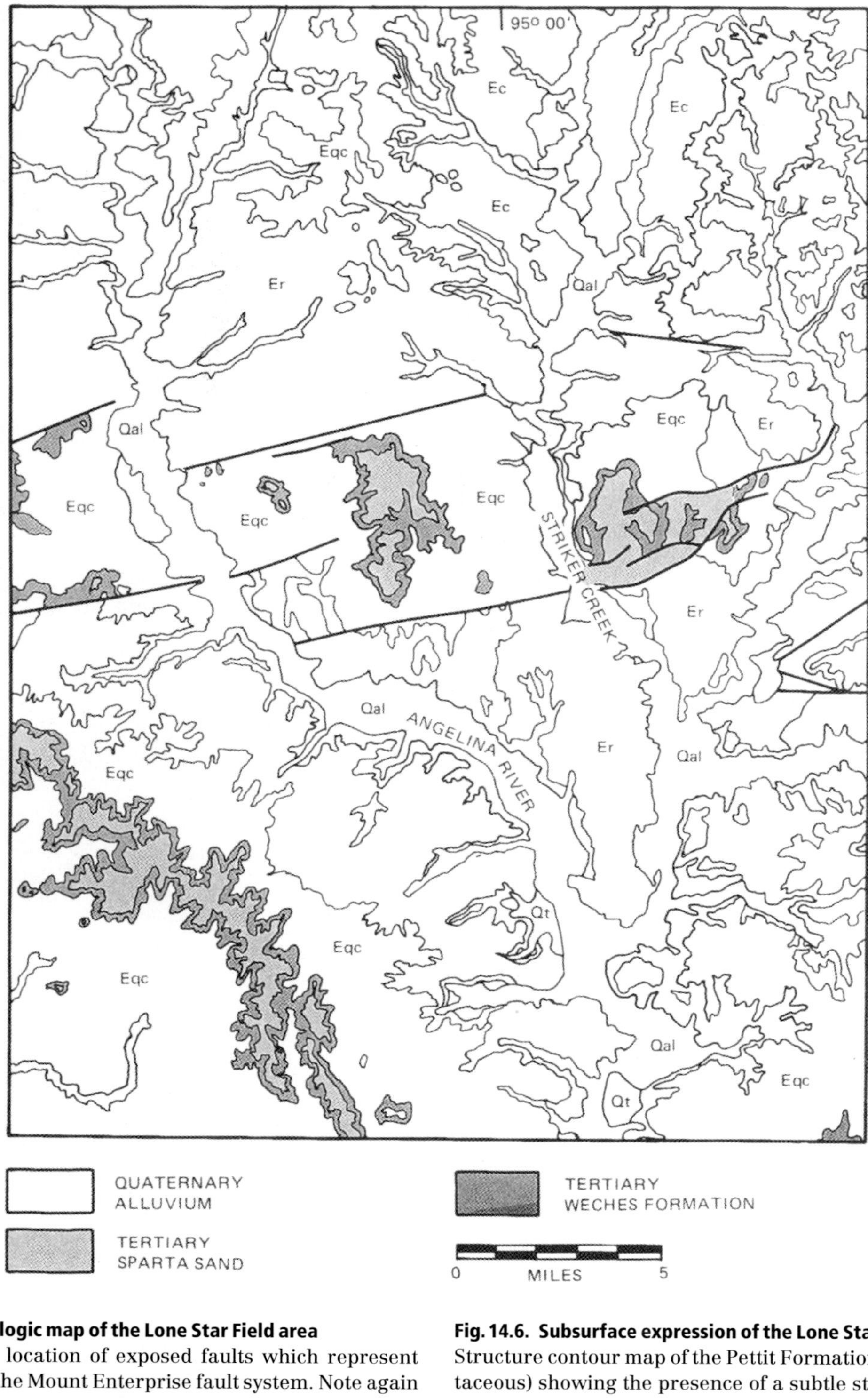

Fig. 14.5. Geologic map of the Lone Star Field area
Showing the location of exposed faults which represent segments of the Mount Enterprise fault system. Note again the relationship between the exposed fault and the abrupt changes in the width of the floodplain. The circular feature shown on the imagery is also noticeable, causing a deflection in the Angelina River which outlines the periphery of the structure in a semicircular pattern. (After Flawn 1968)

Fig. 14.6. Subsurface expression of the Lone Star Field area ▶
Structure contour map of the Pettit Formation (Lower Cretaceous) showing the presence of a subtle structural high which was detected on the imagery. Note that there is no apparent structural manifestation of the cross-trending linear features which were identified on the imagery and were attributed to the regional unconformities. (Courtesy of Exxon Exploration Co.)

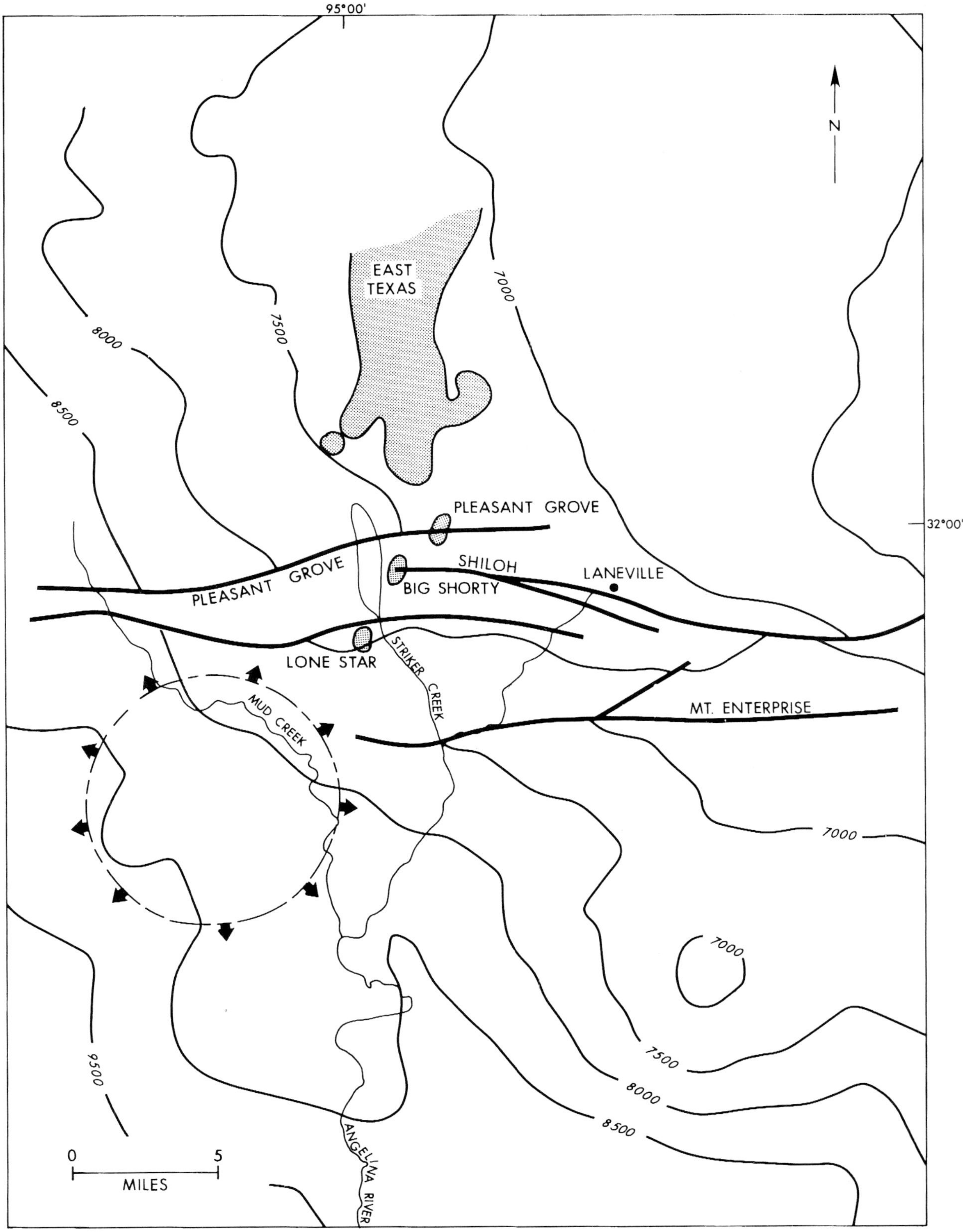
95°00'
32°00'
N
EAST TEXAS
7000
7500
8000
8500
PLEASANT GROVE
PLEASANT GROVE
SHILOH
LANEVILLE
BIG SHORTY
LONE STAR
STRIKER CREEK
MUD CREEK
MT. ENTERPRISE
7000
7000
7500
8000
8500
9500
ANGELINA RIVER
0
5
MILES

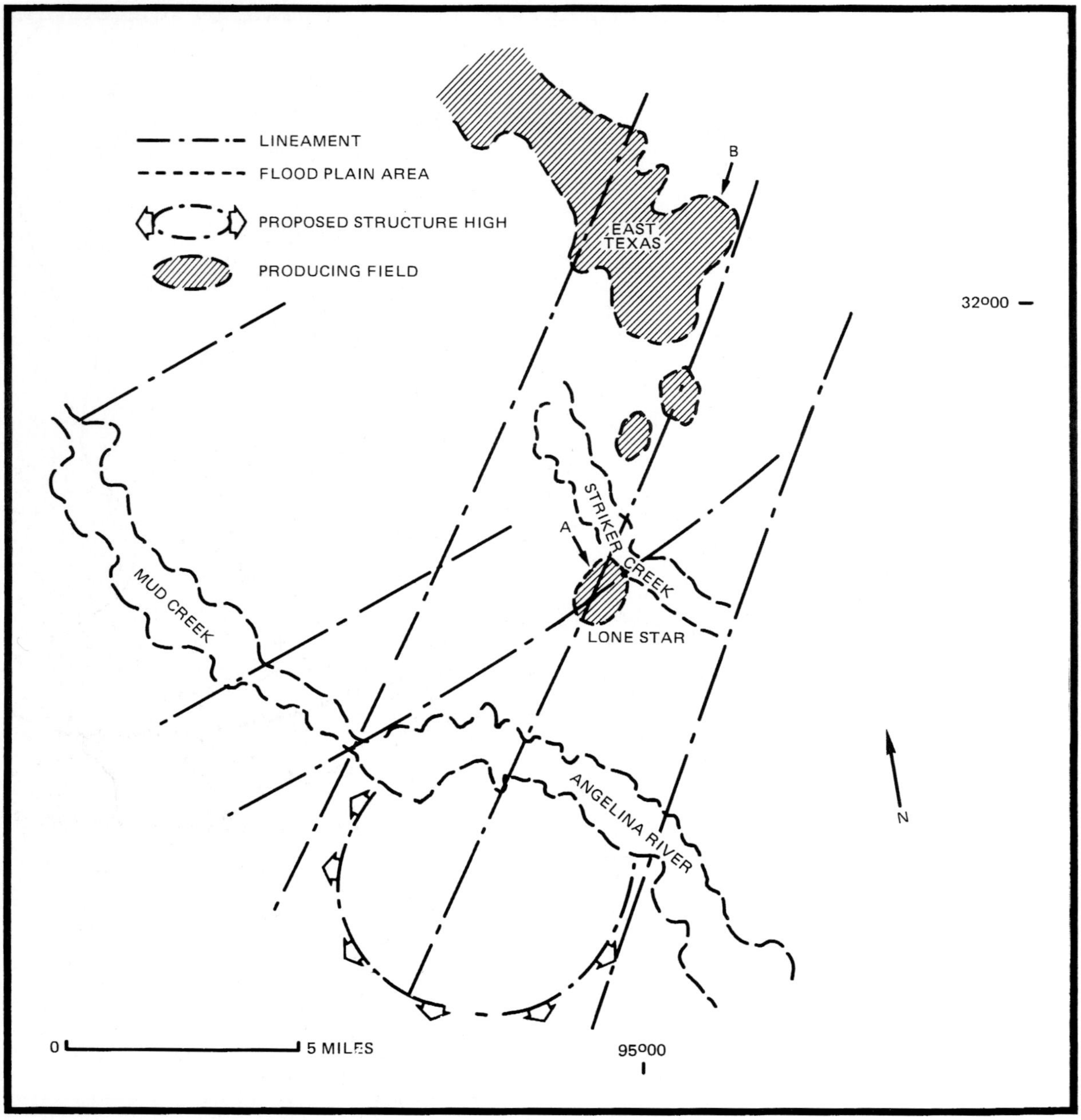

Fig. 14.7. Lead map for the Lone Star area
Generalized map showing the relationship between geological structures interpreted from Landsat and production in the Lone Star Field. Production in this field appears related to traps that occur along the intersection of northeast-trending and north-trending linear features representing faults and unconformities, respectively. Additional prospects are anticipated at similar intersections within this region

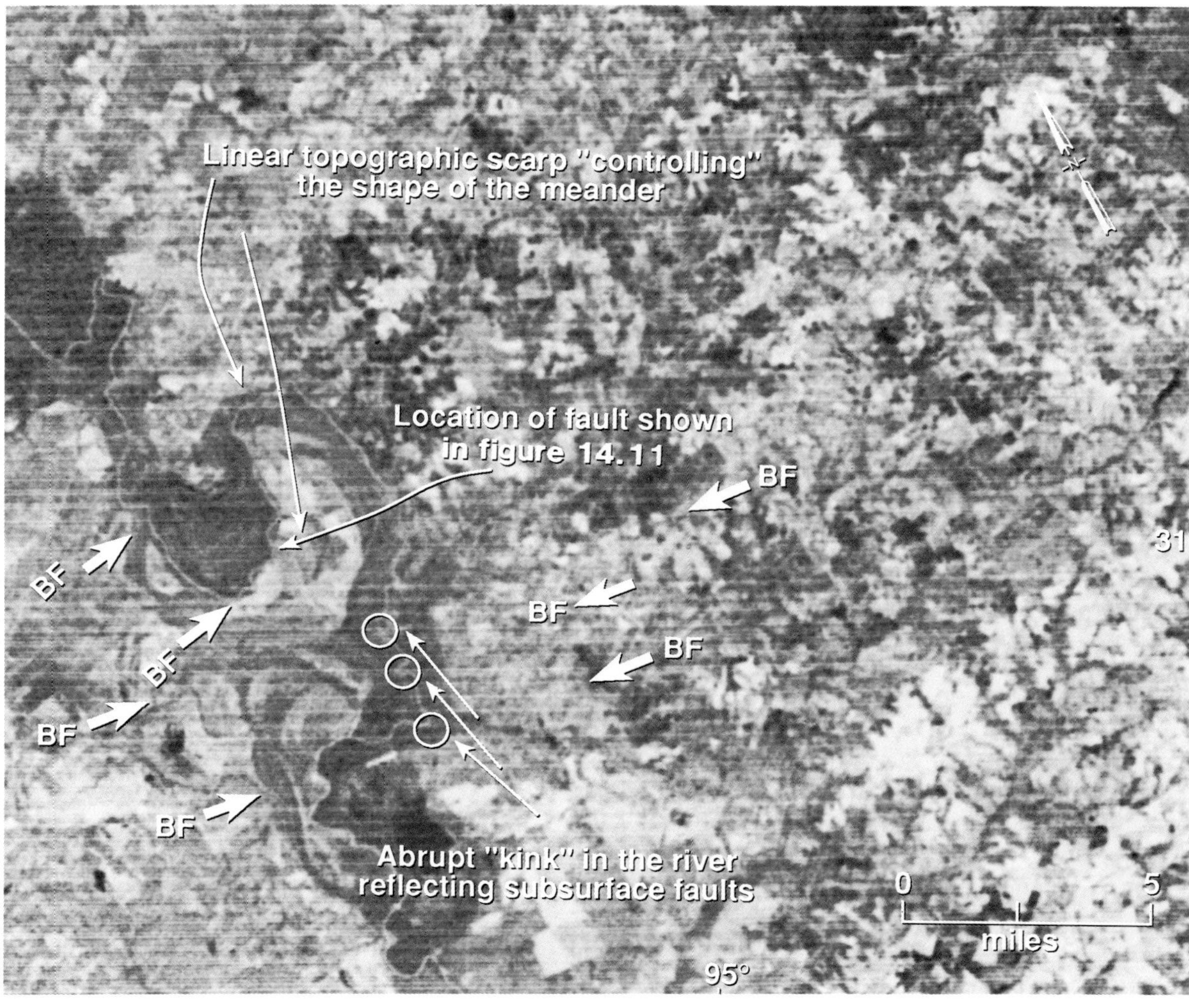

Fig. 14.8. Landsat interpretation of the Elkhart Graben area
Interpretation of a satellite image of the buried extension of the Elkhart Graben area. Linear features which represent the surface expression of subsurface faults exert significant influence on the characteristics of the Trinity River channel orientation and floodplain configuration

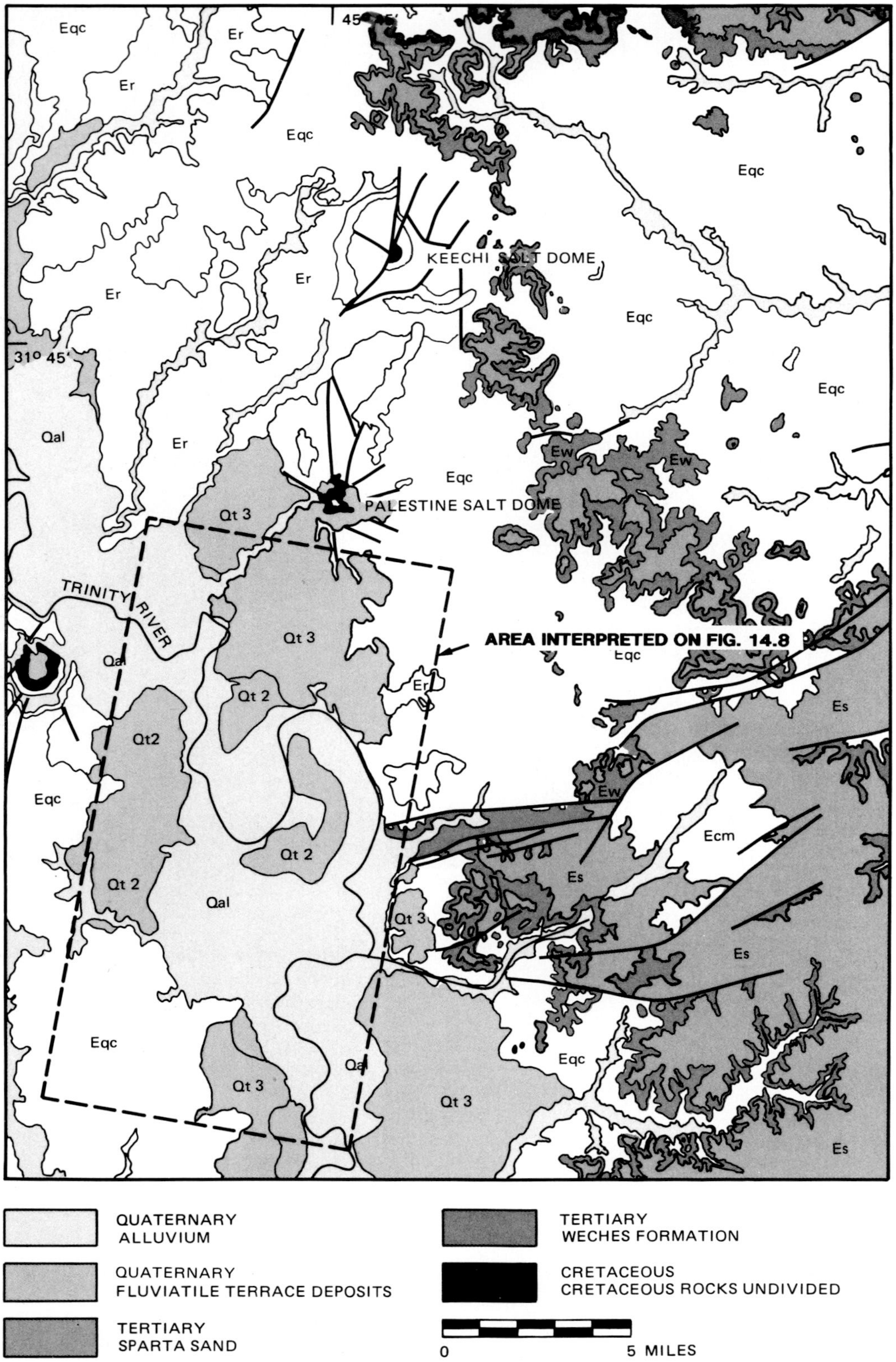

Eqc
Er
Eqc
Eqc
KEECHI SALT DOME
Er
Er
Eqc
31° 45'
Eqc
Qal
Er
Eqc
PALESTINE SALT DOME
Ew
Ew
Qt 3
TRINITY RIVER
Qt 3
AREA INTERPRETED ON FIG. 14.8
Eqc
Qal
Qt 2
Er
Es
Qt2
Ew
Eqc
Ecm
Qt 2
Qt 2
Es
Qal
Qt 3
Es
Eqc
Eqc
Qal
Qt 3
Qt 3
Es
QUATERNARY ALLUVIUM
QUATERNARY FLUVIATILE TERRACE DEPOSITS
TERTIARY SPARTA SAND
TERTIARY WECHES FORMATION
CRETACEOUS CRETACEOUS ROCKS UNDIVIDED
0
5 MILES

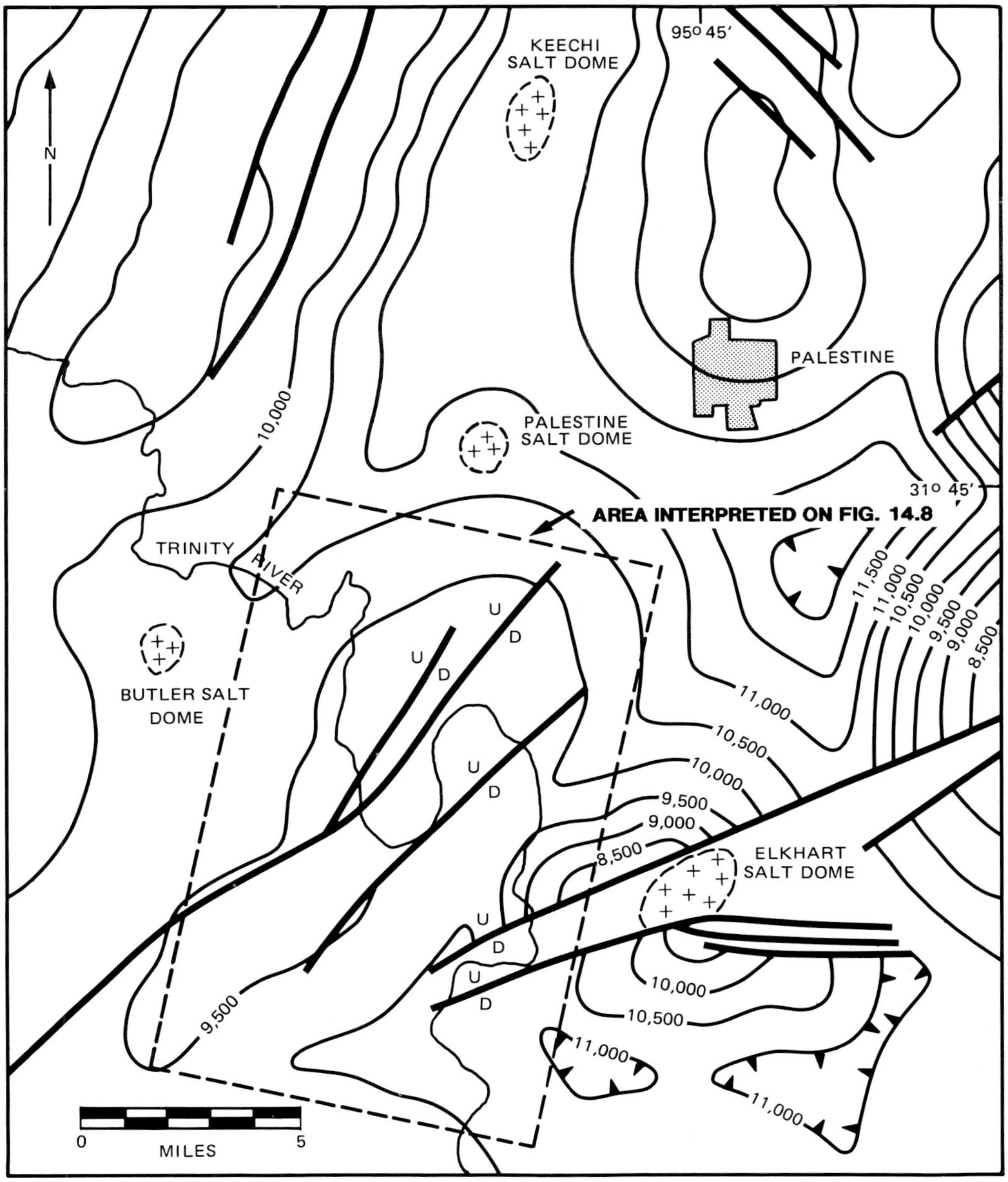

◀ **Fig. 14.9. Geologic map of the Elkhart Graben area**
Showing the surface expression of the Elkhart Graben which represents portions of the Mount Enterprise fault system. Note the absence of faults in the Trinity River floodplain area in spite of their strong surface expression on imagery. (After Flawn 1968)

Fig. 14.10. Subsurface expression of the Elkhart Graben
Surface structure map of the Pettit Formation (Lower Cretaceous) showing the extension of the Elkhart Graben into the Trinity River floodplain area. Note the strong correspondence of mapped faults with the linear features interpreted from Landsat data in Fig. 14.8. (Courtesy of Exxon Exploration Co.)

Fig. 14.11. Surface expression of obscured faults

Outcrop displays faulting of Eocene and Quaternary sediments in the vicinity of the Elkhart Graben area. The photograph, which was taken along the valley wall of the Trinity River, illustrates the presence of faulting in this area which was not depicted by surface maps but was clearly observed on the imagery. See small box in Fig. 14.8. (Photographed by C. A. Dengo, Exxon Exploration Co.)

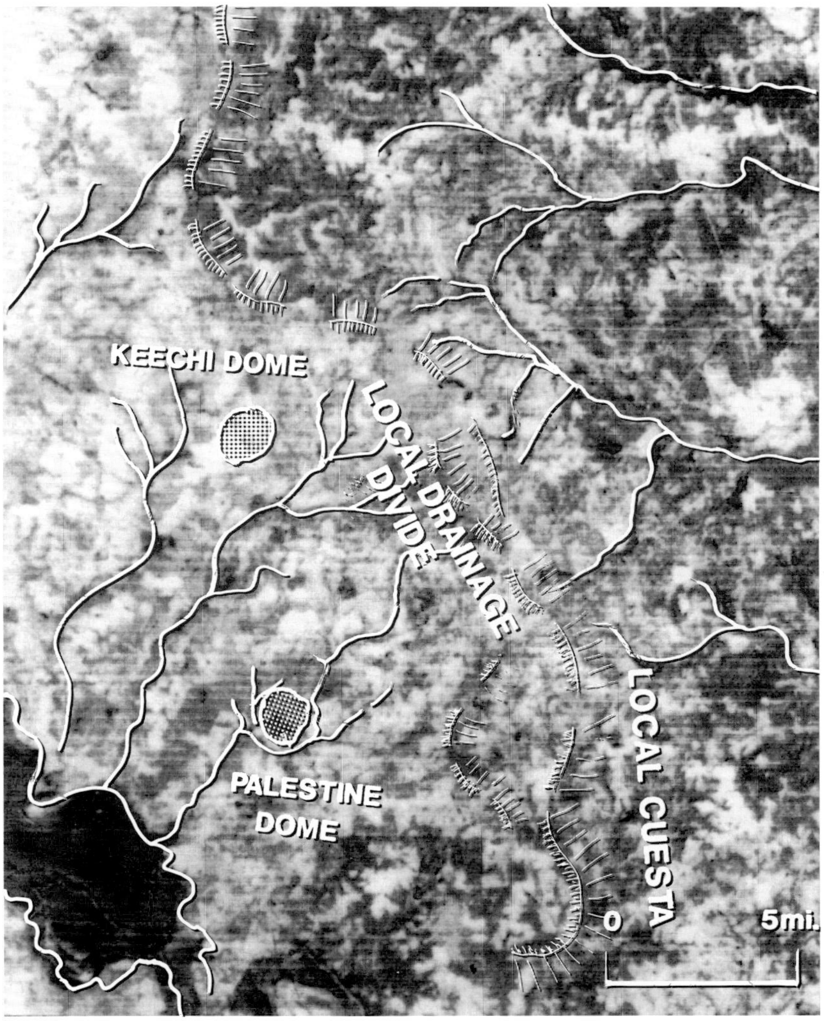

Fig. 14.12. Surface expression of salt domes

Satellite imagery subscene and structural interpretation of the Palestine area. A topographic escarpment forms the eastern limit of an obliterated topography that resulted from intense erosion above the Keechi and Palestine salt domes. The peripheries of the domes are enhanced by the presence of structurally controlled stream valleys. The topographic expression of the escarpment and the salt domes is further illustrated in Fig. 14.14.

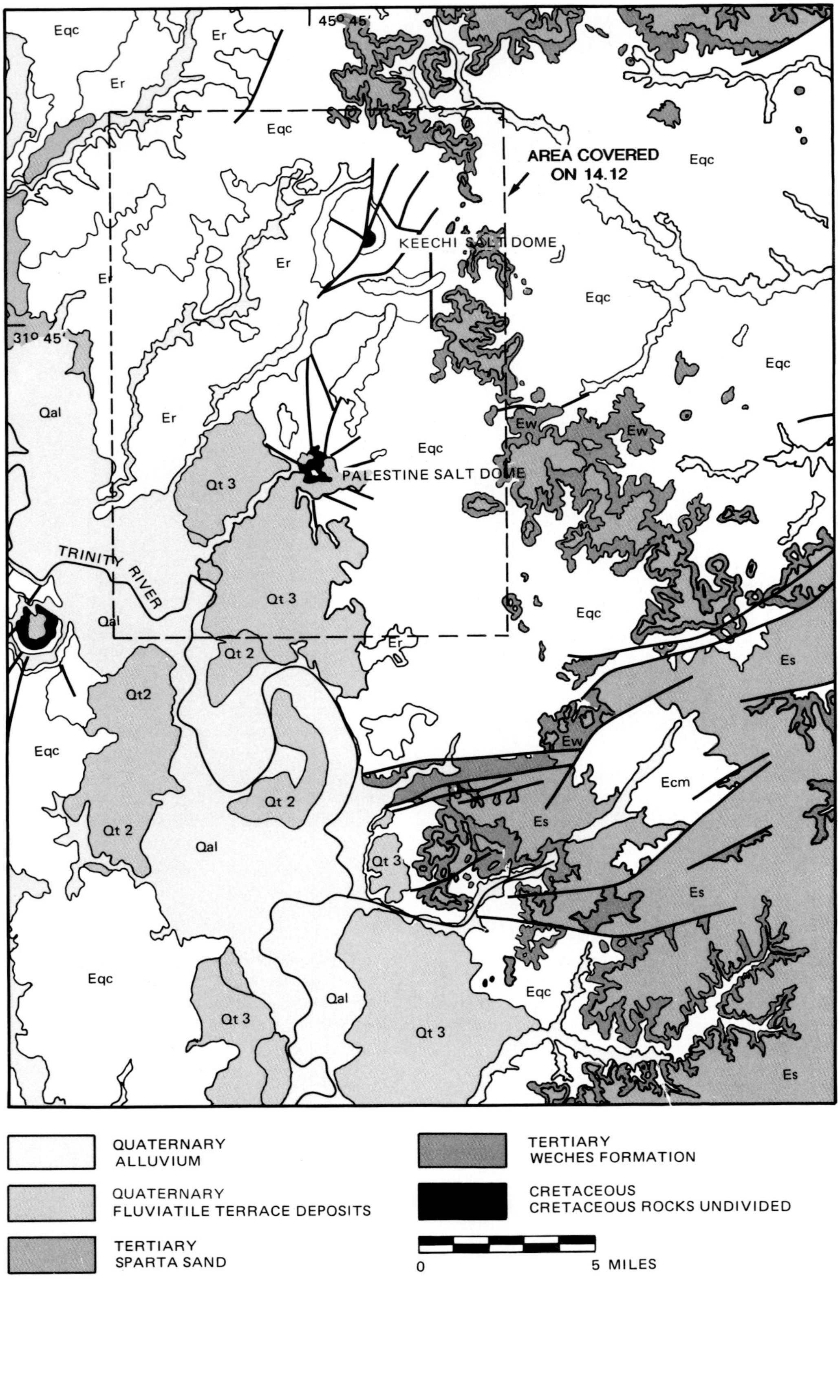
Eqc
Er
45° 45'
AREA COVERED ON 14.12
KEECHI SALT DOME
31° 45'
Qal
Ew
PALESTINE SALT DOME
Qt 3
TRINITY RIVER
Qt 2
Es
Ecm
QUATERNARY ALLUVIUM
QUATERNARY FLUVIATILE TERRACE DEPOSITS
TERTIARY SPARTA SAND
TERTIARY WECHES FORMATION
CRETACEOUS CRETACEOUS ROCKS UNDIVIDED
0
5 MILES

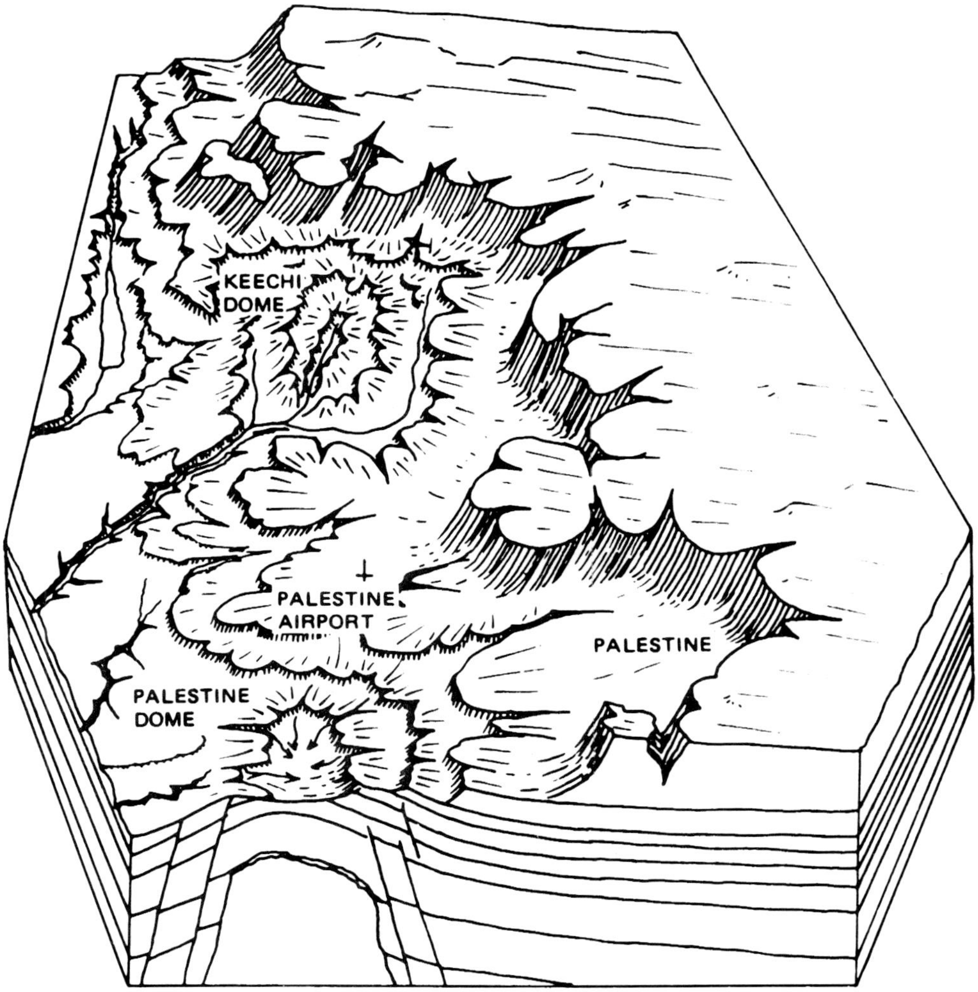

◀ **Fig. 14.13. Geological expression of salt domes**
Geological map of the Palestine area shows the Keechi and Palestine domes and the intense radial faulting associated with their diapirs. Note the receding of the Tertiary Weches formation in this area forming a local erosional embayment which enhances the presence of these salt domes in the area

Fig. 14.14. Topographic expression of salt domes
Drainage patterns and slope assemblages resulting from erosion of the Palestine and Keechi domes. Data was compiled from Landsat imagery, low-altitude aerial photographs and topographic maps. Schematic vertical exaggeration is approximately 10x. (After Berger and Aghassy 1982)

Appendices

A. List of Symbols Used in the Interpretation of Imagery Data

Fig. A.1

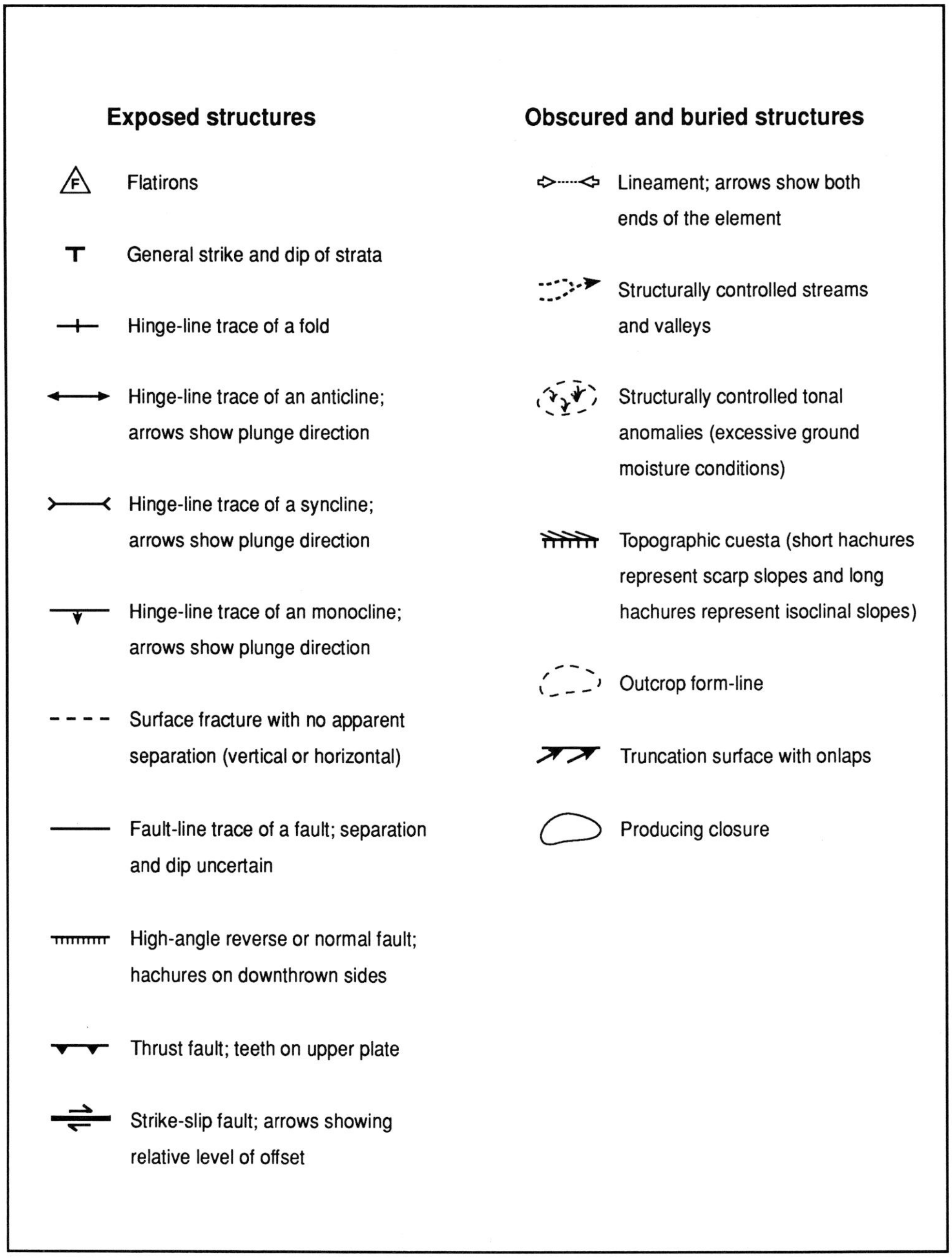

B. Abbreviations Used on Image

Fig. B.1

AD	Anomalous drainage density	IN	Inverted FLT
AF	Alluvial fan	IS	Interrupted slope
AM	Abandoned meander	LB	Lithostratigraphic boundary
BDD	Buried dip direction	LN	Linear negative FLT
BA	Bright anomaly	LO	Look direction
BF	Buried fault	LT	Lineament
BM	Breached monocline	MB	Marker beds
BS	Breached structure	MF	Multidirectional fault
BSV	Box-shaped valley	MM	Man-made feature
BWS	Basement warp structure	MS	Marginal stream
CA	Cold area	NF	Negative FLT
CP	Concentric pattern	OS	Obsequent stream
CS	Cone-shaped valley	PC	Parallel composite FLT
CSP	Compressional splay	PP	Polyphase FLT
CU	Cuesta	PF	Pseudo-FLT
DA	Dark anomally	PN	Plunging nose
DD	Drainage divide	RD	Radial drainage
DL	Dogleg	SC	Structural closure
DP	Drilling pads	SD	Slope direction
DS	Deflected stream	SF	Sinuous FLT
DV	Deflected valley	S&P	Sapping and piping
EB	Erosional breaching	SR	Shuttered ridges
EDD	Exposed dip direction	SS	Subsequent stream
EH	Erosional halo	TB	Turtleback structure
FA	Fold axis	TD	Trapdoor
FB	Fault boundary	TF	Triangular facets
FI	Flatirons	TO	Termination by onlap
FLT	Fault-line trace	TRF	Tear fault
FP	Floodplain	TS	Truncated surface
FR	Fractures	TST	Truncated structure
GD	Groundwater discharge	TV	Transverse valley
HV	Heavy vegetative cover	WA	Warm areas

C. Drainage Patterns

Radial Drainage Patterns (Related to Folds)

Fig. C.1

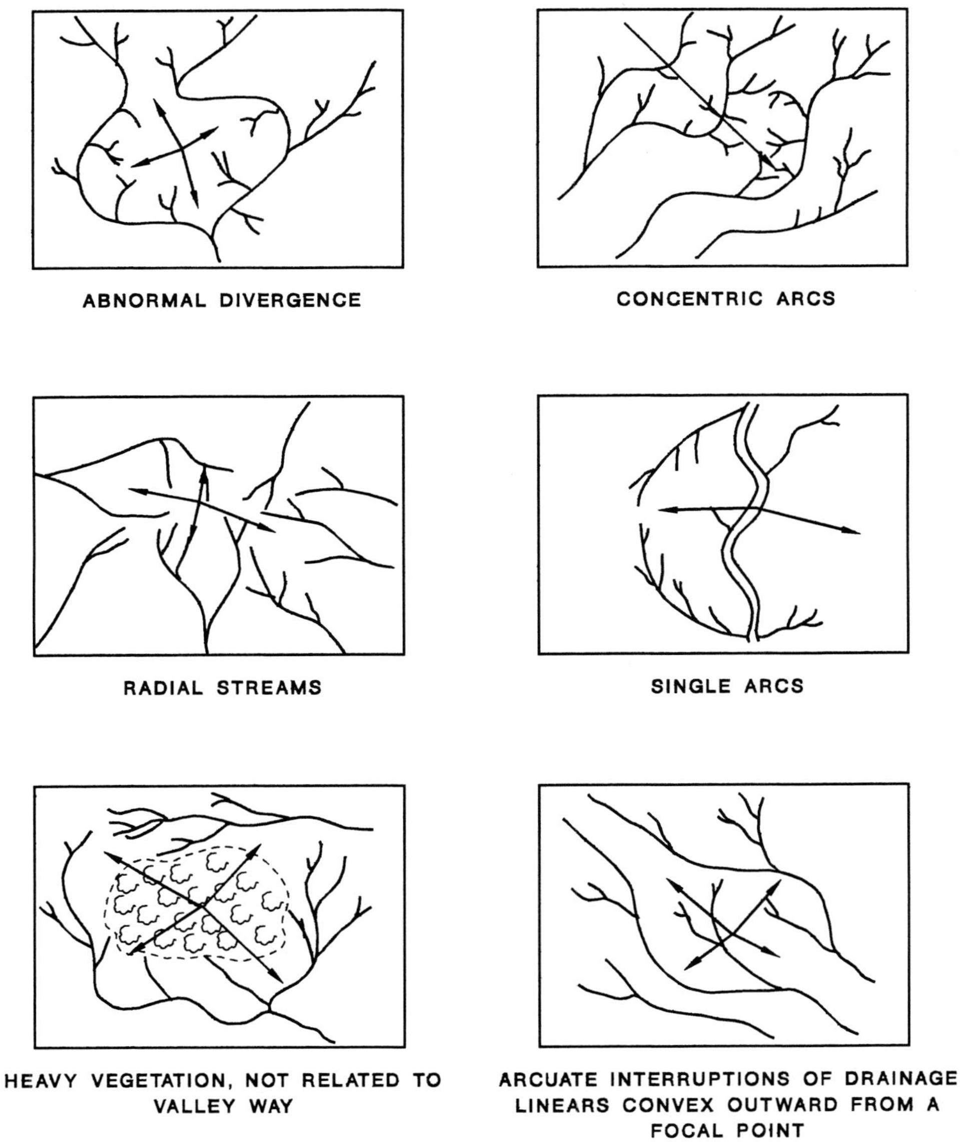

Local Increase in Drainage Densities

Fig. C.2

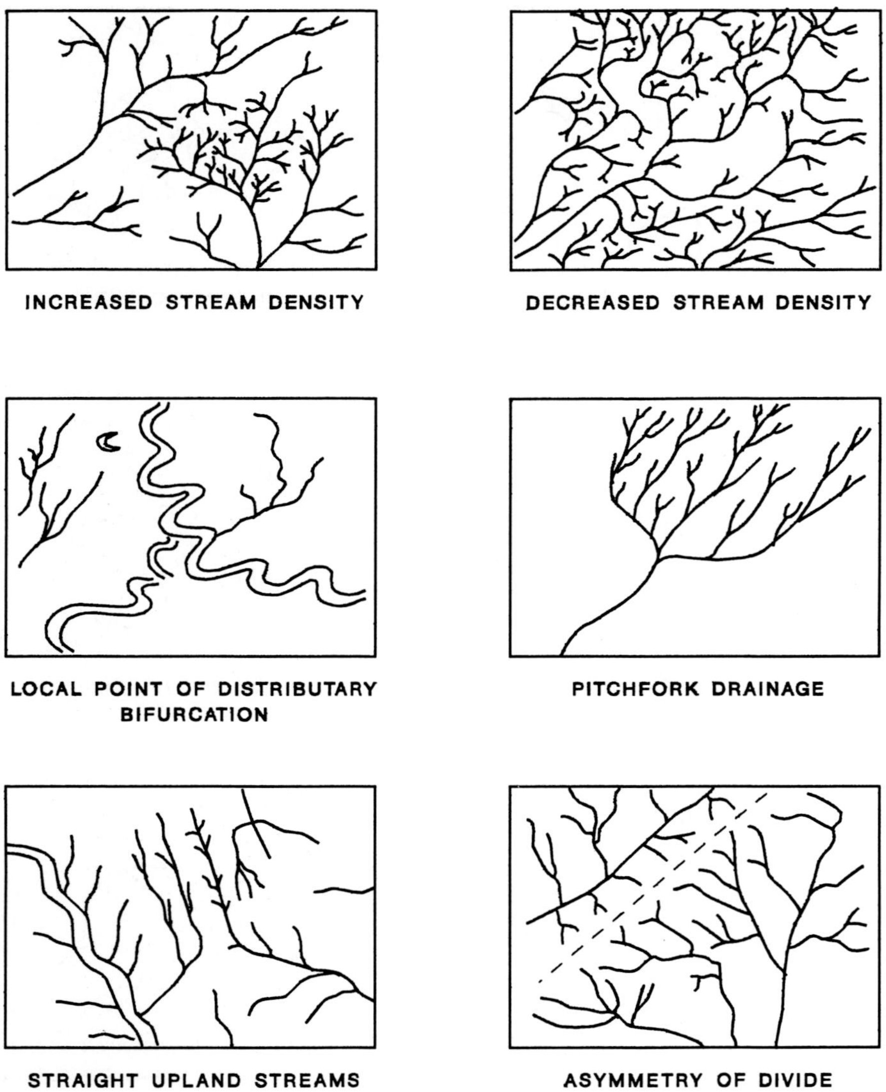

Local Drainage Anomalies (Related to Cross-Trending-Structures)

Fig. C.3

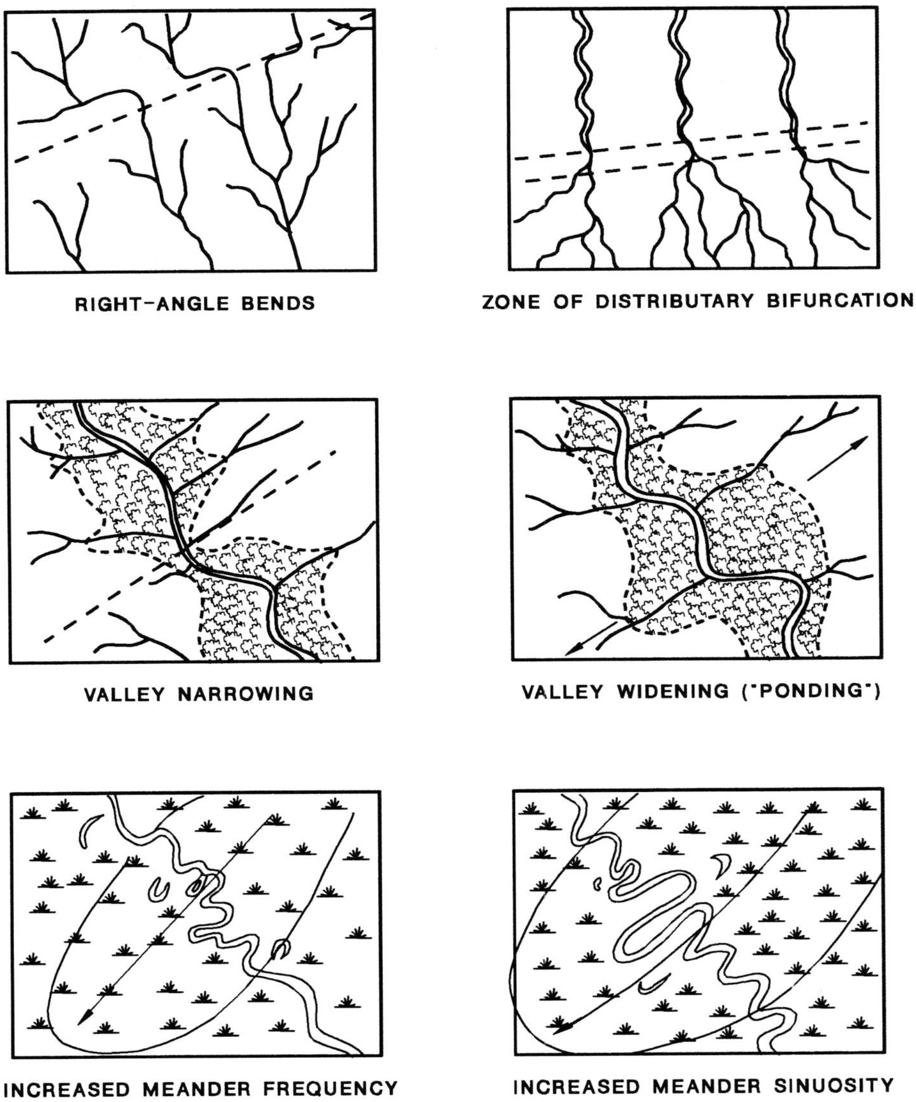

Linear Drainage Patterns (Related to Fault and Fractures)

Fig. C.4

A

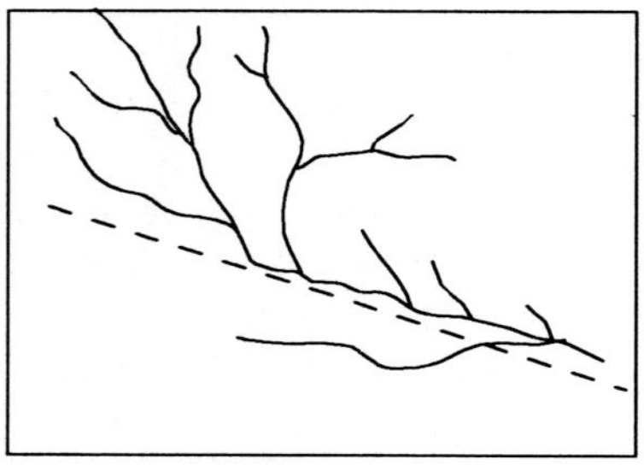

INTERSECTION OF PROMINENT DRAINAGE LINEARS

B

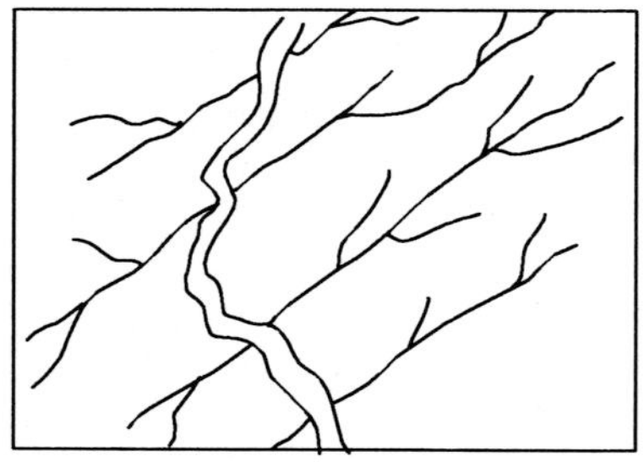

180° TRIBUTARIES

C

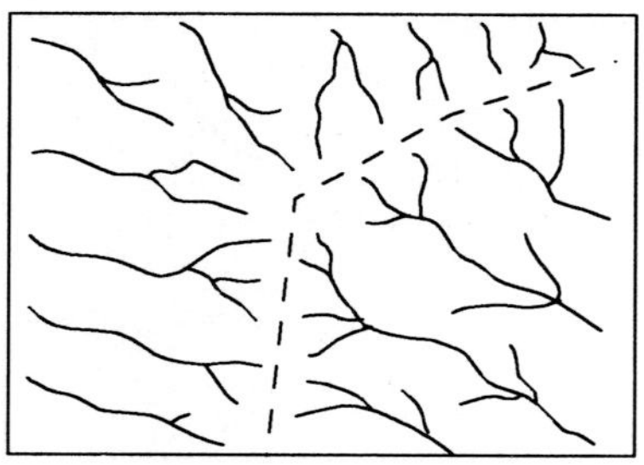

DEVIATION OF DIVIDE

D

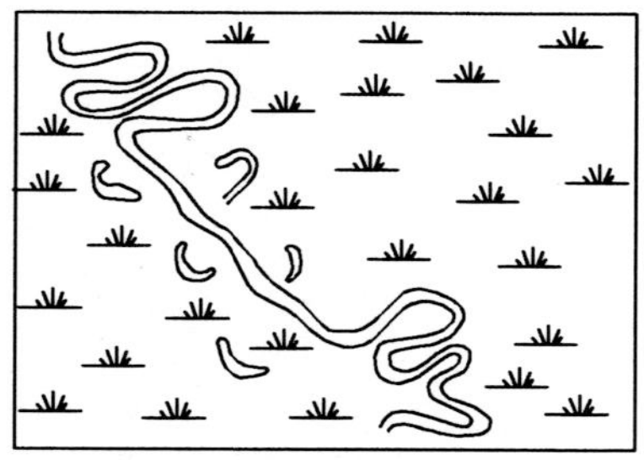

STRAIGHTENED MEANDERS

E

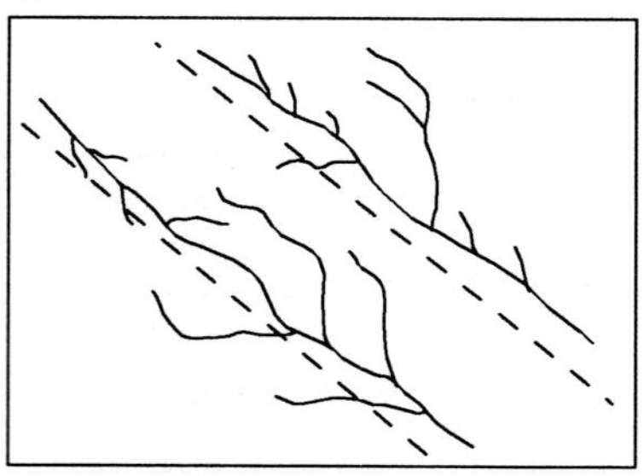

DRAINAGE LINEAR OR LINEARS IN COMBINATION WITH ANOTHER ELEMENT

F

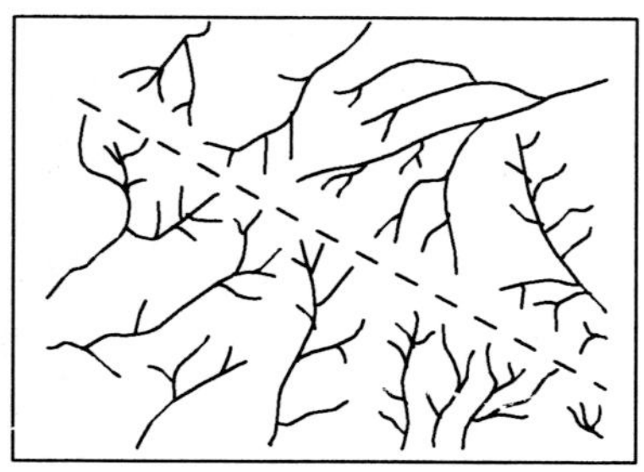

LONG, STRAIGHT DIVIDE

D. Major Sources of Images

EOSAT

In 1984, the Reagan administration successfully introduced legislation to transfer operation responsibility of the Landsat systems from the National Oceanic and Atmospheric Association (NOAA) to a commercial organization. The consortium accepted by the US Congress was the Earth Observation Satellite Company (EOSAT). The extensive library of Landsat data is now available from the EOSAT archive and can be purchased by contacting the Customer Service Department in Lanham, Maryland. A listing of available scenes can also be obtained from EOSAT. This service is free and covers all types of Landsat data (acquired since 1972). To request a search, it is necessary to provide one of the following:

- the latitude and longitude coordinates for the corner points of the study area,
- a specific location such as a city name or prominent geographical feature, or
- a satellite Path/Row number which is available from Worldwide Reference System (WRS) maps or other published Path/Row sources (Fig. D.1). WRS Path/Row maps are available free from EOSAT. (It is important to note that Landsats 1, 2 and 3 followed a different Path/Row system than Landsats 4 and 5.)

It is also possible to specify maximum acceptable cloud cover, minimum acceptable image quality and the time of year preferred. Further information on EOSAT products can be obtained from their Customer Service Department, International Headquaters, 4300 Forbes Boulevard, Lanham, Maryland 20706, USA. Phone: (301) 552-0500 or (800) 344-9933. Fax: (301)552-0572.

Fig. D.1. Sample of a World Reference System Path/Row map (Courtesy of EOSAT)

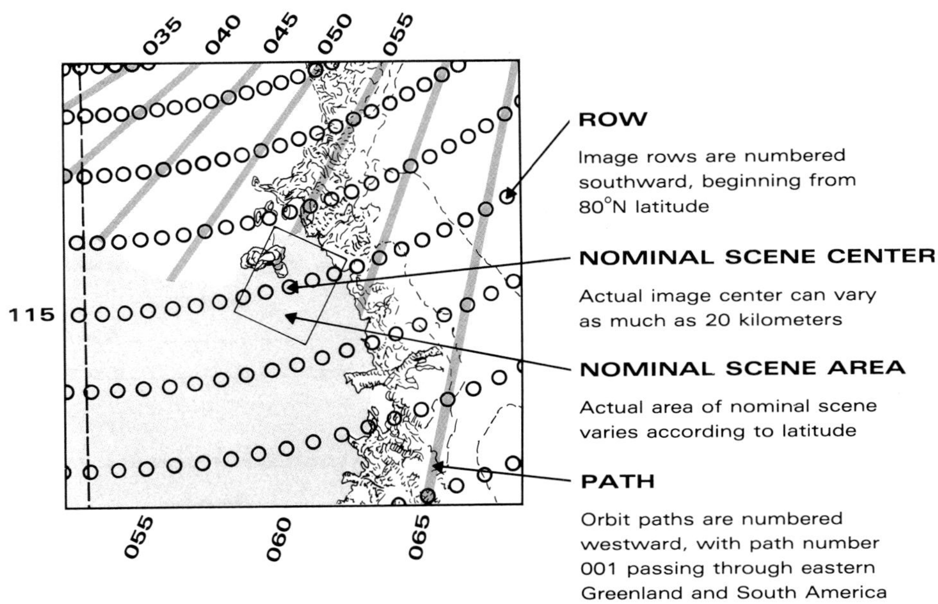

SPOT

The SPOT satellite is owned and operated by the Centre National d'Etudes Spatials (CNES), the French national space agency. Worldwide commercial operations are anchored by private companies including SPOT Image Corporation in the USA, SPOT Image in France, Satimage in Sweden and 60 other distributors from around the world. Product orders are placed by completing a Licence Request Form and specifying the scene parameters (area of coverage, scene characteristics, deliverable medium) for the requested SPOT products. SPOT Image maintains a computerized archive of over 3 million scenes which is updated daily and includes a digital quick-look system called DALI (Device to Access and Look at SPOT Inventory). DALI searches are specified by defining the area of interest as a polygon whose vertices are given as latitude and longitude coordinates. If the required imagery is not on file, users can submit a programming request to SPOT Image. The survey area can be designated using:

- a Grid Reference System (GRS) map (see Figs. D.2, D.3) which divides the Earth's surface into grid sections in which the columns (K) are parallel to the satellite ground tracks and the rows (J) are lines of latitude (the "K, J" designation corresponds to the GRS nodes closest to the center of the scene),
- a polygon whose vertices are given as standard latitude/longitude points, or
- a circle of a specified diameter with a center given by latitude and longitude.

Users also specify:

- the desired date or range of dates for the survey to take place,
- whether images are desired in panchromatic mode (with 10-m resolution) or multispectral mode (with 20-m resolution),
- if images should be captured at near-vertical angles or if off-nadir imaging is acceptable,
- the desired base-to-height ratio (if stereoscopic imagery is required), and
- the level of processing and output format (available choices include computer-compatible tape, CD-ROM and paper prints).

Further information can be obtained from SPOT Image S. A., 5 rue de Satellite, P. O. Box 4359, F31030 Toulouse Cedex, France. Telephone: (33) 63 19 40 40. Fax: (33) 62 19 40 11, or from Spot Image Corporation, 1897 Prestion White Drive, Reston, Virginia 22091, USA. Telephone: (703) 620-2200. Fax: (703) 648-1813.

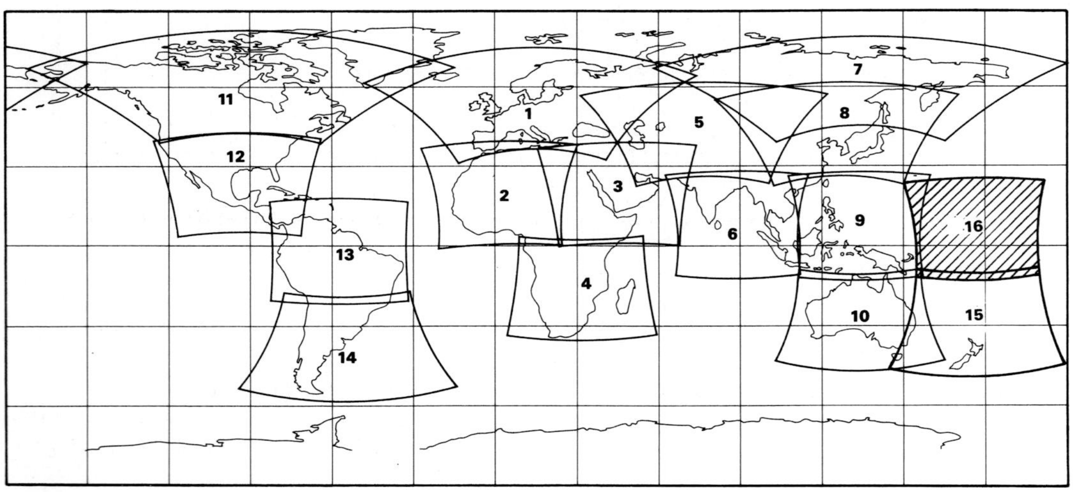

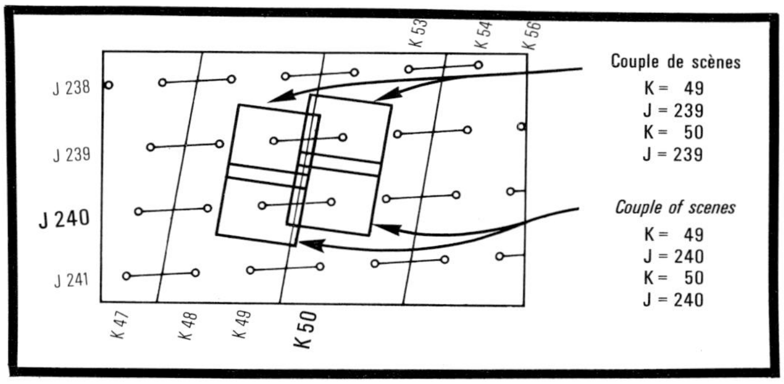

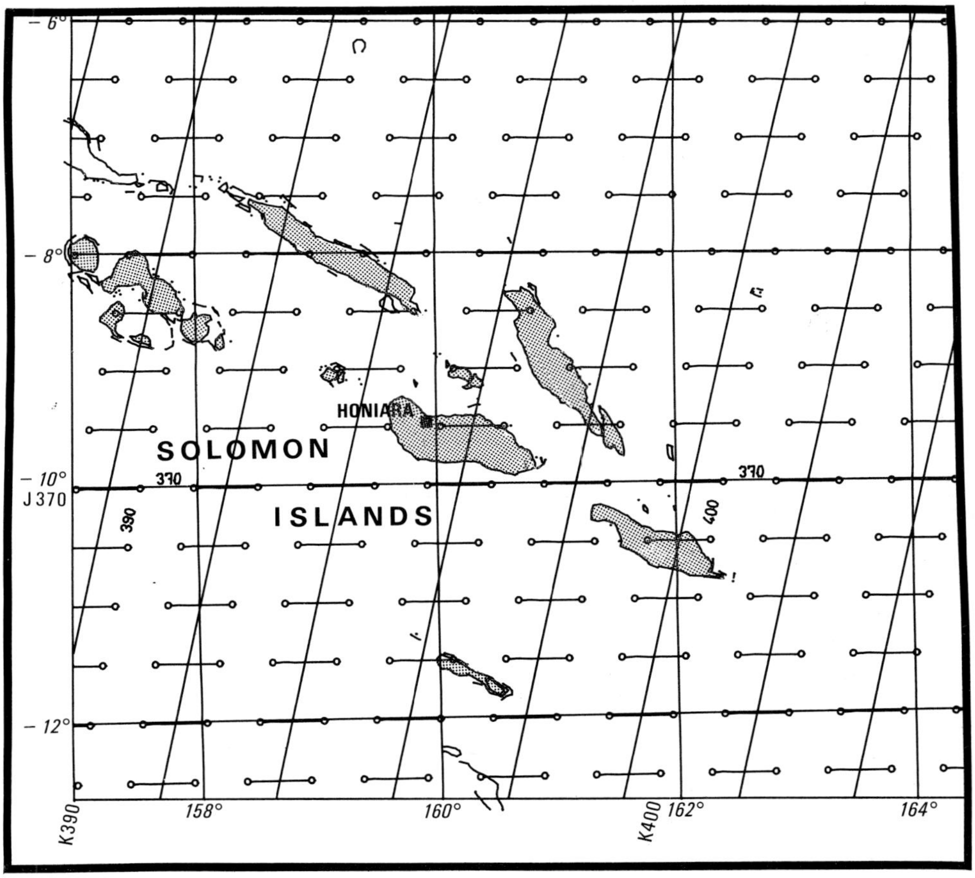

◀ **Fig. D.2. Layout of the SPOT Grid Reference System**
(Courtesy of SPOT Image)

Fig. D.3. Using the SPOT GRS sheets
Above A sample of specifying a scene using J and K coordinates. *Below* a sample of a GRS scheet covering the South Pacific. (Courtesy of Spot Image)

Intera

Intera Information Technologies is the only privately owned remote sensing company that actively collects and processes airborne synthetic-aperture radar imagery for hydrocarbon exploration and other purposes. Intera gathers data using its own STAR-1 and STAR-2 radar systems which are flown on board a Conquest aircraft. Radar images are available in either high-resolution mode (6-m resolution, 23-km swath width) or standard resolution (12- to 25-m resolution, and 46-to 200-km swath width). Data are available, in some key areas, on a nonexclusive basis or can be acquired exclusively for a client in other areas (Fig. D.4). Intera also acquires airborne thermal scanner and aeromagnetic data as well as specialized aerial photography. Further information on Intera's SAR and other remote sensing products can be obtained from Intera Information Technologies, 1000, 645 6th Ave. S.W., Calgary, Alberta, Canada, T2P 4G8. Telephone: (403) 266-0900. Fax: (403)-265-0499.

Fig. D.4. Areas of nonexclusive stereo radar coverage
In some cases, only part of the country indicated is covered. *Asterisk* indicates availability of high-resolution data

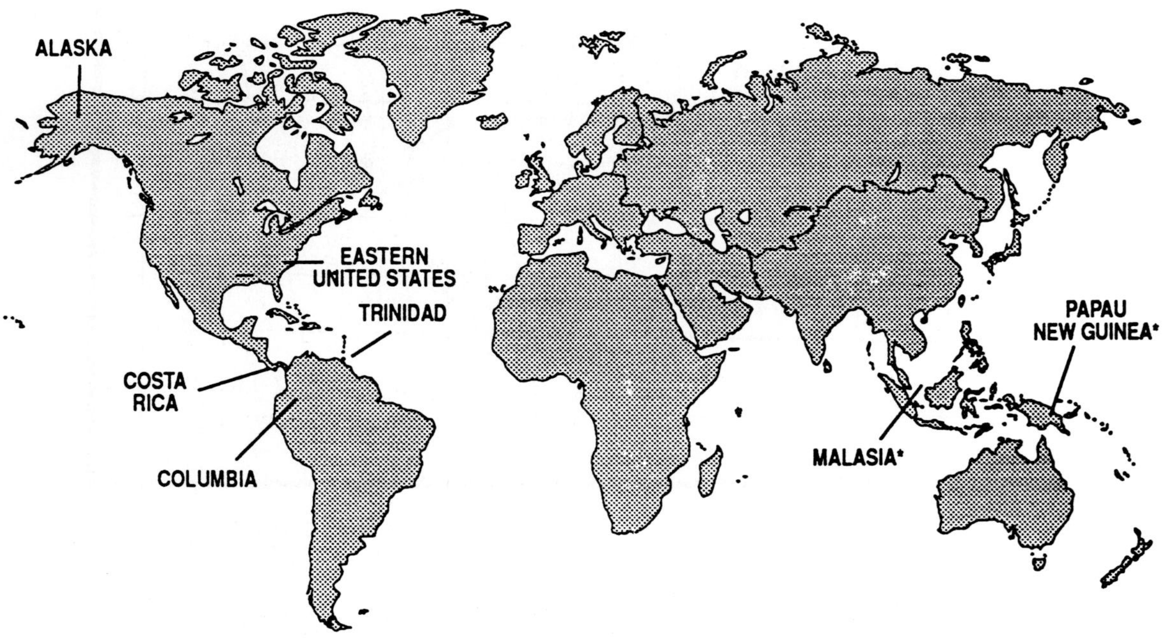

RADARSAT

RADARSAT International (RSI) was established in 1989 by a consortium of Canadian companies interested in the promotion of space and satellite remote sensing technology. RSI will be the commercial distributor of all data from the RADARSAT satellite, due to be launched in 1995 (Fig. D.5). In the interim, RADARSAT participates in the ERS consortium and is responsible for marketing ERS-1 data products in the USA and Canada. (A search on the availability of ERS-1 data can be obtained from RSI free of charge.) As well, RSI is the exclusive distributor of Landsat and SPOT data in Canada.

RSI will be conducting a number of airborne tests simulating RADARSAT's capabilities to help plan the data acquisition agenda for the satellite. Proposals are being accepted for data acquisition projects for both the airborne testing and the satellite itself.

Further information of RSI's operations can be obtained from the Data Centre of RADARSAT International, Inc., 3851 Shell Road, Suite 200, Richmond, British Columbia, Canada, V6X 2W2. Telephone: (604)244-0400. Fax: (604)244-0404, or RADARSAT International, 275 Slater Street, Suite 1203, Ottawa, Ontario, K1P 5H9. Telephone: (613)238-5424. Fax: (613)238-5425.

Fig. D.5. Planned coverage for the RADARSAT satellite RADARSAT will be able to capture any area in Canada within 3 days. Regions at the equator can be imaged every 4 days and regions above 79° lat. (including the North Pole) are covered on a daily basis. RADARSAT will also feature complete Antarctic coverage for two periods during the first 2 years of its mission

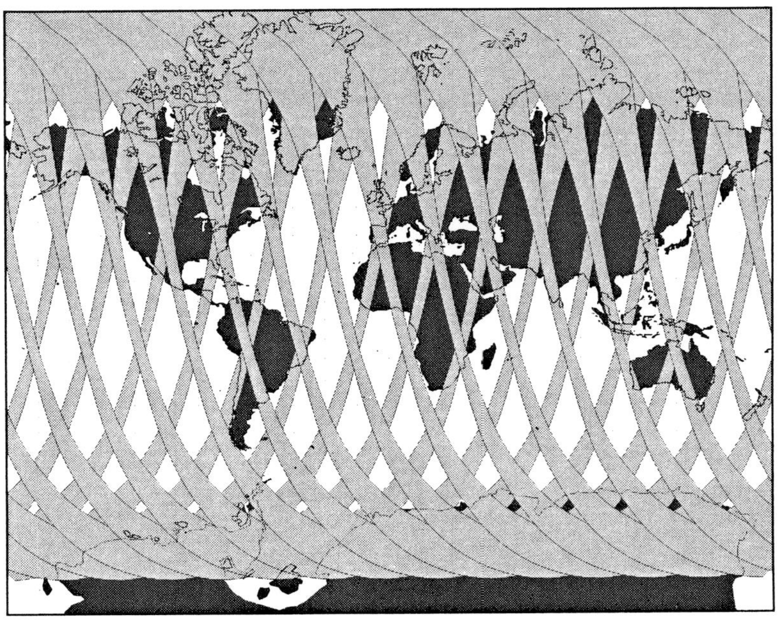

Glossary

active system: A remote sensing system that emits its own radiation and then monitors reflections from that radiation to gather information from the Earth's surface.

air base: The separation between the two observation sites of a stereo survey.

angular field of view: The angle subtended by lines extending from the imaging system to the edges of its ground swath.

apparent object distance: The distance from the eye base at which a projected stereo object appears to be.

atmospheric window: A region of the EM spectrum that is relatively free from scattering by atmospheric gases. Spaceborne imaging systems must be careful to choose bandwidths that correspond to atmospheric windows.

azimuth direction: The direction in which a radar system is moving. Commonly referred to as the flight direction. In SLAR systems, azimuth direction is perpendicular to the look direction.

azimuth resolution: Roughly speaking, the width of the image swath captured by a radar system. Azimuth resolution deteriorates with increasing distance from platform to surface and, thus, becomes an acute problem for a space-based system. Ways of improving azimuth resolution include decreasing the wavelength used, using a longer antenna, or employing synthetic aperture radar.

bandwidth: A certain range of the electromagnetic spectrum. It also often refers to the spectral range used by a satellite.

basement warp structure (BWS): A subtle structure that develops over a basement feature by the combined influence of differential compaction and structural reactivation. Such a structure, although more subtle, mimics the configuration of the underlying basement feature.

box-shaped valley: Deeply incised valley with steep valley walls that are uniquely developed by the process of piping and sapping.

breaching: Erosional process describing the removal of the crest of folded strata.

buried structure: A geological structure whose rock units are completely covered by sediments and thus must be analyzed indirectly by examining its effect on surface features.

charge-coupled detector (CCD): a semiconductor device used in some imaging systems to measure reflectance values.

cone-shaped stream valley: Typical surface expression of an antecedent stream valley which crosses a positive topographic feature such as a dome or a horst block.

contrast stretching: An image enhancement technique that redistributes the intensity values of an image's pixels in order to take advantage of the full radiometric range. As the name implies, this procedure's main result is an improvement in contrast and, hence, interpretability.

convergence angle: The angle subtended by lines drawn from each eyeball to the object in question.

cuesta: A typical erosional expression of an inclined bedrock strata. Expressed as an asymmetric ridge with one long and gentle slope (called the dip or isoclinal slope) and one short, steep slope (called the scarp slope). Also referred to as a questa.

cutoff point: In contrast stretching, the value below which all DNs are set to zero. All data below the cutoff pont are essentially lost by the operation. See saturation cutoff.

decision space: A region of a bivariate (or higher dimension) plot identified as indicating one particular material. Decision space can be assigned manually (in supervised classification) or by computer (in unsupervised classification).

deflected drainage: See perturbed drainage patterns.

depression angle: The angle between a horizontal line and a line extending from the imaging platform to its target. Can be related to the incidence angle if the ground is level.

diagnostic reflectance range: An area of the EM spectrum where the spectral response curve of one material differs greatly from a specified set of other materials, allowing it to be identified.

dielectric constant: An electrical property of a material that influences radar returns.

digital elevation model (DEM): A stereo analysis technique where a computer takes information from a digitized stereo pair and produces one orthographically correct digital map with correlated elevation measurements.

digital number (DN): The value assigned to each pixel to indicate the reflectance level corresponding to the ground area represented by the pixel. In a multispectral system, more than one DN is assigned to each pixel.

digitizing system: A device that takes an image and transforms it into pixel data usable by a computer. It is the first step required for image enhancement if the data were not originally gathered in a computer-compatible format.

dip slope: See isoclinal slope.

disparity angle: Twice the angle between the central axis of the eyeball to the projection point on the retina of the object in question. The central axis extends from the center of the eye's lens to the back of the retina.

DN bivariate plot: A scatter diagram created by correlating two or more DN distribution histograms. Bivariate plots can then be used to define parameters for computerized classification.

DN distribution histogram: A plot of DN values versus their relative occurrence in a specific scene. Histograms can be analyzed to reveal the contrast of the image and the degree of homogeneity. See DN bivariate plot.

dogleg edge pattern: An abrupt change in orientation of the zigzag pattern of high-angle dip-slip faults.

drainage anomaly: Unique pattern of stream channels and valleys that appears anomalous with respect to the surrounding drainage features. Usually reflects local topographic and structural control on the development of such drainage systems.

drainage density: Total length of stream valleys divided by their drainage basin area.

drainage frequency: Total number of stream valleys divided by their drainage basin area.

drainage halo: An area of extremely high-frequency and high-density drainage features which are characterized by box-shaped valleys. Develop by the process of piping and sapping along buried and obscured structures.

early breaching stage: Describes a fold whose upper layers have been removed. Often appears as a series of concentric cuestas.

edge enhancement: An image enhancement technique that applies a mathematical algorithm to all the pixels in a scene attempting to highlight boundaries and edges by maximizing edge contrast difference. Edge enhancement can be either directional (favoring one trending direction over others) or nondirectional.

electromagnetic (EM) radiation: Energy caused by electrical or magnetic disturbances that can travel through empty space at the speed of light (3×10^8 m/s). EM radiation emanates from the sun (as a result of fusion processes) and reflects off of surfaces on the Earth into space. The radiation is altered by the absorptive properties of materials on the Earth and, therefore, can be used by detectors (in a passive mode) to indicate conditions of the surface. Alternatively, detectors can emit their own radiation (an active mode) and monitor subsequent reflections.

electromagnetic (EM) spectrum: The range of observed wavelengths of electromagnetic radiation including radio waves, microwaves, infrared radiation, visible light, ultraviolet light, X-rays and gamma rays. The lower-intensity manifestations of EM radiation are called "waves" because of their more wave-like nature, whereas the higher-intensity manifestations are called "rays" because of they are more particle-like. In reality, all EM radiation has both particle and wave characteristics.

ERS-1: (the European Remote Sensing Satellite) Launched by the European Space Agency in 1991.

exposed structure: A structure that can be recognized and analyzed by the surface expressions of its inclined bedrock strata and fault-line traces.

eye base: The separation distance between the two eyes.

fault-line trace (ELT): The linear feature resulting from the erosion of a fault line. Fault-line traces often preserve important diagnostic surface and structural attributes recognizable on image data.

fault scarp: A steep slope or cliff formed by a fault.

first-order BWS: Regional-scale, subtle feature in the sedimentary cover that develops over basement structures and topography. Includes major arches and domes and large-scale fault systems and basement block boundaries.

flatiron: A typical surface expression of a dissected cuesta. Expressed as a triangularly shaped ridge with an apex that points in the dip direction.

floating dot: A technique used in stereo imagery analysis that presents a dot that can be moved in three dimensions over the apparent topography, providing readings of elevation. Often produced with a parallax bar.

focusing accommodation effect: A cue to depth perception arising from the detection of the tension required in the lens to maintain an object in focus.

foreshortening effect: A distortion on radar images that occurs when the emitted wavefront strikes the top of a vertical feature. Causes the near slope of a symmetrical mountain, for example, to appear shorter than the far slope. The effect is most pronounced in the far range. Cannot be systematically corrected.

form-line surface structure (FSS) map. A stereo mapping technique that reconstructs eroded or buried segments of a key horizon by taking dip and strike measurements from underlying or overlying layers.

framing system: A system that creates an image by capturing an entire scene at once by means of a photographic plate or other recording medium. See scanning system.

frequency: the number of oscillations per unit time or the number of wave crests that pass a specific point per unit time. Given in s^{-1} or Hertz. Can be related to wavelength by the equation *freq.* $\times$ *wavelength = speed of light* and is thus used interchangeably with wavelength.

grey level: A common synonym for brightness value or DN:

ground resolution cell: The smallest area on the ground that can be resolved by an imaging system. Each ground resolution cell becomes one pixel on the image. See pixel.

ground swath: A track of land that is scanned in one pass by an imaging system. Can be in the direction of travel or perpendicular to it.

hazy anomaly: A local distortion on an image appearing as a smudged or erased area. Has been postulated to be related to chimney phenomena and related surface alterations.

highly deformed: Describes relatively complex structural features characterized by steeply inclined and overturned beds.

hogback: An asymmetrical ridge produced by highly tilted strata.

horsetails: A slang term used to describe the dispersion of a master wrench fault into a series of small splay features.

hue: Indicates the dominant wavelength of a color. Hue is usually given as a number between 0 and 255, where 0 represents red, 85 represents green and 170 represents blue. See IHS system.

hydrocarbon chimney: The process of slow leakage of hydrocarbons from an effective hydrocarbon trap to the overlying sediments. Results in the development of unique features which can be detected by surface and subsurface exploration tools.

hydrocarbon surface alteration: Local changes in the mineralogical, spectral, and tonal characteristics of exposed rock units over leaking hydrocarbon-bearing areas.

image enhancement: Any method applied to imagery data to improve its interpretability. Includes contrast stretching, edge enhancement, false-color compositing and IHS transformations.

image restoration: The first step in data enhancement. Removes systematic errors such as periodic line dropouts, random noise and atmospheric scattering effects.

inbound obsequent stream: Flows toward the center of a dome and is usually collected by the central breaching stream following inversion of topography.

incidence angle: The angle between the ground and a line extending from the object to the imaging system. Can be related to depression angle if the ground is level.

incised valley: A deep valley cut by a rejuvenated stream with the course acquired from a previous cycle.

intensity: Indicates the "brightness" value of a color. In imaging systems, the sensitivity of intensity is set by the radiometric resolution. See IHS system.

intensity-hue-saturation (IHS) system: A system for labeling all possible colors where any color is identified as a point on a cylinder. Distance from the axis of the cylinder indicates saturation, angular position indicates hue, and distance along the axis indicates intensity.

intensity-hue-saturation (IHS) transformations: Any image enhancement technique that separates intensity, hue, or saturation components from an image and modifies them. See normalization.

inverted FLT: A fault-line trace that reflects an inverse relationship between the structural movement of the fault and its present topographic expression. Attributed to differential erosion along the fault.

isoclinal slope: The long and gentle slope of a cuesta. The isoclinal slope reflects the dip direction. Also called a dip slope.

JERS-1: (The Japanese Earth Resources Satellite) The first radar imaging satellite from Japan, launched in 1992.

Landsat: A series of satellites used for remote sensing that were deployed by NASA. Formally called the ERTS-1 (-Earth Resource Technology Satellite). Carries the MSS and TM imaging systems.

late breaching stage: Describes a fold that has undergone a complete removal of its crest, but still has its limbs preserved.

layer-cake geology: A slang term for a structural plateau with escarpments and gentle slopes that develop on resistant and nonresistant beds, respectively, and is dissected by dendritic drainage systems.

layover effect: A distortion on radar images that occurs when the emitted wavefront strikes the top of a vertical feature and makes it appear to lean towards the platform. The effect is most severe near the platform.

lineament: Large-scale linear feature in the sedimentary cover reflecting a wide zone of faults and fractures that develop over reactivated basement faults and zones of weakness.

linear density contours: Contour values that reflect the number of linear features per unit area.

linear feature: A small-scale surface feature that reflects individual faults and fractures in the sedimentary cover.

look direction: The direction in which a radar system emits radiation. Can also be called range direction. In SLAR systems, look direction is perpendicular to the azimuth direction and the flight path.

marginal subsequent stream: Forms as preexisting major stream and stream segments and tends to adjust its flow around newly formed obstacles.

mildly deformed: Describes low-amplitude structures with average dips of 0 to 5°. Such structures generally do not manifest surface expression of inclined bedrock strata (i. e. flatirons).

moderately deformed: Described relatively simple structures with well-expressed rims and related flatirons. Dips range from 5° to 60°.

multidirectional positive FLT: A fault-line trace related to a high-angle, dip-slip normal or reverse fault. These FLTs display a positive topographic relief which directly reflects the vertical movement along the fault, making it easy to map from satellite imagery.

multispectral: Refers to an imaging system that records reflectance values in more than one bandwidth.

multispectral scanning system (MSS): The original imaging system used on board the first generation of Landsat satellites.

NASA: The National Aeronautic and Space Administration (USA).

near-polar orbit: An orbit that follows closely the lines of longitude and passes nearly over the poles.

normalization: An image enhancement technique where intensity, hue and saturation values are separated, enhanced and recombined. See IHS system and IHS transformation.

obliterative stage: Describes a fold whose structural relief has been completely eliminated by erosional processes leaving only concentric rims and portions of plunging noses as remnants.

obscured structure: A geological structure whose rock units are partially covered by vegetation or soil and thus must be analyzed indirectly by examining its effect on surface features.

obsequent stream: A stream draining the scarp slope of a cuesta. Obsequent streams are typically shorter than their subsequent stream counterparts and cover the slope with more frequency.

off-nadir imaging: The ability of a satellite to capture an image swath that is not directly below its flight path but rather off to one side or another. Off-nadir imaging allows viewing of the same area more than once per orbit. This capability is useful for surveys of rapidly changing conditions and also for the generation of stereo imagery.

orthographic map: A map with no geometric distortions – distances shown correspond directly with distances on the surface. Images taken in an off-nadir fashion (for the purpose of stereo or otherwise) will not be orthographically correct. Such images can be corrected by creating a digital elevation map.

outbound consequent stream: Forms in a radiating pattern from the center of a dome. It may be either collected by the marginal subsequent components or flow directly outward.

parallax: The displacement of an object caused by a shift in observational position. Because the displacement is related to distance from observer to object, parallax serves as the basis for stereo imagery which uses two images of the same scene from slightly shifted positions to extract elevation information.

parallax bar: A stereo analysis tool that produces a floating dot on the apparent terrain.

parallel composite FLT: A fault-line trace related to a rotated fault block or a listric dip-slip fault. These FLTs appear as wide zones for parallel linear features that form a transition zone between upthrown and downthrown blocks.

passive system: A remote sensing system that relies on naturally occurring radiation from the Earth's surface to create an image. See active system.

periodic line dropout: A common defect on scanning imaging systems that causes a black line to appear across the data at regular intervals representing a loss of data. This error is usually corrected before interpretation begins.

perturbed drainage patterns: Local deflection of stream valleys caused by the presence of topographic obstacles which are related to structural controls.

photostratigraphy: Method of mapping the sequence stratigraphic components of outcrop units with remote sensing data.

piping and sapping: Erosional processes related to the movement of groundwater in the subsurface. Characterized by unique drainage that develops by the process of collapse of caves and pipes.

pixel: The smallest discrete area on an image (from PICTure ELement). DNs are assigned to pixels according to the brightness of each band within the area that the pixel represents.

polyphase FLT: A fault-line trace that exhibits a mixture of diagnostic features which indicate a kinematic history of several distinct tectonic events with varying stress regimes.

positive-relief stage: Describes folds that are at the initial stages of their erosional evolution exhibiting a direct relationship between topography and structures.

preferred look directions: The direction chosen for a radar survey which takes advantage of radar shadows to improve the appearance of features of interest.

pseudo-FLT: A fault-line trace expressed as an alignment of geomorphic features that develop in front of positive FLTs but does not reflect its actual location. For example, alluvial plains or highly eroded hogbacks.

pseudoscopic illusion: The illusion that the topography is inverted – canyons become ridges and vice versa. Can be caused by two things: a reversal of the order of the stereo pairs (left eye sees right image and vice versa) or by viewing a monoscopic image oriented so that shadows fall towards the viewer.

questa: See cuesta.

radar: (RAdio Detection And Ranging) An active imaging system that sends out EM radiation in the microwave region and then measures response times and intensity values from the resulting echoes.

radar altimeter: The simplest implementation of a radar system that measures echo return times and then correlates them to position to gather information about either the ground elevation or the platform altitude, depending on which is known.

radar imaging system: Radar system that combines information on return times and reflection intensity with information on position to produce an image of the area of interest.

RADARSAT: A radar imaging satellite system to be launced by Canada with help from NASA in late 1994. Will include extensive off-nadir imaging capabilities.

radar scatterometer: A radar system that measures the degree to which emitted signals are scattered by the area of interest. Common uses of scatterometers include mapping ocean currents and wind patterns.

radar shadows: Dark regions on a radar image caused when large objects block incoming radiation. Radar shadows become more exaggerated with lower depression angles and can seriously hamper interpretations. They can also, however, highlight linear features that might otherwise be hard to detect.

radiance: A measure of the intensity of incoming radiation. It is measured in watts per square meter. See reflectance.

radiometric resolution: The sensitivity of an imaging system to variations in brightness. Radiometric resolution is given as the number of increments used between the two extemes – completely dark and completely bright. Higher spatial resolution means smaller increments and, therefore, a picture with more information and more likely to benefit from contrast stretching.

range direction: See look direction.

range resolution: The minimum spacing between two objects in the range direction for them to be individually discerned on a radar image. Range resolution is directly related to the length of the EM radiation pulse emitted by the radar system. Also, range resolution improves with decreasing depression angle.

ratio enhancement: An image enhancement technique that attempts to eliminate the distortion of spectral data due to uneven lighting conditions. Because shadow regions will experience a decrease in the same proportion in each band, taking the ratio of two bands and then displaying the result yield an improved image.

real-aperture radar: A radar system that does not employ the synthetic-aperture system. Also called a "brute force" radar.

reconnaissance tools: Gravity, magnetic and satellite image data which are used in exploration for early recognition of geological structures. The term contrasts with prime tools which include well and seismic data.

reflectance value: The ratio of radiant energy to reflected energy of a particular object measured in a particular bandwidth. Reflectance is also often used loosely as a qualitative term specifying the degree of radiance from a particular object. See radiance.

retina: The light sensitive membrane coating the back of the eyeball that interprets incoming light into electrical impulses for the brain.

retinal disparity: The principle that gives binocular vision the ability to perceive depth. When the two eyes are focused on one object, an object at a different distance away will be projected onto each retina at a displacement from the center proportional to the distance of the object.

saturation: Indicates the "purity" of a color or the degree to which all wavelengths correlate to the dominate wavelength (the hue). 0 saturation is a completely "impure" color that appears grayish (and, in effect, has no hue), whereas higher purity indicates "brighter" colors. See IHS system.

saturation cutoff: In contrast to stretching, the value above which all DNs are set to maximum intensity, all data above the saturation cutoff level are essentially lost. See cutoff point.

scanning system: A system that creates an image by sweeping individual sensors across a scene and recording the variations in reflectance. See framing system.

scarp slope: The short and steep slope of a cuesta. The scarp slope indicates the anti-dip direction.

SEASAT: Satellite launched by NASA in 1978. SEASAT was one of the first spaceborne imaging radar systems. Power failure ended its data-gathering capability less than 5 months later.

second-order BWS: Prospect-scale, subtle feature in the sedimentary cover that develops over basement structure and topography. Includes individual uplifted horst blocks, local domal structures and faults.

side-looking airborne radar (SLAR): Refers to airborne radar systems that are mounted such that their field of view is perpendicular to the flight path of the airplane. Can also refer to a spaceborne radar system operating in the same manner.

sinuous positive FLT: A fault-line trace related to low-angle thrust faults. These FLTs usually form distinct topographic boundaries between the upper and lower thrust plates.

SIR-A, SIR-B, SIR-C: (Shuttle Imaging Radar) The radar imaging systems run by NASA on board the space shuttle.

slant-range image: A distortion on radar images caused by the compression of scale in the near range and extension of scale in the far range. Cannot be systematically corrected.

spatial resolution: Roughly speaking, the size of the ground resolution cell or the smallest discernible area by the imaging system. Strictly speaking, the minimum spacing required between two objects on the ground for them to be imaged as two distinct objects. Spatial resolution is usually given in meters. Better spatial resolution yields a sharper picture.

spectral response curve: Shows reflectance values versus frequency (or wavelength).

SPOT (Systéme Probatoire d'Observation de la Terre): A remote sensing satellite launched by the French National Space Agency (Centre National d'Etudes Spatiales) in 1984. SPOT features better spatial resolution than the Landsat satellites and the availibility to produce stereo images through off-nadir imaging.

standard false-color composite (SFC): A standard method of viewing Landsat and other satellite data. When applied to MSS data, bands 4, 5, and 7 are assigned to blue, green, and red, respectively.

stereo imagery: Technique of using parallax properties from two images of the same area to extract elevation information. The two images can be obtained by systems with off-nadir capabilities and can be viewed with a variety of equipment.

stereoscope: A device consisting of lenses and mirrors that allows viewing of stereo pairs.

structural inversion: Describes a process where reactivation of structures caused the inversion of their structural inversion. For example, graben-filled sediments which were inverted and are expressed as a positive fold.

structural style: The unique assemblage of individual structures that reflect their mode of deformation.

structurally controlled stream: Synonymous with drainage anomaly.

subsequent incised valley: A deep valley cut by a rejuvenated stream that circum navigates and follows the strike of an exposed structure. The process indicates that the rate of downcutting is less than the rate of rejuvenation.

subsequent stream: A stream that drains the isoclinal slope of a cuesta. Subsequent streams are typically longer and less frequent than their corresponding obsequent streams.

sun-synchronous orbit: An orbit with a period designed to coincide with the rotation of the Earth so that is passes over any particular latitude of the Earth at the same local time. This leads to similar illumination conditions on each pass and allows the compilation of multiple images into uniform mosaics. Most remote sensing satellites follow sun-synchronous orbits.

superimposed style: Unique assemblages of individual structures that exhibit anomalous trends and structural characteristics in comparison to their surrounding structures. Such phenomena reflect the presence of preexisting structural elements that were reactivated and influenced the development of structures in a region during the latest mode of deformation.

supervised classification: A computer-assisted information extraction technique where decision spaces are assigned by an interpreter. The image is then examined by computer and areas corresponding to the decision spaces are marked accordingly. See unsupervised classification.

surface roughness: The most significant factor contributing to the radar reflectance properties of a particular area. Surface roughness is often discussed qualitatively in terms of specular and nonspecular surfaces. Surface roughness can also be analyzed quantitatively through the use of vertical relief which is the average height of small-scale irregularities in the terrain surface.

synthetic aperture radar (SAR): Method of improving azimuth resolution by analyzing the Doppler shift of echo responses as a particular feature is passed. So named because it simulates the effect of having a much larger antenna.

tear fault: A detached strike-slip fault usually confined to a single thrust sheet in fold and thrust belts. Produce profound surface expression of a negative FLT.

thematic mapper (TM): The second-generation imaging system used on board the Landsat satellite. TM featured improved spectral and spatial resolution.

third-order BWS: Small-scale, subtle linear features in the sedimentary cover which reflect predominately the basement grain.

tonal and spectral anomaly: A remote sensing term used to describe the unique expressions of buried and obscured structures related to their influence on ground moisture conditions.

topographic inversion: Describes an inverse relationship between structure and topography. For example, a breached anticline whose crest is expressed as a topographic low.

transverse incised valley: A deep valley cut by a rejuvenated stream that cuts across a positive structure feature. The process indicates that the rate of downcutting is greater than the rate of rejuvenation.

treetop geology: The ability to map surface topography indirectly by examining variations in the heights of treetops. The effect created by radar's ability to measure altitude changes very accurately which is one of the major adavantages of radar over passive systems.

unsupervised classification: A computer-assisted information extraction technique where decision spaces are assigned automatically by computer after a preliminary examination of the image data. Areas on the image are then assigned classifications according to the decision spaces.

vertical exaggeration: The common stereo phenomenon where objects are made to appear taller than they actually are.

vertical relief: The average height of small-scale irregularities of a surface. Used as an indication of surface roughness in radar imagery.

V-shaped valley: Typical surface expression of stream valleys that develop by fluvial processes. In inclined strata, the apex of the V-shaped valleys point in the anti-dip direction.

wavelength: The distance between successive crests of a wave given in meters. Wavelength can be related to frequency by the equation *freq.* × *wavelength* = *speed of light* and is thus used interchangeably with frequency.

Location Index

Subject Index

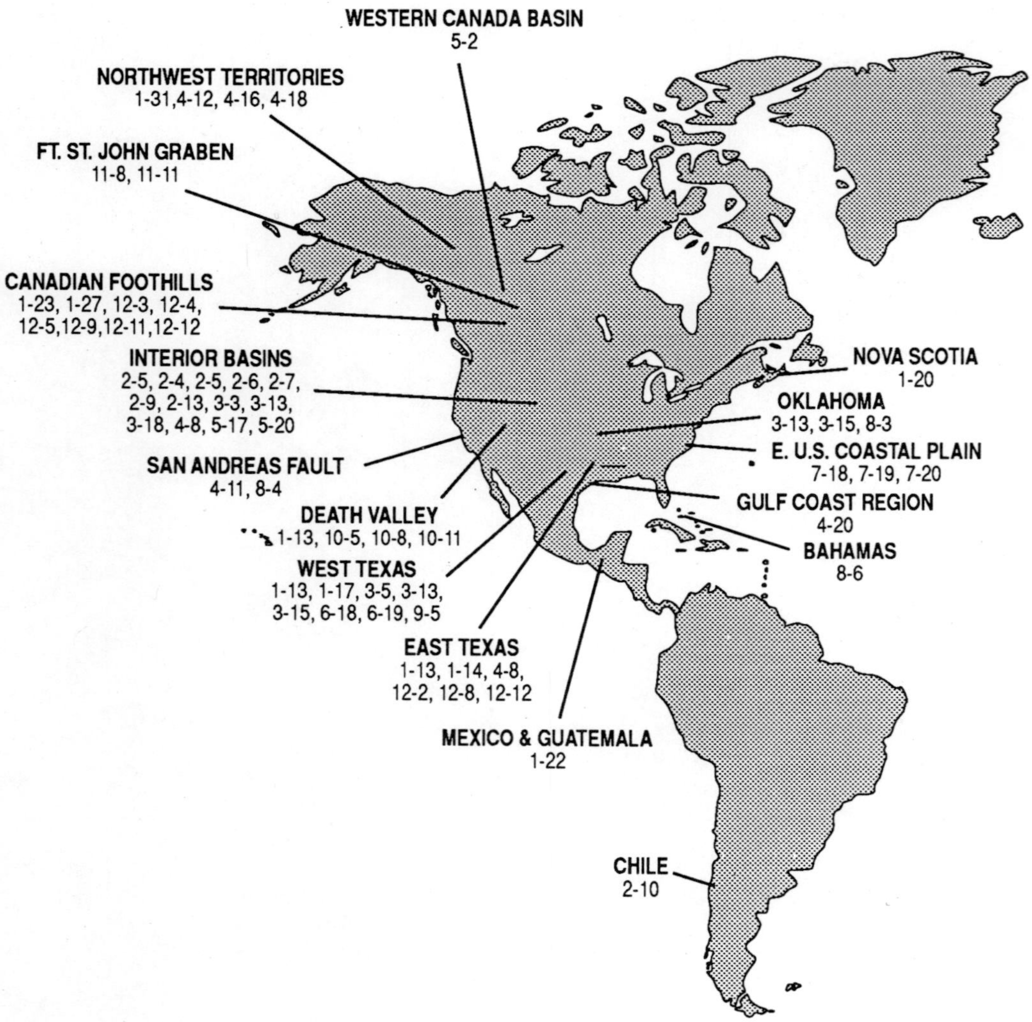
WESTERN CANADA BASIN
5-2
NORTHWEST TERRITORIES
1-31,4-12, 4-16, 4-18
FT. ST. JOHN GRABEN
11-8, 11-11
CANADIAN FOOTHILLS
1-23, 1-27, 12-3, 12-4,
12-5,12-9,12-11,12-12
INTERIOR BASINS
2-5, 2-4, 2-5, 2-6, 2-7,
2-9, 2-13, 3-3, 3-13,
3-18, 4-8, 5-17, 5-20
SAN ANDREAS FAULT
4-11, 8-4
DEATH VALLEY
1-13, 10-5, 10-8, 10-11
WEST TEXAS
1-13, 1-17, 3-5, 3-13,
3-15, 6-18, 6-19, 9-5
EAST TEXAS
1-13, 1-14, 4-8,
12-2, 12-8, 12-12
MEXICO & GUATEMALA
1-22
NOVA SCOTIA
1-20
OKLAHOMA
3-13, 3-15, 8-3
E. U.S. COASTAL PLAIN
7-18, 7-19, 7-20
GULF COAST REGION
4-20
BAHAMAS
8-6
CHILE
2-10